WENNER–GREN CENTER
INTERNATIONAL SYMPOSIUM SERIES

VOLUME 48

RECEPTOR–RECEPTOR INTERACTIONS
A NEW INTRAMEMBRANE INTEGRATIVE MECHANISM

RECEPTOR–RECEPTOR INTERACTIONS

A NEW INTRAMEMBRANE INTEGRATIVE MECHANISM

Proceedings of an International Symposium held at
The Wenner-Gren Center, Stockholm, October 9th – 11th, 1986

Edited by

Kjell Fuxe
Department of Histology
Karolinska Institutet
Stockholm, Sweden

and

Luigi F. Agnati
Department of Human Physiology
University of Modena
Modena, Italy

PLENUM PRESS • NEW YORK AND LONDON

First published 1987

Published in Great Britain by
THE MACMILLAN PRESS LTD
Houndmills, Basingstoke, Hampshire RG21 2XS
and London
Companies and representatives
throughout the world

Published in the United States of America by
PLENUM PUBLISHING CORPORATION
233 Spring Street, New York, NY 10013, USA

ISBN-13: 978-1-4684-5417-8 e-ISBN-13: 978-1-4684-5415-4
DOI: 10.1007/978-1-4684-5415-4
LCCN 87–042721

Contents

Session III
Receptor–Receptor Interactions in Psychoactive Drug Action

The Participants

Ann Acheson
Max Planck-Institute for Psychiatry
Klopferspitz 18 A
D-8033 Martinsried
Munich
F.R.G

Luigi F. Agnati
Department of Human Physiology
University of Modena
Via Campi 287
411 00 Modena
ITALY

Julius Axelrod
Section on Pharmacology
Laboratory of Clinicla Science
NIMH
Bethesda, MD 20205
U.S.A.

Tamas Bartfai
Arrheniuslaboratoriet
University of Stockholm
Svante Arrhenius väg
S-10691 Stockholm
SWEDEN

Guiseppi Bartholini
Synthélabo - L.E.R.S.
58, rue de la Glaciére
75013 Paris
FRANCE

Etienne Baulieu
Département de Chimie Biologie
Hôpital de Bicêtre
78, avenue du Général Leclere
94270 Bicêtre
FRANCE

Michael Berridge
ARC Unit of Insect Neurophysiology
and Pharmacology
Department of Zoology
University of Cambridge
Cambridge CB2 2EJ
ENGLAND

Floyd Bloom
Division of Preclinical
Neuroscience and Endocrinology
Scripps Clinic
La Jolla, CA 92037
U.S.A.

Claus Braestrup
Pharmaceuticals R & D
Novo Industri A/S
DK-2880 Bagsvaerd
DENMARK

Nicoletta Brunello
Institute of Pharmacology
and Pharmacognosy
Via A. del Sarto 21
20129 Milano
ITALY

Richard Cerione
Department of Pharmacology
Cornell University
New York State College of
Veterinary Medicine
Ithaca, NY 14853-6401
USA

Jean-Pierre Changeux
Neurobiologie Moléculaire
Institute Pasteur
75024 Paris Cedex 15
FRANCE

Jacqueline Crawley
Unit on Behavioral Neuropharmacology
Clinical Neuroscience Branch
NIMH
Bethesda, Maryland 20205
U.S.A.

Claudio Cuello
Department of Pharmacology and Therapeutics
McGill University
McIntyre Medical Sci.Bld.
3655 Drummond St.
Montreal
PQ CANADA H3G IY6

Bertil Fredholm
Department of Pharmacology
Karolinska Institute
Box 60400
S-104 01 Stockholm
SWEDEN

Kjell Fuxe
Department of Histology
Karolinska Institute
Box 60400
104 01 Stockholm
SWEDEN

Menek Goldstein
New York University
Medical Center
School of Medicine
550 First Avenue
New York
N.Y. 10016
U.S.A.

Paul Greengard
Laboratory of Molecular and Cellular Neuroscience
The Rockefeller University
1230 York Avenue
New York
N.Y. 10021-6399
U.S.A.

Antonio Gropetti
Department of Pharmacology
University of Milan
Via Vanvitelli, 32
20129 Milan
ITALY

Ernst Helmreich
Department of Physiological Chemistry
University of Würzburg Medical School
D-8700 Würzburg
F.R.G.

Edith Heilbronn
Unit for Neurochemistry and Neurotoxicology
University of Stockholm
106 91 Stockholm
SWEDEN

Paul Herrling
Wander Research Institute
Box 2747
CH-3001 Bern
SWITZERLAND

Barry Hoffer
Department of Pharmacology
University of Colorado
Medical Center
Denver
CO 80262
U.S.A.

Morley Hollenberg
Department of Pharmacology
University of Calgary
Alberta
CANADA 72N 4N1

Anders Härfstrand
Department of Histology
Karolinska Institutet
Box 60400
S-104 01 Stockholm
SWEDEN

Tomas Hökfelt
Department of Histology
Karolinska Institute
Box 60400
104 01 Stockholm
SWEDEN

Karl Jakobs
Department of Pharmacology
University of Heidelberg
D-6900 Heidelberg
FRG

Gösta Jonsson
Department of Histology
Karolinska Institutet
Box 60400
104 01 Stockholm
SWEDEN

Eric Kandel
Center of Neurobiology & Behavior
Columbia University
630 W 168th Street
New York, N.Y. 10032
U.S.A.

S. Langer
Synthélabo-L.E.R.S.
58, Rue de la Glaciere
F-75103
Paris
FRANCE

Rita Levi-Montalcini
Laboratory of Cell Biology
Consiglio Nazionale dell Ricerche
via G. Romagnosi 18/A
00196 Rome
ITALY

Jan Lundberg
Department of Pharmacology
Karolinska Institute
Box 60400
104 01 Stockholm
SWEDEN

Pierre Magistretti
Département de Pharmacologie
Centre Médical Universitaire
1205 Geneva
SWITZERLAND

Robert Moss
Department of Physiology
Health Sciences Center,
Southwestern Medical School,
University of Texas
5323 Harry Hines Boulevard
Dallas, TX 75235
U.S.A.

Lars Olson
Department of Histology
Karolinska Institutet
Box 60400
104 01 Stockholm
SWEDEN

Howard Rasmussen
Dept. of Internal Medicine
and Cell Biology
Yale University
New Haven
Connecticut 06511
U.S.A.

William Rostene
INSERM U 55
Hôpital Saint-Antoine
184, rue du Fg Saint Antoine
75571 Paris Cedex 12
FRANCE

Elaine Sanders-Bush
Dept. of Mental Health and
Mental Retardation
1501 Murfreesboro Road
Nashville, Tennessee 37217
U.S.A.

Agi Schonbrunn
Lab. of Toxicology
Department of Physiology
Harvard School of Public Health
Boston, MA 02115
U.S.A.

George Siggins
Div. Preclin. Neurology and Endocrinology
Research Institute Scripps Clinic
10666 N Torrey Pines Rd.
La Jolla, CA 92037
U.S.A.

P.F. Spano
University of Brescia
Institute of Pharmacology
Faculty of Medicine
I-25124 Berscia
ITALY

Lennart Stjärne
Department of Physiology I
Karolinska Institutet
Box 60400
S-104 01 Stockholm
SWEDEN

J.P. Tassin
Groupe NB, INSERM U 114
Collège de France - Neuropharmacologie
11 Place Marcelin Berthelot
75231 Paris Cedex 05
FRANCE

Lars Terenius
Department of Pharmacology
BMC, Box 573
751 23 Uppsala
SWEDEN

Gino Toffano
Fidia Research Lab.
35031 Abano Terme
Via Ponte della Fabbrica 3/A
ITALY

David U'Prichard
Nova Pharmaceutical Corporation
5210 Eastern Avenue
Baltimore
Maryland 21224-2788
U.S.A.

S. Vicini
The Fidia-Georgetown Institute for the Neurosciences
Georgetown University Med. Center
3900 Reservoir Road, N.W.
Washington, D.C. 20007
U.S.A.

Jean-Didier Vincent
Unité de Recherches de Neurobiologie des Comportements
INSERM U 176
rue Camille Saint Saëns
33077 Bordeaux Cedex
FRANCE

B. Wise
The Fidia-Georgetown Institute for the Neurosciences
Georgetown University Med. Center
3900 Reservoir Road, N.W.
Washington, D.C. 20007
U.S.A.

Corrigendum:

A. Guidotti
The Fidia-Georgetown Institute for the Neurosciences
Georgetown University Med. Center
3900 Reservoir Road, N.W.
Washington, D.C. 20007
U.S.A.

Opening Address

Kjell Fuxe and Luigi F. Agnati

Colleagues and friends,

Professor Agnati and myself and the Wenner-Gren Foundation wish you all very welcome to this Wenner-Gren Center Symposium on Receptor-Receptor Interactions. A new Intramembrane Integrative Mechanism.

Very early some investigators realized that e.g. GABA receptors could be considered as being supramolecular complexes, in which regulatory modulations take place. Thus, it was found by Tallman, Tomas and Gallager (1978) that the addition of GABA to crude membrane preparations increased the affinity of the benzodiazepine binding site. The benzodiazepines were in turn found to increase the density of the high affinity recognition site for GABA as shown by Costa, Guidotti and Toffano (1978). At this time period we were ourselves involved with biochemical and functional work on receptors indicating that such reciprocal regulations may as a matter of fact take place between different types of supramolecular complexes, that is between entirely different receptors. Thus, there was a phenomenon in the membrane waiting to be christened and we called this and other related phenomena "receptor-receptor interactions" (Fig. 1). This title underlines the fact that our attention will be focused on the integrative capabilities of the membrane obviously without dismissing a holistic view of the nerve cell. Very likely we will find out that some sophisticated elaborations of information are performed at the membrane level, via interactions within and among different classes of macromolecules (such as receptors, ion pumps, ion channels). One of the consequences of these phenomena may be that electrically coded and chemically coded information can in this way very effectively modulate each other and interconvert into each other. In fact, in view of the membrane polarization these elaborations can be performed in the best way at the membrane level. Another consequence can be that intracytoplasmatic mechanisms are liberated from certain tasks and can instead work on selected and elaborated inputs. The way we look upon this phenomenon can be illustrated by analogy with the information handling in networks of sensory systems. Thus, e.g. in the case of visual inputs highly sophisticated operations take place in the retina and in the lateral geniculate body, so that selected

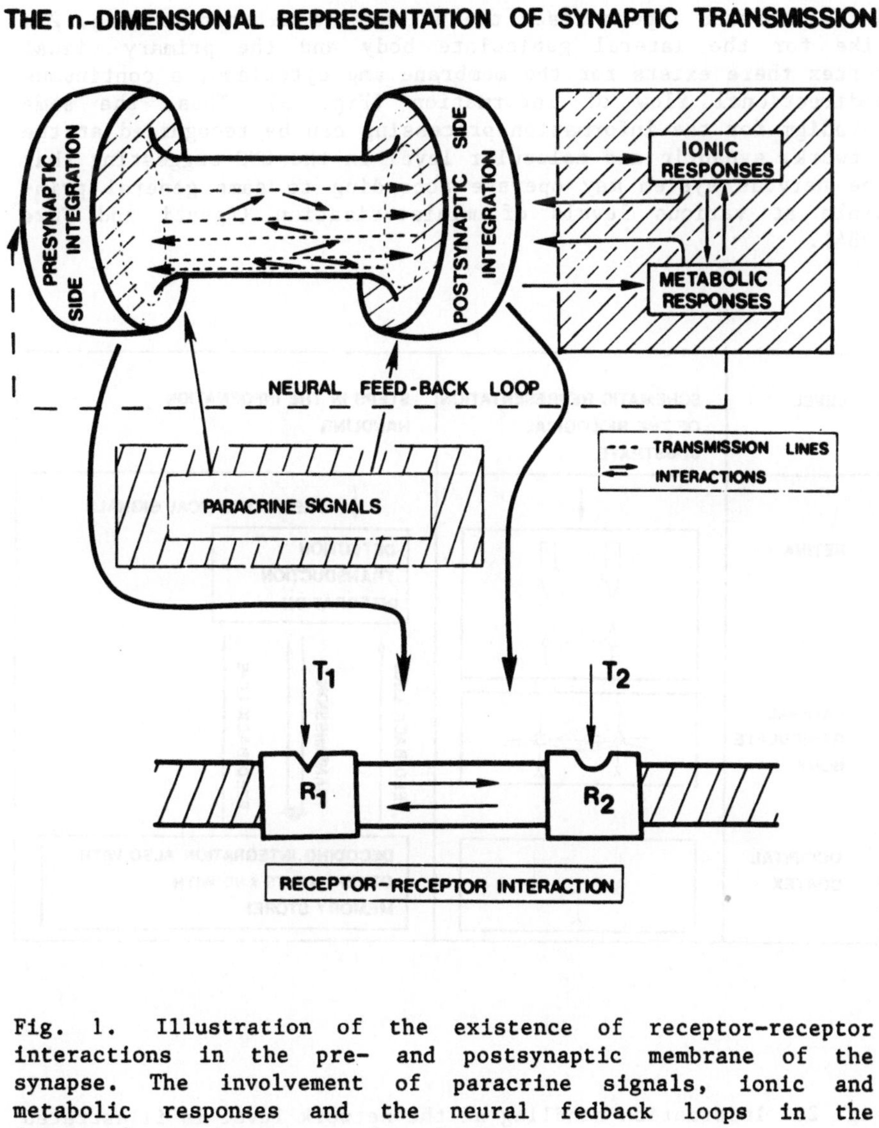

Fig. 1. Illustration of the existence of receptor-receptor interactions in the pre- and postsynaptic membrane of the synapse. The involvement of paracrine signals, ionic and metabolic responses and the neural fedback loops in the integrative processes are also shown.

and elaborated inputs reach the primary visual cortex (Fig. 2). Like for the lateral geniculate body and the primary visual cortex there exists for the membrane and cytoplasm, a continuous bidirectional flow of information (Fig. 3). Thus, the same paradigm for the information processing can be recognized at the network, synaptic and molecular level in the CNS suggesting that the nervous system may operate according to same general principle at various levels of miniaturization (Agnati and Fuxe 1984).

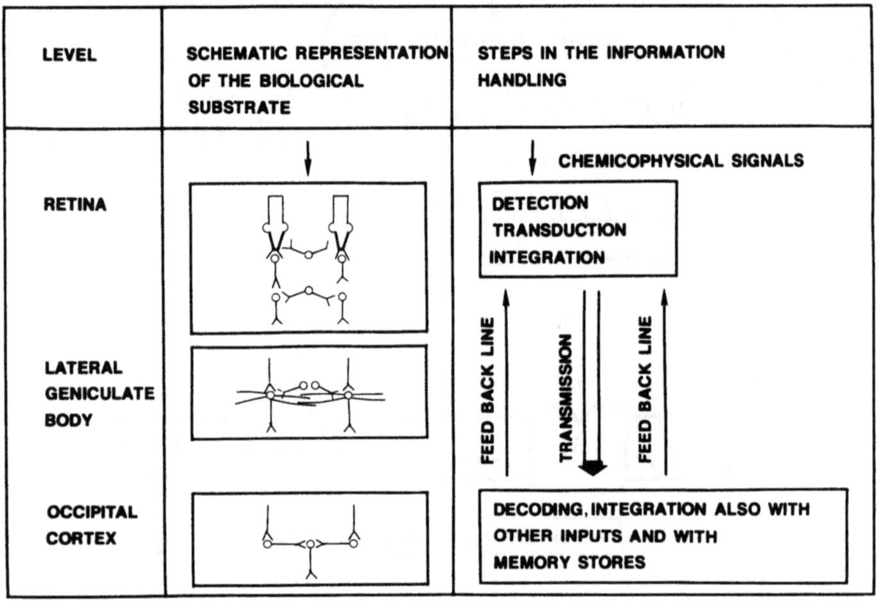

Fig. 2. Information handling at the network level is illustrated in the central visual pathways from the retina to the striate cortex via the lateral geniculate body.

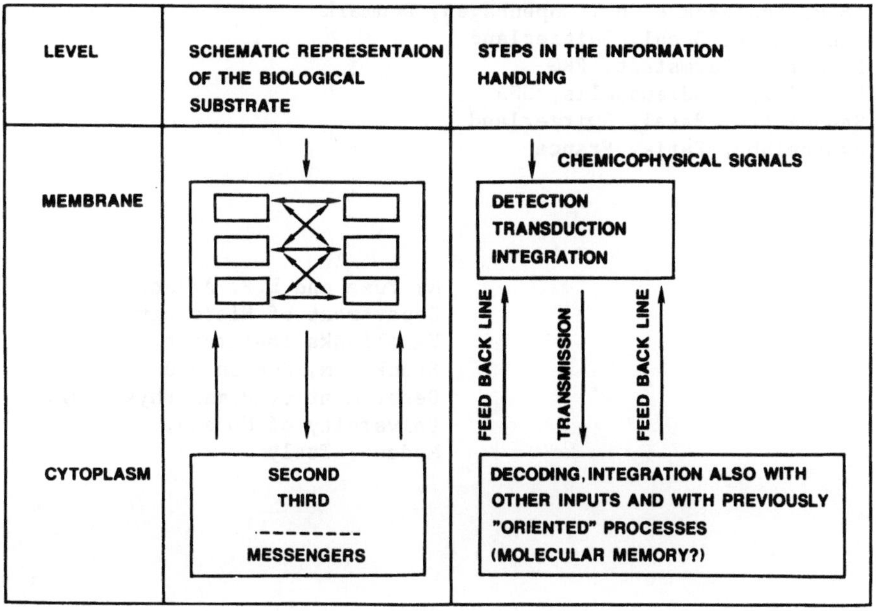

Fig. 3. Illustration of the information handling at the membrane-cytoplasm level and their interactions.

We believe it is important to have been able to bring together so many outstanding scientists from many parts of the world to elucidate the receptor-receptor interactions and other integrative mechanisms in the membrane. We look forward to a penetrating analysis of the problems giving us new inspiration to get on with the work.

We are very much indepted to the Wenner-Gren Center Foundation and its secretary Professor D. Ottoson and Mrs. G. Hultgren whose genuine support in all aspects of this conference made this symposium possible. For generous support we are indepted to the following sponsors.

The Swedish Medical Research Council
Fidia, Laboratoris, Abano-Terme, Italy

Astra Pharmaceuticals, Södertälje, Sweden
AB Ferrosan, Malmö, Sweden

AB H. Lundbeck & Co., Copenhagen, Denmark
Ciba-Geigy, Basel, Switzerland
E. Merck, Darmstadt, FRG
Eli Lilly, Indianapolis, USA
Sandoz Ltd, Basel, Switzerland
Synthelabo, Paris, France

K. Fuxe and L.F. Agnati
Department of Histology,
Karolinska Institutet,
Stockholm, Sweden and
Department of Human Physiology,
University of Modena,
Modena, Italy

REFERENCES

Agnati, L.F., Fuxe, K., Zini, I., Lenzi, P. and Hökfelt, T. (1980) Aspects on receptor regulation and isoreceptor identification. Med. Biol. 58, 182-187.

Agnati, L.F. and Fuxe, K. (1984). New concepts on the structure of the neuronal networks: The miniaturization and hierarchical organization of the central nervous system. Biosci. Rep. 4, 93-98.

Guidotti, A., Toffano, G. and Costa, E. (1978). An endogenous protein modulates the affinity of GABA and benzodiazepine receptors in rat brain. Nature 257, 553-555.

Tallman, J.F., Thomas, J.W. and Gallager, D.W. (1978). GABA-ergic modulatin of benzodiazepine binding site sensitivity. Nature 274, 384-385.

Session I
Effects of Messengers on Membrane Receptors

Part A
Effects of Transmitters and Modulators

Chairman: T. Hökfelt

1
Interactions between Convergent Afferent Transmitter Systems

Floyd E. Bloom, John Morrison, Elena Battenberg, David Lewis
and Michael Campbell

INTRODUCTION

In many respects, the subject of transmitter interactions recalls questions more than a decade old: how does one transmitter substance influence the responses of a cell to its other presumptive afferent transmitters. Such concerns were the obsessions of early iontophoretic experiments (see Bloom, 1974), when one sought patterns in the responsiveness to sets of transmitters as possibly significant attributes of specific neurons. The physiological response question then shifted to neurochemical and neuroanatomical determinants as to which transmitters and which source neurons a given target cell might normally need to recognize. Those data allowed us (see Foote et al., 1983; Siggins and Gruol, 1986) and others to evaluate synaptically released transmitters, rather than to deal with exogenously applied transmitter candidates. Many of the answers to those structural questions were provided by our hosts for this symposium, led by Dr. Fuxe and his collaborators.

Eventually, it became possible to ask more substantive questions of the actions of different transmitters on a given neuronal population. In many cases, it is now possible to resolve the differing mechanisms by which transmitters can directly affect their targets, and how these various response mechanisms could integrate signals of transmitters that converge onto common targets (see Siggins and Gruol, 1986). This form of cellular interaction through convergent afferents was generally termed <u>heterosynaptic regulation.</u>

Clearly, our interests in transmitter-transmitter interactions now are heavily motivated by a possibly related regulation through multiple transmitters contained within a single synaptic afferent, which we might denote as <u>homosynaptic coordination.</u> Such concepts arose through the work of Hokfelt and collaborators (see Hokfelt, this volume). The functional implications of homosynaptic coordination through co-existing transmitters is indeed an interesting and provocative re-statement of the earlier concerns over heterosynaptic transmitter-transmitter interaction, and one that has many aspects suitable for investigation in isolated preparations in vitro (see Magistretti, this volume).

Nevertheless, electrophysiological documentation of regulatory interactions between transmitters, whether they co-exist, or whether they terminate as separate circuits on common targets, still requires the demonstration by cytochemical microscopic methods that the transmitters presumed to interact actually make functional connections with the neurons on which the interactions have been evaluated. Therefore, given that the organizers of this meeting were gracious enough to include in the program two of my collaborators whose reports will develop the electrophysiologic (see Siggins this volume) and biochemical (see Magistretti, this volume) themes of the work we began together, this discussion will be focussed on three accumulations of our anatomic data pertinent to these interpretive concerns.

Cortical Organization of Interacting Transmitters: Noradrenergic (NA) and VIP-Containing Circuits.

The NA coeruleo-cortical system in the rat has two major characteristics: 1) a rich network of NA innervation is oriented longitudinally through the grey matter and branches widely; 2) the NA fibers permeate all layers of the dorsal and lateral cortical regions with a relatively uniform laminar pattern, (c.f. Morrison et al. 1984,1985). Thus, the tangential afferent NA innervation of rodent neocortex is organized quite differently than the highly localized, radial thalamo-cortical afferents.

Although the light microscopic details of the neocortical monoaminergic systems has progressed substantially in the past 15 years, the nature of the cells to which these fibers project in cortex has only begun to clarify after a period of extreme contentiousness. The interpretations of Molliver (see Molliver et al., 1982) are quite similar to ours (see Bloom, 1979; Foote et al., 1983): both 5-HT (also see Azmitia and Gannon, 1986) and the two cortical catecholamines typically do form specialized contacts with their target neurons in frequencies that suggest that nearly all their cortical boutons do so.

We hypothesized (c.f., Morrison and Magistretti, 1983; Magistretti, this volume) that the anatomic overlap between the tangentially organized noradrenergic system and the radially organized intrinsic bipolar neurons containing Vasoactive Instestinal Polypeptide (VIP) provided a structural basis for significant interactions across transmitter systems. Subsequently, major biochemical interactions in vitro were indeed reported (see Magistretti, this volume) and supported by extracellular pharmacologic interactions in vivo (see Ferron et al., 1985; Siggins, this volume). Those data strongly imply that adrenergic and VIP-ergic systems converge physiologically on pyramidal neurons. From electron microscopy and golgi analysis Peters and Kimerer (1981) concluded that bipolar neurons synapsed mainly with assymetric junctions, upon the dendritic spines of pyramidal cells. However, by subsequent immunocytochemical studies (see Connor and Peters, 1984), VIP immunoreactive terminals were found to form mainly symmetrical axo-spinous junctions. Despite the minor controversy in the nature of the junction, evidence for VIP actions on glia and on the cerebral vasculature have also been proposed (cf Morrison and Magistretti, 1983).

Since the cellular target and synaptic type of the VIP terminals are critical to our data interpretations, we recently began to study VIP terminations in rat visual cortex. Using our own modifications of a pre-embedment, immunoperoxidase strategy for peptide localization (see Figs 1-4), we have identified a large number of terminals that contain VIP immunoreactivity. As would be predicted from the light microscopy, both dendrites (Fig 2) and boutons (Figs 1,3,4) were immunoreactive. In favorable sections, boutons with specialized junctions were quite common, of which more than 75% are symmetrical and virtually all are on small dendritric spines. The immunoreactive terminals contain mainly small electronlucent vesicles, with occasional large granular vesicles away from the synaptic zone. In addition, presumptive dendrites, containing few conventional vesicles, and rare large granular vesicles, are also immunoreactive for VIP, occasionally exhibiting synaptic structures similar to the reciprocal synapses of the olfactory bulb. We tentatively conclude that VIP bipolar cells in the rat cortex contact spiny neurons exclusively to form specialized symmetric junctions, and that they show no consistent relationship to glia or blood vessels. The symmetric contacts occupy those cortical layers in which pyramidal dendrites are prominent (see Connor and Peters, 1984). Thus, while more definitive evidence of the target cell type is required, we can conclude that there is a strong presumptive body of evidence favoring a physiological interpretation of the VIP-noradrenaline interaction in the rodent cortex.

Definition of Common or Unique Target Cells - A Virtue of Primate Cortical Complexity

More recent studies of monoamines in the far more highly differentiated, gyrencephalic primate neocortex have shown the need for considerable refinement in concepts of monoamine circuit specificity. Thus, the specific density and pattern of NA or 5-HT innervation in a particular cortical locus varies systematically with the cytoarchitectonic region and its hierarchical order

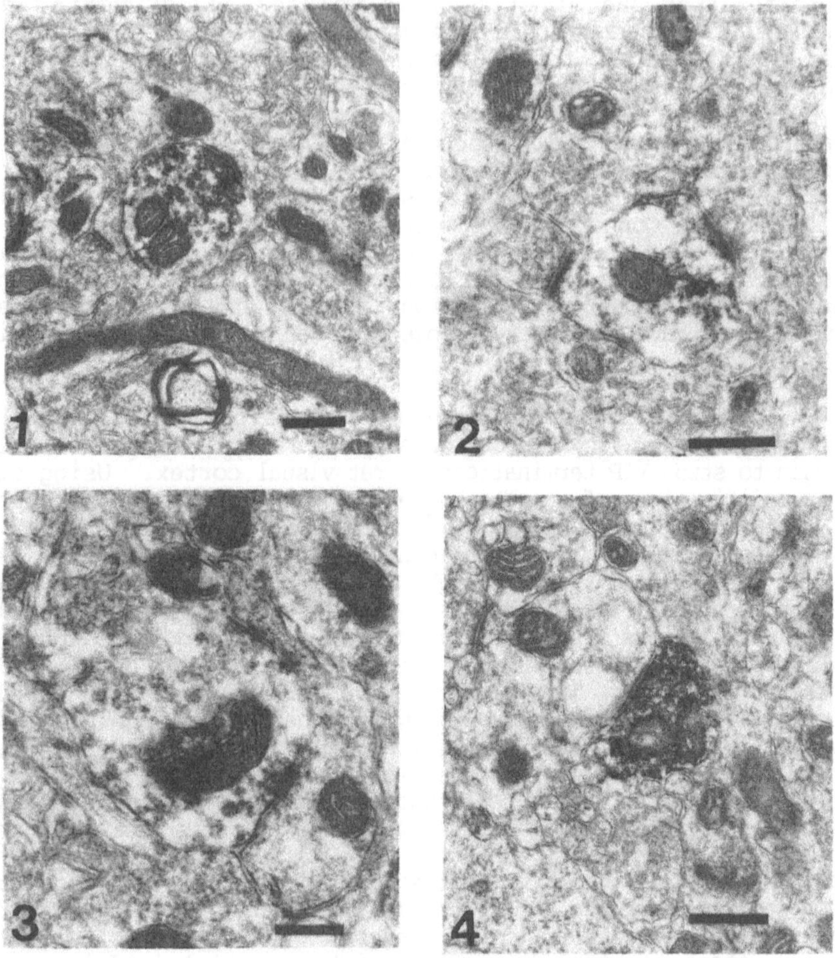

Figures 1-4. Ultrastructural localization of vaso-active intestinal polypeptide (VIP) by immunoperoxidase methodology in rat visual cortex. Rat brains were fixed and prepared as described for light microscopic localization of VIP (see refs in Magistretti and Morrison, 1983) except that triton was omitted, and cryoprotection was achieved with 10% dimethylsulfoxide in saline for rapid freeze-thaw exposures to facilitate antibody penetrance in vibratome sectioned slabs. Figs 1,3 and 4 show typical small VIP immunoreactive terminals with small dendritic spines. Those in Figs 1 and 4 are of the symmetric type, while that seen in Fig 3 shows a more pronounced post-synaptic specialization. Large granular vesicles can be seen in the large reactive bouton in Fig 3. In Fig 2, a typical immunoreactive dendritic profile bearing reactive synaptic vesicles can be seen receiving contact from at least 6 non-reactive boutons. Calibration bar= 0.5μ.

within functional systems as well as by the lamina, the species and the age of the primate (see Morrison et al., 1984; Morrison and Foote, 1986). Recent detailed examinations of several primate cortical regions substantiate unique innervation patterns for each of the monoamines within regions with cytoarchitectonically specified laminar specializations.

In squirrel monkey primary visual cortex, the fiber systems containing specific markers for serotonin or noradrenaline exhibited a high degree of complementary laminar specialization: layers V and VI receive a moderately dense NA projection and a sparse 5-HT projection, whereas layers IVa and IVc receive a very dense 5-HT projection and are largely devoid of NA fibers (c.f. Morrison and Foote, 1986). These patterns of innervation again offer hypothetical transmitter system interactions: the raphe-cortical 5-HT projections should regulate geniculo-striate input to spiny stellate cells with the coeruleo-cortical NA projection directed predominantly at pyramidal output cells.

This interpretation of the light miscroscopic laminar patterns also rests upon ultimate answers through ultrastructural immunocytochemistry. Although neither direct (permanganate fixed) or immunocytochemical detection of primate noradrenergic terminals seems feasible at this time, immunocytochemical detection of 5-HT terminals has been achieved by several groups (see Azmitia and Gannon, 1986). We have been analyzing such immunoreactive terminals within the primate visual cortex for the past 3 years. We find that within layers III-V, a very high density of 5-HT immunoreactive terminals can be found that approaches 1% of the total boutons. These boutons generally make symmetrical synapses on small dendritic spines that frequently receive other non-reactive axo-spinous terminals. The 5-HT boutons vary considerably in their diameter, ranging from a few tenths of a micron to more than ten times that size (see Figs. 5-7). Again, far more than half the boutons in a single random section exhibit specialized contact zones, occasionally assymetric in nature, and very rarely making axosomatic junctions on small intracortical neurons. When boutons are found in serial sections (see Figs. 6,7), their specialized contact zone occupies a substantial portion of the bouton-spine interface. Interestingly, these immunoreactive 5-HT terminals show few if any of the large granular vesicles observed by others in presumptive rodent 5-HT cortical terminals. Thus, the important structural interaction presumed from light microscopy can be supported by this preliminary ultrastructure.

Anticipating Transmitter Interactions from Laminar Innervation Patterns

The extensive innervation of the primate visual cortex by NA and 5-HT represents two distinct global systems independently

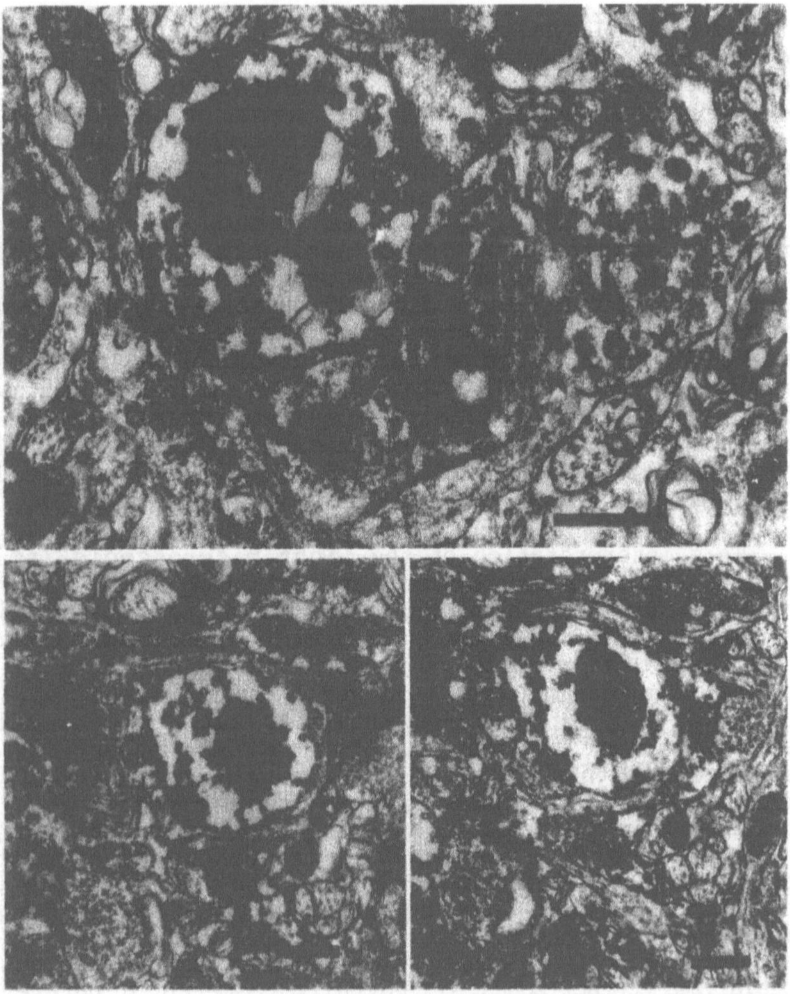

Figures 5-7. Ultrastructural localization of 5-HT by immunoperoxidase methodology in cynomolgous primary visual cortex. Tissue was fixed and prepared as for light microscopy (see Morrison and Foote, 1986) with omission of triton. Fig 5 portrays a relatively large immunoreactive terminal that is making a specialized (assymetric) contact on its upper right surface with a dendritic spine which is less that 1/5 the diameter of the bouton. Immunoreactivity within the bouton is seen on small electron lucent vesicles, and one possible large granular vesicle. Figs 6 and 7 represent serial sections through the same immunoreactive terminal, around whose perimeter are several very small dendritic spines. The one spine located at approximately "7 o'clock" on the perimeter contains a prominent assymetric post-synaptic specialization in both sections, estimated to be 2-300 nM apart. Calibration bar = 0.5μ.

suited to selective transmitter interaction. The primate auditory cortex (see Fig.8) offers another potentially heuristic system for transmittter-transmitter interactions. Here there are extensively specialized and highly unique laminar innervation patterns for acetylcholine-, noradrenaline-, dopamine- and serotonin-containing fibers (Campbell, Lewis and Morrison, in preparation). Dopamine containing terminals are present throughout primate neocortex in numbers and density patterns far exceeding any seen in rodent cortex (Lewis et al., 1986). Because the targets of cortical and

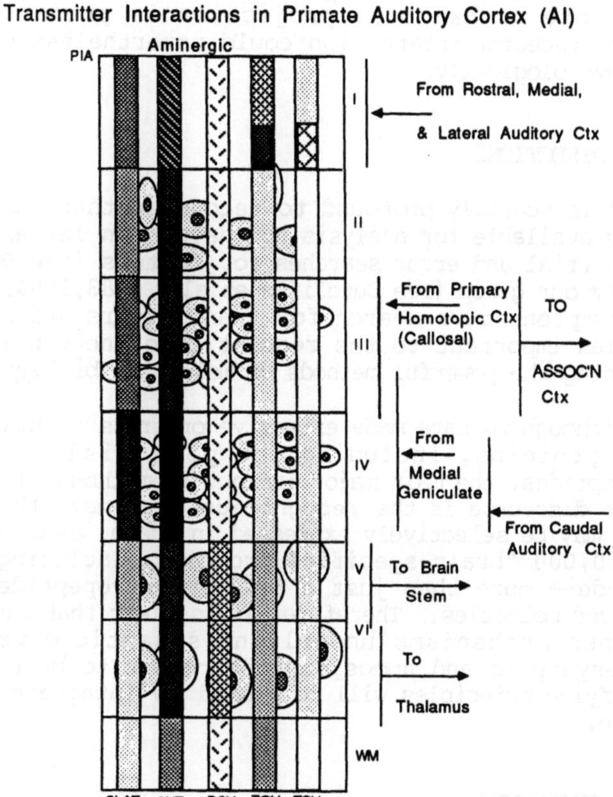

Figure 8: Graphic schematic of the lamina by lamina estimates of the terminal field innervation patterns in cynomolgous primary auditory cortex of the 4 major monoamine immunoreactivity markers (ChAT= Choline Acetyltransferase; DβH= Dopamine-β-hydroxylase; and TOHr and TOHc indicate tyrosine hydroxylase in the rostral and caudal portions of this cortical field). The monoamine patterns, indicated by grey scales proportional in darkness to the innervation density, are superimposed on icons representing the major cell classes of this cortical field, whose maker afferent and efferent connections are indicated at right. (Monoamine localization data and connectivity assessments from Campbell, Lewis and Morrison, in preparation.)

sub-cortical afferents to this region are well known, the laminar patterns of the monoaminergic innervation suggests which monoamines should be able to interact with which afferent systems to this region. These predictions are directly testable, albeit with some effort. Thus, by evaluating the relative effects of stimulating a specific monoaminergic source nucleus, it should be possible to test the presumption that acetylcholine and serotonin would regulate medial geniculate input, and that all 4 systems might produce significant interactions on responses to callosal input. While the identity of the transmitter for either the geniculate or callosal projection is presently unknown, receptor-receptor interaction could nevertheless be documented here physiologically.

FUTURE DIRECTIONS

It is scarcely profound to recognize that the transmitter systems available for analysis now have been largely discovered through trial and error searches for factors (see Bloom, 1984). Recently our group (see Sutcliffe et al., 1983,1984) initiated a more comprehensive search for transmitters and other unique molecules important to the regulation of neuronal activity by exploiting the powerful methods of molecular biology.

Although we have made extensive progress in identifying novel brain proteins, including one potential new family of neuropeptides, the main major insight from those studies for the present discourse is the recognition that more than 60% of the genome may be selectively expressed in brain, amounting to no less than 30,000 "brain-specific" proteins, including-- one must conclude-- more than just a few new neuropeptides, and other messenger molecules. Therefore, it is clear that many signals and response mechanisms underlying synaptic diversity, both heterosynaptic and homosynaptic, remain to be found and that simplifying principles will be needed to integrate this complex picture.

ACKNOWLEDGEMENTS

This research is supported by the McNeil Pharmaceutical Company, the Whittier Foundation and NINCDS 22347. We thank Madeline Cappello for preparation of ultrastructural material, and Nancy Callahan for manuscript preparation.

REFERENCES

Azmitia, E.C. and Gannon, P.J. (1986). The primate serotonergic system: A review of human and animal studies and a report on macaca fascicularis. In Advances in Neurology, Vol. 43:

Myoclonus. (eds. S. Fahn et al.). Raven Press, New York, p. 407.

Bloom, F.E. (1974). To spritz or not to spritz: The doubtful value of aimless iontophoresis. Life Sci. 14, 1819-1834.

Bloom, F.E. (1979). Central noradrenergic neurons: Structure-function considerations. In Catecholamines: Basic and Clinical Frontiers, (ed. E. Usdin). Pergamon Press, pp. 609-618.

Bloom, F.E. (1984). The functional significance of neurotransmitter diversity. Am. J. Physiol., 246, C184-C194.

Connor, J.R. and Peters, A. (1984). Vasoactive intestinal polypeptide-immunoreactive neurons in rat visual cortex. Neuroscience 12, 1027-1044.

Ferron, A., Siggins, G.R. and Bloom, F.E. (1985). Vasoactive intestinal polypeptide acts synergistically with norepinephrine to depress spontaneous discharge rates in cerebral cortical neurons. Proc. Natl. Acad. Sci. U.S.A. 82, 8810-8812.

Foote, S.L., Bloom, F.E., and Aston-Jones, G. (1983). Nucleus locus ceruleus: new evidence of anatomical and physiological specificity. Physiological Reviews 63, 844-914.

Lewis, D.A., Campbell, M.J., Foote, S.L., Goldstein M., and Morrison, J.H. (1986). The distribution of tyrosine hydroxylase-immunoreactive fibers in primate neocortex is widespread but regionally specific. J. of Neuroscience, in press.

Molliver, M.E., Grzanna, R., Lidov, G.H.W., Morrison, J.H., and Olschowka, J.A. (1982). Monoamine systems in the cerebral cortex. In Cytochemical Methods in Neuroanatomy. Alan R. Liss, Inc., New York. pp. 255-277.

Morrison, J.H. and Foote, S.L. (1986). Noradrenergic and serotonergic innervation of cortical, thalamic, and tectal visual structures in old and new world monkeys. J. Comp. Neurol. 243: 117-138.

Morrison, J.H. and Magistretti, P.J. (1983). Monoamines and peptides in cerebral cortex: Contrasting principles of cortical organization. Trends in Neurosciences 8: 146-151.

Morrison, J.H., Foote, S.L. and Bloom, F.E. (1984). Laminar, regional developmental and functional specificity of monoaminergic innervation patterns in monkey cortex. In Monoamine Innervation of Cerebral Cortex, (eds. L. Descarries, T.A. Reader, and H.H. Jasper). Alan R. Liss, Inc., pp. 61-75.

Peters, A. and Kimerer, L.M. (1981). Bipolar neurons in rat visual cortex: a combined Golgi-electron microscope study. J. Neurocytology 10, 921-946.

Siggins, G.R. and Gruol, D.L. (1986). Synaptic mechanisms in the vertebrate central nervous system. In Handbook of Physiology Volume on Intrinsic Regulatory Systems of the Brain, (ed. F.E. Bloom). The American Physiological Society, Bethesda, Maryland, pp. 1-114.

Sutcliffe, J.G., Milner, R.J., Shinnick, T.M., and Bloom, F.E. (1983). Identifying the protein products of brain specific genes with antibodies to chemically synthesized peptides. Cell 33, 671-682.

Sutcliffe, J.G., Milner, R.J., Gottesfeld, J.M., and Reynolds, W. (1984). Control of neuronal gene expression. Science 225, 1308-1315.

2

Interactions of Peptides and Monoamines in Central Neurons: Role of Second Messengers

George R. Siggins, André Ferron, Jorge Mancillas, Samuel Madamba and Floyd E. Bloom

INTRODUCTION

Our interest in receptor-receptor interactions and the involvement of second messengers started in the late 1960s, with the observation that noradrenaline (NA) and cyclic AMP augment the amplitude of excitatory postsynaptic potentials, while at the same time directly hyperpolarizing cerebellar Purkinje cells (Siggins et al., 1971; Hoffer et al., 1973). This finding was later extended by the studies of Foote et al. (see 1983) and Woodward et al. (1979), showing that NA, although directly inhibitory, also could enhance both inhibitory and excitatory responses to other transmitters. This augmentation of transmitter responses was termed "neuromodulation" by some but "enabling" by Bloom (1979). Because neuromodulation implies lack of a direct effect by the "neuromodulator", whereas NA clearly has a direct inhibitory effect on most central neurons, we prefer to use the "enabling" term (see Siggins and Gruol, 1986).

In the time since these early observations, many reports have appeared showing similar interactive phenomena between neurotransmitters, and especially between monoamines and neuropeptides. The important observations of the coexistence of monoamines and peptides in nerve fibers (see Lundberg et al., this volume) have now added even greater incentive for investigation of such transmitter interactions. The mechanisms of these interactions are not yet well-defined, but could include: 1) interactions of nearby (either subsynaptic or extrasynaptic) receptors; 2) alteration by one transmitter of the biophysical properties of the postsynaptic membrane, thus indirectly altering the response to the second transmitter; 3) the interaction of second messengers generated by the transmitters in the target cell.

In this chapter we will discuss three recent observations of unusual amine-peptide interactions in the CNS. The first two, studied in vivo, involve what appear to be the expression of converging inputs to a single target neuron (see Bloom et al., this volume) and the third exemplifies a potential interaction involving coexisting substances.

Synergism Between NA and VIP in Cerebral Cortex.

A transmitter role for vasoactive intestinal polypeptide (VIP) in cerebral cortex is supported by several biochemical, histochemical and electrophysiological studies (see Siggins and Gruol, 1986; Bloom et al., this volume). Both VIP and NA stimulate cyclic AMP formation and glycogenolysis in cortical slices (see Magistretti, this volume). As suggested by Bloom et al.(this volume), NA- and VIP-containing neurites in cortex may converge on a common target neuron, the pyramidal cell. Interestingly, VIP and NA synergistically increase cyclic AMP in cerebral cortex (Magistretti, this volume). It may be relevant that electrophysiological studies of identified cortical pyramidal neurons have suggested that the depressions of discharge evoked by iontophoresis of NA may be mediated by cyclic AMP (Stone and Taylor, 1977). Therefore, we (Ferron et al., 1985) applied VIP and NA iontophoretically to rat cortical neurons to determine if this interaction was detectable at the single cell level.

We first tested the direct effects of VIP alone on the spontaneous firing rate of sensorimotor cortical neurons. VIP depressed 24%, excited 20%, had biphasic effects on 2% and had no apparent effect on 54% of over 100 cells. Fourteen cells identified as pyramidal neurons by antidromic activation showed the same heterogeneity of response as unidentified cells. As previously reported (see Foote et al., 1983), NA alone had a predominantly depressant action. When a neuron responded to NA (5-100 nA), the current was reduced until little or no NA effect was detectable (usually 1-5 nA) and possible interactions between NA and VIP were examined. In over half of 47 neurons studied, ejection of VIP during subthreshold NA administration resulted in pronounced reduction of firing rate (see fig 1 of Ferron et al., 1985). Such synergistic inhibitions resulted whether VIP alone had previously elicited excitatory, inhibitory, or negligible effects on the test neuron. When VIP alone had a depressant action, administration of NA markedly enhanced this depressant effect. Even when VIP alone was excitatory, concurrent subthreshold NA treatment reversed the VIP effect to inhibition (6 of 9 cells).

NA alone activates cyclic AMP synthesis in cortex via β-adrenergic receptors, yet the synergism between VIP and NA in activating cyclic AMP production is mediated by an alpha adrenoceptor (Magistretti; this volume). Therefore, we examined the effect of phenylephrine, an alpha receptor agonist, on neuronal responses to VIP. In 10 cells showing an interaction between VIP and NA, 9 revealed comparable, pronounced interactions with phenylephrine. Thus, the interaction of VIP and NA at the cellular level may also involve alpha receptor activation.

Our results suggest that the direct response to VIP could depend upon the extent of ongoing activation of adrenoreceptors by endogenous NA, or that VIP enhances NA-induced inhibitions so that a direct VIP excitatory effect is masked by these inhibitions. The cortical influence of VIP-containing bipolar neurons could be viewed as bi-directionally conditional. That is, under minimal noradrenergic tonus these neurons might transmit excitatory information, whereas during periods of enhanced NA release (for example, upon behavioral arousal (Foote et al., 1983) they could become powerfully inhibitory.

Our demonstration of an interaction between VIP and NE at the cellular level is consonent with the biochemical data on cortical slices using cyclic AMP generation as the end-point (Magistretti, this volume). Such parallel findings may point to the mediation by cyclic AMP of NA and VIP evoked depressions of neuronal firing in cortex, as previously suggested by Stone and Taylor (1977) for NA. The reported enhancement by NA of synaptic and other transmitter responses, including inhibitory ones, (see Foote et al., 1983) may be related phenomena. However, it is not yet clear how these metabolic and electrophysiologic events are actually linked. If NA and VIP-containing fibers do indeed converge on the same cortical target cell, it is feasible that cyclic AMP is the intracellular mediator of their synergistic interaction.

Interactions of Somatostatin and Acetylcholine Effects.

It is now evident that the tetradecapeptide somatostatin (SS14) has a widespread distribution not only in the hypothalamus but throughout the central and peripheral nervous systems. Although there is considerable evidence for its role as a neurotransmitter in the central nervous system (see Siggins and Gruol, 1986), the effect of SS14 on neuronal activity is of some debate. In hippocampus and cerebral cortex, where somatostatin-containing neurons have been identified (see Morrison et al., 1982,1983), SS14 was first reported to depress neuronal activity in vivo (Renaud et al., 1977). Although studies in our laboratory have also revealed predominantly depressant or hyperpolarizing effects of SS14 in hippocampus (see Pittman and Siggins, 1981), other authors have reported excitatory or depolarizing actions (Dodd and Kelly, 1978; Olpe et al., 1980; Delfs and Dichter, 1983). The reason for these differences is unclear.

Recently, we observed that SS14 had inhibitory effects on the spontaneous firing of hippocampal pyramidal cells in vivo unless acetylcholine (ACh) was included in the multibarrel iontophoresis assembly. Therefore, we (Mancillas et al., 1986) investigated this phenomenon in more detail in the dorsal hippocampus and parietal cortex, where fields of somatostatin-containing (Morrison et al., 1982,1983) and cholinergic (see, e.g., Houser et al., 1985) nerve terminals overlap.

In halothane anesthetized rats, SS14 applied by electroosmosis depressed spontaneous discharge rate in about 90% of all 58 cells tested in either hippocampus (17 CA1 and 19 CA3 pyramidal cells) or cortex (22 cells). None of the cells showed increased firing with SS14 alone. The SS14-induced inhibitions were consistently slow in onset and persisted beyond the offset of application (see Fig. 1A of Mancillas et al., 1986). By contrast, ACh (applied by iontophoresis) facilitated firing with a relatively fast time-course, as previously reported (see Siggins and Gruol, 1986). When brief repetitive iontophoretic pulses of ACh were applied during a period of sustained exposure (2-3 mins) to SS14, ACh responses were markedly (up to 100%) enhanced (see Fig 1B of Mancillas et al., 1986). This effect of SS14 was observed in 20 of 26 neurons tested, including 8 of 10 in cortex, 5 of 8 in CA1 and 7 of 8 in CA3. In contrast, SS14 typically depressed glutamate responses as well as baseline firing, in both cortex and hippocampus. The SS14 enabling effect appeared to be dose-dependent: low currents of SS14 depressed the basal firing between ACh-induced rate increases, but did not affect the magnitude of those increases, whereas greater SS14 currents evoked a clear increase in the magnitude of ACh responses. At still higher currents, SS14 initially enhanced responses to ACh, and then immediately depressed them and basal firing.

When the temporal relationship between the application of SS14 and ACh was reversed (that is, with short iontophoretic pulses of SS14 applied to neurons whose firing rate was tonically increased by continuous application of small, "leak" amounts of ACh) an increase in neuronal firing resulted. These apparent "excitatory" responses to SS14 pulses during tonic ACh activation would seem to be dependent on an activation of a muscarinic cholinergic receptor, as they were blocked by the muscarinic antagonist scopolamine.

Our results indicate that iontophoretic application of SS14 can yield different, but consistent, effects on postsynaptic cell activity, depending on what other neurotransmitters are being tested simultaneously. This interaction appears to be selective for SS14 and ACh, as there was no enhancement of responses to glutamate. It does not depend on the sequence of tests.

We do not yet understand the molecular or cellular mechanisms that underlie the interactions between SS14 and ACh. We recognize that the two SS14 actions could arise from any of several sites (e.g., pre- or postsynaptic) where ACh is known to act (see Siggins and Gruol, 1986). Further studies in hippocampal slice preparations are now being pursued to elucidate the mechanisms. Nevertheless, the observations reported here offer several implications. First, they raise the possibility that the previously reported excitatory actions of SS14 could have resulted from unrecognized influences of exogenous or endogenous ACh. Second, if the effects of iontophoretic SS14 accurately reflect those of synaptically released SS14, the peptide could be conceived to have at least two types of conditional effects on hippocampal and cortical neurons. Its "primary" effect, in the absence of ACh, is to inhibit spontaneous discharge, but in the presence of ACh, the functional consequences may depend on the concentration and temporal context of the release of the two substances. Thus, the SS14/ACh interaction in cortex and hippocampus also might be considered bidirectionally conditional. The possible role of second messengers in this interaction has not been explored directly but is suggested by studies indicating an involvement of cyclic nucleotides, Ca^{++} or the triphosphoinositol system in responses to SS14 or ACh individually (see, e.g., Reisine and Guild, 1985; Malenka et al., 1986).

Somatostatin and GABA Responses in the Hippocampal Slice.

In the hippocampus and other brain regions, it is now recognized that some form of somatostatin probably coexists with gamma-aminobutyric acid (GABA) (Somogyi et al., 1984). Because many terminals surrounding the CA1 pyramidal cell layer (and similar to GABAergic "baskets") are immunohistochemically reactive for certain prosomatostatin-derived peptides (Morrison et al., 1982), it seems likely that GABA could be co-released with somatostatin(s) onto pyramidal neurons. Therefore, we have begun studies designed to test whether SSs alter responses of CA1 pyramidal neurons to GABA, using the hippocampal slice preparation and intracellular voltage recording. The hippocampal slice was maintained completely submerged in continuously flowing artificial CSF (see Pittman and Siggins, 1981). Pulses of GABA (5-15 sec long) were applied by pressure from a micropipette to the pyramidal cell layer before, during and after superfusion of SS14 or SS28 (the prohormone precursor for SS14 with potent somatostatinergic actions).

Our results to date provide an interesting contrast to those conditional ones obtained with NA/VIP and SS14/ACh. As described previously (Pittman and Siggins, 1981), superfusion of SS14 or SS28 alone (0.1 to 1 μM) produced long-lasting inhibitory and hyperpolarizing effects (Fig 1), accompanied by 10-15% decreases in input resistance signifying a slightly increased conductance (probably to K^+ ions). GABA applied by pipette also often hyperpolarized CA1 pyramidal cells when K-acetate was used in the recording micropipette, although pure depolarizations and biphasic depolarizing/hyperpolarizing responses were also observed. The hyperpolarizations were accompanied by large (40-75%) decreases in input resistance; since no hyperpolarizations were seen when Cl^- was injected into the cells via KCl-containing pipettes, this probably represents an increased conductance to Cl^- ions. The depolarizations obtained with K-acetate filled recording pipettes were accompanied by smaller apparent decreases (10-40%) in input resistance.

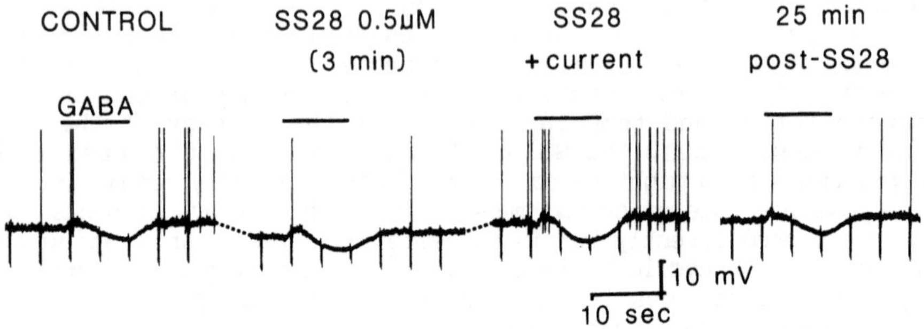

Figure 1. Superfusion of SS28 hyperpolarizes a CA1 pyramidal cell and augments GABA-evoked hyperpolarizations. GABA administered by pressure (10 psi) from a micropipette (bars above record). Dotted line indicates 4 mV of hyperpolarization by SS28 (2nd panel). Then, positive current was used (3rd panel) to return the membrane potential back to resting levels (-68 mV), whereupon the size of the GABA hyperpolarization increased to 175% of control. Also note the SS28 potentiation of the GABA-evoked reduction in the periodic hyperpolarizing pulses monitoring input resistance: thus, the GABA-evoked conductance increase is also potentiated by SS28.

Interactions of SS and GABA were subtle. In 2 of 4 pyramidal neurons tested to date there was enhancement by SS14 or SS28 of the GABAergic hyperpolarization. The remaining 2 cells showed no change in GABA hyperpolarizations. However, the hyperpolarization produced by the SS would likely move the membrane potential closer to the equilibrium potential for Cl^-, thus tending to reduce the GABA hyperpolarization. Indeed, when we artifically hyperpolarized the membrane by injection of negative current from the recording pipette, the GABA-evoked hyperpolarizations were reduced, abolished or reversed to depolarizations. Therefore, we tested the effects on the SS/GABA interaction of "clamping" the membrane potential at various levels. As seen in figure 1, whereas SS28 superfusion hyperpolarized the neuron and only slightly enhanced the size of the GABA-evoked hyperpolarization, when the membrane potential was brought back to the original resting potential by positive current injection, a clear SS28 potentiation of the GABA hyperpolarization occurred.

In the light of the mechanisms responsible for SS and GABA effects, such subtle interactions make functional sense. Under some circumstances (e.g., during hyperexcitability) there may be need for stronger inhibitory controls than obtained with release of GABA alone. It is likely that large amounts of GABA will have a self-limiting effect due to: 1) shunting effect of the large increase in Cl^- conductance, 2) hyperpolarization toward the Cl^- equilibrium potential, and 3) desensitization to GABA. Indeed, we have noted that GABA superfusion can reduce the hyperpolarizing response to GABA applied by pipette. If, as in some peripheral systems (see Lundberg, this volume), the co-localized peptide is released only with trains of presynaptic volleys, this might augment and extend (by virtue of the longer action of SS) the shorter GABAergic response. The use of K^+ channel activation by SS thus may provide a synergistic adjunct to the Cl^- channel activation by GABA.

However, we recognize that there are multiple effects of GABA, mediated by different GABA receptor subtypes (see Siggins and Gruol, 1986). We have focussed on the hyperpolarizations associated with large conductance increases most likely involving Cl^- (via GABA-A receptor activation). Further studies are needed to determine the effect of SS on GABA-B responses (also probably hyperpolarizing but involving K^+) and the depolarizing dendritic responses to GABA. The possible involvement of second messengers in this interaction is untested, but it seems relevant that SS14 may have an inhibitory influence on cyclic AMP generation (via the N_i G-protein) and block the ability of cyclic AMP to raise intracellular Ca++ levels (see Reisine and Guild, 1985).

CONCLUSIONS AND FUTURE DIRECTIONS

In this account we have described three amine-peptide interactions, two that may result from converging (or heterosynaptic) inputs and that appear to be bidirectionally conditional in effect, and one resulting from a co-existing (or homosynaptic--see Bloom, this volume) input that is "enabling" or additive in effect. However, it must be emphasized that the interactive studies described above (and indeed almost all others reported elsewhere) involve the testing of transmitter candidates applied exogenously and not released synaptically. Therefore, interpretations based on these results are subject to the criticisms of the methods used to apply the transmitters (see Siggins and Gruol, 1986). Future studies of such interactions must attempt to correlate these early findings with those obtained with synaptically released transmitters, as has been possible in the peripheral nervous system (see Lundberg, this volume).

In the past, most electrophysiological studies have dealt principally with each possible synapse or neurotransmitter in isolation, as if it transmitted a single, immutable message from one neuron to another under all conditions. However, it is becoming evident that such an idealized one-on-one signal holds strictly only for a very few models, such as the motor endplate on skeletal muscle. In the CNS, the receptive field of each dendrite or soma must receive thousands of synaptic inputs, using an untold number of different messengers. Obviously, the chances for simultaneous or near simultaneous discharge of these inputs must be high and interactions of their effects highly probable, especially for transmitters that are long-acting. Thus, it is no longer sufficient simply to test a single neurotransmitter.

A given neuronal region may see a 'cocktail' mixture of many neuromessengers, each in a wide range of concentrations, depending upon the discharge rate of the synaptic terminals releasing them. The question of what concentrations of each transmitter should be tested confounds the issue: in general the peptides seem more potent than the other transmitters, but does this mean that they should be co-tested in much lower amounts than the others? More data are needed on amounts or proportions of transmitters actually released, and the concentrations seen by the target cell receptors. It is not even certain that the lowest effective concentration is the most physiologically meaningful: perhaps the most sensitive responses only reflect overflow of excessive transmitter to more remote sites that might not usually 'see' the messenger except in very low concentrations.

The various types of interactions of transmitters now becoming apparent may actually help us answer a major question on the need for a diversity of neurotransmitters (see Siggins and Gruol, 1986). Some neurobiologists estimate that there may be fifty or more neuromessengers, and the number increases rapidly with the discovery of each new peptide. What then is the need for all of these substances, if many of them, tested singly, act like other transmitters. The answer may lie in the types of interactions elicited by the appearance of two or more substances together. Thus, each transmitter may provide a unique pattern or "fingerprint" of holistic actions derived from a combination of its direct effects with its various forms of interactions with other transmitters. These unique interactions might well involve various permutations and combinations of the several types of second messengers.

This research was supported by USPHS Grant AA-06420 and AM-26741.

REFERENCES

Bloom, F.E. (1979). Chemical integrative processes in the central nervous system. In Neurosciences - Fourth Intensive Study Program. (eds. Schmitt, F.O. and Worden, F.G.). MIT Press, Cambridge, pp. 51-58.

Delfs, J.R. and Dichter, M.A. (1983). Effects of somatostatin on mammalian cortical neurons in culture: physiological actions and unusual dose-response characteristics. J. Neurosci. 3, 1176-1188.

Dodd, J. and Kelly, J.S. (1978). Is somatostatin an excitatory transmitter in the hippocampus? Nature 273, 674-675.

Ferron, A., Siggins, G.R. and Bloom, F.E. (1985). Vasoactive intestinal polypeptide acts synergistically with noradrenaline to depress spontaneous discharge rate in cerebral cortical neurons. Proc. Natl. Acad. Sci. U.S.A. 82, 8810-8812.

Foote, S.L., Bloom, F.E. and Aston-Jones, G. (1983). Nucleus locus ceruleus: new evidence of anatomical and physiological specificity. Physiol. Rev. 63, 844-914.

Hoffer, B. J., Siggins, G. R., Oliver, A. P., and Bloom, F. E. (1973). Activation of the pathway from locus coeruleus to rat cerebellar Purkinje neurons: Pharmacological evidence of noradrenergic central inhibition. J. Pharmacol. Exp. Therap. 184, 553-569.

Houser, C.R., Crawford, G.D., Salvaterra, P.M. and Vaughn, J.E. (1985). Immunocytochemical localization of choline acetyltransferase in rat cerebral cortex: A study of cholinergic neurons and synapses. J. Comp. Neurol. 234, 17-34.

Malenka, R.C., Madison, D.V., Andrade, R. and Nicoll, R.A. (1986). Phorbol esters mimic some cholinergic actions in hippocampal pyramidal neurons. J. Neurosci. 6(2), 475-480.

Mancillas, J.R., Siggins, G.R. and Bloom, F.E. (1986). Somatostatin selectively enhances acetylcholine-induced excitations in rat hippocampus and cortex. Proc. Natl. Acad. Sci. U.S.A., in press, 1986.

Morrison, J.H., Benoit, R., Magistretti, P.J. and Bloom, F.E. (1983). Immunohistochemical distribution of pro-somatostatin-related peptides in cerebral cortex. Brain Res. 262: 344-351.

Morrison, J.H., Benoit, R., Magistretti, P.J., Ling, N. and Bloom, F.E. (1982). Immunohistochemical distribution of prosomatostatin-related peptides in hippocampus. Neurosci. Lett. 34, 137-142.

Olpe, H-R., Balcar, V.J., Bittiger, H., Rink, H. and Sieber, P. (1980). Central actions of somatostatin. Eur. J. Pharmacol. 63, 127-133.

Pittman, Q.J. and Siggins, G.R. (1981). Somatostatin hyperpolarizes hippocampal pyramidal cells in vitro. Brain Res. 221, 402-408.

Reisine, T. and Guild, S. (1985). Pertussis toxin blocks somatostatin inhibition of calcium mobilization and reduces the affinity of somatostatin receptors for agonists. J Pharm. Exper. Therap. 235, 551.

Renaud, L.P., Pittman, Q.J., Blume, H.W., Lamour, Y. and Arnauld, E. In Central Nervous System Effects of Hypothalamic Hormones and Other Peptides (eds. Collu, et al.). Raven Press, New York pp. 147-161.

Siggins, G. R., Oliver, A. P., Hoffer, B. J., and Bloom, F. E. (1971). Cyclic adenosine monophosphate and norepinephrine: Effects of transmembrane properties of cerebellar Purkinje cells. Science 171, 192-194.

Siggins, G.R. and Gruol, D.L. (1986). Mechanisms of transmitter action in the vertebrate central nervous system. In Handbook of Physiology, Volume on Intrinsic Regulatory Systems of the Brain. (ed. F.E. Bloom). The American Physiological Society, Bethesda, Maryland, pp. 1-114.

Somogyi, P., Hodgson, A.J., Smith, A.D., Nunzi, M.G., Gorio, A. and Wu, J-Y. (1984). Different populations of GABAergic neurons in the visual cortex and hippocampus of cat contain somatostatin or cholecystokinin-immunoreactive material. J. Neurosci. 4, 2590-2603.

Stone, T.W. and Taylor, D.A. (1977). Microiontophoretic studies of the effects of cyclic nucleotides on excitability of neurones in the rat cerebral cortex. J. Physiol. 266, 523-543.

Woodward, D.J., Moises, H.C, Waterhouse, B.D., Hoffer, B.J. and Freedman, R. (1979). Modulatory actions of norepinephrine in the central nervous system. Federation Proc. 38, 2109-2116.

3
Multiple Transduction Mechanisms for Dopamine D2 Receptors in Anterior Pituitary Cells

M. Memo, C.Missale and P.F. Spano

A large body of experimental evidence points to the existence of multiple classes of dopamine (DA) receptors in mammalian brain and peripheral tissues which can be distinguished by their pharmacological specificity and localization and by their transduction mechanisms (Spano et al., 1978; Kebabian and Calne, 1979). Specifically, two types of DA receptors whose stimulation affects cellular cyclic AMP have been recently characterized. Stimulation of the so-called D1 DA receptors increases cyclic AMP formation, whereas stimulation of the D2 DA receptors reduces the formation of cyclic AMP (Spano et al., 1978; Stoof and Kebabian, 1981). Recent observations, however, rise the question whether the transduction mechanism of the DA signal may involve second messengers other than cyclic AMP (Simmonds and Strange, 1985; Memo et al, 1986). The studies on the transduction mechanisms induced by DA receptor stimulation have taken great advantage from the identification in the anterior pituitary of a single class of DA receptors, i.e. the D2 type (Spano et al., 1978; Caron et al., 1978).Since this peculiar property, the anterior pituitary has been taken as a viable model for studying both intracellular modifications and functional expression induced by D2 DA receptor stimulation.

The anterior pituitary contains a particular type of endocrine cells, termed lactotrophs, which are specialized in the synthesis and release of prolactin (PRL). Like other pituitary cells, lactotrophs are excitable and show both spontaneous and evoked electrical activity consisting of calcium-dependent large amplitude spikes. Both secretory and electrical activities of lactotrophs are regulated by the brain through stimulatory and inhibitory substances which are released from nerve endings in the median eminence and interact with specific receptors located in the cell membrane. Different polypeptides, including thyrotropin-releasing hormone (TRH), vasoactive intestinal peptide (VIP), angiotensin II (A-II) and neurotensin (NT), have been described to stimulate PRL release (see Muller and MacLeod,1982, as review). Some of them, such as TRH and NT, depolarize excitable cells (Osawa and Kimura, 1979; Stanzione and Zieglgansberger, 1983). On the other hand, it is now generally accepted that DA, through stimulation of D2 receptors and the following hyperpolarization of the target cells, inhibits PRL release. The biological mechanisms involved in linking the receptor binding of these agents to the modulation of both electrical activity and PRL release are still not well understood.

PRL release appears to be regulated by the intracellular levels of at least two different second messengers: cyclic AMP and calcium. The evidence for an involvement of cyclic AMP in PRL release is that compounds which increase intracellular cyclic AMP levels may increase PRL release (Nagasawa and Yanay, 1972; Tam and Dannies, 1981; Schettini et al., 1983). It has been also shown that DA inhibits both basal and VIP-stimulated adenylate cyclase activity in different types of PRL-secreting cells thus suggesting that these intracellular events may be linked with the inhibition of hormone secretion (Onali et al., 1981; Enjalbert and Bockaert, 1983). However, several observations are consistent with the hypothesis that intracellular calcium plays a relevant role in the secretion of PRL (Thorner et al., 1980). In this respect, high potassium levels or sodium channel activators, which both depolarize pituitary cells and increase calcium influx, enhance PRL release. In addition, different peptides, such as TRH, A-II, and bombesin, appear to induce PRL release by activation of phospholipase C which leads to an increase of phosphatidylinositol breakdown and intracellular calcium concentration (Sutton and Martin, 1982, Enjalbert et al., 1986). Thus, PRL secretion appears to be promoted by activation of two distinct membrane transduction mechanisms which involve a rise in the intracellular concentrations of either cyclic AMP or calcium.

Among PRL releasing factors that do not affect adenylate cyclase system but do increase both phosphatidylinositol breakdown and intracellular calcium levels is NT (Goedert et al., 1984; Memo et al., 1985). This peptide releases PRL in different anterior pituitary cell preparations (Enjalbert et al., 1982, Memo et al., 1984). Our recent report indicates that the mechanism by which NT induces PRL release involve an activation of calcium channels resulting in an increase of intracellular calcium concentrations (Memo et al., 1986a). Since we found that DA was capable to inhibit the release of PRL induced by NT (Memo et al., 1986b), the NT receptors associated with calcium channels have been taken as a model to ascertain whether the DA receptors present in lactotrophs are functionally linked with calcium channels.

In this report, we shall briefly review our recent data on the characterization of NT receptors associated with calcium influx and PRL release and the identification of a DA receptor type which is functionally linked with NT-sensitive calcium channels.

CHARACTERIZATION OF NEUROTENSIN RECEPTORS ASSOCIATED WITH CALCIUM INFLUX AND PROLACTIN RELEASE

NT is a tridecapeptide which was first isolated from bovine hypothalamus by Carraway and Leeman in 1973. According to the neurochemical studies, large amounts of NT and its receptors are found in the brain and in gastrointestinal organs where this peptide has a wide physiological activity (Nemeroff and Prange, 1982). NT is present in the synaptosomal fraction of brain homogenates and this peptide is released from brain tissue in a calcium-dependent manner (Iversen et al., 1978). These findings suggest that NT may play the role of neurotransmitter in the central and peripheral nervous system.

TABLE 1

PROPERTIES OF ^{3}H-NEUROTENSIN BINDING TO RAT PITUITARY CELL MEMBRANE PREPARATIONS

binding parameters	
-High affinity	Kd = 2.6 nM
-Saturable	Hill coeff. = 0.96
-Definite number of binding sites	Bmax = 126 fmol/mg prot.
-Same affinity for kinetic and equilibrium constants	Kd (kinetic) = 2.3 nM Kd (equilib) = 2.6 nM
-Pharmacology	stereospecifically displaced
-Sensitivity to GTP	decrease in affinity constant

We found that both NT and its derivative L-Phe-NT increase PRL release from primary culture of anterior pituitary cells in continuously superfused columns in a dose- and time-dependent manner. On the contrary, D-Phe-NT was completely inactive in releasing PRL at any concentration used.This is clearly depicted in the representative dose-response curves in fig. 1. The ED-50

FIG. 1

EFFECTS OF DIFFERENT CONCENTRATIONS OF NEUROTENSIN AND ITS ANALOGS ON PRL RELEASE FROM PITUITARY CELLS

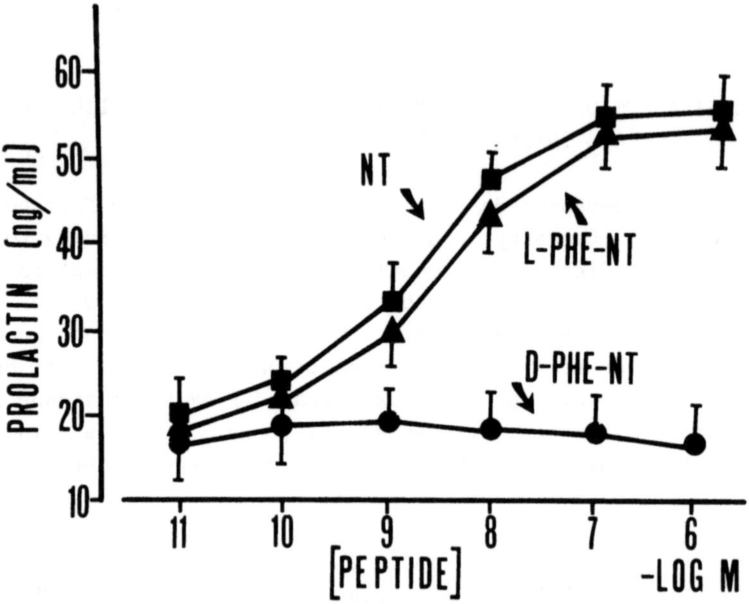

calculated from the results reported in fig. 1 were 1.5 nM for NT and 2.5 nM for L-Phe-NT. NT-stimulated PRL release appeared to be calcium-dependent since it was unaffected by pretreatment of the cells with the sodium channel blocker tetrodotoxine but virtually abolished by either depletion of cellular calcium with EGTA or calcium channel blockade by inorganic calcium entry blockers (data not shown).

On these bases, we characterized the NT receptors associated with PRL release in dispersed cells from the anterior lobe of rat pituitary and investigated on the second messengers involved in this action.

The ^{3}H-NT binding to membranes from anterior pituitary dispersed cells was found to be saturable and stereospecific. Scatchard analysis of the data gave a straight line indicating a Bmax value of 126 fmol/mg prot. and a Kd value of 2.6 nM. The calculated IC-50 for inhibiting ^{3}H-NT binding to hypophysial membrane preparations were 5.8 nM, 7.8 nM, and 3,000 nM, for NT, L-Phe-NT and the pharmacological inactive form D-Phe-NT, respectively. The kinetic and pharmacological characteristics of ^{3}H-NT binding to pituitary

FIG. 2

EFFECTS OF DIFFERENT CONCENTRATIONS OF NEUROTENSIN AND ITS ANALOGS ON CALCIUM INFLUX INTO PITUITARY CELLS

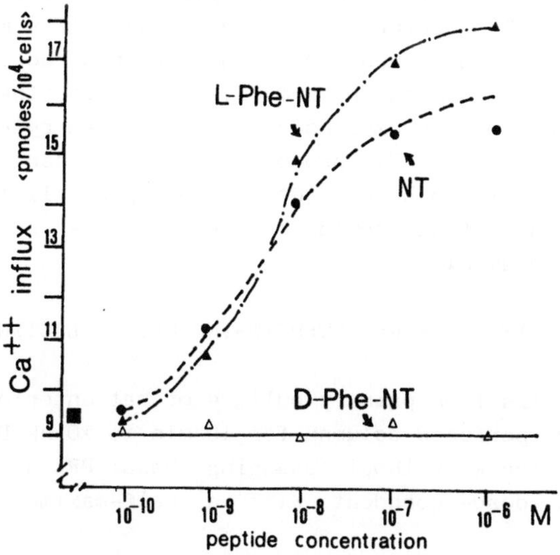

cell membranes are summarized in table 1.

To evaluate whether the NT-induced PRL release was cyclic AMP-mediated, we measured the effects of this peptide on both adenylate cyclase activity and cyclic AMP accumulation in different tissue preparations. We found that NT, up to the concentration of 1 uM, did not affect the cyclic AMP generating system in homogenates of anterior pituitary from either male or lactating female rat. The same pattern of results was obtained measuring cyclic AMP formation in intact cells. It is evident from these results that NT receptors present in pituitary cells are not coupled either in stimulation or in inhibtion with the adenylate cyclase system.

Since lowering the extracellular concentrations of calcium led to a marked reduction of the PRL release induced by NT, we tested the possibility that NT receptor stimulation allows a change in calcium channel permeability. Indeed, we found that NT and its analogs stereospecifically enhanced the influx of calcium into dispersed cells from rat anterior pituitary. The time required for NT to elicit a maximal increase in calcium influx was 10 min. A dose-related stimulation of calcium influx into the cells was obtained with a range of peptide concentrations from 0.1 nM to 1,000 nM. The results are illustrated in fig. 2. The concentrations required to elicit the half-maximal stimulation were 4.2 nM for NT, 5.8 nM for L-Phe-NT, and more than 1,000 nM for D-Phe-NT.

In summary, stimulation of NT receptors by NT or its analogs in dispersed cells from the anterior lobe of rat pituitary enhances the release of PRL and the influx of calcium in a stereospecific manner. The significant correlation of the effective concentrations of these peptides in binding and biological activities indicates that occupancy of NT receptors of anterior pituitary modifies the permeability of calcium channels resulting in an accumulation of intracellular calcium which presumably initiates the cascade of events that are ultimately expressed as an enhanced release of PRL. The mechanism by which NT promotes an increase in calcium permeability has still to be elucidated. One possibility is that NT-induced calcium influx could be mediated by activation of phosphoinositol breakdown.

DOPAMINE INHIBITION OF NEUROTENSIN-INDUCED PROLACTIN RELEASE

Exposure of cells from primary culture of rat anterior pituitary in continuously perfused columns for 10 min to 10 nM DA prevented NT-induced PRL release without changing basal PRL secretion. The effect of DA was dose-dependent with a half-maximal inhibitory

TABLE 2

INHIBITORY EFFECTS OF DOPAMINE AND VARIOUS DOPAMINERGIC AGNISTS ON THE INCREASE OF CALCIUM INFLUX AND PROLACTIN SECRETION BOTH INDUCED BY NEUROTENSIN

	Calcium influx IC-50 (nM)	PRL secretion IC-50 (nM)
Bromocriptine	1.0	0.8
Lisuride	2.5	1.5
Dopamine	7.0	5.0
N-propyl-norapomorphine	7.0	15
Apomorphine	10	20
Dihydroergotoxine	50	30
Epinephrine	100	500
SKF 38393	1,000	1,000
Norepinephrine	---	2,000
Phenylephrine	10,000	10,000
Isoproterenol	10,000	10,000

Values are the mean of three experiments run in triplicate with at least five concentrations of the tested drug.

concentration of 5 nM. Like DA, various DAergic agonists, such as bromocriptine, lisuride, and apomorphine, inhibit NT-induced PRL release.

In the same experimental conditions, 10 nM DA completely suppresses NT-induced calcium influx into pituitary cells. The IC-50 of DA for this effect was 7 nM. The inhibitory action of DA was mimicked by bromocriptine, lisuride, and apomorphine and stereospecifically reversed by sulpiride. The data obtained for DAergic inhibition of both PRL release and calcium influx elicited by NT are summarized in table 2. (Memo et al., 1986b).

The close similarity in the concentrations required to inhibit PRL secretion and calcium influx supports the hypothesis that DA, as well the various DAergic agonists, prevents NT-induced PRL release by impairing the capability of NT to open calcium channels.

It has been reported that DA and various DAergic agonists inhibit VIP stimulation of PRL release in a variety of cell preparations (Enjalbert et al., 1982; Memo et al., 1984). The effects of VIP are generally believed to be mediated by a cyclic AMP-dependent pathway and DA appears to act modulating in an inhibitory way VIP stimulation of adenylate cyclase activity (Onali et al., 1981; Nicosia et al., 1982). Thus, DA receptors negatively coupled with adenylate cyclase have been proposed to be the DA receptors involved in inhibiting PRL secretion (Enjalbert and Bockaert, 1983). Unlike VIP, NT do not stimulate adenylate cyclase activity, thus rising the question whether or not the inhibitory action of DA on NT-induced calcium influx was cyclic-AMP mediated.

TABLE 3

EFFECTS OF DOPAMINE ON NEUROTENSIN AND VIP STIMULATED PROLACTIN SECRETION FROM CONTROL AND PERTUSSIS TOXIN TREATED PITUITARY CELLS

	Prolactin (ng/ml)	
	control	pertussis toxin
Basal	375 + 24	405 + 35
VIP	875 + 85*	915 + 85*
VIP + DA (100 nM)	580 + 60	890 + 85*
Neurotensin	810 + 80*	825 + 85*
Neurotensin + DA (5 nM)	520 + 75	615 + 70

* $p < 0.01$ vs basal values.

VIP, 100 nM; neurotensin, 10 nM.

We found that DA, in a concentration range from 0.1 nM to 10 nM, did modify neither adenylate cyclase activity in homogenates nor cyclic AMP formation in pituitary cells from male or lactating female rats. In addition, this amine, alone or in combination with NT i.e. in a condition where the intracellular concentration of ionized calcium is enhanced, was unable to alter the cyclic AMP generating system in cultured pituitary cells (data not shown). This observation indicated that very low concentrations of DA were capable to inhibit NT-induced PRL release without modifying cyclic AMP levels.

The hypothesis of an uncoupling between DA receptors and adenylate cyclase system was further tested exposing the pituitary cells to pertussis toxin. Pertussis toxin, an exotoxin produced by Bordetella Pertussis, abolished receptor-mediated inhibition of adenylate cyclase in several tissues (Boyer et al., 1984, Hughes et al., 1984). The effect of the toxin seems to be due to the ADP-ribosylation of a particular GTP-binding protein, putatively Ni. In our experimental conditions, we found a complete loss of DA inhibition of both PRL release and cyclic AMP formation induced by VIP. As shown in tables 3 and 4, at the time at DA receptors negatively coupled with inhibition of adenylate cyclase were not operative, DA was still fully effective in suppressing NT stimulation of both PRL release and calcium influx.

TABLE 4

EFFECTS OF DOPAMINE ON BOTH CALCIUM INFLUX INDUCED BY NEUROTENSIN AND CYCLIC AMP ACCUMULATION INDUCED BY VIP IN CONTROL AND PERTUSSIS TOXIN-TREATED PITUITARY CELLS

	control	pertussis toxin
Calcium influx		
---	7.5 + 0.7	7.0 + 1.0
NT (10 nM)	17.2 + 1.3*	17.0 + 1.6*
DA (10 nM)	7.1 + 0.5	7.4 + 0.5
NT + DA	9.1 + 0.9	9.5 + 1.0
Cyclic AMP accumulation		
---	6.1 + 0.5	7.3 + 0.5
VIP (100 nM)	13.8 + 1.0*	14.9 + 1.2*
DA (100 nM)	5.5 + 0.4	7.5 + 0.5
VIP + DA	9.0 + 1.0	13.5 + 1.0*

* $p < 0.005$ vs the correspondent basal values.
Calcium influx values are expressed as pmoles of $^{45}Ca/10^4$ cells. Cyclic AMP values are expressed as pmol/mg prot./min.

In summary, the suggestion that DA acts independently of cyclic AMP in the control of PRL secretion induced by NT comes from two observations: first, very low concentrations of DA (from 0.1 nM to 10 nM) did not alter the cyclic AMP generating system but inhibit NT-induced PRL secretion; second, abolishement of receptor-mediated inhibition of adenylate cyclase by pertussis toxin-treatment allowed unchanged the capability of DA to inhibit NT-induced both calcium influx and PRL release. The transduction mechanism involved in this DA action has still to be elucidated. Some recent reports suggest that DA inhibits phosphatidylinositol breakdown in different tissue preparations (Simmonds and Strange, 1985; Pizzi et al., 1986), thus proposing the phospholipase C as the transducer at the D2 DA receptor supramolecular complex.

CONCLUSION

Opening of calcium channels in response to NT receptor stimulation allows calcium ions to move down their electrochemical gradient into the cytoplasm. This calcium influx not only transfers depolarizing charge into excitable cells but also carries a specific message to be decoded by calcium receptor proteins. The signal leads to the initiation of the cascade of events that are ultimately expressed as PRL hormone secretion.

DA inhibits NT-induced PRL secretion by preventing the increased permeability of calcium channels. These effects appear to be independent of adenylate cyclase modulation but functionally related with calcium channels.

These data propose the existence in pituitary cells of multiple transduction mechanisms for D2 DA receptors, one of which is related with inhibition of adenylate cyclase activity whereas the other possibly involves ion fluxes across the membrane and/or phosphatidylinositol turnover changes.

The existence of multiple transduction mechanisms for D2 DA receptors could be explained by the so-called "mobile receptor" theory. The mobile receptor model stands that neurotransmitter-receptor complex may interact with a number of "effector" substituents in the plane of the membrane. This hypothesis evolved from two sources. First, observations with fluorescence-labelled antibodies directed at cell surface constituents demonstrated that membrane macromolecules could diffuse freely in the plane of the membrane (Edidin et al., 1976). Second, it was observed that a number of receptor-specific agonists could independently stimulate adenylate cyclase activity in a manner indicative of a unique adenylate cyclase enzyme responding to a variety of hormonal stimuli, thus implying that each independent hormone-receptor

FIG. 3
INTERACTION OF DOPAMINE RECEPTOR WITH MULTIPLE EFFECTORS

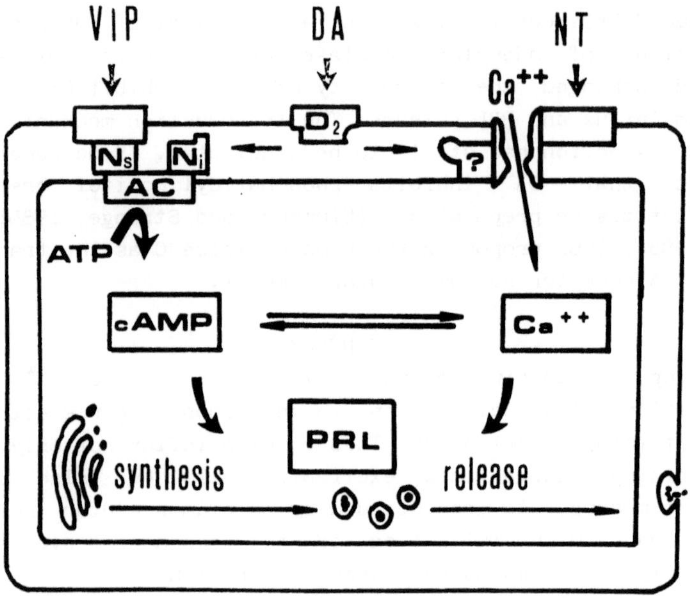

moiety can freely compete for the effector (adenylate cyclase) in the plane of the membrane (Hollemberg, 1985). The concept of DA receptors as freely diffusible membrane constituents that, upon binding a ligand, can interact with other membrane effectors, is illustrated in fig. 3.

According to the mobile or floating receptor model, the protein containing the recognition site for DA, upon binding its ligand, changes its molecular conformation. The conformation of the DA-receptor complex is now capable of interacting in the plane of the membrane with two separate membrane effector molecules which can be identified as the G-protein-adenylate cyclase moiety complex and the NT-sensitive calcium channel. D2 DA receptor stimulation in pituitary cells could thus control the intracellular concentrations of the two second messengers involved in PRL secretion, cyclic AMP and calcium.

REFERENCES

Boyer JL, A Garcia, C Posadas, JA Garcia-Sainz (1984) Differential effects of pertussis toxin on the affinity state for agonists of renal alpha 1 and alpha 2 adrenoceptors. J Biol Chem 259:8076-8079

Caron MG, M Beaulieu, V Raymond, B Gagne', J Drouin, RJ Lefkowitz, F Labrie (1978) Dopaminergic receptors in the anterior pituitary gland. J Biol Chem 33:2244-2253.

Carraway R, SE Leeman (1973) The isolation of a new hypothalamic peptide, neurotensin, from bovine hypothalami. J Biol Chem 248:6854-6861

Edidin M, Y Zaganaski, TJ Lardner (1976) Measurement of membrane protein lateral diffusion in single cell. Science 19:466-468

Enjalbert A, S Arancibia, M Ruberg, M Priam, MT Bluet-Pajot, WH Rotztejn, C Kordon (1980) Stimulation of in vitro prolactin release by vasoactive intestinal peptide. Endocrinology 31:200-204.

Enjalbert A, S Arancibia, M Priam, MT Bluet-Pajot, C Kordon (1982) Neurotensin modulation of prolactin secretion in vitro. Neuroendocrinol 34:95-98.

Enjalbert A, J Bockaert (1983) Pharmacological characterization of the D2 dopamine receptors negatively coupled with adenylate cyclase in rat anterior pituitary. Mol Pharmacol 23:576-584.

Enjalbert A, F Sladeczek, G Guillon, P Bertrand, C Shu, J Epelbaum, A Garcia-Sainz, S Jard, C Lombard, C Kordon, J Bockaert (1986) Angiotensin II and dopamine modulate both cAMP and inositol phosphate productions in anterior pituitary cells. J Biol Chem 261:4071-4075

Goedert M, RD Pinnock, CP Downes, PW Mantyh, PC Emson (1984) Neurotensin stimulates inositol phospholipid hydrolysis in rat brain slices. Brain Res 323:193-197

Hollemberg MD (1985) Receptor models and the action of neurotransmitters and hormones: some new perspectives. In: Neurotransmitter receptor binding (HI Yamamura, SJ Enna, MJ Kuhar, eds) Raven Press, pp 1-40

Hughes AR, MW Martin, TK Harder (1984) Pertussis toxin differentiates between two mechanisms of attenuation of cyclicAMP accumulation by muscarinic cholinergic receptors. Proc Natl Acad Sci 81:5680-5684.

Kebabian JW, DB Calne (1979) Multiple receptors for dopamine. Nature 227:93-96.

Iversen LL, DS Iversen, FE Bloom, C Douglas, M Brown, W Vale (1978) Calcium dependent release of somatostatin and neurotensin from rat brain in vitro. Nature 273:161-163.

Memo M, E Carboni, M Trabucchi, PF Spano (1984) Dopamine modulation of different intracellular changes induced by prolactin releasing agents. Adv Biosci 48:155-161.

Memo M, E Carboni, M Trabucchi, MO Carruba, PF Spano (1985) Dopamine inhibition of neurotensin-induced increase of calcium influx into rat pituitary cells. Brain Res 347:253-257

Memo M, L Castelletti, A Valerio, C Missale, PF Spano (1986a)

Identification of neurotensin receptors associated with calcium channels and prolactin release in rat pituitary. J Neurochem 47:1682-1688

Memo M, L Castelletti, C Missale, A Valerio, MO Carruba, PF Spano (1986b) Dopaminergic Inhibition of prolactin release and calcium influx induced by neurotensin in anterior pituitary is independent of cyclic AMP System. J Neurochem 47:1689-1695

Memo M, C Missale, MO Carruba, PF Spano (1986) D2 dopamine receptors associated with inhibition of dopamine release from rat neostriatum are independent of cyclic AMP. Neurosci Lett 71:192-196

Muller EE and RM MacLeod, eds (1982) Neuroendocrine perspectives. Elsevier Biomedical vol 1

Nagasawa L, R Yanay (1972) Promotion of pituitary prolactin release in rats by dibutyryl adenosine 3,5-monophosphate. J Endocrinol 55:215-219

Nemeroff CB, AJ Prange jr, eds (1982) Neurotensin a brain and gastrointestinal peptide. Ann N Y Acad Sci vol 400

Nicosia S, A Spada, C Borghi, L Cortellazzi, G Giannatasio (1982) Effects of vasoactive intestinal polypeptide (VIP) in human prolactin (PRL) secreting pituitary adenomas. Stimulation of PRL release and activation of adenylate cyclase. FEBS Lett 112:159-162.

Onali P, JP Schwartz, E Costa (1981) Dopaminergic modulation of anterior pituitary. Proc Natl Acad Sci 78:6531-6534.

Ozawa S, N Kimura (1979) Membrane potential changes caused by thyrotropin-releasing hormone in the clonal GH3 cell and their relationship to secretion of pituitary hormone. Proc Natl Acad Sci 76:6017-6020

Pizzi M, F d'Agostini, M Da Prada, PF Spano, WE Haefely (1986) Dopamine D2 receptor stimulation decreases the inositol 1,4,5-trisphosphate level of rat striatal slices. Nature submitted

Schettini G, MJ Cronin, RM MacLeod (1983) Adenosine 3,5-monophosphate (cAMP) and calcium-calmodulin interrelation in the control of prolactin secretion: evidence for dopamine inhibition of cAMP accumulation and prolactin release after calcium mobilization. Endocrinology 112:1801-1807

Simmonds SH, PG Strange (1985) Inhibition of inositol phospholid breakdown by D2 dopamine receptors in dissociated bovine anterior pituitary cells. Neurosci Lett 60:267-272

Spano PF, S Govoni, M Trabucchi (1978) Studies on pharmacological properties of dopamine receptors in various areas of the central nervous system. Adv Biochem Psychopharmacol 19:155-165.

Stoof JC, JW Kebabian (1981) Opposing roles for D1 and D2 dopamine receptors in efflux of cyclic AMP from rat neostriatum. Nature 294:366-368

Stanzione P, W Zieglgansberger (1983) Action of neurotensin on spinal cord neurons in the rat Brain Res 268:111-118.

Sutton CA, TFJ Martin (1982) Thyrotropin-releasing hormone (TRH) selectively and rapidly stimulates phosphatidylinositol turnover in GH pituitary cells a possible second step of TRH action. Endocrinology 110:1273-1280.

Tam SW, PS Dannies (1981) The role of adenosine 3,5-monophosphate in dopaminergic inhibition of prolactin release in anterior pituitary cells. Endocrinology 109:403-408

Thorner MO, JT Hackett, F Murad, RM MacLeod (1980) Calcium rather than cyclic AMP as the physiological intracellular regulator of prolactin release. Neuroendocrinology 31:390-402.

4
Receptor Reserves at α_2-adrenergic and Dopaminergic Autoreceptors

M. Goldstein, E. Meller and C.H. Adler

Repeated observations that the maximal response elicited by an agonist can be obtained when only a fraction of receptors are occupied has led to the concept of "spare receptors" (or "receptor reserve"). A large pool of "spare receptors" has been found in many peripheral tissues and partial agonists with substantial, but submaximal, intrinsic efficacy often produce maximal response (Nickerson 1956; Stephenson 1956; Ariens et al., 1960). The law of mass action relates fractional occupancy to ligand concentration and it follows that a lower agonist concentration will suffice to elicit a particular level of response the greater the excess of functional receptors (larger receptor reserve). In order to delineate the relationship between receptor occupancy and closely coupled physiologic responses at α_2-adrenergic and dopaminergic autoreceptors, we have inactivated a portion of these receptors with the recently described irreversible antagonist N-ethoxycarbonyl-2-ethoxy-1,2-dihydroquinoline (EEDQ) (Meller et al., 1985). This compound was found to dose-dependently inactivate central α_2-adrenrgic and dopaminergic receptors *in vivo*.

Since α_2-adrenergic and dopaminergic autoreceptors are involved in the regulation of their corresponding neuronal activities and since their responsiveness depends on the concentration of the released transmitter, we reasoned that variability in physiologic responsiveness for a particular receptor type in different brain regions or at different loci might be due to differences in the number of "spare receptors". In the present report we will summarize evidence supporting the presence of "spare receptors" or "receptor reserve" at the α_2-adrenergic autoreceptors which mediate the release of norepinephrine (NE), and at the dopamine (DA) autoreceptors which regulate the synthesis of the transmitter DA.

α_2-ADRENERGIC AUTORECEPTORS

Dose-response curves for the inhibition of ^{3}H-NE release from rat occipitoparietal cortex slices by the full α_2 agonist UK 14,304 (Cambridge 1981) were generated using both continuous and repetitive stimulation methods (Frankhuyzen and Mulder, 1982). The EC_{50}'s for UK 14,304 inhibition of ^{3}H-NE release were not significantly different (12 and 24 nM) using the repetitive and continuous methods, respectively. Following inactivation of α_2-adrenoceptors by treatment of rats with EEDQ, the EC_{50} value for UK 14,304 was shifted 7-fold to the right (control = 23.9 nM; EEDQ = 169 nM) and there was a decrease of 21.5% in the maximal response. The "q" value (the fraction of receptors remaining active) and the K_A (activation) constant were calculated from double reciprocal plots of equieffective concentrations of UK 14,304 required to inhibit the release of ^{3}H-NE from controls and from EEDQ treated animals (Furchgott, 1966). In addition we also determined the number of receptor binding sites (B_{max}) by measuring the binding of ^{3}H-RX-781094, a selective α_2-adrenoceptor antagonist, to cerebral cortical membranes at various time intervals after administration of EEDQ (Adler et al., 1985). It is evident from the results presented in Table 1 that by 48 hours after administration of EEDQ repopulation of only 5.6% of the functional receptors had occurred, yet 70% of maximum inhibition of ^{3}H-NE release was obtained. By 96 hours after administration of EEDQ 24.3% and at 192 hrs 36.8% of the functional receptors had returned, whereas the percentage of maximum inhibition of ^{3}H-NE release at these time intervals was 87 and 95, respectively. The data in Table 1 also show that at various time intervals after administration of EEDQ the binding of ^{3}H-RX 781094 to cerebral cortical membranes is significantly decreased.

Table 1

Comparison of α_2 Adrenergic Autoreceptor Function and Calculated Receptor Number with Radioligand Binding Densities

Time Post-EEDQ (hr)	Inhibition of Release (% of Maximum)	q x 100 (%)	[^{3}H]RX 781094 Binding (% of B_{max})
48	70	5.6	35
96	87	24.3	55
192	95	38.8	77

At the indicated times following EEDQ-treatment (1.6 mg/kg, i.p.), evaluation of UK-14304-inhibition of [^{3}H]NE release in rat occipitoparietal cortex slices and [^{3}H]RX 781094 binding to rat cerebral cortical membranes (Adler et al., 1985) was performed. The q values were calculated as indicated in the text. The maximal effect for UK-14304-inhibition of [^{3}H]NE release was 76.2% [^{3}H]RX 781094 binding was in Tris HCL and had a K_D = 0.92 ± 0.03 nM and B_{max} = 2.81 ± 7 fmol/mg protein (n = 27) (Adler et al., 1985).

Repopulation kinetics analysis (Mauger et al., 1982) of these results shows that the recovery of 50% of the maximal response of the autoreceptor is 2.4 days (Adler et al., 1985), while the recovery of half of the total number of functional receptors is approximately 10 days. Thus, the recovery of α_2 receptor function is more rapid than recovery of the estimated fraction of active receptors (q). Repopulation kinetics analysis of α_2-receptor binding sites labeled with ^{3}H-RX-781094 was found to be monoexponential with a $t_{1/2}$ of 4.1 days. A comparison of the "q" values with the B_{max} values obtained from the binding studies indicates that the number of active receptors determined from functional studies differs from the number of receptor binding sites. Such discrepancies have also been found in other receptor systems (Minneman and Abel, 1984), and it might suggest that the functional receptors involve only a discrete subpopulation of the total receptor binding sites.

DA AUTORECEPTORS

The effects of various DA agonists on DA autoreceptors which regulate the synthesis of DA in the striatum were investigated by measuring their potencies to inhibit the enhanced synthesis of L-DOPA elicited by administration of γ-butyrolactone (GBL) (Roth, 1979). The potencies of full and partial DA agonists at DA autoreceptors were determined following partial inactivation of these receptors by treatment of rats with EEDQ. Apomorphine (Apo) and N-propylnorapomorphine (NPA) dose-dependently and completely reversed the enhanced synthesis of striatal L-DOPA elicited by GBL administration in control animals. Following treatment with low doses of EEDQ the dose response curves were shifted to the right without significant reductions in maximal response. Larger or multiple doses of EEDQ produced increasing rightward shifts with progressive reductions in maximal response. The fraction (q) of receptors remaining active after irreversible blockade of DA receptors with EEDQ was determined, as described earlier, from double-reciprocal plots of equieffective doses of Apo or NPA required to elicit the same level of response before and after treatment with the inactivator. Apo and NPA elicit 50% of the maximal response in presence of 17% and 5.8% of active receptors, respectively (Meller et al., 1986; Meller et al, to be published). It was estimated that maximal response by NPA could still be obtained when 38% of the receptors remained active. The partial DA agonist EMD 23,448 (Seyfried et al., 1982; Goldstein et al., 1984) and (+) 3-PPP (Clark et al., 1985) also dose-dependently and completely reversed the enhanced synthesis of striatal DOPA induced by GBL, while the partial DA agonist (-) 3-PPP dose-dependently, but not completely, reversed the enhanced L-DOPA synthesis. These results were used to calculate fractional occupancy of the receptor as a function of administered dose.

Subsequent comparison of receptor occupancy vs. response showed a steep hyperbolic relationship for NPA and much shallower relationships for EMD 23,448 and (+) 3-PPP. The relationship for (-) 3-PPP was linear (Meller et al., to be published). The results presented in Table 2 show the relationship between response and occupancy and the calculated relative efficacies for various tested DA agonists. Although EMD 23,448 and (+) 3-PPP show weak relative efficacy, nevertheless, due to the large receptor reserve at the DA autoreceptors, they exert full intrinsic activity (i.e. the tissue response in controls is indistinguishable from that elicited by the full agonist NPA).

DISCUSSION

Our conclusions that there are central "spare" α_2-adrenergic and DA autoreceptors are based on the findings that following partial inactivation of these receptors with EEDQ, the dose-response curves of the corresponding full agonists UK 14-304, Apo and NPA were shifted to the right without significant reductions in the maximal response. However, the maximal responses of such weak partial DA agonists such as EMD 23,448 and (+) and (-) 3-PPP are drastically reduced following partial inactivation of the striatal DA autoreceptors. These findings are in agreement with the theory of receptor reserve, which predicts that for a full agonist, only a small proportion of the receptors need to be occupied to obtain a maximal tissue response (R.P. Stephenson, 1956; M. Nickerson, 1956). Indeed our results show that only about 1.5% of the α_2-receptors need to be occupied by the full α_2 agonist UK 14-304 to achieve 50% of the maximal inhibition of ^{3}H-NE release from cerebral cortical slices, and 3.8% of the receptors need to be occupied by the full DA agonist NPA to achieve 50% inhibition of the enhanced striatal L-DOPA synthesis elicited by GBL administration. The presence of "spare" α_2-adrenergic and DA autoreceptors in the CNS could have important physiological and pharmacological implications. The sensitivity

Table 2

Relationships Between Response and Occupancy and Calculated Relative Efficacies of Various Dopamine Agonists

DA agonists	% receptor occupancy	% maximal response	relative efficacy
NPA	3.8	50	1.0
EMD 23,448	18	50	0.19
(+) 3-PPP	30	50	0.12
(-) 3-PPP	100	50	0.05

The relative efficacy was calculated from the ratio of the fraction of receptors required to be occupied by NPA over that fraction required to be occupied by each of the other agonists at the 50% response level.

of neurons to endogenous and exogenous agonists might be controlled by the magnitude of the receptor reserve. Thus, the greater the excess of functional receptors present, the lower is the concentration of agonist required to produce an effect in a specific brain region. Furthermore, the threshold for the development of tolerance to an agonist might be greater if there is a large receptor reserve.

The presence of "receptor reserve" has to be taken into account in the interpretation of the result of various studies on the mode of action of drugs. For example, conflicting data on the effects of antidepressants on α_2-adrenergic receptors have been reported. The reported discrepancy between antidepressant-induced decreases in the number of α_2-adrenergic receptors (Sugrue, 1983) and an apparent lack of functional alterations (Schoffelmeer et al., 1984) could be explained by the existence of a "receptor reserve." The findings that antihypertensive drugs such as - methyl dopa or clonidine do not show functional down-regulation might also be related to the presence of "receptor reserve" at the α_2-adrenergic receptors. Thus, the "receptor reserve" might be of therapeutic importance in mediating drug action.

In a separate study we have also shown that a moderate receptor reserve exists for the α_2-adrenergic heteroreceptors which regulate 5-HT release (Adler, Ph.D. thesis, 1986). It seems possible that autoreceptors in general may have large fractions of spare receptors which enable the autoregulation of release and/or synthesis to occur at low concentrations of the transmitter in the synapse. This might be of importance for central epinephrine neurotransmission since its release at the α_2 adrenoceptors is probably regulated by low concentrations of the transmitter. The selective ability of some partial DA agonists to stimulate DA autoreceptors and supersensitive, but not normosensitive, postsynaptic DA receptors could also be explained by the existence of a relatiavely large receptor reserve at pre- vs. postsynaptic sites. The ability of an agonist to produce its effect at the receptor site depends on the density of the receptor and on the efficiency of coupling. Thus, low efficacy agonists may produce responses only in tissues which have a sufficient density of the receptor and an efficient coupling mechanism. In this study we have shown that low efficacy partial DA agonists such as EMD 23,448 and (+) 3-PPP produce maximal responses at DA autoreceptors which regulate DA synthesis. These DA agonists produce maximal responses presumably because the coupling of the receptor to the effector component is efficient and therefore the "receptor reserve" is large. It has been previously suggested that low levels of agonist occupancy at DA autoreceptors and supersensitive postsynaptic DA receptors induces a conformational change resulting in an increased responsivenes of the receptors to the agonist (Carlsson, 1983). This hypothesis could be further extended and modified by considering that it has been shown that reversible coupling between the receptor and the effector components is promoted by the agonist; thus low agonist occupancy at these receptors would shift the equilibrium to the noncoupled state. We postulate that the sustained absence of coupling promotion by endogenous DA results in increased efficiency of receptor-effector coupling by exogenous DA agonists.

The increased efficiency of DA receptor-effector coupling and the presence of a large "receptor reserve" might explain the increased sensitivity of DA autoreceptors to agonists as compared to normosensitive postsynaptic DA receptors. This hypothesis could be applied to supersensitive postsynaptic DA receptors, and the antiparkinsonian activity of partial DA agonists (Goldstein et al., 1984) may result from the increased population of receptor reserve. Finally, the lack of correspondence between functional responsiveness and increased total DA receptor binding sites (B_{Max}) observed in animals with central supersensitive receptors could be explained by the findings that maximal function requires the occupancy of only a fraction of the active receptors by the DA agonists.

ACKNOWLEDGEMENTS

These studies were supported in part by the R. and J. Bendheim Fellowship awarded by the Parkinson's Disease Foundation, NIMH 02717 and NINCDS 06801.

REFERENCES

Adler, C.H., Meller, E. and Goldstein, M. (1985). Recovery of $alpha_2$-adrenoceptor binding and function after irreversible inactivation by N-ethoxycarbonyl-2-ethoxy-1,2-dihydroquinoline (EEDQ). Eur. J. Pharmacol. 116, 1178-178.

Ariens, E.J., Van Rossum, J.M. and Koopman, P.C. (1960). Receptor reserve and threshold phenomena. Arch. Int. Pharmacodyn Ther. 127, 459-478.

Cambridge, D. (1981). UK-14,304, a potent and selective $alpha_2$-agonist for the characterization of alpha-adrenoreceptor subtypes. Eur. J. Pharmacol. 72, 413-415.

Carlsson, A. (1983). Dopamine receptor agonists: intrinsic activity vs. state of receptor. J. Neur. Trans. 57, 309-315.

Clark, D., Hjorth, S. and Carlsson, A. (1985). Dopamine receptor agonists: mechanisms underlying autoreceptor selectivity. II. Theoretical considerations. J. Neur. Trans. 62, 171-207.

Frankhuyzen, A.L. and Mulder, A.H. (1982). A cumulative dose-response technique for the characterization of presynaptic receptors modulating [^{3}H]noradrenaline release from rat brain slices. Eur. J. Pharmacol. 78, 91-97.

Furchgott, R.F. (1966). The use of beta-haloalkylamines in the differentiation of receptors and in the determination of dissociation constants of receptor-agonist complexes. Adv. Drug Res. 3, 21-55.

Goldstein, M., Helmer, E. and Fuxe, K. (1984). Effects of dopamine agonists on presynaptic normosensitive and on postsynaptic supersensitive DA receptors. IUPHAR 9th International Congress of Pharmacology, Abstr. 1155P.

Goldstein, M., Lieberman, A. and Battista, A.F. (1984). The therapeutic potential of centrally acting dopamine agonists. Trends in Pharmacol. Sci. 5, 227-230.

Mauger, J.P., Sladeczek, F. and Bockaert, J. (1982). Characteristics and metabolism of $alpha_1$ adrenergic receptors in a nonfusing muscle cell line. J. Biol. Chem. 257, 875-879.

Meller, E., Bohmaker, K., Goldstein, M. and Friedhoff, A.J. (1985). Inactivation of D_1 and D_2 dopamine receptors by N-ethoxycarbonyl-2-ethoxy-1,2-dihydroquinoline in vivo. J Pharm. Exp. Ther. 233, 656-662.

Meller, E., Helmer-Matyjek, E., Bohmaker, K., Adler, C.H., Friedhoff, A.J. and Goldstein, M. (1986). Receptor reserve at striatal dopamine autoreceptors: Implications for selectivity of dopamine agonists. Eur. J. Pharmacol. 123, 311-314.

Minneman, K.P. and Abel, P.W. (1984). Spare $alpha_1$-adrenergic receptors and the potency of agonists in rat vas deferens. Mol. Pharmacol. 25, 56-62.

Nickerson, M. Receptor occupancy and tissue response. (1956). Nature 178, 697-698.

Roth, R.H. (1979). Dopamine autoreceptors: pharmacology, function and comparison with post-synaptic dopamine receptors. Commun. Psychopharmacol. 3, 49.

Schoffelmeer, A.N.M., Hoorneman, E.M.D., Sminia, P. and Mulder, A.H. (1984). Presynaptic α_2 and postsynaptic B-adrenoreceptor sensitivity in slices of rat neocortex after chronic treatment with various antidepressant drugs. Neuropharm. 23, 115-120.

Seyfried, C.A., Fuxe, K., Wolf-H.P. and Agnati, L.D. (1982). Demonstration of a new type of dopamine receptor agonist: an indolyl-3-butylamine. Actions at intact versus supersensitive dopamine receptors in the rat forebrain. Acta Physiol. Scand. 116, 465-468,

Stephenson, R.P. (1956). A modification of receptor theory. Br. J. Pharmacol. 11:379-393.

Sugrue, M.F. (1983). Chronic antidepressant therapy and associated changes in central monoaminergic receptor functioning. Pharmacol. Ther. 21, 1-33.

Part B
Effects of Trophic Factors and Gangliosides

Chairman: L. Olson

5
Potentiation of NGF-mediated Effects of Laminin

Ann Acheson and Hans Thoenen

Nerve growth factor (NGF) is a well-defined protein which is necessary for the development and maintainance of differentiated properties of peripheral sympathetic and neural crest-derived sensory neurons (for review see Thoenen et al., 1985 and refs therein). Many morphological and biochemical effects of NGF have been described in both the physiological target cells of NGF and a clonal cell line (PC12) derived from a rat pheochromocytoma. NGF is known to bind to specific cell-surface receptors, an event which is followed by internalization of the ligand-receptor complex; however, the mechanism of signal transduction after NGF's interaction with its receptor is unknown.

The effects of soluble neurotrophic factors such as NGF on neuronal survival and neurite outgrowth are known to be influenced by the extracellular matrix. For example, laminin, a glycoprotein component of the extracellular matrix, potentiates both of these effects of NGF in embryonic chick sympathetic neurons (Edgar et al., 1984), while having no survival-promoting effect by itself. Laminin may thus facilitate the analysis of NGF's biochemical signalling mechanism.

We chose to study the ability of laminin to potentiate NGF's effects in bovine adrenal chromaffin cells. Like sympathetic neurons, adrenal chromaffin cells are derived from the neural crest, synthesize and release catecholamines and have NGF receptors (Naujoks et al., 1982). In calf chromaffin cells, 48 hr NGF treatment results in a selective increase in the levels of tyrosine hydroxylase (TH) (Naujoks et al., 1982; Acheson et al., 1984). These cells also respond to substrate-bound laminin *in vitro* with an increase in TH levels (Acheson et al., 1986a), indicating the presence of laminin receptors on these cells. Laminin, however, does not potentiate the long-term increase in TH levels brought about by NGF (Acheson et al., 1986a). In addition to its long-term effects, substrate-bound laminin also brings about a rapid activation of TH (Acheson et al., 1986a).

Recent reports have indicated that, in PC12 cells, a variety of the effects of NGF are blocked by inhibitors of the enzyme S-adenosylhomocysteine (SAH) hydrolase. These drugs block methyltransferase reactions by end-product inhibition due to increased intracellular SAH levels. Effects of NGF which are blocked by SAH hydrolase inhibitors include membrane ruffling (Seeley et al., 1984), increased transcription of the proto-oncogene c-*fos* (Greenberg et al., 1985) and changes in the state of phosphorylation of a protein associated with the cytoskeleton (Landreth and Rieser, 1985). All these effects occur after only minutes of NGF exposure, the same time frame as is required for TH activation in PC12 cells (Greene et al., 1984). We have therefore examined the effect of NGF on TH activity in calf chromaffin cells, as well as the ability of laminin to potentiate and SAH hydrolase inhibitors to block this effect.

SHORT-TERM EFFECT OF NGF ON TH ACTIVITY

Calf chromaffin cells, cultured at low density (10,000 cells/cm^2) on either polyornithine- or laminin-coated dishes, were treated with NGF for 2-60 min. When polyornithine was used as a substrate, NGF brought about an increase in TH activity which became apparent after 5 min, and stabilized at a maximal level after 30-60 min of NGF exposure (Table 1). With a laminin substrate, there was a much larger increase in TH activity, which was maximal after 5 min, and declined thereafter to a somewhat lower plateau, remaining elevated relative to cells on polyornithine (Table 1). The effect of laminin alone to activate TH was apparent in the baseline values, as reported previously (Acheson et al., 1986a).

Table 1. Short-term effect of NGF treatment on calf chromaffin cell TH activity.

Time of NGF exposure	TH Specific Activity[1] (nmoles/min/mg) **Cells plated on:** *Polyornithine*[2]	*Laminin*
0 min	2.25 ± 0.30	3.45 ± 0.25
5 min	3.95 ± 0.75	18.3 ± 2.25
15 min	4.13 ± 0.30	10.8 ± 0.90
60 min	5.10 ± 0.15	10.1 ± 0.75
60 min (no NGF)	2.13 ± 0.17	4.03 ± 0.20

1 Data are expressed as mean ± S.E.M. of values obtained from 5 independent experiments.

2 Calf chromaffin cells were cultured at low density for 1 day and were then harvested and replated at the same low density on either polyornithine- or laminin-coated dishes at the same low density. Three hr post-replate, NGF (800 ng/ml) was added for the times indicated. Cells were then harvested and the specific activity of TH was determined.

SHORT-TERM BIOCHEMICAL EFFECTS OF INHIBITORS

We next examined the ability of two SAH hydrolase inhibitors, 5'-deoxy-5'-methylthioadenosine (MTA) and 9-beta-arabino-furanosyladenine (ara-A), to block protein synthesis (^{35}S-methionine incorporation into TCA-precipitable material) versus methylation (^{3}H-methyl-methionine incorporation). Both MTA and ara-A had been reported to be potent methyltransferase inhibitors in other systems (Schanche et al., 1984). Since the specific incorporation of label was approximately 7-fold higher for ^{3}H-methyl-methionine as compared to ^{35}S-methionine (9.41 ± 0.47 versus 1.32 ± 0.12 pmoles label/mg protein), a selective decrease in ^{3}H-methyl-methionine incorporation was taken as indicating a specific action of the drug as a methyltransferase inhibitor.

Neither MTA nor ara-A affected protein synthesis, whereas both drugs produced a >60% inhibition of methylation (Table 2). MTA and ara-A were therefore considered to act selectively to decrease methylation.

Table 2. Incorporation of ^{35}S-methionine and ^{3}H-methyl-methionine into cellular protein after short-term drug treatment.

Treatment[1]	% Control Incorporation of:[2]	
	^{35}S-methionine	^{3}H-methyl-methionine
MTA (3 mM)	97 ± 3	28 ± 1
ara-A (10 uM)	112 ± 2	42 ± 2

1 Calf chromaffin cells were incubated in the presence of the drugs (final concs as indicated) for 2.5 hr, then in the continued presence of drug with ^{35}S-methionine or ^{3}H-methyl-methionine for an additional hr. Total cellular proteins were solubilized and TCA precipitated.

2 Data were expressed as CPM/mg protein, and are shown as a percentage of control values, obtained from DMSO-treated cells. Values are means ± S.E.M. from 3 independent experiments.

EFFECT OF INHIBITORS ON NGF-MEDIATED TH ACTIVATION

Cells plated on laminin which had been preincubated for 3 hr with the drugs were exposed to NGF for 5 min, and the extent of TH activation was determined. MTA and ara-A, which had been shown to block methylation, both inhibited TH activation (Table 3), whereas cycloheximide, an inhibitor of protein synthesis, was without effect.

Table 3. Effect of inhibitors on NGF-mediated TH activation

Treatment[1]	TH Specific Activity[2] (nmoles/min/mg)	
Polyornithine	3.9 ± 0.83	
+ Laminin	17.2 ± 1.8	(100%)
+ Laminin + MTA (3 mM)	2.5 ± 0.15	(14%)
+ Laminin + ara-A (10 uM)	3.4 ± 0.17	(20%)
+ Laminin + cycloheximide (1.4 mM)	15.8 ± 1.6	(92%)

1 Cells were harvested and replated onto either polyornithine- or laminin-coated dishes. Cells were then preincubated with the drugs as indicated for 3 hr, at which time NGF (800 ng/ml) was added to all of the dishes. After 5 min of NGF exposure, cells were harvested and specific TH activity was determined.

2 Data are expressed as the mean ± S.E.M. of 3 independent experiments. Values shown in parentheses are taken as a percent of the effect on laminin alone.

EFFECT OF INHIBITORS ON 8-BROMO-cAMP-MEDIATED TH ACTIVATION

In order to evaluate possible non-specific effects of these drugs, we next examined their ability to block 8-br-cAMP-induced TH activation. cAMP-dependent TH activation shows roughly the same time course as that seen for NGF-mediated TH activation in cells plated on laminin, i.e. the activation is maximal after 5-15 min (data not shown). Cells were incubated with the various drugs for 3 hr, then were exposed to 8-br-cAMP (1 mM) for 5 min. The extent of TH activation was then determined. None of the drugs blocked cAMP-mediated TH activation (Table 4).

Table 4. Effect of inhibitors on the short-term activation of TH by 8-br-cAMP

Treatment[1]	TH Specific Activity[2] (nmoles/min/mg)
No drug	1.42 ± 0.08
8-br-cAMP (1 mM)	3.20 ± 0.19
MTA (3 mM) + 8-br-cAMP	3.37 ± 0.17
ara-A (10 uM) + 8-br-cAMP	3.27 ± 0.20
Cycloheximide (1.4 mM) + 8-br-cAMP	3.63 ± 0.14

[1] Cells plated on polyornithine were pre-treated with the drugs as indicated for 3 hr. 1 mM 8-br-cAMP was added to the cells for 5 min, and cells were then harvested and specific TH activity determined.

[2] Data are expressed as the mean ± S.E.M. of 5 replicate dishes in a single experiment.

Additional experiments showed that the long-term effect of NGF to increase TH levels described earlier was also blocked by MTA (data not shown). However, the long-term effect of 8-br-cAMP, which also produces an increase in TH levels (Acheson et al., 1984), was not affected by MTA treatment (data not shown).

EFFECT OF MTA ON NGF-MEDIATED NEURONAL SURVIVAL

It is not known whether the most important developmental effect of NGF, the selective regulation of neuronal survival, is brought about by the same signalling mechanism as NGF-induced phenomena that occur in cells which do not require NGF for survival, as is the case for both chromaffin and PC12 cells. It was thus of interest to determine whether drugs which block NGF's effects in PC12 and chromaffin cells could also inhibit the survival-promoting action of NGF.

Studies were therefore carried out on embryonic chick sympathetic neurons. As with chromaffin cells, the ability of laminin to potentiate NGF's effects provided the necessary tools to assess potential biochemical signalling mechanisms. When sympathetic neurons are plated on a laminin substrate, >80% of the neurons survive (as opposed to 30% on polyornithine; Edgar et al., 1984), so that essentially all cells with NGF receptors survive. Neurite outgrowth is also extremely rapid, allowing it to be quantitatively assessed after only several hours (Edgar et al., 1984). An additional advantage of this system is that elevated concentrations of K^+ (35 mM) support the survival of the same population of sympathetic neurons as does NGF, and high K^+-mediated survival is also potentiated by laminin (Wakade et al., 1983). Indeed, it is possible to switch between NGF and high K^+ *in vitro* without any significant neuronal death (Wakade and Thoenen, 1984), so that high K^+ can be used as a control when studying NGF-mediated survival of sympathetic neurons. An additional short-term effect of NGF on sympathetic neurons is the rapid (maximal after 15 min) and specific dephosphorylation of a 70 Kd protein, p70 (Por and Huttner, 1984). Studying p70 has the advantage that it is a major neuronal phosphoprotein which is affected by both NGF and high K^+ in the same way, i.e. its state of phosphorylation is decreased by both agents.

Our recent studies have shown that MTA selectively blocks NGF- but not high K^+-mediated survival of sympathetic neurons. In these studies, embryonic chick neurons were cultured for 2 days in the presence of high K^+, then half of them were switched to receiving NGF as a survival factor. Two days later, in the presence of NGF or high K^+ alone, 72% and 84% of the neurons survived, respectively (data are expressed as % of the total number of cells plated at the beginning of the experiment). When high K^+ was the survival-promoting agent, the addition of MTA did not alter survival (76%). However, only 17% of the cells which received NGF together with MTA survived after 2 days (Acheson et al., 1986b). The ability of neurons plated on laminin to rapidly respond to NGF with neurite outgrowth was also blocked by MTA, such that after 5 hr *in vitro*, 75% of the neurons with NGF alone bore neurites greater than 5 cell diameters long, whereas only 22% of those receiving NGF plus MTA had neurites (Acheson et al., 1986b). MTA's effect on neurite outgrowth was reversible upon removal of the drug, indicating its lack of general toxicity. MTA was also shown to block all NGF-induced changes in protein phosphorylation in these cells, including the decrease in p70 phosphorylation induced by NGF, but not that by high K^+ (Acheson et al., 1986b). We concluded from these data that the early molecular events of NGF-mediated neuronal survival differ from those of high K^+-mediated survival.

We conclude that MTA inhibits an early biochemical step in the cascade of reactions initiated by the interaction of NGF with its receptor. In the series of experiments presented above, the controls for the specificity of the drugs included: measurement of protein synthesis, both activation and induction of TH brought about by treatment of the cells with 8-br-cAMP, high K^+-mediated neuronal survival, and reversibility of the blockade of neurite outgrowth. Results from all these control experiments indicated that MTA treatment did not produce a general, non-specific inhibitory effect in the cells. For example, while NGF-mediated neuronal survival was blocked, the same end effect mediated by high K^+ was not altered. Similarly, in the presence of MTA, 8-br cAMP still activated TH.

Although MTA blocks protein methylation, it is not certain that this effect is causally related to the blockade of NGF's effects. Given the nature of the drug (an adenosine analog) and the high concentrations necessary to block NGF's effects, it may be that other effects of the drug are responsible for its selective inhibition of NGF-mediated events. The role of protein methylation in eukaryotic cells has not yet been shown to serve a regulatory function, as it has in bacterial systems. Indeed, N-methylation of proteins is irreversible and can be excluded as a regulatory post-translational modification. The eukaryotic protein carboxyl methylating enzyme has been shown to affect only modified proteins, in which isoaspartate/D-aspartate residues are substrates (O'Connor et al., 1984; Aswad, 1984; Clarke, 1985). A carboxyl methylating system analogous to that in bacteria has not been demonstrated in eukaryotic cells (Clarke, 1985). The substoichiometric nature of carboxylmethylation of proteins such as calmodulin in intact or semi-purified systems (Johnson et al., 1985) also speaks against a regulatory function for this post-translational modification, and, in fact, it has been shown that when the normal aspartate residues are converted to isoaspartate in these proteins by base treatment at 37 ^oC, carboxylmethylation becomes stoichiometric (Johnson et al., 1985). Carboxylmethylation has thus been hypothesized to serve as a tag for protein degradation or repair (Clarke, 1985). Nonetheless, our data (Acheson and Thoenen, 1986; Acheson et al., 1986b) provide further evidence for the selectivity of MTA to inhibit NGF's signalling mechanism in a variety of different systems. MTA and related compounds are thus useful tools for the elucidation of the molecular mechanism of signal transduction following the interaction of NGF with its receptor.

REFERENCES

Acheson, A., Naujoks, K. and Thoenen, H. (1984) Nerve growth factor-mediated enzyme induction in primary cultures of bovine adrenal chromaffin cells: specificity and level of regulation. J. Neurosci. 4: 1771-1780.

Acheson, A., Edgar, D., Timpl, R. and Thoenen, H. (1986a) Laminin increases both levels and activity of tyrosine hydroxylase in calf adrenal chromaffin cells. J. Cell Biol. 102: 151-159.

Acheson, A., Vogl, W., Huttner, W.B. and Thoenen, H. (1986b) Methyltransferase inhibitors block NGF-regulated survival and protein phosphorylation in sympathetic neurons. EMBO J. in press.

Aswad, D. (1984) Stoichiometric methylation of porcine adrenocoticotropin by protein carboxyl methyltransferase requires deamidation of asparagine 25. J. Biol. Chem. 259: 10714-10721.

Edgar, D., Timpl, R. and Thoenen, H. (1984) The heparin-binding domain of laminin is responsible for its effects on neurite outgrowth and neuronal survival. EMBO J. 3: 1463-1468.

Clarke, S. (1985) Protein carboxyl methyltransferases: distinct classes of enzymes. Ann. Rev. Biochem. 54: 479-506.

Greenberg, M.E., Greene, L.A. and Ziff, E.B. (1985) Nerve growth factor and epidermal growth factor induce rapid transient changes in proto-oncogene transcription in PC12 cells. J. Biol. Chem. 260: 14101-14110.

Greene, L.A., Seeley, P.J., Rukenstein, A., DiPiazza, M. and Howard, A. (1984) Rapid activation of tyrosine hydroxylase in response to nerve growth factor. J. Neurochem. 42: 1728-1734.

Johnson, B.A., Freitag, N.E. and Aswad, D.W. (1985) Protein carboxyl methyltransferase selectively modifies an atypical form of calmodulin. J. Biol. Chem. 260:

10913-10916.

Landreth, G.E. and Rieser, G.D. (1985) Nerve growth factor- and epidermal growth factor-stimulated phosphorylation of a PC12 cytoskeletally associated protein in situ. J. Cell Biol. 100: 677-683.

Naujoks, K.W., Korsching, S., Rohrer, H. and Thoenen, H. (1982) Nerve growth factor-mediated induction of tyrosine hydroxylase and neurite outgrowth in cultures of bovine adrenal chromaffin cells: dependence on developmental stage. Dev. Biol. 92: 365-379.

O'Connor, C.M., Aswad, D.W. and Clarke, S. (1984) Mammalian brain and erythrocyte carboxyl methyltransferases are similar enzymes that recognize both D-aspartyl and L-isoaspartyl residues in structurally altered protein substrates. Proc. Natl. Acad. Sci. USA 81: 7757-7761.

Por, S.B. and Huttner, W.B. (1984) A M_r 70,000 phosphoprotein of sympathetic neurons regulated by nerve growth factor and by depolarization. J. Biol. Chem. 259: 6526-6533.

Schanche, J.-S., Schanche, T., Ueland, P.M., Holy, A. and Votruba, I. (1984) The effect of aliphatic adenine analogues on S-adenosylhomocysteine and S-adenosylhomocysteine hydrolase in intact rat hepatocytes. Mol. Pharmacol. 26: 553-558.

Seeley, P.J., Rukenstein, A., Connolly, J.L. and Greene, L.A. (1984) Differential inhibition of nerve growth factor and epidermal growth factor effects on the PC12 pheochromocytoma line. J. Cell Biol. 98: 417-426.

Thoenen, H., Korsching, S., Heumann, R. and Acheson, A. (1985) Nerve growth factor. *In*: Growth Factors in Biology and Medicine (CIBA Foundation Symposium 116), Pitman, London, pp. 113-128.

Wakade, A.R. and Thoenen, H. (1984) Interchangeability of nerve growth factor and high potassium in the long-term survival of chick sympathetic neurons in serum-free culture medium. Neurosci. Lett. 45: 71-74.

Wakade, A.R., Edgar, D. and Thoenen, H. (1983) Both nerve growth factor and high K^+ concentrations support the survival of chick embryo sympathetic neurons. Exp. Cell Res. 144: 377-384.

6
Gangliosides as Modulators of Neuronotrophic Interactions

G. Toffano, R. Dal Toso, L. Facci, G. Ferrari, D. Benvegnù, A. Consolazione, M. Favaron and A. Leon

INTRODUCTION

Basic neurobiological research has recently reconsidered the capability of the adult mammalian central nervous system (CNS) neurons to undergo adaptive functional and morphological modifications in response to external noxious perturbations (Tsukahara, 1981; Cotman and Nieto-Sampedro, 1984). This new interest mainly stems from lesion and transplantation experiments documenting an unprecedented inherent plasticity of the mature CNS neurons following brain damage. The lesioned axons are now known to possess the ability to regrow when growth-promoting signals and substrates are introduced into their environment (Richardson et al., 1980; Kromer et al., 1981). In addition, a growing number of studies have provided evidence for a naturally occurring sprouting ability of intact axons following partial deafferentation (Cotman and Nieto-Sampedro, 1984). Yet loss of neuronal connectivity and function are still today common consequences of brain damage in the adult. What is critically needed is the comprehension of the cellular and molecular mechanisms underlying mature CNS plasticity following injury and its relationship to repair. Hopefully, this will in the near future provide novel ways for ameliorating the outcome following brain injury.

Among several aspects, the problem of lesion-induced CNS neuronal plasticity and repair presently converges on the study of growth-promoting signals capable of affecting neuronal cell survival and/or fiber outgrowth in vitro and in vivo. The ultimate goal, once these substances have been identified, is to obtain information as concerns the mechanisms regulating the expression of these molecules in the mature brain and the ability of the mature neurons to respond to these molecules. In this regard, neuronotrophic factors (NTFs) and gangliosides, in particular monosialoganglioside GM_1 (nomenclature according to Svennerholm, 1963) are now being analyzed with increasing attention.

BRAIN INJURY AND NEURONOTROPHIC FACTORS

As adult mammalian CNS injury is often associated with neuronal cell damage and death, a fundamental prerequisite for improving functional repair processes is obviously the identification and study of the environmental signals affecting survival of the adult normal and damaged CNS neurons. NTF(s) have long been known to regulate during development both the number of surviving neurons and of axons which ultimately maintain stable functional synaptic contacts with the innervation field (Cowan et al., 1984). Recently, a similar role has also been attributed to such molecules throughout adulthood especially following injury (Varon et al., 1984).

In particular, neuronotrophic factors are now known to occur in the adult normal and lesioned mammalian brain (Nieto-Sampedro et al., 1983; Whittemore et al., 1985; Toffano and Leon, 1986). Furthermore recent reports concerning the only well known NTF-nerve growth factor (NGF)-have documented that the early intracerebral administration of NGF is effective in reducing both neuronal cell death and behavioral deficits associated with injury to CNS cholinergic neurons in adult rodents (Will and Hefti, 1985; Hefti, 1986; Varon et al., 1986). Together these evidences have been interpreted as indicating that NTF(s) are useful in preventing cell death, if sufficiently operative at early post-lesion stages. Presumably the endogenous occurring NTF(s) are per se inadequate soon after injury (perhaps due to an increase in trophic need of the damaged neurons) and, even though NTF activity has been reported to increase following brain injury, this increase occurs at a time subsequent to most neuronal cell death.

Thus, given that NTF(s) are essential and effective components regulating neuronal cell survival and hence better conditions for facilitating functional repair, it is feasible that their efficacy during early post-lesion stages depends not only on their availability but also on the ability of the neurons to respond to these signals. As concerns the latter, gangliosides have been recently shown to be functionally associated with growth factor receptors (Bremer et al., 1984). In addition a growing body of literature, briefly summarized below, is available indicating that the GM_1 ganglioside is effective in facilitating neuronotrophic interactions in vitro and, most important, mature neuronal cell survival and repair following CNS injury in vivo (see Ledeen et al., 1984 for a recent review; Toffano et al., 1986).

GM_1 AND CNS NEURONAL CELL PLASTICITY: INVOLVEMENT IN NEURONOTROPHIC INTERACTIONS.

Gangliosides, sialic acid-containing glycosphingolipids have long been recognized as characteristic cell surface constituents of mammalian cells, in particular neuronal cells. Although these molecules have been identified as receptors for bacterial toxins and viruses, relatively little is known about their normal function in membranes (for review see Ledeen et al., 1984). In this regard, several lines of evidence are now available linking gangliosides, in particular GM_1 (see Figure 1), to plasticity of developing and mature CNS neurons.

In particular recent studies have described changes in GM_1 ganglioside expression with changes in the patterns of differentiation of CNS tissue (Willinger and Schachner, 1980). These findings have led to postulate that GM_1 may play a specific role in axon sprouting and growth. This conclusion is consistent with reports indicating that anti-GM_1 antibodies produce long lasting morphological and behavioral abnormalities when administered to developing animals (Kasarskis et al., 1981). In addition, whilst exogenously supplied GM_1 has been reported to affect neuritic outgrowth from a variety of clonal and primary (including CNS) neuronal cells in vitro, (see Figure 2), anti-GM_1 antibodies have been documented to inhibit neurite regeneration both in vitro and in vivo (Spirman et al., 1982; Ferrari et al., 1983; Leon et al., 1984; Facci et al., 1984; Sparrow et al., 1984; Skaper et al., 1985; Doherty et al., 1985; Leon et al., 1986).

Related to the above, another growing body of literature indicates that most probably GM_1 is involved in the regulation of neuronal cell responsiveness to neuronotrophic factors (Skaper et al., 1985; Leon et al., 1986). Many of the in vitro neuritogenic effects of the exogenously supplied GM_1 are associated with enhanced neuronal cell survival and/or development. In most cases, the attainment of these GM_1 effects is strictly dependent on the presence of adequate NTF(s) and associated with enhanced neuronal cell responsiveness to the NTF.

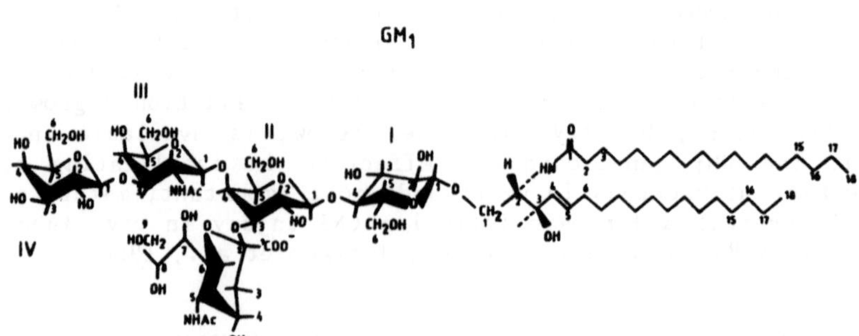

Figure 1

Chemical structure of GM_1 monosialoganglioside.

For example, both fetal chick (E8) dorsal root ganglionic (DRG) and pheochromocytoma (PC12) cells are known to necessitate NGF for neuritic outgrowth. In both cases, the exogenous GM_1-induced facilitation of neuritic outgrowth was shown to absolutely necessitate NGF. In fact the GM_1 alone was totally ineffective (Leon et al., 1984; Ferrari et al., 1983). Noteworthy is that the exogenously supplied GM_1 has been reported to result in stable incorporation of the ganglioside with the surface of many cell types in culture (Radsak et al., 1981; Facci et al., 1984) and to behave, in some instances, like the endogenously occurring gangliosides (Schwartzmann et al., 1983). These findings, together with the evidence that anti-GM_1 antibodies and choleragenoid inhibit respectively NGF-induced neurite outgrowth from DRG explants (Schwartz and Spirman, 1982) and from primed PC12 cells (Ferrari et al., in preparation), indicate that both the endogenous and exogenous GM_1 molecules play a significant role in the transduction mechanisms of the NGF signal. In addition, as analogous results were obtained in primary neuronal cell culture systems, e.g. ciliary ganglia and CNS mesencephalic neurons, known to necessitate trophic support other than NGF, the GM_1 potentiation of NTF efficacy on neuronal cells is not limited to NGF.

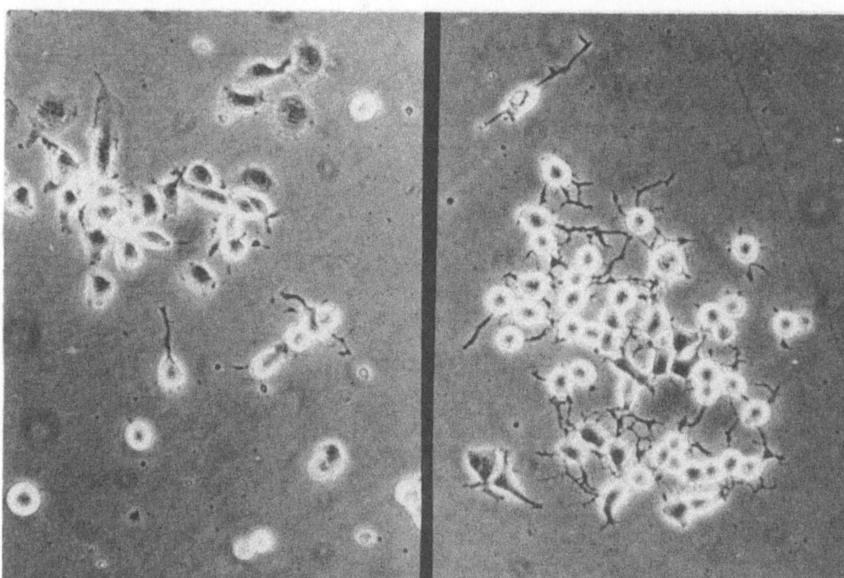

Figure 2

Effect of GM_1 addition to neuroblastoma N_2A cells.
A) Control cells, B) GM_1-treated cells.
N_2A cells were exposed for 24 hours to GM_1 (10^{-4}M) in Dulbecco's modified Eagle's medium plus 10% fetal calf serum.

CORRELATION BETWEEN GM_1 EFFECTS IN VITRO AND IN VIVO: FUTURE PERSPECTIVES

As previously outlined and exemplified in Figure 3, it is conceivable that neuronotrophic deficit at early post-lesion stages may well be one of the underlying causes of progressive neuronal cell death following brain injury in adult mammals. Feasibly, agents or conditions capable of ameliorating early post-lesion neuronal cell response to endogenously occurring NTF(s) may well be effective in ameliorating prognosis.

GM_1 ganglioside is now known to be able to favourably modulate neuronal cell behavior in response to NTF(s) in vitro. Furthermore, the systemic administration of GM_1 has been shown to improve outcome following experimentally induced brain damage to adult rodents. In many instances, this effect has been associated with enhanced neuronal cell survival (Toffano et al., 1986). Interestingly the intracerebral administration of NTF(s) has been reported to produce a similar effect. Together these findings have led to the working hypothesis that GM_1 effects in vivo are, as in vitro, related to the modulation of NTF activity, per se inadequate soon after the injury.

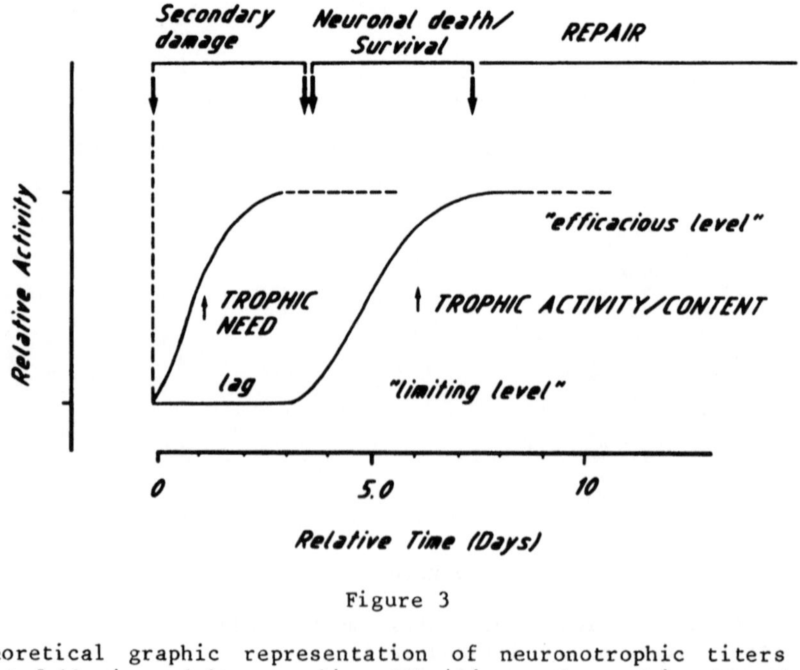

Figure 3

Theoretical graphic representation of neuronotrophic titers and need following adult mammalian CNS injury: the working hypothesis on GM_1 effects in vivo (see text for further elucidations).

Although much work is still necessary so as to establish the validity of the above mentioned hypothesis, initial in vivo experiments are promising. Should this prove to be a reality, it re-emphasizes the importance of understanding cellular and molecular mechanisms underlying adult mammalian lesion-induced neuronal plasticity. Their knowledge is a fundamental prerequisite for deriving novel pharmacological strategies aiming at reducing damage and/or restoring function following trauma or stroke in humans. The GM_1 ganglioside is an example.

REFERENCES

Bremer, E.G., Hakomori, S.I., Bowen-Pope, D., Raines, E. and Ross, R. (1984). Ganglioside-mediated modulation of cell growth, growth factor binding and receptor phosphorylation. J. Biol. Chem., 259, 6818-6825.

Cotman, C.W. and Nieto Sampedro, M. (1984). Cell biology of synaptic plasticity. Science, 225, 1287-1294.

Cowan, W.M., Fawcett, J.W., O'Leary, D.D.M. and Stanfield, B.B. (1984). Regressive events in neurogenesis. Science, 225, 1258-1265.

Doherty, P., Dickson, J.G., Flanigan, T.P. and Walsh, F.S. (1985). Ganglioside GM_1 does not initiate, but enhances neurite regeneration of nerve growth factor-dependent sensory neurones. J. Neurochem., 44, 1259-1265.

Facci, L., Leon, A., Toffano, G., Sonnino, S., Ghidoni, R. and Tettamanti, G. (1984). Promotion of neuritogenesis in mouse neuroblastoma cells by exogenous ganglioside. Relationship between the effect and the cell association of ganglioside GM_1. J. Neurochem., 42, 299-305.

Ferrari, G., Fabris, M. and Gorio, A. (1983). Gangliosides enhance neurite outgrowth in PC12 cells. Devl. Brain Res., 8, 215-221.

Hefti, F. (1986). Nerve growth factor (NGF) promotes survival of septal cholinergic neurons after fimbrial transections. J. Neurosci., 6, 2155-2162.

Kasarskis, E.J., Karpiak, S.E., Rapport, M.M., Yu, R.K. and Bass, N.H. (1981) Abnormal maturation of cerebral cortex and behavioral deficit in adult rats after neonatal administration of antibodies to gangliosides. Devel. Brain Res., 1, 25-35.

Kromer, L.F., Björklund, A. and Stenevi, U. (1981). Regeneration of the septo hippocampal pathways in adult rats is promoted by utilizing embryonic hippocampal implants as bridges. Brain Res., 210, 173-200.

Ledeen, R.W., Yu, R.K., Rapport, M.M. and Suzuki, K. (1984). Ganglioside Structure, Function and Biomedical Potential. Plenum Press, New York.

Leon, A., Benvegnù, D., Dal Toso, R., Presti, D., Facci, L., Giorgi, O. and Toffano, G. (1984). Dorsal root ganglia and nerve growth factor: a model for understanding the mechanism of GM_1 effects on neuronal repair. J. Neurosci. Res., 12, 277-287.

Leon, A., Benvegnù, D., Dal Toso, R., Giorgi, O., Presti, D., Tettamanti, G. and Toffano G. (1986). Neuronal cell cultures and monosialoganglioside: a model for comprehension of mechanisms underlying central nervous system repair. In Experimental Brain Research, Supplementum 13: Processes of Recovery from Neural Trauma. (eds. G.M. Gilad, A. Gorio and G.W. Kreutzberg). Springer-Verlag, Berlin.

Nieto-Sampedro, M., Manthorpe, M., Barbin, G., Varon, S. and Cotman, C.W. (1983). Injury-induced neuronotrophic activity in adult rat brain. Correlation with survival of delayed implants in a wound cavity. J. Neurosci., 3, 2219-2229.

Radsak, K., Schwarzmann, G. and Wiegandt, H. (1981). Studies of the cell association of exogenously added sialoglycolipids. Z. Physiol. Chem., 363, 263-272.

Richardson, P.M., Mc Guinnes, U.M. and Aguayo, A.J. (1980). Axons from CNS neurons regenerate into PNS grafts. Nature, 284, 264-265.

Schwartz, M. and Spirman, N. (1982). Sprouting from chicken embryo dorsal root ganglia induced by nerve growth factor is specifically inhibited by affinity purified antiganglioside antibodies. Proc. Natl. Acad. Aci. USA, 79, 6080-6083.

Schwartzmann, G., Hoffmann-Bleihauer, P., Shubert, J., Sandhoff, K. and Marsh, D. (1983). Incorporation of ganglioside into analogues into fibroblast cell membranes. A spin label study. Biochemistry, 22, 5041-5048.

Skaper, S.D., Katoh-Semba, R. and Varon, S. (1985). GM_1 ganglioside accelerates neurite outgrowth from primary peripheral and central neurons under selected culture conditions. Devl. Brain Res., 23, 19-26.

Sparrow, J.R., Mc Guinnes, C., Schwartz, M. and Grafstein, B. (1984). Antibodies to gangliosides inhibit goldfish optic nerve regeneration in vivo. J. Neurosci. Res., 12, 233-243.

Spirman, N., Sela, B.A. and Schwartz, M. (1982). Antiganglioside antibodies inhibit neuritic outgrowth from regenerating goldfish explants. J. Neurochem., 39, 874-877.

Svennerholm, L. (1963). Chromatographic separation of human brain gangliosides. J. Neurosci., 10, 613-623.

Toffano, G. and Leon, A. (1986). Central nervous system repair and nerve cell cultures. Pharmacol. Res. Commun., 18, 187-201.

Toffano, G., Agnati, L.F. and Fuxe, K.G. (1986). The effect of the ganglioside GM_1 on neuronal plasticity. Int. J. Devl. Neurosci., 4, 97-100.

Tsukahara, N. (1981). Synaptic plasticity in the mammalian central nervous system. Ann. Rev. Neurosci., 4, 351-379.

Varon, S., Manthorpe, M. and Williams, L.R. (1984). Neuronotrophic and neurite-promoting factors and their clinical potentials. Dev. Neuroscience, 6, 73-100.

Varon, S., Williams, L.R. and Gage, F.H. (1986). Exogenous administration of neuronotrophic factors in vivo protects CNS neurons against axotomy-induced degeneration. Progress in Brain Res., In Press.

Whittemore, S.R., Nieto-Sampedro, M., Needels, D.L. and Cotman, C.W. (1985). Neuronotrophic factors for mammalian brain neurons: injury induction in neonatal, adult and aged brains. Develop. Brain Res., 20, 169-178.

Will, B. and Hefti, F. (1985). Behavioural and neurochemical effects of chronic intraventricular injections of nerve growth factor in adult rats with fimbria lesions. Behav. Brain Res., 17, 17-24.

Willinger, M. and Schackner, M. (1980). GM_1 ganglioside as a marker for neuronal differentiation in mouse cerebellum. Develop. Biol., 74, 101-107.

7
Influence of Gangliosides and Nerve Growth Factor on the Plasticity of Forebrain Cholinergic Neurons

A. Claudio Cuello, D. Maysinger, L. Garofalo, P. Tagari, P.H. Stephens, E. Pioro and M. Piotte

INTRODUCTION

Neurons from the medial septal nucleus and nucleus of the vertical limb of the diagonal band of Broca provide an important cholinergic input to the hippocampus (Lewis and Shute, 1967; Oderfeld-Nowak et al, 1974; Meibach and Siegel, 1977). The cortex receives a widespread distribution of cholinergic fibres, the majority of which seem to originate from the nucleus basalis magnocellularis (NBM) (Johnston et al, 1981; Fibiger, 1982; Cuello and Sofroniew, 1984). From immunohistochemical studies it would appear that a topographic representation exists for this projection (Mesulam et al, 1986; Ingham et al, 1985). These fibers represent approximately 70% of the total cholinergic component of the cortex (Lehman et al, 1982), the remainder deriving from local circuit neurons (Sofroniew et al, 1982; Johnston et al, 1981). This participation of forebrain cholinergic neurons has been emphasized in recent years since a decrease in choline acetyltransferase (ChAT) activity has been reported to occur in the cortex and in the NBM of patients with Alzheimer's disease Bowen et al, 1983; Davies and Maloney, 1976; Perry et al, 1977; Rossor et al, 1982; Sims et al, 1983). Furthermore, a reduced number of cells in the latter area has been reported as a feature of Alzheimer's disease (Whitehouse et al, 1982). On the basis of these findings it has been suggested that cell loss in the NBM is a primary factor in the aetiology of Alzheimer's disease, analogous to the loss of nigral dopaminergic neurons observed in Parkinson's disease. We have speculated instead on the possiblity that the involvement of forebrain cholinergic neurons in Alzheimer's disease is a consequence of a primary cortical lesion. Consistent with this proposal is the fact that in the rat, ChAT immunoreactive neurons shrink and lose processes following desvacularising lesions of the cortex (Sofroniew et al, 1983) and that a specific depletion of ChAT activity is detected in the microdissected nucleus basalis, but not in other CNS areas, of cortically lesioned animals (Stephens et al, 1985). We have been interested in the possiblity of preventing pharmacologically the secondary, retrograde, subcortical involvement of cholinergic neurons following a primary cortical lesion. Repair and regenerative changes can take place in the CNS in response to injury, with a multitude of known and unknown factors participating in those processes. Our studies were directed towards examining the effects of the monosialo-ganglioside (GM_1) and nerve growth factor (NGF) on subcortical cholinergic neurons following removal of the hippocampus or desvacularizing lesions of the neocortex.

EXPERIMENTAL EVIDENCE FOR A PROTECTIVE ROLE OF GANGLIOSIDES AND NGF ON FOREBRAIN CHOLINERGIC NEURONS

In the rat, signs of retrograde degeneration of cholinergic neurons of the nucleus basalis magnocellularis (NBM) can be attained by compromising the terminal blood supply to part of the neocortex. This leads to necrosis of the subadjacent cortical area whilst leaving the blood supply to subcortical structures intact. Similarly, retrograde involution and loss of septal nucleus cholinergic neurons results from the removal of the hippocampus (Figure 1a-c). The surgical interruption of cortical pial vessels results in specific morphological (Sofroniew et al, 1983) (Figure 1a, d and e) and biochemical (Stephens et al, 1985) alterations of the cholinergic component of the NBM which projects to the neocortex. These alterations are expressed as a reduction of the cross sectional area of cholinergic somata, a diminution of neuronal processes (Sofroniew et al, 1983) and a depletion of choline acetyltransferase activity in the NBM but not other CNS areas (Stephens et al, 1985).

Figure 1: Morphological consequences of lesions of terminal fields on basal forebrain cholinergic neurons.

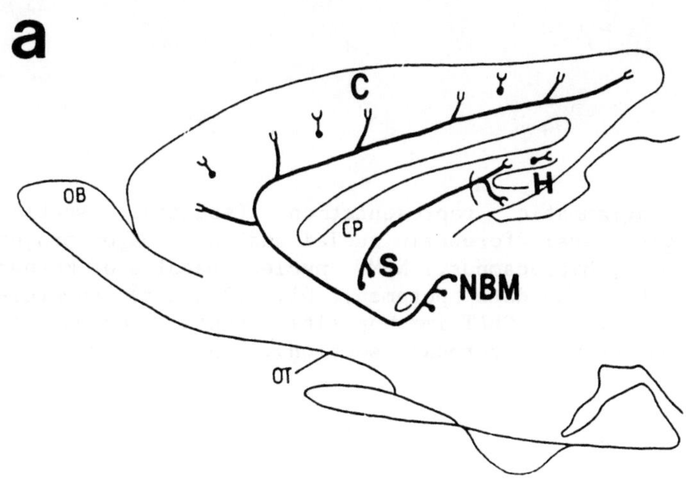

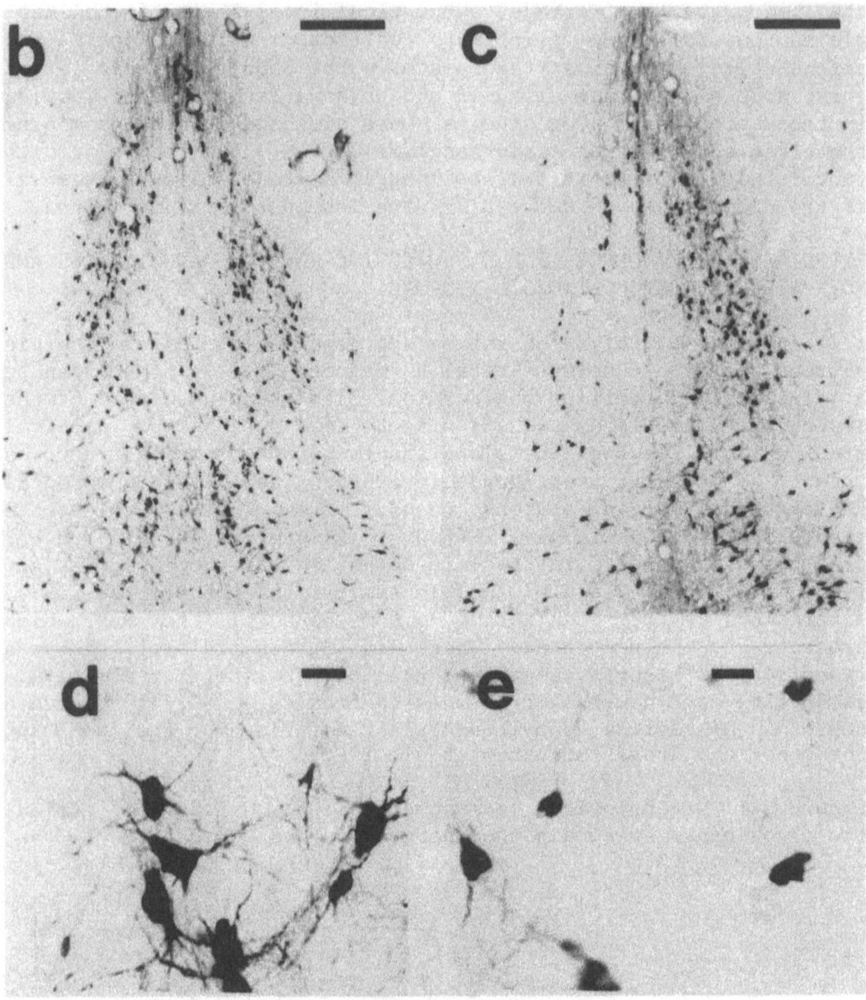

a) A diagrammatic representation (sagittal section) of cholinergic basal forebrain nuclei and their major projections; S, septum; H, hippocampus,; NBM, nucleus basalis of Meynert; C, cortex; CP, caudate putamen; OT, olfactory tubercle; OB, olfactory bulb. b) ChAT immunopositive septal neurons from an unlesioned animal (coronal section). Scale bar = 1mm. c)

Ipsilateral degeneration with loss of immunostained cholinergic septal neurons was apparent 10 weeks after a suction lesion of the left hippocampus. Scale bar = 1mm. d) ChAT immunopositive neurons in the NBM of an unoperated animal. Scale bar = 50 micrometres. e) Retrograde changes in ipsilateral NBM cholinergic neurons 30 days after a unilateral cortical lesion resulted in a shrunken appearance and loss of cell processes. Scale bar = 50 micrometres.

In young animals (30 days old at surgery), both the biochemical and morphological changes reach a maximum at the 30th post-operative day. There is, however, no apparent recovery from cell shrinkage in these animals up to 120 post-operative days, although a full recovery of the ChAT enzymatic activity in young and mature animals (120 days old at time of operation) has occurred at this point (see Figure 2) (Stephens et al, submitted).

Figure 2: Retrograde biochemical effects on the ChAT activity of the microdissected nucleus basalis of the rat following extensive unilateral cortical lesions.

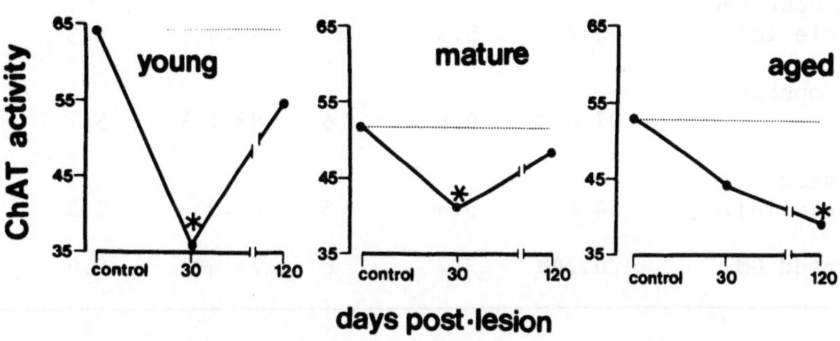

Note that a marked depletion in ChAT activity is observed at all ages. This loss of activity is followed by spontaneous recovery in the young and mature animals but not in the aged rats. Choline acetyltransferase activity (ordinate, nmols ACh/mg protein/hour) was measured in microdissected homogenates of NBM ipsilateral to the cortical lesion in young, mature and aged animals at various times after the operation (abcissa, days).

As monosialo-ganglioside GM_1 displays apparent trophic neuronal functions (see Toffano, this volume), its effects were investigated on the model of retrograde degeneration of forebrain cholinergic neurons mentioned above. In our first set of studies the daily intraperitoneal administration of 30 mg/kg of GM_1 to young rats bearing cortical lesions completely prevented the ipsilateral decrease in ChAT activity otherwise seen. When compared to vehicle injected controls, treatment with GM_1 caused an increase in ChAT enzymatic activity in both sham operated and cortically damaged animals (Cuello et al, 1986) (Table 1). The effects of long term ganglioside administration in the population of ChAT immunoreactive neurons of the medial septal nucleus and vertical limb of the diagonal band following unilateral hippocampal removal (Sofroniew, et al, in press) were similarly examined. In this system, ganglioside administration arrested the cell losses observed in the ipsilateral side of these structures. Table 2 summarizes these findings in the medial septum (see also Figures 1b and c).

TABLE 1: The effect of chronic ganglioside GM_1 treatment on ChAT activity in microdissected NBM from animals with unilateral (right) cortical lesions.

	Left			Right		
	ChAT	S.E.M.	n	ChAT	S.E.M.	n
Sham operated Vehicle inj.	52.7	5.3	5	49.4	4.3	5
Sham operated GM_1	97.4**	6.6	6	83.1**	5.7	6
Operated Vehicle inj.	54.5	5.4	5	43.0*	2.5	5
Operated GM_1	70.7**	3.7	6	75.4**	5.0	6

Choline acetyltransferase activity (nmols ACh/mg protein/hour) was measured in the ipsilateral (right) and contralateral (left) nucleus basalis from cortically lesioned or sham operated young animals after treatment for 30 days with GM_1 or vehicle (saline) *$p < 0.01$; ** $p < 0.001$

To elucidate further the protective effects of GM_1 in retrograde neuronal involvement, we explored circumstances related to both the age of the recipient and the regime of ganglioside administration. Successful protection of forebrain cholinergic neurons of the nucleus basalis was attained in only young and mature rats. Furthermore, relatively long term treatment with gangliosides (30 mg/kg daily for 30 days) failed to revert the discrete but significant change in nucleus basalis ChAT activity in aged animals (Stephens et al, in press).

We have also observed that time of initiation of ganglioside administration is crucial for the effective protection of forebrain cholinergic neurons. Thus, in young rats, when the initiation of the ganglioside treatment was delayed for 10 days no recovery in the ChAT activity of the nucleus basalis was observed. Based on these observations, we have administered 30 mg/kg of GM_1 for seven days starting from the first post-operative day and, in another series, commencing at the time of surgery. Only in the latter circumstance did ChAT enzymatic activity remain at control levels in the NBM following cortical lesions.

TABLE 2: Effects of unilateral (left) hippocampal removal and ganglioside

Treatment	ChAT immunoreactive cell number (mean ± s.e.m.) ipsilateral	contralateral	% Difference
unoperated + vehicle	682.3 ± 41.3	687.3 ± 22.7	- 0.7
unoperated + GM_1	647.7 ± 23.6	645.5 ± 36.6	+ 0.4
lesioned + vehicle	170.0 ± 25.8	637.3 ± 45.8	-73.3*
lesioned + GM_1	639.0 ± 30.3	725.5 ± 20.6	-14.7

ChAT immunoreactive cells were counted at 5 matched levels in each of 3 animals from each group after 30 days of treatment with GM_1 (30 mg/kg daily IP) or vehicle. *$p < 0.01$ by unpaired t-test.

In 30 day old animals, ganglioside administration (30 mg/kg daily i.p. for 30 days) was not initiated until 10 days after unilateral (left) cortical lesions. A significant decrement in ChAT activity was observed in the ipsilateral (left) basalis, which could not be reverted by the delayed administration of large doses of gangliosides.

We found that short term application of ganglioside resulted in an equally effective protection of cholinergic neurons from retrograde degeneration. The possibility of reducing the amount of the compound required was further explored. At the same time the experiments were assigned to elucidate the possible site of ganglioside actions, i.e., peripheral vs central effects. For this we compared the effects of large (30 mg/kg) and small (5 mg/kg) doses of gangliosides on cholinergic neurons of the NBM following decortication. Figure 4 summarizes these results which indicated that 5 mg/kg of GM_1, delivered intracerebroventricularly via osmotic mini-pumps, for only seven days, rendered a protection of cholinergic neurons comparable to that of 30 mg/kg, intraperitoneally. However, these small doses of ganglioside did not prevent cholinergic involution when applied systemically, suggesting a preferential central site of action of this compound.

Figure 3: The effect of short-term GM_1 treatment, initiated immediately following unilateral cortical lesion.

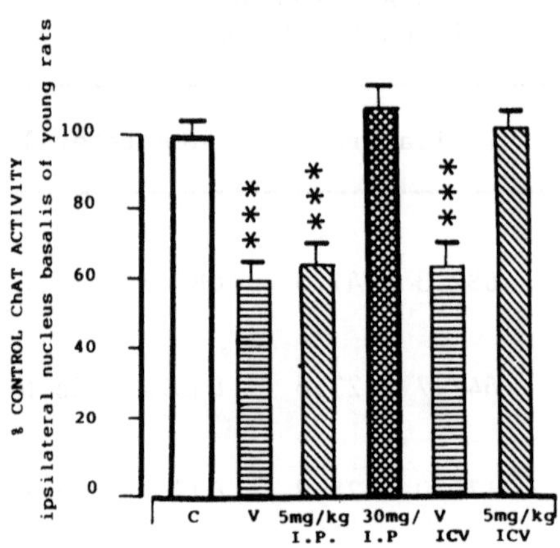

Ganglioside GM_1 was administered intraperitoneally (i.p. 5 mg/kg (n=6) or 30 mg/kg (n=10) daily for 7 days) or centrally (i.c.v., 5 mg/kg (n=11) by continuous infusion into the contralateral cerebral ventricle by osmotic mini-pumps commencing immediately after surgery. Vehicle was administered intraperitoneally (n=12) and intracerebroventricularly (n=5) to the lesioned animals and controls (n=9).

The short treatment with small doses of GM_1 applied via the lateral ventricle also appeared to conserve the morphology of the ChAT immunoreactive neurons as judged by immunocytochemistry and image analysis. Thus, the distribution of the cross-sectional area of ChAT immunoreactive neurons in the NBM of lesioned animals was shifted towards a smaller size population after the lesion, indicating shrinkage of somata. However, in animals receiving GM_1 for 7 days commencing at the time of surgery, the distribution was more comparable with control values, suggesting a degree of morphological protection (Figure 4).

Figure 4: The effect of short-term GM_1 treatment on the area of NBM ChAT immunoreactive neurons after cortical lesions.

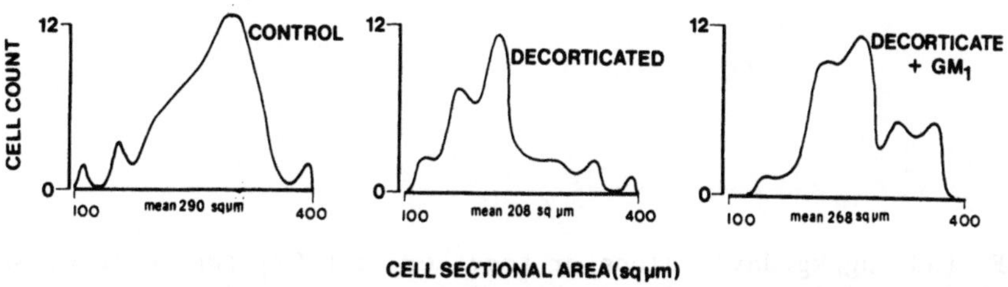

Matched sections from two animals from each group were immunostained for choline acetyltransferase ChAT (Eckenstein and Thoenen, 1982) and horseradish peroxidase (Cuello et al, 1984). A shift towards smaller somata was noted in lesioned animals (centre), which was largely averted by intraperitoneal ganglioside treatment for 7 days (30 mg/kg daily commencing at operation, right).

As the above effects of gangliosides resemble in vivo effects of nerve growth factor (NGF) (see Levi-Montalcini, this volume), we consider it important to compare the effects of GM_1 with those of NGF in the present cortex-basalis model, and further, explore for possible interactions. Our results in this direction indicate that doses of NGF (intracerebroventricular) which produce stimulation of ChAT activity in septo-hippocampal neurons after partial transection of the fimbria (Hefti et al, 1984) produce an analogous protection of the cholinergic neurons in the nucleus basalis while the simultaneous administration of NGF and GM_1 resulted in an unexpected potentiation of these effects. In these experiments the values of ChAT activity of the microdissected NBM rose above normal in the lesioned side (Figure 5a). Furthermore, in the remaining neocortex, ipsilateral to the lesion, ChAT activity rose to 237% of control values (see Figure 5b).

Figure 5: Effect of NGF and GM_1 on ChAT activity in the microdissected ipsilateral NBM and ipsilateral cortex of mature rats following unilateral decortication.

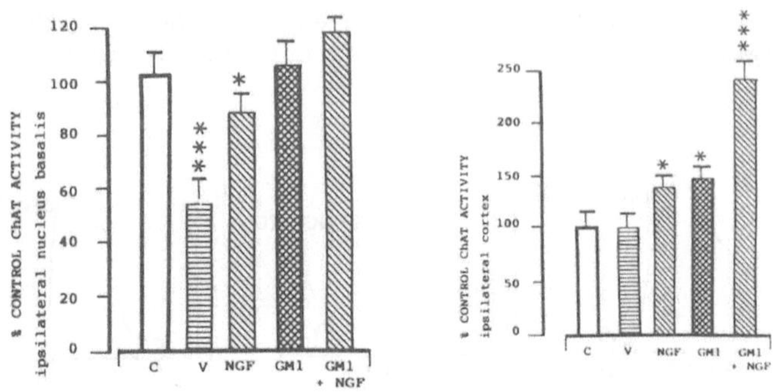

NGF (80 µg/kg/day) alone or together with GM_1 (or vehicle) was administered i.c.v. via osmotic minipumps for 7 days commencing at surgery in cortically lesioned mature rats. ChAT activity is expressed at a % of control (unoperated) values. *$p < 0.05$, ***p < 0.001) by unpaired t-test on untransformed data.

DISCUSSION

The present experimental model for retrograde degeneration of forebrain cholinergic neurons supports the idea that the subcortical cholinergic involvement observed in Alzheimer's disease and related conditions might be caused by a primary cortical lesion. This view is also substantiated by evidence of losses of cortical somatostatin immunoreactivity (Davies et al, 1980; Beal et al, 1986), which, in the neocortex, is apparently restricted to local circuit neurons and the demonstration of somatostatin's involvement in early plaques (Morrison et al, 1985; Roberts et al, 1985). Assuming that the present experimental model parallels, with a different time course, a retrograde deterioration of cholinergic forebrain neurons such as may occur in Alzheimer's disease, it is of interest to investigate the possibilities of arresting or reversing such a process. In this regard lesions in the neocortex or ablation of the hippocampus represent valid models to study the retrograde involution or losses of forebrain cholinergic cells.

It has recently been observed that central cholinergic neurons are able to transport retrogradely nerve growth factor (NGF) (Korsching and Thoenen, 1983) and that they possess receptors to it (Taniuchi et al, 1986; Richardson et al, 1986). Furthermore, the application of NGF in newborn rats leads to an increment in CNS ChAT activity (Heumann et al, 1984; Mobley et al, 1986). NGF has, however, no such effects on the dopaminergic nigrostriatal pathway (Schwab et al, 1979).

The particular responsiveness of CNS cholinergic neurons offers certain hopes that trophic factors might promote growth and repair of these neurons. Thus, we have observed a remarkable recovery of the neurotransmitter biosynthetic enzyme ChAT in the nucleus basalis (see Figure 5). Similar recoveries in biochemical parameters of cholinergic function have also been reported in the reverse situation, i.e., after stereotaxic lesions of the NBM (Wenk et al, 1984; Casamenti et al, 1985). In our model for retrograde degeneration, this capacity for spontaneous recovery is absent in aged animals, a finding which is perhaps to be expected, as the production of endogenous trophic factors in response to CNS injuries is diminished in aged animals (Nieto Sampedro et al, 1982).

In the cortex-basalis model discussed here a dissociation is observed between morphological (cell size) and biochemical (ChAT activity) parameters in the nucleus basalis four months after lesions of the neocortex. A differential participation of multiple, uncharacterized endogenous factors could explain such dissociation between biochemical and morphological signs of recovery at 120 post-operative day in young and mature animals. Evidence for the existence of separate factors regulating biochemical and morphological parameters in cholinergic neurons has been produced by Smith and collaborators (1983; 1985; 1986).

Large doses of gangliosides administered intraperitoneally are able to prevent the dramatic loss of cholinergic cell neurons subsequent to hippocampal ablation (Sofroniew et al, in press). This is a situation similar to that brought about by the intracerebroventricular administration of nerve growth factor. Similarly, intracerebroventricular administration of NGF has been reported to prevent the loss of septal cholinergic neurons following hippocampal lesions (Williams et al, 1986). The fact that gangliosides can prevent such cholinergic cell losses is remarkable, as in this case the target has been removed and the operated animal is presumably unable to release the trophic factors required to maintain the cell body at axonal projection target sites. Gangliosides have similarly been shown to stimulate the reinnervation of the hippocampus after partial lesions of the fimbria (Oderfeld-Nowak et al, 1984), or septal nucleus (Wojcik et al, 1982).

Our results with the model of cholinergic cortex-basalis degeneration reveal that the time of ganglioside administration is important in attaining full protection of these neurons from retrograde involution. It was seen that small amounts of GM_1 are capable of preventing degenerative changes when applied from the onset of cortical lesion. Thus, the initiation and duration of ganglioside treatment and the total amount administered are relevant to the development of the therapeutic potential of GM_1. A delay of just one day in the low doses short treatment scheme or 10 days delay in the case of high doses prolonged treatment greatly attenuates ganglioside effectiveness. The importance of an early treatment with gangliosides concurs with the concept of a critical period of low availability of endogenous trophic factors immediate after lesions of the nervous system (see also Toffano, this volume). Thus, the existence of critical times of initiation of treatment, duration and doses indicate that, after an injury of the CNS, a delicate balance is established between exogenously administered GM_1 and endogenous trophic factors. This is probably esssential for securing the protection of neuronal elements after direct or indirect cellular disruption. It is conceivable that there exist a variety of CNS trophic factors acting on subsets of transmitter specific neurons. Such factors are apparently released when a CNS lesion occurs, the amount being dependent on the time of lesion onset and the animal's age (Nieto Sampedro _et al_, 1982). This is in line with our observations that the protective effects of GM_1 in the present model of retrograde degeneration of cholinergic neurons are also dependent on the age of the recipient. Besides their scientific interests, these aspects (age, dosage, initiation and duration of treatment) should be carefully considered in the possible development of ganglioside therapeutic strategies for the treatment of Alzheimer's or other neurodegenerative diseases.

The mechanisms by which gangliosides bring about these protective effects on cholinergic neurons are far from clear. The co-operativity found in our studies with the simultaneous administration of GM_1 and NGF is suggestive of interactions with trophic factors. There is the distinct possiblity that GM_1 facilitates the interactions between trophic factors and their corresponding membrane receptors. Exogenous gangliosides can be incorporated into membranes of neural membrane preparations (Toffano _et al_, 1980). Gangliosides inserted in cell membranes could elicit a number of receptor-receptor or ligand-receptor interactions. Along these lines, great deal of attention was raised by the observation that cholera toxin, which is a "non-specific" activator of adenylate cyclases, binds rather specifically to the ganglioside GM_1 (Heyningen, 1973; Holmgren _et al_,1975). Nevertheless, while the beta-subunit of cholera toxin is responsible for this specific ganglioside binding, it is the alpha-subunit which activates the cyclases (Gill and King, 1975; Bitensky _et al_, 1975). Thus, exogenous gangliosides, once inserted into neural membranes, could bind to cholera toxin leaving the biologically active site free. A similar mechanism might be operative in the case of specific trophic factors and hormones. In the latter case, there is documentation for the interaction of thyrotropin and of the follicle stimulating hormone with the ganglioside GM_1 (Mullin _et al_, 1976; Deleers, 1979), both cases depending on homologies of aminoacid sequences of these hormones with the beta-subunit of choleratoxin (Kurosky _et al_, 1977). Further knowledge of the nature of endogenous factors released as a consequence of neural injury and their receptors could provide a firmer base for the understanding of these cell membrane molecular interactions.

ACKNOWLEDGEMENTS

This communication is dedicated to Dr. Rita Levi-Montalcini on the occasion of her 1986 Nobel Award in Medicine.

The authors wish to thank Drs. Rita Levi-Montalcini, P. Calissano and L. Aloe for the generous provision of nerve growth factor. We would also like to thank Sharon MacMillan for secretarial assistance, and Drs. Thoenen and Eckenstein and Boehringer-Mannheim for the provision of monoclonal antibodies against choline acetyltransferase. A.C. Cuello acknowledges financial support from the Office of the Dean of Medicine, McGill University, the Medical Research Council (Canada) and Fidia Research Laboratories.

REFERENCES

Beal, M.F., Benoit, R., Mazurek, M.F., Bird, E.D. and Martin, J.B. (1986). Somatostatin-28_{1-12}-like immunoreactivity is reduced in Alzheimer's disease cerebral cortex. Brain Res., 368, 380-383.

Bitensky, M.W., Wheeler, M.A., Mehta, H. and Miki, N. (1975). Cholera toxin activation of adenylate cyclase in cancer cell membrane fragments. Proc. Natl. Acad. Sci. U.S.A., 72, 2572-2576.

Bowen, D.M., Allen, S.J., Benton, J.S., Goodhart, M.J., Haan, E.A., Palmer, A.M., Sims, N.R., Smith, D.D.T., Spillane, J.A., Esiri, M.M., Neary, D., Snowdon, J.B., Wilcock, G.K. and Davison, A.N. (1983). Biochemical assessment of serotonergic and cholinergic dysfunction and cerebral atrophy in Alzheimer's disease. J. Neurochem., 41, 266-272.

Casamenti, F., Bracco, L., Bartolin, L. and Faper, G. (1985). Effects of ganglioside treatment in rats with a lesion of the cholinergic forebrain nuclei. Brain Res., 338, 45-52.

Cuello, A.C., Sofroniew, M.V. (1984). The anatomy of the CNS cholinergic neurons. TINS, 7, 74-78.

Cuello, A.C., Milstein, D., Wright, B., Bramwell, B., Priestley, J.V. and Jarvis, J. (1984). Development and application of a monoclonal rat peroxidase anti-peroxidases (PAP) immunocytochemical reagent. Histochemistry, 30, 257-261.

Cuello, A.C., Stephens, P.H., Tagari, P.D., Sofroniew, M.V. and Pearson, R.C.A. (1986). Retrograde changes in the nucleus basalis of the rat, caused by cortical damage, are prevented by exogenous ganglioside GM_1. Brain Res. 376, 373-377.

Davies, P. and Maloney, A.J.F. (1976). Selective loss of central cholinergic neurons in Alzheimer's Disease. Lancet, II, 1403.

Davies, P., Katzman, R. and Terry, R.D. (1980). Reduced somatostatin-like immunoreactivity in cerebral cortex of cases of Alzheimer disease and Alzheimer senile dementia. Nature, 288, 279-280.

Deelers, N., Chatelain, P., Poss, A. and Ruysschaert, J.M. (1979). Specific interaction between follitropin and GM_1 ganglioside incorporated into lipid membranes. Biochem. Biophys. Res. Comm., 89, 1102-1106.

Eckenstein, F. and Thoenen, H. (1982). Production of specific antisera and monoclonal antibodies in choline acetyltransferase. Characterization and use for identification of cholinergic neurons. EMBO J. 1, 363-368.

Fibiger, H.C. (1982). The orgnaization and some projections of cholinergic neurons of the mammalian forebrain. Brain Res. Rev., 4, 327-388.

Gill, D.M. and King, C.A. (1975). The mechanism of action of cholera toxin in pigeon erythrocyte lysates. J. Cell Biol., 250, 6424-6432.

Hefti, F.A., Dravid, A. and Hartikka, J. (1984). Chronic intraventricular injections of nerve growth factor elevate hippocampal choline acetyltransferase activity in adult rats with partial septo-hippocampal lesions. Brain Res., 293, 305-309.

Heumann, R., Schwab, M.E., Merkl, R. and Thoenen, H. (1984). Nerve growth factor (NGF) mediated induction of choline acetyltransferase (ChAT) in PC12 cells: evaluation of the site of action of NGF and the involvement of lysosomal degradation products of NGF. J. Neurosci., 4, 3039-3050.

Holmgren, J., Lonnroth, I., Mansson, J.E. and Svennerholm, L. (1975). Interaction of cholera toxin and membrane GM_1 ganglioside of small intestine. Proc. Natl. Acad. Sci. U.S.A., 72, 2520-2524.

Ingham, C.A., Bolam, J.P., Wainer, B.H. and Smith, A.D. (1985). A correlated light and electron microscopic study of identified cholinergic basal forebrain neurons that project to the cortex in the rat. J. Comp. Neurol., 239, 176-192.

Johnston, M.V., McKinney, M. and Coyle, J.T. (1981). Neocortical cholinergic innervation: A description of extrinsic and intrinsic components in the rat. Brain Res., 43, 159-172.

Korshing, S. and Thoenen, H. (1983). Quantitative demonstartion of the retrograde axonal transport of endogenous nerve growth factor. Neurosci. Lett., 39, 1-4.

Kurosky, A., Markel, D.E., Peterson, J.W. and Fitch, W.M. (1977). Primary structure of cholera toxin β-chain: A glycoprotein hormone analog? Science, 195, 299-301.

Lehman, J., Nagy, J.I., Almadja, S. and Fibiger, H.C. (1982). The nucleus basalis magnocellularis; the origin of a cholinergic projection to the neocortex in the rat. Neuroscience, 5, 1161-1174.

Lewis, P.R. and Shute, C.C.D. (1967). The cholinergic limbic system: Projections to hippocampal formation, medial cortex, nuclei of the ascending cholinergic reticular system and the subfornical organ and supra-optic crest. Brain, 90, 521-540.

Meibach, R.C. and Siegel, A. (1977) Efferent connections of the hippocampal formation in the rat. Brain Res., 124, 197-224.

Mesulam, M., Mufson, E.J. and Wainer, B.H. (1986). Three-dimensional representation and cortical projection topography of the nucleus basalis (Ch4) in the macaque: concurrent demonstration of choline acetyltransferase and retrograde transport with a stabilised tetramethylbenzidine method for horseradish peroxidase. Brain Res., 367, 301-308.

Mobley, W.C., Rutkowski, J.L., Tennekoon, G.I., Gemski, J., Buchanan, K. and Johnston, M.V. (1986) Nerve growth factor increases choline acetyltransferase activity in developing basal forebrain neurons. Mol. Brain Res., 1, 53-62.

Morrison, J.H., Rogers, J., Scherr, S., Benoit, R. and Bloom, F.E. (1985). Somatostatin immunoreactivity in neuritic plaques of Alzheimer's patients. Nature, 314, 90-92

Mullin, B.R., Fishman, P.H., Lee, G., Aloj, S.M., Ledley, F.D., Winand, R.J., Kohn, L.D. and Brady, R.O. (1976). Thyrotropin-ganglioside interactions and their relationship to the structure and function of thyrotripin receptor. Proc. Natl. Acad. Sci. U.S.A., 73, 842-846.

Nieto-Sampedro, M., Lewis, E.R., Cotman, C.W., Manthorpe, E.M., Skaper, S.D., Barbin, G., Longo, F.M. and Waron, S. (1982). Brain injury causes a time dependent increase in neuron trophic activity at the lesion site. Science 217, 860-861.

Oderfeld-Nowak, B., Narkiewicz, O., Bialowas, J., Wieraszko, A. and Gradkowska, M. (1974). The influence of septal nuclei lesions on activity of acetylcholinesterase and choline acetyltransferase in the hippocampus of the rat. Acta Neurobiol. Exp. 34, 583-601.

Oderfeld-Nowak, B., Skup, M., Ulas, J., Jezierska, M., Gradkowska, R. and Zaremba, M. (1984). Effect of GM_1 ganglioside treatment on post lesion responses of cholinergic neurons in rat hippocampus after various partial deafferentations. J. Neurosci. Res., 12, 409-420.

Perry, E.R., Perry, R.H., Blessed, G. and Roth, M. (1977). Necropsy evidence of central cholinergic deficits in senile dementia. Lancet, I, 189.

Richardson, P.M., Verge-Issa, V.M.K. and Riopelle, R.J. (1986). Distribution of neuronal receptors for nerve growth factor in the rat. J. Neurosci., 6, 2312-2321.

Roberts, G.W., Crow, T.J. and Polak, J.M. (1985). Location of neuronal tangles in somatostatin neurons in Alzheimer's disease. Nature, 314, 92-94.

Rossor, M.N., Garret, N.J., Johnson, A.L., Mountjoy, D.Q., Roth, M. and Iversen, L.L. (1982b). A post-mortem study of the cholinergic and GABA systems in senile dementia. Brain, 185, 313-330.

Schwab, M.E., Otten, U., Agid, Y. and Thoenen, H. (1979). Nerve growth factor (NGF) in the rat CNS: absence of specific retrograde axonal transport and tyrosine hydroxylase induction in locus coeruleus and substantia nigra. Brain Res., 168, 473-483.

Sims, N.R., Bowen, D.M., Allan, S.J., Smith, C.C.T., Neary, D., Thomay, D.J. and Davison, A.N. (1983). Presynaptic cholinergic dysfunction in patients with dementia. J. Neurochem., 401 503-509.

Smith, R.G. and Appel, S.H. (1983). Extracts of skeletal muscle increase neurite outgrowth and cholinergic activity of fetal rat spinal motor neurons. Science, 219, 1079-1081.

Smith, R.G., McManaman, J. and Appel, S.H. (1986). Trophic effects of skeletal muscle extracts on ventral spinal cord neurons in vitro separation of a protein with morphologic activity from proteins with cholinergic activity. J. Cell Biol., 101, 1608-1621.

Smith, R.G., Vaca, K., McManaman, J. and Appel, S.H. (1985). Selective effects of skeletal muscle extract fractions on motoneuron development in vitro. J. Neurosci., 6, 439-447.

Sofroniew, M.V., Eckenstein, F., Thoenen, H. and Cuello, A.C. (1982). Topography of choline acetyltransferase-containing neurons in the forebrain of the rat. Neurosci. Lett., 33, 7-12.

Sofroniew, M.V., Pearson, R.C.A., Cuello, A.C., Stevens, P.H. and Tagari P. (1986). The effect of parenterally administered GM_1 ganglioside on retrograde degeneration of cholinergic cells of the basal forebrain of the rat. Brain Res., 398, 393-396.

Sofroniew, M.V., Pearson, R.C.A., Eckenstein, F., Cuello, A.C. and Powell, T.P.S. (1983). Retrograde changes in cholinergic neurons in the basal forebrain of the rat following cortical damage. Brain Res., 289, 370-374.

Stephens, P.H., Cuello, A.C., Sofroniew, M.V., Pearson, R.D.A. and Tagari, P. (1985). The effect of unilateral decortication upon choline acetyltransferase and glutamate decarboxylase activities in the nucleus basalis and other areas of the rat brain. J. Neurochem., 45, 1021-1026.

Stephens, P.H., Tagari, P.C. and Cuello, A.C. (1987). Cholinergic neurons in aged rats: changes after cortical lesions. Neurobiol. Aging (submitted).

Tanuichi, M., Schweitzer, J.B. and Johnson, E.M. (1986). Nerve growth factor receptor molecules in rat brain. Proc. Natl. Acad. Sci., 83, 1950-1954.

Toffano, G., Benvegnu, D., Bonetti, A.C., Facci, L., Leon, A., Orlando, P., Ghidoni, R. and Tettamanti, G. (1980). Interactions of GM_1 ganglioside with crude rat brain neuronal membranes. J. Neurochem., 35, 861-866.

Van Heyningen, S. (1974). Cholera toxin: Interaction of subunits with ganglioside GM_1. Science, 183, 656-666.

Wenk, G.L. and Olton, D.S. (1984). Recovery of neocortical ChAT activity following ibotenic acid injection into the nucleus basalis of Meynert in rats. Brain Res., 293, 184-186.

Whitehouse, P.J., Price, D.L., Strubie, R.G., Clark, A.W., Coyle, J.T. and DeLong, M.R. (1982). Alzheimer's disease and senile dementia loss of neurons in the basal forebrain. Science, 215, 1237-1239.

Williams, L.R., Peterson, G.M., Varon, S. and Gage, F.H. (1986). Continuous infusion of NGF prevents non-cholinergic as well as cholinergic neuronal death in the medial septum after fimbria formix transection. Society for Neuroscience 16th Annual Meeting, Washington, November 9-14, 1986, Abstr. 219.3, page 787.

Wojcik, M., Ules, J. and Oderfeld-Nowak, B. (1982). The stimulating effect of ganglioside injections on the recovery of choline acetyltransferase and acetylcholinesterase activities in the hippocampus of the rat after septal lesions. Neuroscience, 7, 495-499.

8
Hippocampal Stimulation of Cholinergic Expression in Primary Cultures of Basal Forebrain Neurons

Bradley C. Wise and Michel B. Emerit

A major goal of development neurobiology is to define the cellular and molecular mechanisms which regulate the survival, growth and synaptogenesis of central nervous system neurons. A role for specific neurotrophic factors in CNS development has been postulated based on described processes occurring in the peripheral nervous system such as naturally occurring neuronal death (Cowan et al., 1984). Such a process might serve to match the size of the afferent neuronal population with that of the target area of innervation. A factor determining neuronal death might be neuronal competition for a limited quantity of trophic factors present in the target region. The identification, characterization and our present understanding of the physiology of the nerve growth factor (NGF) emphasizes the importance of target-drived molecules on the survival, growth and differentiation of an afferent neuronal population (Thoenen and Barde, 1980; Yanker and Shooter, 1982). In mature neurons the maintenance of a critical number of terminal boutons may also require trophic support, albeit at a much lower level than is needed during development. Neurodegenerative diseases, such as senile and presenile dementias, may be caused in part by abnormalities in the supply of neurotrophic factors specific for certain neurons (Appel, 1981).

Our studies are concerned with the trophic influence that hippocampal neurons exert on their afferent cholinergic innervation originating from basal forebrain neurons. The approach we have taken is to grow basal forebrain neurons in culture and examine the effects of hippocampal cells on cholinergic phenotypic expression in the co-cultures. Our results indicate that hippocampal neurons exert a positive trophic effect on the survival and/or maturation of their afferent septal neurons.

EXPERIMENTAL PROCEDURES

The preparation of basal forebrain neuronal primary cultures (often referred to as septal cells) from fetal rats is a modification of the procedure of Hefti et al. (1985). Instead of adding horse and rat sera to the final culture media, the following serum-free supplements (Bottenstein and Sato, 1979) are added (final concentrations in parenthesis): insulin (10 μg/ml), transferrin (20 μg/ml), putrescine (0.1 mM), progesterone (20 nM) and sodium selenite (30 nM). The cells can survive up to 13 days in this serum-free medium and are usually used after 5-7 days in vitro (DIV). Primary cultures of cells from hippocampus and other fetal brain areas are prepared by a method similar to that described for the septal cells.

Acetylcholine (Ach) synthesis in intact cells is measured by a modification of the method of Johnson and Pilar (1980). The neurons are washed, incubated with 60 mM K^+, washed again, and then incubated with ^{3}H-choline (12.5-50 nM) for 10 min. The reaction is terminated by rapid washing followed by the addition of perchloric acid. ^{3}H-Choline and ^{3}H-Ach are extracted, precipitated and separated by reverse-phase HPLC exactly as described by Eva et al. (1984).

RESULTS

Characteristics of Basal Forebrain Neuronal Cultures

Basal forebrain cultures prepared from 17-day old rat fetuses and grown under serum-free conditions in the absence of anti-mitotic agents consist predominantly of neurons (>98% of cell population are neurons) as revealed by neurofilament and neuron-specific enolase immunocytochemistry. The remaining cells (<2% of total) are stained by antibodies against glial fibrillary acidic protein, a marker for astrocytes. Thus these cultures and presumably those from other fetal brain regions (see below) can be considered as nearly pure neurons ruling out glial cells as sources of neurotrophic support.

Basal forebrain cells possess a high affinity choline uptake process that is partially Na^+-dependent (50% decrease in uptake under Na^+-free conditions), is partially inhibited (40%) by hemicholinium-3 (a choline uptake blocker), and is temperature dependent (no uptake seen at 4°C). The cells can synthesize, in situ, acetylcholine (Ach) from the radiolabeled choline and Ach synthesis is inhibited when choline uptake is studied in the presence of hemicholinium-3 (10 μM). ^{3}H-Ach is measured by reverse-phase HPLC with an enzyme-loaded post column reactor and electrochemical detection (Eva et al., 1984). In situ Ach synthesis by septal cultures is a specific marker for cholinergic neurons. In fact hippocampal cells maintained under culture conditions identical to septal neurons fail to exhibit Ach synthesis although they possess a choline uptake process inhibited by hemicholinium-3.

Co-cultures of Basal Forebrain and Hippocampal Neurons

When basal forebrain cells are co-cultured with their target hippocampal neurons Ach synthesis is greatly stimulated (up to 10-fold) (Fig. 1). The increase in Ach synthetic capacity is most likely localized to the septal cells since hippocampal cells cultured alone have negligible Ach synthesis (Fig. 1). The

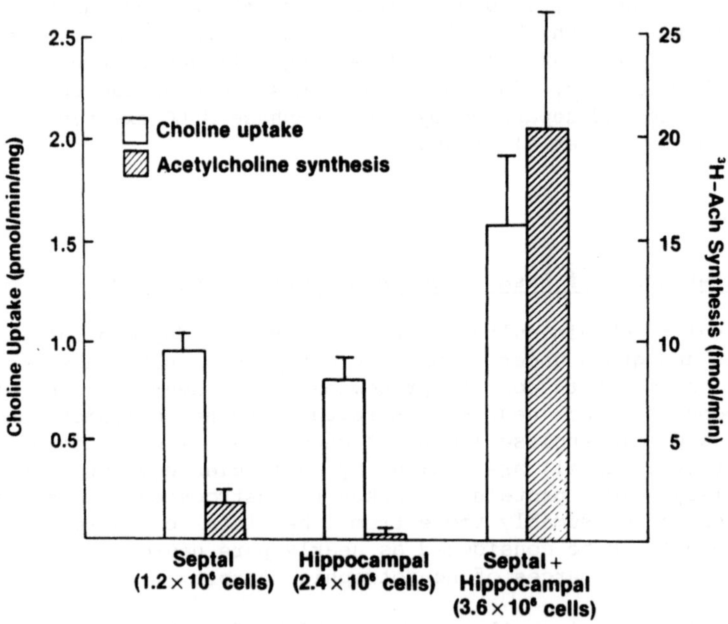

Figure 1. Co-culturing septal and hippocampal cells stimulate Ach synthesis. Septal and hippocampal cells are cultured for 6 days at the indicated densities and culture conditions. Cells were labeled with ^{3}H-choline (12.5 nM) for 10 min. and synthesized ^{3}H-Ach extracted, separated by reverse-phase HPLC and quantitated by liquid scintillation spectroscopy as described in experimental procedures section. Values are Means $\pm$ S.E. of 4 experimental determinations.

hippocampal cells are prepared and cultured identically to the septal cells and, therefore, consist of predominantly neurons. This minimizes a putative role of glial cells in stimulating Ach synthesis. Choline uptake is present in both septal and hippocampal cells and the co-cultures demonstrate an additive effect (Fig. 1).

The stimulatory effect of hippocampal cells on Ach synthesis is independent from potential increases in cholinergic cell density resulting from an enhanced cell survival. Septal cells plated at a density of 3.6×10^6 cells/dish have a lower rate of ^{3}H-Ach synthesis than co-cultures at a similar cell density (Fig. 2). It is also seen that the stimulation of Ach synthesis by hippocampal cells is dependent on the number of target cells (Fig. 2).

The specificity of the hippocampus to stimulate septal cell Ach synthesis was also studied by co-culturing basal forebrain cells with cells from other fetal brain regions (Table 1). An

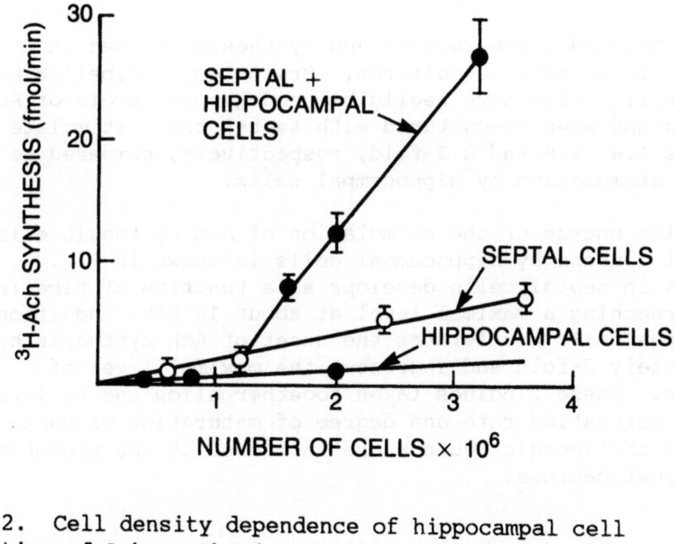

Figure 2. Cell density dependence of hippocampal cell stimulation of Ach synthesis. Septal, hippocampal and co-cultures were maintained in culture for 7 days at the indicated densities of cells. Synthesis of Ach was performed as described in Fig. 1. The values represent Mean $\pm$ S.E. of 4 experimental determinations.

Table 1. Effect of fetal cells from various brain areas on ^{3}H-Ach synthesis in septal cells

Cultures	^{3}H-Ach Synthesis Co-Cultures (fmol/min)	Individual Cultures (fmol/min)
Septal Cells (1.2 X 10^6)	2.10 ± 0.42	2.10 ± 0.42
+Hippocampal Cells (1 X 10^6)	8.58 ± 0.51	0.37 ± 0.06
+Striatal Cells (1 X 10^6)	3.98 ± 0.24	1.05 ± 0.17
+Mesencephalic Cells (1 X 10^6)	4.83 ± 0.36	0.12 ± 0.01
+Cerebellar Cells (1 X 10^6)	3.82 ± 0.18	0.18 ± 0.02
+Brainstem Cells (1 X 10^6)	2.87 ± 0.42	0.21 ± 0.02

Primary cultures of the various fetal rat brain regions were prepared as described in experimental procedures and cultured for 7 days at the indicated cell densities. Cells were either cultured individually or in co-cultures with septal cells. Ach synthesis was assayed as described in Figure 1 and the values are Means ± S.E. of 3 determinations.

approximately additive rate of Ach synthesis is seen in septal-striatal cell co-cultures. Brainstem, cerebellar and mesencephalic cells have negligible endogenous levels of Ach synthesis and when co-cultured with septal cells stimulate Ach synthesis 1.4, 1.8 and 2.3-fold, respectively, compared to a 4.1-fold stimulation by hippocampal cells.

The time course of the stimulation of Ach synthesis elicited in septal neurons by hippocampal cells is shown in Fig. 3. Ach synthesis in septal cells develops as a function of time in culture reaching a maximal level at about 10 DIV. Addition of hippocampal cells accelerates the onset of Ach synthesis by approximately 2-fold and increases the maximal level of synthesis. These findings taken together allow one to infer that the maturation rate and degree of maturation of basal forebrain cholinergic neurons are increased by the presence of their target neurons.

Effects of Conditioned Media and Soluble Extracts on Basal Forebrain Neuronal Maturation

Treatment of basal forebrain cultures with conditioned media from hippocampal cell cultures results in a small but significant stimulation of Ach synthesis (Table 2). Soluble

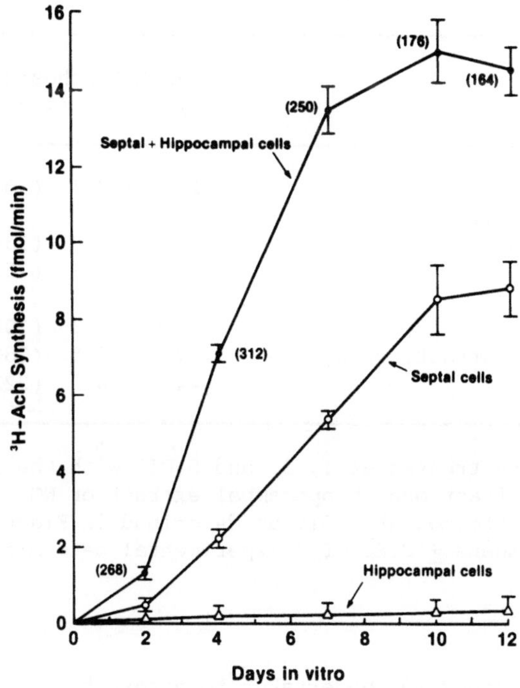

Figure 3. Time course of the hippocampal cell stimulation of septal cell Ach synthesis. Septal (1 X 10^6), hippocampal (1 X 10^6) and co-cultures (2 X 10^6 cells) were maintained in culture for the indicated times and then assayed for Ach synthesis as described in Fig. 1 except that labeling was performed with 50 nM ^{3}H-choline. The values represent Means $\pm$ S.E. of 4 experimental determinations.

extracts of hippocampal cells also increase Ach synthesis in septal cells (Table 2). The stimulatory effect of conditioned media and cell extracts is lower than that seen in the presence of intact hippocampal cells. This may be a consequence of the effective concentration of active material in the three different preparations. The results indicate that at least part of the stimulatory effect of hippocampal cells is mediated by a soluble factor(s) released by the cells. This is borne out in experiments demonstrating a stimulatory action of soluble tissue extracts of hippocampus on septal cell Ach synthesis (Table 2). NGF is known to stimulate Chat activity in these cultures (Hefti et al., 1985) and it also stimulates Ach synthesis (Table 2).

Table 2. Effect of various additions on ^{3}H-Ach synthesis in cultured septal cells.

Addition	^{3}H-Ach Synthesis fmol/min/mg	(% control)
A. Control	20.5 ± 4.1	(100)
Hippocampal Cell		
Conditioned media, 175 µg	27.8 ± 1.0	(136)
Extract, 25 µg	31.2 ± 1.7	(152)
B. Control	19.9 ± 2.1	(100)
Hippocampus extract, 90 µg	57.0 ± 9.5	(286)
NGF, 100 ng	44.2 ± 4.0	(222)

Septal cells were treated at 1, 3, and 5 DIV with the indicated amounts of cell fractions, hippocampal extract or NGF. Ach synthesis was performed at 7 DIV as described in Figure 1. The values are the Means ± S.E. of 4 experimental determinations.

DISCUSSION

The present study was undertaken to assess the influence of target hippocampal cells on the development and phenotypic expression of their afferent septal neurons in dissociated cell culture. Serum-free culture conditions were chosen to eliminate as much as possible the various influences of glial cells on neuronal development and to detect potential cellular responses to neurotrophic factors that may be masked in the presence of serum. Thus, it is essential to use a simplified system to study target-afferent neuronal interactions. The basal forebrain cholinergic neurons can survive and express differentiated properties when cultured in serum-free medium. Proliferation of glial cells is for the most part inhibited under serum-free conditions; less than 2% of the total cell population are astrocytes as detected by GFAP immunohistochemistry.

The principal finding of our studies is that target hippocampal neurons regulate the phenotypic expression of afferent cholinergic neurons evaluated by increases in the rate of Ach synthesis. Acetylcholine synthesis measures the net synthesis, storage and degradation of the neurotransmitter formed from the labeled choline precursor and thus is a physiologically relevant measure of cholinergic cell activity. Preliminary data (not shown) also indicates that hippocampal cells stimulate choline acetyltransferase (Chat) activity in septal neurons although to a lesser extent. The time course of septal Ach synthesis stimulated by hippocampal cells strongly suggests that the rate of maturation of the cholinergic neuron, at least in vitro, is enhanced in the presence of its target neuron. An additional effect on cell survival cannot be ruled out. Immunostaining for Chat or acetylcholinesterase (AchE) histochemistry will in the future be used to examine whether cholinergic neuronal survival or morphology, for example neurite growth and branching, is altered by hippocampal neurons.

The importance of target-derived neurotrophism in CNS development is evident from other in vitro studies. Organotypic co-cultures of septal and hippocampal explants reveal the outgrowth of AchE fibers from the septal explant into the hippocampal slice (Gahwiler and Hefti, 1984; Rimvall, 1985) with the potential establishment of functional cholinergic synapses (Gahwiler and Brown, 1985). However, interpretation of these results is complicated by the fact that glial cells are present in the explant cultures and may contribute to the fiber growth. Embryonic ciliary ganglion neurons when cultured with their target striated muscle cells exhibit an enhanced Ach synthetic capacity (Tuttle et al., 1983). Co-culturing striatal cells with mesencephalic neurons stimulates the dopaminergic characteristics of the cultures (di Porzio et al., 1980; Hemmendinger et al., 1981) while norepinephrine uptake in brainstem cultures is stimulated by the addition of target cerebellar or striatal neurons (di Porzio and Estenoz, 1984). These studies suggest that, in vitro, the phenotypic expression of various types of afferent neurons is regulated in a positive manner by the presence of their respective target cells.

The mechanism behind the hippocampal stimulation of the cholinergic phenotypic expression in our dissociated cell co-cultures is unclear at the present time. Conditioned media from, and soluble extracts of, dissociated hippocampal neuronal cultures stimulate Ach synthesis in basal forebrain neurons suggesting an involvement of soluble molecule(s) that are released from the hippocampal cells. Axonal growth and guidance may be due to both selective cell adhesion and concentration gradients of soluble molecules (Campenot, 1982; Letourneau, 1983). Neuron-neuron recognition and adhesion which are mediated by membrane-bound proteins may also be operative in the basal forebrain co-cultures. Perhaps soluble hippocampal factors act initially to stimulate septal cell axonal growth followed by cell-cell contact and formation of a functional synapse. Alternatively, the increased level of Ach synthesis may be due to an increased extent of neuritic arborization of cholinergic neurons. Whatever the mechanism may be, it seems to be specific for hippocampal cells. Cells prepared from brain regions that do not receive a heavy cholinergic innervation do not stimulate Ach synthesis as effectively as hippocampal cells.

In summary, basal forebrain cholinergic neurons can be maintained in primary cultures in the absence of serum. The expression of cholinergic properties in situ is stimulated by the target hippocampal cells and the effect seems to be an enhancement of cholinergic neuron maturation. Soluble trophic factors are implicated in the mechanism of the target-derived neurotrophism, but may not be the sole molecules responsible for the hippocampal activity. In general, the results indicate that target neurons in the absence of glial cells regulate the phenotypic expression of the afferent neuronal population.

REFERENCES

Appel, S.H. (1981). A unifying hypothesis for the cause of amyotrophic lateral sclerosis, parkinsonism, and Alzheimer disease. Ann. Neurol., 10, 499-505.

Bottenstein, J.E., and Sato, G.H. (1979). Growth of a rat neuroblastoma cell line in serum-free supplemented medium. Proc. Natl. Acad. Sci. USA, 76, 514-517.

Campenot, R.B. (1982). Development of sympathetic neurons in compartmentalized cultures. II. Local control of neurite survival by nerve growth factor. Dev. Biol., 93, 13-21.

Cowan, W.M., Fawcett, J.W., O'Leary, D.D.M., and Stanfield, B.B. (1984). Regressive events in neurogenesis. Science, 225, 1258-1265.

Di Porzio, U., Daguet, M.C., Glowinski, J., and Prochiantz, A. (1980). Effect of striatal cells on in vitro maturation of mesencephalic dopaminergic neurones grown in serum-free conditions. Nature, 288, 370-373.

Di Porzio, U., and Estenoz, M. (1984). Positive control of target cerebellar cells on norepinephrine uptake in embryonic brainstem cultures in serum-free medium. Dev. Brain Res., 16, 147-157.

Eva, C., Hadjiconstantinou, M., Neff, N.H., and Meek, J.L. (1984). Acetylcholine measurement by high-performance liquid chromatography using an enzyme-loaded postcolumn reactor. Anal. Biochem., 143, 320-324.

Gahwiler, B.H., and Brown, D.A. (1985). Functional innervation of cultured hippocampal neurones by cholinergic afferents from co-cultured septal explants. Nature, 313, 577-579.

Gahwiler, B.H., and Hefti, F. (1984). Guidance of acetylcholinesterase-containing fibres by target tissue in co-cultured brain slices. Neurosci., 13, 681-689.

Hefti, F., Hartikka, J., Eckenstein, F., Gnahn, H., Heumann, R., and Schwab, M. (1985). Nerve growth factor increases choline acetyltransferase but not survival or fiber outgrowth of cultured fetal septal cholinergic neurons. Neurosci., 14, 55-68.

Hemmendinger, L.M., Garber, B.B., Hoffman, P.C., and Heller, A. (1981). Target neuron-specific process formation by embryonic mesencephalic dopamine neurons in vitro. Proc. Natl. Acad. Sci. USA, 78, 1264-1268.

Johnson, D.A., and Pilar, G. (1980). The release of acetylcholine from post-ganglionic cell bodies in response to depolarization. J. Physiol., 299, 605-619.

Letourneau, P.C. (1983). Axonal growth and guidance. Trends in Neurosci., 6, 451-455.

Rimvall, K., Keller, F., and Waser, P.G. (1985). Development of cholinergic projections in organotypic cultures of rat septum, hippocampus and cerebellum. Dev. Brain Res., 19, 267-278.

Thoenen, H., and Barde, Y.A. (1980). Physiology of nerve growth factor. Physiol. Rev., 60, 1284-1335.

Tuttle, J.B., Vaca, K., and Pilar, G. (1983). Target influences on [^{3}H]Ach synthesis and release by ciliary ganglion neurons in vitro. Dev. Biol., 97, 255-263.

Yankner, B.A., and Shooter, E.M. (1982). The biology and mechanism of action of nerve growth factor. Ann. Rev. Biochem., 51, 845-868.

Part C

Effects of Neurosteroids and Classical Steroids

Chairman: J.-Å. Gustafsson

9
Neurosteroids: Pregnenolone and Dehydroepiandrosterone in the Brain

Etienne-Emile Baulieu, Paul Robel, Oliver Vatier, Marc Haug, Claude Le Goascogne and Eliane Bourreau

SUMMARY

Brain testosterone and corticosteroids arise from peripheral sources, whereas pregnenolone (Δ5-P) and dehydroepiandrosterone (DHA), the 3β-hydroxy-Δ5 derivatives of cholesterol which serve as precursors of steroid hormones in steroidogenic glands, accumulate in the brain by proper mechanism(s), even after surgical or pharmacological suppression of endocrine glands (adrenals, gonads). They are found in definite proportions as free steroids, or sulfate and fatty acid esters. Two enzymes involved in side-chain cleavage, cytochrome $P\text{-}450_{scc}$ and adrenodoxin, have been immunohistochemically localized in white matter, suggesting a possible modulatory/trophic general function ; they were also present in a few neurons of the olfactory bulb, entorhinal cortex and cingulum, evoking an olfactory pathway. The "neurosteroid" concept is based on changes observed in a variety of physiological situations, including diurnal rhythm, development, and heterosexual exposure of male to female rats. Pharmacologically, DHA decreased a particular type of male mice agressive behavior linked to lactating female signal. The mode of action of Δ5-P and DHA is yet unknown. It may include their transformation to classical steroid hormones acting on a paracrine mode, or directly their binding to unknown membrane or intracellular receptors, or even the control of neuronal functions by their insertion into membranes.

INTRODUCTION

Steroid hormones, as for example testosterone and glucocorticosteroids, are produced by peripheral endocrine glands, circulate in blood, and cross the blood brain barrier to reach the brain where they bind, in specific regions, to receptors mediating their activities (reviews in Fuxe et al., 1981). Specific regions of the brain exhibit steroid metabolizing activities, e.g. transformation of testosterone to estrogens, that may be of physiological importance.

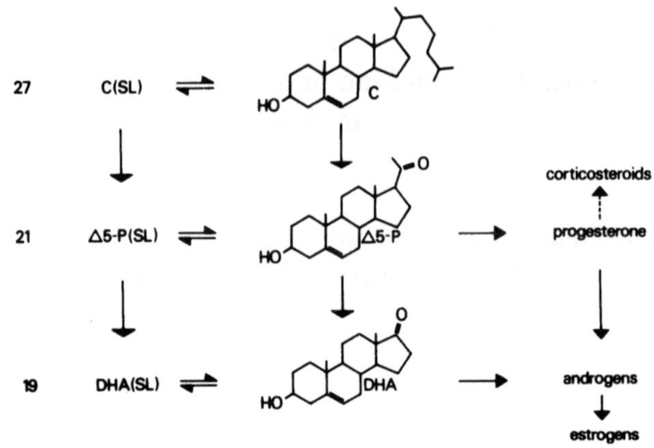

Fig. 1. Pregnenolone (Δ5-P) and dehydroepiandrosterone (DHA) synthesis and metabolism. In the text, when Δ5-P or DHA is sulfated or esterified by fatty acid or both, we indicate S, L or SL according to the case. When we measure together free and sulfate or free, sulfate and lipoidal, we note steroid(S) or steroid(SL), respectively.

We have characterized pregnenolone (Δ5-P) and dehydroepiandrosterone (DHA) in the brain of several mammalian species (rat, mouse, pig, monkey, human) (Corpéchot et al., 1981a, 1983). These steroids are 3β-hydroxy-Δ5 compounds structurally related to cholesterol from which they normally derive, in steroidogenic glands, by successive side chain cleavages. Specific cytochrome P-450 hydroxylase systems are involved, the mitochondrial cytochrome $P\text{-}450_{scc}$ for the conversion of cholesterol into Δ5-P, and the cytochrome $P\text{-}450_{17\alpha}$ for the conversion of Δ5-P into DHA (Lieberman et al., 1984) (Fig. 1). In fact, we found Δ5-P and DHA not only as free, non conjugated steroids, but also as 3-sulfate esters and 3-fatty acid esters (lipoidal derivatives). The mammalian brain contains enzymes responsible for the formation and the hydrolysis of these conjugates. Other yet unknown 3β-hydroxysteroid conjugates may also be present, (although in small amounts) (Robel and Baulieu, 1985). We initially discovered the presence of DHA(S) in the rat brain (Corpéchot et al., 1981a) to our surprise, since rodent adrenals do not produce sizable amounts of this steroid, contrary to primates (Baulieu et al., 1965). Moreover, we observed that, besides the large difference between brain and blood concentrations, the cerebral values were apparently independent of adrenal and testis sources, which were suppressed surgically and/or pharmacologically. This contrasted with classical steroid hormones, testosterone and corticosterone, whose concentrations in the brain not only were below those in plasma, but also readily disappeared after removal of the corresponding endocrine glands. Therefore we studied also Δ5-P, the immediate biochemical precursor of DHA, and the same results essentially were found (Corpéchot et al., 1983).

Limited experiments were conducted in other species, including monkeys (see later). The results were interesting enough to justify further studies which are presented in this summary of our recent work. We obtained confirmating data for the independence of brain Δ5-P and DHA from peripheral sources, and despite the lack of a definitive demonstration of Δ5-P and DHA biosynthesis along the cholesterol pathway*, the discovery of specific locations for mitochondrial cytochrome $P\text{-}450_{scc}$ and ferredoxin highly suggests the synthesis of the steroids in brain : we call them <u>neurosteroids</u>, with the understanding that their physiology is different from that of the "imported" steroidal derivatives which eventually act in the brain. It is proposed that there are different steroid pools in the brain, steroids of peripheral origin and neurosteroids, respectively, corresponding to endocrine and paracrine functions.

ACCUMULATION OF BRAIN PREGNENOLONE AND DEHYDROEPIANDROSTERONE INDEPENDENTLY OF ENDOCRINE GLANDS SUPPLY

The adult male rat

The procedures for the identification and measurement of Δ5-P and DHA, either non conjugated or released from sulfate or fatty acid esters, have been described previously (Corpéchot et al., 1981a, 1983 ; Robel et al., 1986a). We did not measure Δ5-androstene-3β,17β-diol, an estrogenic DHA reduced metabolite. We did not find measurable amounts of 17-hydroxy-pregnenolone. The quantitation of rat Δ5-P and DHA appeared to depend on several environmental and methodological factors. Some have been documented as age, lighting schedule, hour of sacrifice, housing conditions (number of males per cage, female rats in the same room), and stress after castration (ORX) plus adrenalectomy (ADX), or the corresponding sham operation (sham). Seasonal variations, with lower levels in winter, were also detected (but not systematically studied). The work up conditions for the removal and initial processing of brain tissue were also critical.

We foun. differences in Δ5-P and DHA concentration between selected brain areas. Steroids were constantly higher in hypothalamus and olfactory bulb than in any other brain structure, in spite of a large range of values between different experiments. The brain concentrations of Δ5-P and DHA were always severalfold larger than the corresponding plasma concentrations. The concentration of DHAS was even larger in brain than in adrenal.

<u>Effects of combined adrenalectomy and castration on the brain steroids of adult male rats</u>. Neither Δ5-P(S) nor DHA(S) disappeared from rat brain 15 d after combined ADX + ORX (Corpéchot et al., 1981a, 1983). The concentrations of Δ5-P tended to be lower than in intact males, but were identical in ADX and in sham operated animals. This contrasted with the results obtained with testoste-

rone, which decreased immediately after castration and was undetectable in the brain 1 and 7 d after operation.

Effects of stress and adrenal stimulation or suppression. Brain DHA(S) was unchanged after administration of ACTH or dexamethasone for 3 d. Stress conditions, prevailing 2-3 d after ADX + ORX or the corresponding sham operation, resulted in a significant increase of brain DHA(S) and Δ5-P(S) concentrations. Thereafter, they tended to decrease, but no difference was observed between sham and really operated animals. Regeneration of adrenals was excluded as a cause for the maintenance of plasma Δ5-PL and brain Δ5-P, Δ5-PS and Δ5-PL.

Is the accumulation of brain Δ5-P and DHA of peripheral origin ? ^{3}H-Δ5-P was injected by the intracardiac route (25 µCi in 500 µl of 10 % propanediol in isotonic saline) to several adult male rats and the tritium content was measured in two brain areas, pituitary, adrenals and plasma. Concentrations of radioactivity was severalfold larger in brain than in plasma, with the highest levels in hypothalamus. However, the clearance from brain was as rapid as from plasma (Robel et al., 1986b and O. Vatier, unpublished results).

^{3}H-Δ5-P also was injected stereotaxically in the lateral ventricles or in olfactory bulb. It was cleared as rapidly as after peripheral injections. Also, aminophyllin (75 mg/kg) produced a rapid, large, but transient increase of Δ5-P in plasma and brain. The increase of brain Δ5-P was no more observed after adrenalectomy. In these experiments, no significant change of DHA was recorded. Therefore, there is certainly a brain compartment for Δ5-P(S), and to a smaller extent for DHA(S), which is dependent on the delivery of circulating steroids. However there is another pool or compartment, apparently autonomous, and responsible for the long term persistence of brain Δ5-3β-hydroxysteroids. The available results do not favor the concept of a lipoidal reserve of steroids.

The adult female monkey (macaca fascicularis)

Two control adult spayed females and two dexamethasone-treated spayed females have been studied (Robel et al., 1986b). Effectiveness of adrenal suppression was indicated by the decrease of plasma cortisol to undetectable levels 20 h after the last injection of dexamethasone and by a marked decrease of plasma DHA(S) (Table 1).

Δ5-P(SL) and DHA(S) were found in all tested brain regions. Concentrations were larger in hypothalamus, hippocampus and rhinencephalon than in cerebellar or cerebral cortex. Δ5-P(S,L) concentration was much larger in brain than in plasma, and not markedly modified by adrenal suppression, with the possible exception of hippocampus. Conversely, the concentration of DHA(S) in brain was

Table 1. Effect of dexamethasone on plasma (ng/ml) and brain (ng/g) steroids of Macaca fascicularis[a]

	Plasma				Brain[b]			
	Control		Dex[c]		Control		Dex	
	1	2	3	4	1	2	3	4
DHA(S)	117	54	5	19	27	20	11	8
Δ5-P(S)	2	1	1	1	11	8	8	8
Δ5-PL	-[d]	0	-	0	16	36	18	20
Cortisol	240	72	0	0	-	-	-	-
Age (years)	8	~ 20	13	~ 20				

[a] Macaca fascicularis females, castrated for > 6 months.
[b] Average of measurements performed on hypothalamus, hippocampus, cerebral and cerebellar cortex and rhinencephalon.
[c] Dexamethasone, 4 mg i.m. daily for 3 d.
[d] Not determined.

smaller than in plasma, a result which is explained by the high concentration of DHAS in primate blood (Baulieu et al., 1965 ; Cutler et al., 1978), and the impermeability of blood brain barrier to steroid sulfates (Kishimoto and Hoshi, 1972). Adrenal suppression resulted in decreased brain DHA(S) concentrations, but to a much smaller extent than in plasma, thus suggesting that the accumulation of DHA(S) in monkey brain is partially independent of adrenal production.

METABOLISM OF 3β-HYDROXY-Δ5-NEUROSTEROIDS

Pregnenolone formation from sterol precursors

Since a fraction of brain Δ5-P and DHA was independent of the steroidogenic activity of endocrine glands, and since no evidence was obtained of either extraglandular source(s) or release from brain storage derivatives, many attempts were made to demonstrate the side-chain cleavage of cholesterol by brain slices, total homogenates or mitochondrial preparations. They essentially remain unsuccessful so far. However, we conclusively demonstrated ^{3}H-Δ5-P formation in an established cell line of neural origin, the rat C6 glioma, incubated with ^{3}H-mevalonolactone or ^{3}H-cholesterol. It was inhibited by aminoglutethimide, an inhibitor of side chain cleavage. In contrast, no radioactive Δ5-P was detected after identical incubations of two neuroblastoma cell lines (PC12 and Neuro2A), including in culture conditions favoring neuronal differentiation. No Δ5-P was formed either by mouse fibroblast L-929 cells. Failure to demonstrate so far the biosynthesis of neurosteroids in brain could be explained by the unknown nature of the putative precursor, or by the still undefined location of steroid producing cells. Some attempts to detect Δ5-P with specific antibodies or by autoradiography were unsuccessful.

We have recently pursued another research line, which consists in the immunohistochemical localization of specific steroidogenic enzymes. The presence of ferredoxin and of a cytochrome P-450 of undefined type have been previously demonstrated in bovine brain mitochondria (Oftebro et al., 1979). The side chain cleavage cytochrome P-450 ($P\text{-}450_{scc}$) has been purified from bovine adrenal mitochondria, and used to generate specific antisera (Seybert et al., 1979 ; Du Bois et al., 1981). We have used the corresponding IgGs (kindly provided by M. Waterman) to set up an immunohistochemical technique for the detection of cytochrome $P\text{-}450_{scc}$ in rat tissues (unpublished results). We have verified that the bovine IgGs were able to stain the specific cytochrome in the rat adrenal cortex and ovaries. The optimal conditions thus defined were applied to the adult male rat brain. Specific immune staining was detected in the white matter wherever it was looked for in the brain, including in the cerebellum (Fig. 2). Some neurons were also positively stained, in the entorhinal cortex, in the cingulum, and in the olfactory bulb. Therefore, we work now with the hypothesis

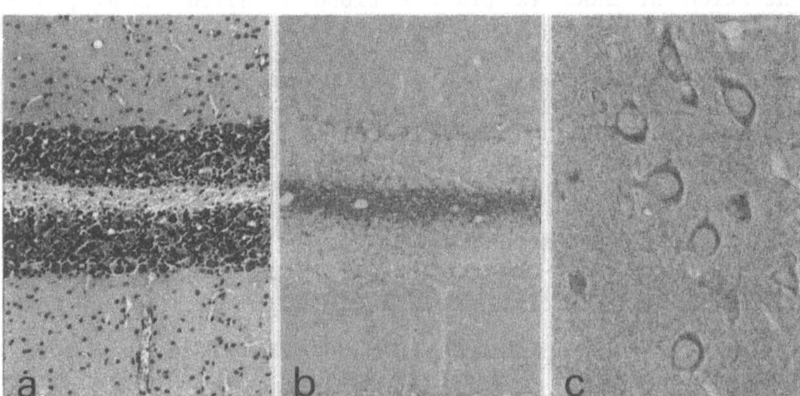

Fig. 2. Immunoperoxydase staining of the rat brain with IgGs to cytochrome $P\text{-}450_{scc}$. a and b, cerebellum (X 120) ; c, entorhinal cortex (X 500) ; a, histological staining with Masson's trichrome, showing the white matter delineated on both sides by the dark granular layers and more externally by the molecular layers ; b, the immunoperoxidase staining exhibits a positive reaction in the zone corresponding to the white matter. c ; immunoreactive cells in the entorhinal cortex.

that oligodendrocytes and a few specific neurons are involved in the production of neurosteroids. No neurons outside the brain have been studied so far.

Incidently, DHA(S) and Δ5-P(S) have been detected in several organs of the adult male rat. In the spleen and kidney, the concentrations of Δ5-P and Δ5-PS were severalfold larger than in plasma (Robel and Baulieu, 1985). It is not excluded that steroids may be formed in these organs, but in kidney the immunohistochemical staining of cytochrome $P\text{-}450_{scc}$ was negative.

Pregnenolone metabolism in the brain

Incubation of rat brain minces with ^{3}H-Δ5-P yielded ^{3}H-progesterone and ^{3}H-Δ5-PL. The yield of progesterone formed per 100 mg tissue was in the ~ 1 % range in hypothalamus, amygdala, olfactory bulb, and other regions of the brain under the experimental condition used, with the exception of the cerebellum and frontal cortex where virtually no metabolism was observed (Clarke et al., 1984). The acyl transferase responsible for the formation of Δ5-PL was further studied (Robel et al., 1986a). A high level of synthetic activity was found in the brain of young male rats (1-3 week old). Palmitate, oleate, and linoleate were the major fatty acid esters. ^{3}H-DHA gave small amounts of ^{3}H-androstenedione and ^{3}H-DHAL. So far we have not detected conversion of Δ5-P to DHA, and of both steroids to their sulfate esters, although steroid sulfokinase activity has been previously reported in mammalian brain (Knapstein et al., 1986). Since steroid sulfates hardly cross the blood brain barrier, those found in the brain are probably formed from their corresponding unconjugated moiety, or from cholesterol sulfate (Lieberman et al., 1984). In the brain of myelin deficient jimpy and quaking mice, DHA sulfoconjugation is markedly decreased (Corpéchot et al., 1981b). The myelin of these affected mice also is characterized by lack of sulfatides. The role of polar and charged sulfates and of low polarity and neural fatty acid esters is not known. No result suggests reciprocal variations of the concentration of unconjugated and conjugated steroids forms which could be responsible for the slow turnover of neurosteroids.

VARIATIONS OF PREGNENOLONE AND DEHYDROEPIANDROSTERONE ACCORDING TO PHYSIOLOGICAL SITUATIONS, SUPPORTING THE CONCEPT OF METABOLIC INDEPENDENCE OF BRAIN 3β-HYDROXY-Δ5-STEROIDS

The large variability observed in our early measurements of brain 3β-hydroxy-Δ5-steroids was later on explained, at least in part, by the major role of several factors such as stress, diurnal variations and intersexual exposure. They are physiologically relevant to regulations of Δ5-P and DHA proper to the brain.

Table 2. Ontogenesis of corticosterone and Δ5-P(S) in rat plasma (ng/ml) and brain (ng/g). Effects of ACTH and dexamethasone[a]

	before birth[b]	days 2-10	day 6 ACTH	day 6 DEX	Adult 10a.m.[c]
<u>Plasma</u>					
Corticosterone	240	5-25	190	0	25
Δ5-P(S)	14	1-2	5	2	1
<u>Brain</u>					
Corticosterone	99	3-7	180	0	10
Δ5-P(S)	27	8-22	12	10	9

[a] Measurements were made on multiple pools of laboratory bred Sherman rats, killed at 10 p.m.

[b] Fetuses were removed immediately after delivery of the first pup.

[c] Adult Sprague Dauley males (11-12 week old) killed 2h after lights on. Mean of 6 determinations.

<u>Ontogenesis</u>

Δ5-P(S), DHA(S) and corticosterone were measured in rat brain between birth and day 22. In newborns of both sexes, Δ5-P(S) and corticosterone were high at delivery in brain and plasma, confirming that adrenal cortical activity is high in the newborn animal (Corbier and Roffi, 1978) (Table 2). DHA(S) was elevated neither in brain nor in plasma. Brain Δ5-P(S) decreased steadily during the first day of life, and on the following days it remained in the 8-22 ng/g range, while brain DHA(S) was very stable during the whole period, in the 1.8-3.4 ng/g range. Interestingly, brain corticosterone decreased rapidly to insignificant levels between day 1 and day 10 (2-5 ng/g), in accordance with the known period of low adrenal activity in the rat (Robel et al., 1986a). Contrary to corticosterone, DHA(S) and Δ5-P(S) concentrations in the brain were always much larger than the corresponding plasma values. In addition, brain Δ5-P(S) was unchanged when 4 d old rats were treated with ACTH or dexamethasone for 2 d. It therefore appears that, with the exception of the short perinatal stress period, brain Δ5-P(S) accumulates through a selective mechanism unrelated to adrenal activity.

Circadian rhythm of 3β-hydroxy-Δ5-steroids and glucocorticosteroid in rat brain and plasma

Male adult rats (11-12 week old) were housed in triads for 2 weeks with lights on between 7 am and 8 pm. Corticosterone, Δ5-P(S) and DHA(S) were measured at 3 h intervals. They underwent circadian variations in both plasma and brain (Synguelakis et al., 1985). As indicated by the cosinor method, the acrophases of Δ5-P(S) in brain and of DHA(S) in plasma significantly preceded the acrophase of corticosterone. The asynchrony of 3β-hydroxy-Δ5-steroid and corticosterone rhythms brings an additional argument in favor of separate regulatory mechanism. To explore the role of adrenal secretion, circadian variations were again studied in groups of rats castrated and adrenalectomized 15 d before killing. Ablation of endocrine glands was followed by a dramatic decrease of plasma and brain corticosterone, but Δ5-P(S) and DHA(S) did not disappear from operated rat brain. Circadian variations of corticosterone in plasma and brain, and of DHA in plasma were no more observed. Contrawise, a significant rhythm of brain DHA(S) persisted with an acrophase at the beginning of the dark period, and a similar trend was observed for Δ5-P(S), although not statistically significant (Robel et al., 1986c).

The persisting circadian rhythm of brain DS after the removal of steroidogenic peripheral glands confirmed that there is a central modulatory neuroendocrine system controlling 3β-hydroxy-Δ5-steroid accumulation. This does not mean necessarily a varying cerebral biosynthesis, since, for example, a rhythmic conversion from the fatty acid ester pool and/or the circadian variation of a binding protein could be involved (Robel et al., 1986a).

Pregnenolone decreases in olfactory bulb of male rats exposed to female siblings

Young adult Sprague-Dawley male (M) rats (285-300 g body weight) were exposed for seven days to the scent of a group of cycling females (F), in absence of visual or tactile contact (M/F), and compared to males similarly exposed to other males (M/M). They were killed at 2 pm and the olfactory bulbs (OB), amygdalae (AMY) and hypothalami (HYP) were collected, as well as the rest of the brain, plasma, spleen and adrenals. Δ5-P(S) and testosterone (T) were measured and M/M vs M/F values were compared (Fig. 3) (Corpéchot et al., 1985 and O. Vatier, P. Brazeau and E.E. Baulieu, to be published).

Heterosexual exposure resulted in a significant decrease of Δ5-P(S) concentrations in OB and to a lesser extent in HYP, but not in any other tissue sample. The concentrations of T in the brain structures, plasma, and adrenals were either slightly increased or unchanged, as were those of DHA(S).

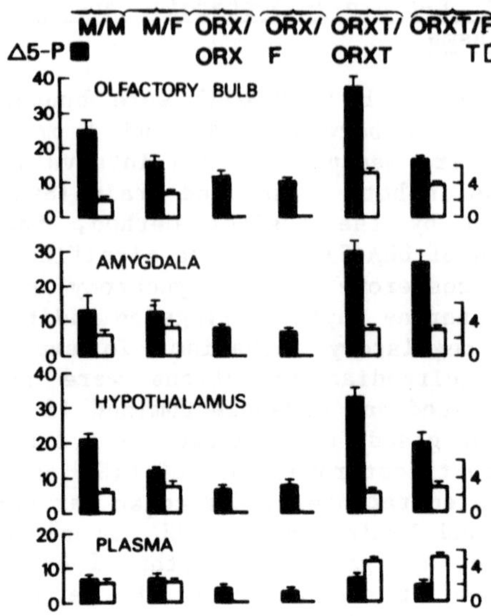

Fig. 3. Variations of Δ5-P(S) and testosterone in the brain and plasma of male rats exposed to siblings of either sex. Adult male Sprague Dawley rats (M) (275-300 g body weight) were housed in individual cages and were exposed for 7 d either to the scent of other males (M/M) or of groups of 4 cycling females (M/F). Groups of males were castrated (ORX), and half of them were treated with testosterone propionate in oil (1 mg/d for 7 d) (ORXT).

When male rats were castrated (for > 7 d, ORX) and exposed to the scent of castrated males (ORX/ORX), T decreased to undetectable levels in plasma and brain, as expected. Castration also had a pronounced effect on Δ5-P(S) concentration in the brain (50 % decrease), but not in plasma and adrenals. Exposure to the scent of females (ORX/F) produced no further decrease of Δ5-P(S) concentration in any tissue investigated. DHA(S) concentration in brain was unchanged by ORX and identical in ORX/ORX and ORX/F groups.

The effect of testosterone in ORX males (ORXT) was remarkable. One mg of testosterone propionate in sesame oil was injected s/c daily for 7 d (controls : vehicle alone), so as to establish normal or even supranormal concentrations of T in plasma and brain. Most interestingly, while Δ5-P(S) in plasma and adrenal was not affected, Δ5-P levels in OB, AMY and HYP exceeded those of intact males of the M/M group. Even more surprising was the major decrease of Δ5-P(S) recorded in OB and HYP of ORX males exposed to intact females, but not in AMY and other brain structure.

Finally, when females were castrated 7 d before the intersexual exposure experiments (OVX), no decrease of Δ5-P was observed in the brain of exposed males (M/OVX). Conversely, when intact females were all in estrus (EST), after an appropriate PMSG-hCG treatment, the decrease of brain Δ5-P(S) in male brain after a 24 h exposure (M/EST) was as important as after exposure for 7 d to the scent of randomly cycling females (M/F).

In conclusion, intersexual exposure of males to cycling females provokes a profound decrease of Δ5-P(S) in the olfactory bulb and hypothalamus (apparently not in other parts of the brain, nor in the blood and adrenals). The overall level of brain Δ5-P(S) is dependent on testosterone, as is also the down-regulation mechanism operating when males are exposed to the scent of females. However, castrated females no more emit the proper signal.

Table 3. Effects of DHA, DHAS, androst-5-ene-3β,17β-diol (Δ5-diol), and 3β-methyl-androst-5-en-17-one ($D-CH_3$) an agression by castrated male mice towards lactating intruders

Treatment groups (trials)	Mean latency to first bite (sec) ± S.D.	mean number of biting attacks/15min ± S.D.
DHA (9)	189 ± 117[a]	16 ± 10[a]
DHAS (9)	115 ± 144	27 ± 15
Δ5-diol (9)	52 ± 50	29 ± 20
Controls (9)	34 ± 50	36 ± 19
DHA (17)	154 ± 218	21 ± 15[a]
$D-CH_3$ (19)	309 ± 340[a]	18 ± 16[a]
Controls (8)	59 ± 108	36 ± 11

Controls were castrated, vehicle injected male Swiss mice.
Steroids were given at the daily dose of 280 nmol for 15 d.
[a] Significantly different from control ($p < 0.03$).

Dehydroepiandrosterone inhibits the agressive behavior of castrated male mice

Group-housed triads of castrated mice attack lactating female intruders. Previous reports indicated that testosterone and estradiol inhibit this agressive behavior (Haug and Brain, 1979). Male mice of the Swiss strain were castrated (ORX) or sham operated (M) when they were 7 week old. One month later, triads were randomly assigned to experimental groups injected each day subcutaneously with 280 nmol of DHA, DHAS, Δ5-androstene-3β,17β-diol vehicle alone for 15 d. One lactating female was then introduced or into each cage 2 h after the last injection, and the latency and number of attacks were recorded for 15 min. DHA inhibited the agressive behavior of castrated males in a dose-dependent manner (Haug et al., 1983). DHA(S) or androstenediol treated animal showed no statistically significant differences with controls, suggesting low metabolic transformation of these compounds to DHA. Transformation of the latter into testosterone was small, as judged by the increased concentration of testosterone in the brain ($p < 0.01$) after DHA injection, and remained far below the physiological values measured in intact males. However, we cannot exclude testosterone accumulation in some specific brain regions. Therefore we injected a DHA analogue, 3β-methyl-androst-5-en-17-one, which is not converted into testosterone (or other Δ4-3-ketonic or phenolic hormones). This compound is not demonstrably estrogenic in rodents (Pashko et al., 1984). Nevertheless, it inhibited the agressive behavior of castrated mice at least as efficiently as DHA, suggesting an inhibitory effect of the DHA molecule by itself on this very particular type of agressive behavior (Table 3).

THE MECHANISM(S) OF ACTION OF 3β-HYDROXY-Δ5-NEUROSTEROIDS

At present, the precise localization and activity of neurosteroids are almost completely unknown, and we can only discuss several tentative models, by analogy with other steroidal systems and ligand induced effects.

3β-hydroxy-Δ5-steroids may be precursors of Δ4-3-keto steroid "hormones" in the brain. We have shown some transformation of Δ5-P into progesterone and of DHA into Δ4-androstenedione in selected brain regions, but no conversion of the former into corticosterone or of the latter into testosterone was detected so far. However, the available results have been obtained in very crude experiments, and formation of hormones restricted to a few neurons would not be detected, whatever its biological significance. Indeed, if there is formation of classical steroid hormones, they probably act paracrinally or autocrinally, within a very limited area. Whether the potential action of Δ4-3-ketonic steroids would be mediated by "ordinary" intracellular receptors or by a membrane mechanism (see later) is a subsequent question.

Δ5-P and DHA do not bind to any intracellular steroid hormone receptor, but so far no specific steroid binding study has seriously be undertaken. Low affinity Δ5-P binding proteins have been described in other tissues, and the concentration of Δ5-P in brain is quite compatible with binding to such protein(s).

DHA(S) and Δ5-P(S) are amphipathic molecules. As such they could constitute integral components of cell membranes, and indeed DHA and Δ5-P themselves may also be considered as "mini-cholesterol" and compete for cholesterol itself for insertion in membrane. Cholesterol sulfate has been shown to affect the fluidity of red blood cell plasma membranes (Bleau et al., 1974). The hydrophobic fatty acid esters of Δ5-P and DHA are also good candidates for integration into plasma membrane. Indeed, more than 50 % of Δ5-PL is found in purified myelin preparations (E. Bourreau, unpublished data). In other words, there are possible models which would account for a "structural" in membrane mechanism in order to explain 3β-hydroxy-Δ5 steroids activities.

Another possibility is that neurosteroids bind to a membrane receptor. The only well documented case for such an interaction of steroid hormone at the cell surface is the initiation of meiosis of Xenopus laevis oocytes by progesterone. Photoaffinity labelling experiments have demonstrated a ~ 30kDa binder of specific characteristics with reference to affinity and hormone specificity (Blondeau and Baulieu, 1984). Several steroidal compounds administered by iontophoresis and/or pressure modify the action potential of locally recorded neurones (Carette et al., 1979). This approach has been used for 3β-hydroxy-Δ5-steroids, in adult female guinea pig (Carette and Poulain, 1984). Neurons in the septo-preoptic area exhibited an excitatory response. DHAS and Δ5-PS always produced a similar effect on the same neuron, while when DHA and DHAS were tested, some neurons were excited by both and others by DHAS only. Responses displayed a short latency in onset and offset, suggesting an action at the membrane level. Very recently Δ5-PS has been shown to interact with the GABA receptor, with characteristics of an antagonist of GABA-ergic neurotransmission (Majewska and Schwartz, 1986).

CONCLUSIONS

Δ5-P and DHA, free and in sulfate and fatty ester forms, are found in the rat brain (and present also in the brain of several other species). Endocrine manipulation and behavioral experiments have indicated that large fractions of these steroids are accumulated in the brain independantly of steroidogenic gland contribution. Indeed, their concentration appears to be controlled by regulatory mechanisms proper to the brain, which in turn may depend upon gonadal hormones. Cytochrome $P\text{-}450_{scc}$ and adrenodoxin, enzymes implicated in cholesterol to pregnenolone formation, are found in the white matter (presumably in oligodendrocytes), and in a few neurons (presently detected in the olfactory bulb, entorhinal cortex and cingulum). The functionnal significance of these observations remains to be studied.

ACKNOWLEDGEMENTS

We acknowledge the help and generosity of Drs. M. Axelson, N. Baumann, P. Brazeau, C. Corpéchot, F. Dray, R. Emiliozzi, R. Guillemin, F. Halberg, J. Martel, J. Sjövall, M. Synguelakis, S. Talha, R. Vikho and M. Waterman at several stages of this work.

*After the Stockholm meeting, direct biochemical evidence for the synthesis of Δ5-P from cholesterol has been obtained and will be soon reported.

REFERENCES

Baulieu, E.E, Corpéchot, C., Dray, F., Emiliozzi, R., Lebeau, M.C., Mauvais-Jarvis, P. and Robel, P. (1965). An adrenal-secreted "androgen" : dehydroisoandrosterone sulfate. Its metabolism and a tentative generalization on the metabolism of other steroid conjugates in man. Rec. Progr. Hormone Res., 21, 1167-1178.

Bleau, G., Bordley, F.H., Longpré, J., Chapdelaine, A., Roberts, K.D. (1974). Cholesterol sulfate. I. Occurrence and possible biological function as an amphipathic lipid in the membrane of the human erythrocyte. Biochem. Biophys. Acta, 352, 1-9.

Blondeau, J.P. and Baulieu, E.E. (1984). Progesterone receptor characterized by photoaffinity labelling in the plasma membrane of Xenopus laevis oocytes. Biochem. J., 219, 785-792.

Carette, B., Barry, J., Linkie, D., Ferin, M., Mester, J. and Baulieu, E.E. (1979). Effets de l'oestradiol-7-acide butyrique au niveau des cellules hypothalamiques. C.R. Acad. Sci. Paris, 288, 631-634.

Carette, B. and Poulain, P. (1984). Excitatory effect of dehydroepiandrosterone, its sulfate ester and pregnenolone sulfate, applied by iontophoresis and pressure, on single neurones in the septo-preoptic area of the guinea pig. Neurosci. Lett., 45, 205-210.

Clarke, C., Groyer, A., Baulieu, E.E. and Robel, P. (1984). Pregnenolone metabolism in the rat brain. 7th Int. Congr. Endocrinol. Quebec, Canada, abstract n°395. Excerpta Medica International Congress Series 652, Amsterdam.

Corbier, P. and Roffi, J. (1978). Increased adrenocortical activity in the newborn rat. Biol. Neonate., 33, 72-79.

Corpéchot, C., Robel, P., Axelson, M., Sjövall, J. and Baulieu, E.E. (1981a). Characterization and measurement of dehydroepiandrosterone sulfate in the rat brain. Proc. Natl. Acad. Sci. USA, 78, 4704-4707.

Corpéchot, C., Robel, P., Lachapelle, F., Baumann, N., Axelson, M., Sjövall, J. and Baulieu, E.E. (1981b). Déhydroépiandrostérone libre et sulfo-conjuguée dans le cerveau de souris dysmyéliniques. C.R. Acad. Sci. Paris, 292, 231-234.

Corpéchot, C., Synguelakis, M., Talha, S., Axelson, M., Sjövall, J., Vihko, R., Baulieu, E.E. and Robel, P. (1983). Pregnenolone and its sulfate ester in the rat brain. Brain Res., 270, 119-125.

Corpéchot, C., Leclerc, P., Baulieu, E.E. and Brazeau, P. (1985). Neurosteroids : regulatory mechanisms in male rat brain during heterosexual exposure. Steroids, 45, 229-234.

Cutler, G.B., Jr., Glenn, M., Bush, M., Hodgen, G.D., Graham, C.E. and Loriaux, D.L. (1978). Adrenarche : a survey of rodents, domestic animals, and primates. Endocrinology, 103, 2112-2118.

Du Bois, R.N., Simpson, E.R., Kramer, R.E. and Waterman, M.R. (1981). Induction of synthesis of cholesterol side chain cleavage cytochrome P-450 by adrenocorticotropin in cultured bovin. Adrenocrotical cells. J. Biol. Chem., 256, 7000-7005.

Fuxe, K., Gustafsson, J.A. and Wetterberg, L. (1981). Steroid Hormone Regulation of the Brain. Pergamon Press, Oxford.

Haug, M. and Brain, P.F. (1979). Effects of treatment with testosterone and oestradiol on the attack by groups of gonadectomized male and female mice towards lactating intruders. Physiol. Behav. 23, 397-400.

Haug, M., Spetz, J.F., Schlegel, M.L. and Robel, P. (1983). La déhydroépiandrostérone inhibe le comportement agressif de souris mâles castrées. C.R. Acad. Sci. Paris, 296, 975-977.

Kishimoto, Y. and Hoshi, M. (1972). Dehydroepiandrosterone sulfate in rat brain : incorporation from blood and metabolism in vivo. J. Neurochem., 19, 2207-2215.

Knapstein, P., David, A., Wu, C., Archer, D.F., Flickinger, G.L. and Touchstone, J.C. (1986). Metabolism of free and sulfoconjugated DHEA in brain tissue in vivo and in vitro. Steroids, 14, 483-498.

Lieberman, S., Greenfield, N.J. and Wolfson, A. (1984). A heuristic proposal for understanding steroidogenic processes. Endocrine Rev., 5, 128-148.

Majewska, M.D., Schwartz, R.D. (1986). Pregnenolone-sulfate : an endogenous antagonist of the γ-aminobutyric acid receptor complex in brain ? Brain Res., in press.

Oftebro, H., Stormer, E.C. and Pedersen, J.I. (1979). The presence of an adrenodoxin-like ferredoxin and cytochrome P-450 in brain mitochondria. J. Biol. Chem. 254, 4331-4334.

Pashko, L.L., Rovito, R.J., Williams, J.R., Sobel, E.L. and Schwartz, A.G. (1984). Dehydroepiandrosterone (DHEA) and 3β-methyl-androst-5-en-17-one : inhibitors of 7,12-dimethylbenz-(a)-anthracene (DMBA)-initiated and 12-O-tetradecanoyl phorbol-13-acetate (TPA)-promoted skin papilloma formation in mice. Carcinogenesis, 5, 463-466.

Robel, P. and Baulieu, E.E. (1985). Neuro-steroids : 3β-hydroxy-Δ5-derivatives in the rodent brain. Neurochem. Int., 7, 953-958.

Robel, P., Corpéchot, C., Clarke, C., Groyer, A., Synguelakis, M., Vourc'h, C. and Baulieu, E.E. (1986a). Neuro-steroids : 3β-hydroxy-Δ5-derivatives in the rat brain. In Neuroendocrine Molecular Biology. (eds. G. Fink, A.J. Harmar and K.W. McKerns). Plenum Press, New York.

Robel, P., Bourreau, E., Corpéchot, C., Dang, C.D., Halberg, F., Clarke, C., Haug, M., Schlegel, M.L., Synguelakis, M., Vourch', C. and Baulieu, E.E. (1986b). Neuro-steroids : 3β-hydroxy-Δ5-derivatives in the rat and monkey brain. J. Ster. Biochem., in press.

Robel, P., Synguelakis, M., Halberg, F. and Baulieu, E.E. (1986c). Persistance d'un rythme circadien de la déhydroépiandrostérone dans le cerveau, mais non dans le plasma, de rats castrés et surrénalectomisés. C.R. Acad. Sci. Paris, 303, 235-238.

Seybert, D.W., Lancaster, J.R., Jr., Lambeth, J.D. and Kamin, H. (1979). Participation of the membrane in the side chain cleavage of cholesterol. Reconstitution of cytochrome $P\text{-}450_{scc}$ into phospholipid vesicles. J. Biol. Chem., 254, 12088-12098.

Synguelakis, M., Halberg, F., Baulieu, E.E. and Robel, P. (1985). Evolution circadienne de Δ_5-3β-hydroxystéroïdes et de glucocorticostéroïdes dans le plasma et le cerveau de rat. C.R. Acad. Sci. Paris, 301, 823-826.

10
Estrogenic and Antiestrogenic Modulation of Neuronal Membrane Sensitivity*

Robert L. Moss, Carol A. Dudley, Yang-In Kim and Misty Eaton

INTRODUCTION

Although studies on the action of steroids on the central nervous system (CNS) have a long history, the discovery of estradiol-concentrating neurons in the brain greatly accelerated experiments designed to determine the electrophysiological consequences of estradiol administration at the level of the single brain cell (For review see Morrell et al., 1984; Moss & Dudley, 1984). The utilization of the in vitro slice preparation as well as the application of estrogen directly to the neuronal membrane in vivo has contributed enormously to our understanding of the cellular mechanism(s) of steroid actions in the brain. The aim of this presentation is twofold: the first is to provide a brief review of the actions of estrogen on CNS neuronal activity and the second aim is to present some new contributions concerning the mechanism(s) by which estrogen affects neuronal activity.

BACKGROUND

Utilizing the technique of steroid autoradiography, select neurons of the mammalian brain have been shown to concentrate estradiol (Morrell et al., 1984). These estradiol concentrating nerve cells are thought to play a critical role as mediators and/or modulators in a wide variety of reproductive as well as nonreproductive functions (Beatty, 1979; Adler, 1981; Feder, 1981; Van Hartesvedt & Joyce, 1986). The mechanism by which estradiol acts in the CNS to mediate nonreproductive function is not as clearly understood as that for reproductive function and behavior. The latter is thought to involve the binding of estradiol to intracellular receptors which, in turn, triggers RNA dependent protein synthesis (Goldberger & Deeley, 1980). It has been demonstrated that the initiation of such neuroendocrine events as gonadotropin release, ovulation and sexual behavior require a minimum of a several hours of estrogen exposure. This time period

*Supported by NIH Grants NS10434 and HD09988

is consistent with a long-term genomic action of the steroid (Etgen, 1979; Etgen & Whalen, 1981/82; McEwen et al., 1982). However, over the past decade, it has also been suggested that estrogen influences neural electrical activity by a different, short latency, non-genomic action (Moss & Dudley, 1984; Drouva et al., 1983; Kelly, 1982). Recently, biochemical evidence has also appeared that supports this notion and suggests that estrogen may act directly on surface membrane receptors to initiate a physiological function (Baulieu, 1979; Godeau et al., 1978; Inaba and Kamata, 1979; Towle and Sze, 1983; Drouva et al., 1983). The concept that estradiol may act at a non-genomic level, i.e., on surface membrane receptors or recognition sites to alter the electrical activity of nerve cells and/or to alter the other cellular functions may constitute an important fundamental intermediary step that is necessary for the CNS action of steroids.

Changes in electrical activity have been detected following both long-term and acute exposure to estrogen. Experiments designed to determine the effect of long-term exposure of estradiol on CNS neurons, that is, pretreating animals with estradiol 24 to 48 hours prior to the electrophysiological recording, have demonstrated estrogen's ability to modulate neuronal firing rate as well as to modulate putative neurotransmitter function (Lincoln, 1967; Yagi, 1970; Yagi and Sawaki, 1973; Kubo, Gorski and Kawakami, 1975; Bueno & Pfaff, 1976; Dufy et al., 1976; Haskins and Moss, 1983; Lofstrom et al., 1977; Ani et al., 1980; Van Hartesvedt and Joyce, 1986). Similarly experiments designed to determine the effect of short-term exposure of estradiol on CNS neurons, that is, intravenous injection or local application of estradiol directly to neurons, have demonstrated estrogen's ability to initiate short latency, rapid effects on the electrical activity of neurons located in areas of CNS that have been shown to concentrate estradiol (Kelly et al., 1976; Kelly et al., 1978; Pfaff, 1980: Kelly, 1982; Moss and Dudley, 1984).

In contrast to the long latency, genomic-like actions, the short latency, rapid effects of estradiol occur within seconds to a few minutes and appear to vary in latency with the route of administration. For instance, systemic injection of estradiol has been shown to increase or decrease the electrical activity of hypothalamic neurons within several minutes (Lincoln and Cross, 1967; Yagi and Sawaki, 1973; Kubo, Gorski and Kawakami, 1975; Bueno and Pfaff, 1976; Dufy and Vincent, 1980). On the other hand, more rapid changes in neural activity have been reported after iontophoretic administration of 17-ß-estradiol to medial preoptic neurons (Kelly et al., 1976; 1977a; 1977b; 1978; Carrette et al., 1979). The aforementioned _in vivo_ experiments suggest a direct, non-genomic action of estradiol on the post and/or presynaptic membrane of hypothalamic neurons.

In subsequent electrophysiological experiments, the emphasis has shifted to the use of _in vitro_ preparations to clarify the role of 17-β-estradiol on the neuronal membrane. Experiments involving the *in vitro* slice preparation have suggested that perfusion application of 17-β-estradiol to hippocampal (Teyler et al., 1980) and to parvocellular-ventromedial hypothalamic and arcuate neurons (Kelly et al., 1980; 1984) initiates a hyperpolarization within a few seconds of application. In a more recent publication (Nabekura et al., 1986), it was demonstrated via intracellular recordings in the slice preparation that 17-β-estradiol initiates a brief depolarization followed by a longer lasting hyperpolarization in amygdala neurons. In addition, the hyperpolarization was shown to be a result of an increase in potassium conductance. This effect persisted after elimination of synaptic input and after suppression of protein synthesis. This laboratory has also recently reported the effects of 17-β-estradiol on ventromedial hypothalamic neurons (Oomura et al., 1986). The results from these experiments suggest that estradiol causes a marked and pronounced depolarization and an increase in electrical activity of hypothalamic neurons.

The aforementioned data taken collectively from *in vivo* and *in vitro* studies suggest that 17-β-estradiol acts directly on the neuronal membrane to alter potassium ionic conductances of some neurons, while in others, 17-β-estradiol acts directly on the postsynaptic membrane to initiate depolarization and to alter voltage sensitive sodium channels.

ELECTROPHYSIOLOGIC ACTIONS OF ESTRADIOL AND ANTIESTROGEN: *IN VIVO* PREPARATION

As previously noted, certain diencephalic nerve cells do possess the necessary neuronal machinery to alter their electrical signature in response to long-term as well as short-term exposure to estrogen (For review see Moss & Dudley, 1984). For instance, changes in single cell extracellular electrical activity of medial preoptic - medial septal neurons have been correlated with [1] the variations in plasma levels of estrogen that occur during the estrous cycle; [2] the subcutaneous as well as intravenous administration of estrogen; and [3] the local application of estrogen directly to the neuronal membrane. Taken together these data support the notion that estrogen has both genomic and non-genomic actions on nerve cells and raises the question as to whether antiestrogens can antagonize the estrogenic-induced changes in neural activity.

The role of non-steroidal antiestrogens, such as nitromophene citrate (CI-628) and triphenylethylene (tamoxifen), as estrogen antagonists has been established in the uterus, pituitary gland and in the central nervous system. Antiestrogens have been demonstrated to antagonize uterine growth (Clark et al., 1974; Katzenellenbogen & Ferguson, 1975), CNS estrogen binding, certain enzymatic activities, gonadotropin secretion and sexual behavior

(Chazel et al., 1975; Landau, 1976, 1977; Whalen et al., 1975; Wade and Blaustein, 1978; Gogan et al., 1980). The antiestrogenic effects may be related to its ability to compete with estrogen for binding sites in the target tissue. However, it should be noted that antiestrogens can, under certain conditions (especially in the absence of estrogen) act as an estrogen agonist (Clark et al., 1974; Drouin et al., 1976).

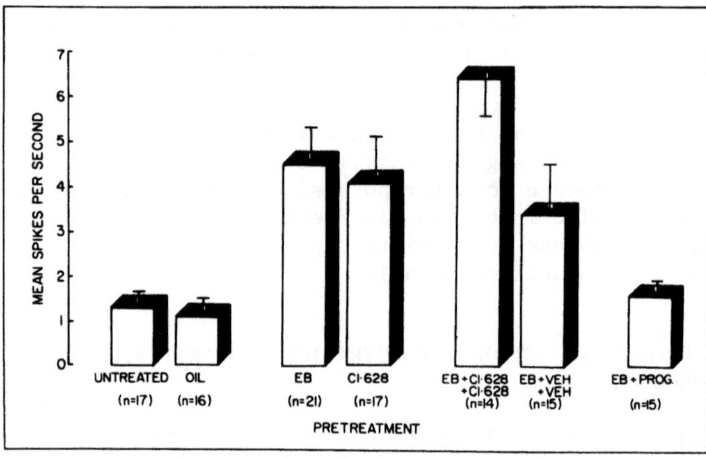

Figure 1:
Mean firing rate of neurons in the preoptic-septal (PO-S) area recorded following various estrogen (EB) and/or anti-estrogen (CI-628) treatments. VEH=vehicle; PROG=progesterone. Pretreatment 24 hr. prior to recording with EB, CI-628, EB + CI-628 or EB + VEH was found to significantly increase ($p<0.005$) the electrical activity of PO-S neurons as compared to the activity recorded in untreated, oil and EB + PROG treated animals. From Moss and Dudley, 1984.

Recently in our laboratory, we have utilized this information on antiestrogen and have addressed the question of whether long-term exposure to antiestrogen can antagonize the estrogen-induced changes in electrical activity. In the first series of experiments, the ability of the antiestrogen, CI-628, to alter the spontaneous electrical activity of preoptic-hypothalamic nerve cells and the membrane sensitivity of these nerve cells to the local application of estradiol was investigated in ovariectomized female rats anesthetized with urethane (as previously described: Moss and Dudley, 1984). Animals were treated (subcutaneously) 24 hours prior to recording with one of the following: [1] control (either oil or no treatment); [2] estradiol benzoate (EB, 4 μg); [3] CI-628 (20 mg/kg of body weight); [4] EB + CI-628; [5] EB + Oil; and [6] EB + Progesterone (2.5 mg; 6 hours prior to recording). The firing rate of single preoptic-hypothalamic neurons was significantly increased with the administration of EB; CI-628; and EB + CI-628 as compared to the firing rate recorded in ovariectomized animals pretreated with oil or receiving no pretreatment (control animals), as well as compared to animals receiving EB + P (Figure 1). However, no significant differences in the spontaneous firing rate were observed following pretreatment with either EB; CI-628; or EB + CI-628 (Figure 1). This suggests that EB as well as CI-628 potentiates neuronal activity and that CI-628 in combination with EB does not antagonize the estrogen-induced increase in neuronal activity.

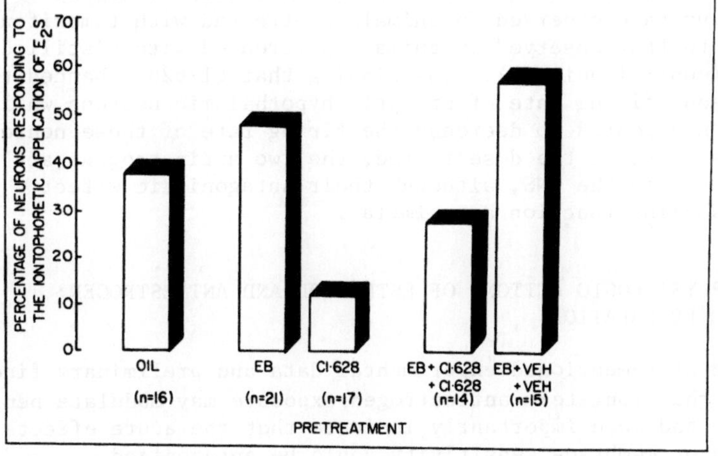

Figure 2: The influence of various estrogen (EB) and/or anti-estrogen (CI-628) treatments (24 hr. prior to recording) on the percentage of preoptic-septal neurons responsive to iontophoretically applied 17-β-E_2S. From Moss and Dudley, 1984.

Further analysis of the results revealed that unlike the spontaneous firing rate, the membrane sensitivity to the iontophoretic application of 17-β-estradiol hemisuccinate (10^{-3}M) directly to the nerve cell surface varied as a function of the pretreatment. No significant differences were observed in the percentage of neurons responding to the iontophoretic application of 17-β-estradiol hemisuccinate (an index of membrane sensitivity) in animals pretreated with oil, EB, or EB + oil (Figure 2). However, the percentage of neurons responding to iontophoretically applied 17-β-estradiol hemisuccinate was significantly reduced in animals pretreated with CI-628 and EB in combination with CI-628 (Figure 2). These findings demonstrate that 24 hours of exposure to the antiestrogen, CI-628, alters the estrogen-induced changes in membrane sensitivity, without altering the estrogen-induced changes in the spontaneous firing rate of preoptic-hypothalamic nerve cells.

These initial experiments with the antiestrogen CI-628 provided less than conclusive results. On one hand, CI-628 potentiated neuronal firing very much like estrogen treatment, while on the other hand, CI-628 decreased the membrane sensitivity to subsequent estradiol applied directly to the cell surface.

In a series of preliminary experiments, the administration of the antiestrogen tamoxifen (40 μg subcutaneously administered 24 hours prior to recording) was observed to suppress the spontaneous firing rate of preoptic-hypothalamic neurons as compared to the firing rate observed in animals pretreated with either EB or CI-628. The firing rate observed in animals pretreated with tamoxifen was similar to that observed in animals pretreated with distilled water or oil (control animals). The finding that CI-628 enhanced the spontaneous firing rate of preoptic-hypothalamic neurons while tamoxifen appeared to decrease the firing rate of these neurons suggests that, at the dose tested, the two antiestrogens act differently in the CNS, although their antagonistic effect neuroendocrine function are similar.

ELECTROPHYSIOLOGIC ACTIONS OF ESTRADIOL AND ANTIESTROGEN: *IN VITRO* PREPARATION

The aforementioned experimental data and preliminary findings suggest that long-term antiestrogen exposure may modulate neuronal activity and more importantly indicate that the acute effects of estrogen on membrane sensitivity could be antagonized.
While the question of how antiestrogen can antagonize the estrogen-induced changes in electrical activity is extremely important, the methods utilized, i.e. the *in vivo* preparation, may be too limiting in scope to provide insight into mechanistic questions. Therefore, we turned to the *in vitro* preparation. The sensitivity of nerve cells in the slice preparation to a variety of transmitters, putative transmitters, and drugs can be tested under more controlled conditions of concentration than those possible *in vivo* (McIlwain and Bachelard, 1971; McIlwain, 1972; Hatton et al., 1980; Hatton, 1984). In addition, *in vitro* slice preparations, defined as thin tissue sheets less than 0.5 mm thick, have been derived from many CNS regions, and well characterized regarding their metabolic, biosynethetic, transport, and electrophysiological properties (For review see: Orrego, 1979; Dingledine, 1984). All these functions

have been demonstrated to persist for several hours of *in vitro* incubation in balanced saline media supplemented with glucose and oxygen (Orrego, 1979). The remainder of this chapter will deal with the preliminary findings observed in our laboratory concerning the antagonistic actions of the antiestrogen tamoxifen on estradiol-induced changes in the spontaneous electrical activity of medial septal-medial preoptic neurons recorded in the *in vitro* preparation.

These preliminary studies were performed on 3-5 month-old female Sprague-Dawley albino rats. At the time of experiment, the animal was decapitated and the brain rapidly removed and placed in cold medium. Frontal sections (400 microns thick) were made through the septal-preoptic regions of the diencephalic aspect of the brain utlizing a Oxford Vibratome or a McIlwain tissue chopper and subsequently maintained at 33±1°C and superfused with a standard Yamamoto's medium. The medium used was composed of 116 mM NaCl, 5 mM KCl, 2.5 mM $CaCl_2$, 1.2 mM $MgCl_2$, 1.2 mM NaH_2PO_4, 25 mM $NaHCO_3$, and 10 mM glucose. The medium was bubbled continuously with 95% O_2 - 5% CO_2 and was maintained at a pH of 7.40.

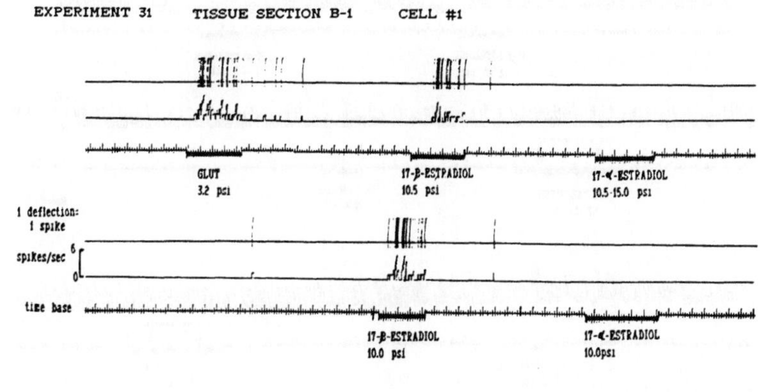

Figure 3: Strip chart record of extracellular activity from a medial preoptic-medial septal area neuron recorded in a slice preparation. Application of glutamate (GLUT) or 17-β-estradiol increased the firing rate of this neuron whereas application of 17-α-estradiol had no effect. The interval between each large deflection on the time base tracing is 5 seconds.

Extracellular recordings were made from 15 medial septal-medial preoptic neurons utilizing the center barrel of a glass multi-barrelled micropipette. The center barrel contained 4M NaCl while outer barrels contained one of the following solutions [1] 10^{-6}M 17-β-estradiol, [2] 10^{-6}M 17-α-estradiol (a weak estrogen isomer), [3] 0.5 mM OH-tamoxifen, and [4] 0.5 M glutamate. The solutions in the outer barrels were applied to the neuronal membrane of recorded cells via the micropressure ejection technique (pressure ranging between 0.5 and 20.0 psi was used to apply the solutions).

We have selected the data obtained from two recorded and tested neurons in order to demonstrate two primary findings that were observed in these preliminary experiments. The first of these findings is shown in Figure 3. This figure is a polygraph record from a medial septal-medial preoptic neuron recorded from an *in vitro* slice preparation. The spontaneous activity of this neuron was extremely slow but could be potentiated by the application of glutamate via micropressure ejection. The spontaneous electrical activity of this neuron was also increased with the application of 17-β-estradiol while the application of 17-α-estradiol had no observable effect on cell activity. Thus in the *in vitro* septal-hypothalamic slice preparation, the application of 17-β-estradiol facilitated the generation of action potentials while a weak estrogen isomer (17-α-estradiol) tested on the same neuron had no effect; this demonstrates the specificity of action of the 17-β-estradiol on the neuronal membrane.

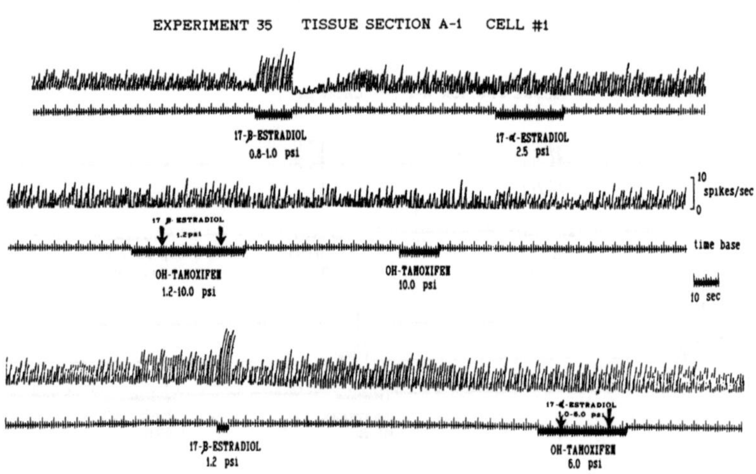

Figure 4: Strip chart record of extracellular activity from a medial preoptic-medial septal neuron recorded in a slice preparation. Micropressure application of 17-β-estradiol increased the firing rate of this neuron whereas 17-α-estradiol had no effect. The increase in firing rate obtained during application of 17-β-estradiol was prevented by application of OH-tamoxifen. OH-tamoxifen alone had no effect on firing rate. Time interval as in Figure 3.

The second finding is shown in Figure 4. This figure displays a polygraph record from another medial septal-medial preoptic neuron that was recorded from the *in vitro* slice preparation. This neuron exhibits a response pattern to 17-β-estradiol and 17-α-estradiol similar to that seen in Figure 3 : that is 17-β-estradiol increased while 17-α-estradiol had no observable effect on the spontaneous electrical activity of the recorded neuron. Moreover, the administration of OH-tamoxifen is shown to antagonize the 17-β-estradiol-induced increase in the spontaneous activity. These findings suggest that specific membrane recognition sites exist on the neurons that are sensitive to both estrogen and antiestrogen.

CONCLUDING REMARKS

While the results concerning the action of CI-628 are difficult to interpret, two conclusions can be drawn concerning the action of tamoxifen: [1] long-term exposure to tamoxifen, unlike EB and CI-628, does not increase unti activity and [2] the local application of tamoxifen suppresses estradiol's action on the electrical activity of medial-septal-medial preoptic neurons in the female rat. The finding that locally applied estrogen can induce a short latency change in extracellular activity coupled with the finding that this membrane response can be blocked by tamoxifen suggests that neuronal membrane recognition and/or receptor sites may exist for 17-β-estradiol. These sites may mediate estrogen-induced changes in electrical activity by opening or closing specific membrane gates (channels). The findings further suggest that when these specific sites are occupied by tamoxifen rather than estrogen, the gates remain unaffected by estrogen. It should be emphasized that this is a proposed mechanism of action of estrogen and that much more definitive research is needed to establish its true non-genomic action. In any case, estradiol may be a true neuromodulator in that it is secreted from a peripheral organ, the ovary, and readily enters the nervous system where it is concentrated by a specific subset of neurons. Furthermore, along with its classical, long latency genomic action in the regulation of endocrinological and reproductive processes, estrogen can rapidly enhance or diminish the electrical activity of central nervous system neurons.

REFERENCES

Adler, N.T. (1981). Neuroendocrinology of Reproduction, Physiology and Behavior. Plenum Press, New York.

Ani, M., Butterworth, P.J. and Thomas, P.J. (1980). Effect of Estradiol on Neurotransmitter Sensitive Adenylate Cyclase. Its Possible Role in 'Sexual Differentiation'. Brain Res., 183, 341-353.

Baulieu, E.-E. (1979). Aspects of Steroid Hormone-Target Cell Interactions. In: Steroid Hormone Receptor Systems. (eds. W.W. Leavitt and J.H. Clark). Plenum Press, New York.

Beatty, W.W. (1979). Gonadal Hormones and Sex Differences in Non-Reproductive Behaviors in Rodents: Organizational and Activational Influences. Hormones Behav., 16, 565-568.

Bueno, J. and Pfaff, D.W. (1976). Single Unit Recording in Hypothalamus and Preoptic Area of Estrogen-Treated and Untreated Ovariectomized Female Rats. Brain Research, 101, 67-78.

Carrette, B., Barry, J., Linkie, D., Ferin, M., Mester, J. and Baulieu, E.-E. (1979) Effets de l'oestradiol-7-α Acide Butyrique Au Niveau de Cellules Hypothalamiques C.R. Acad. Sci. (Paris), Ser. D, 36, 631-634.

Chazal, G., Faudon, M., Gogan, F. and Rotsztejn, W. (1975). Effects of Two estradiol Antagonists Upon Estradiol Uptake in the Rat Brain and Peripheral Tissue. Brain Res., 89, 245-254.

Clark, J.H., Peck, E.J. and Anderson, J.N. (1974). Oestrogen Receptors and Antagonism of Steroid Hormone Action. Nature (Lond.). 251, 446-448.

Dingledine, R. (ed.) (1984). Brain Slice. Plenum Press, New York.

Drouin, J., Lagace, L. and Labrie, F. (1976). Estradiol Induced Increase of the LH Responsiveness to LH Releasing Hormone (LH-RH) in Rat Anterior Pituitary. Endocrinol., 99, 1477-1481.

Drouva, S.V., LaPlante, E. and Kordon, C. (1983). Effects of Ovarian Steroids on in vitro Release of LHRH from Mediobasal Hypothalamus. Neuroendocrinol., 37, 336-341.

Dufy, B. and Vincent, J.D. (1980). Effects of Sex Steroids on Cell Membrane Excitability: A New Concept for the Action of Steroids on the Brain. In: Hormones and the Brain. (eds. D. De Weid and P.A. van Keep). University Park Press, Baltimore, MD.

Dufy, B., Partouche, C., Poulain, D., Dufy-Barbe, L. and Vincent, J.D. (1976). Effects of Estrogen on the Electrical Activity of Identified and Unidentified Hypothalamic Units. Neuroendocrinol., 22, 38-47.

Etgen, A.M. (1979). Antiestrogens: Effects of Tamoxifen, Nafoxidine, and CI-628 on Sexual Behavior, Cytoplasmic Receptors, and Nuclear Binding of Estrogen. Hormones Behavior, 13, 97-112.

Etgen, A.M. and Whalen, R.E. (1981/82). Kinetic Analysis of Estrogen and Antiestrogen Competition for Hypothalamic Cytosol Estrogen Receptors. Evidence for Noncompetitive Ligand-Receptor

Interactions. J. Recept. Res., 2, 531-553.

Feder, H.H. (1981). Experimental Analysis of Hormone Actions on the Hypothalamus, Anterior Pituitary, and Ovary. In: Neuroendocrinology of Reproduction, Physiology and Behavior. (ed. N.T. Adler). Plenum Press, New York.

Godeau, F., Schorderst-Slatkine, S., Hubert, P. and Baulieu, E.-E., (1978). Induction of Maturation in Xenopus Laevis Oocytes by a Steroid Linked to a Polymer. Proc. Natl. Acad. Sci. (U.S.A.), 75, 2353-2357.

Goldberger, R.F. and Deeley, R.G. (1980). Gene Regulation by Steroid Hormone. Springer-Verlag, New York.

Gogan, F., Rotsztejn, W.H. Couturier, L., Chazal, G., Beattie, I. and Kordon, C. (1980). Interaction of the Anti-Estrogen CI-628 and Estradiol on Plasma LH and Hypothalamic LH-RH in the Female Rat. Brain Res., 184, 109-118.

Hatton, G.I., Doran, A.D., Salm, A.K. and Tweedle, C.D. (1980). Brain Slice Preparation: Hypothalamus. Brain Res. Bull., 5, 405-414.

Hatton, G.I. (1984). Hypothalamic Neurobiology. In: Brain Slices. (ed. R. Dingledine). Plenum Press, New York.

Haskins, J.T. and Moss, R.L. (1983). Action of Estrogen and Mechanical Vaginocervical Stimulation on the Membrane Excitability of Hypothalamic and Midbrain Neurons. Brain Res. Bull., 10, 489-496.

Inaba, M. and Kamata, K. (1979). Effect of Estradiol-17β and Other Steroids on Noradrenaline and Dopamine Binding to Synaptic Membrane Fragments of Rat Brain. J. Steroid Biochem., 11, 1491-1497.

Katzenellenbogen, B.S. and Ferguson, E.R. (1975). Anti-Estrogen Action in the Uterus: Biological Ineffectiveness of Nuclear Bound Estradiol After Antiestrogen. Endocrinol., 97, 1-12.

Kelly, M.J., Moss, R.L. and Dudley, C.A. (1976). Differential Sensitivity of Preoptic-Septal Neurons to Microelectrophoresed Estrogen During the Estrous Cycle. Brain Res., 114, 152-277.

Kelly, M.J., Moss, R.L., Dudley, C.A. and Fawcett, C.P. (1977a). The specificity of the Response of Preoptic-Septal Neurons to Estrogen: 17α-Estradiol versus 17β-Estradiol and the Response of Extrahypothalamic Neurons. Exp. Brain Res., 30, 43-52.

Kelly, M.J., Moss, R.L. and Dudley, C.A. (1977b). The Effects of Microelectrophoretically Applied Estrogen, Cortisol, and Acetylcholine on Medial Preoptic-Septal Unit Activity Throughout the Estrous Cycle of the Female Rat. Exp. Brain Res., 30, 53-64.

Kelly, M.J., Moss, R.L. and Dudley, C.A. (1978). The Effect of Ovariectomy on Responsiveness of Preoptic-Septal Neurons to Microelectrophoresed Estrogen. Neuroendocrinol., 25, 204-211.

Kelly, M.J., Kuhnt, U. and Wuttke, W. (1980). Hyperpolarization of Hypothalamic Parvocellular Neurons by 17β-Estradiol and Their Identification Through Intracellular Staining with Procion Yellow. Exp. Brain Res., 40, 440-447.

Kelly, M.J. (1982). Electrical Effects of Steroids in Neurons. In: Brain Peptides: Structure and Function. (eds. K.W. McKerns and V. Pantic). Plenum Press, New York.

Kelly, M.J., Ronnekleiv, O.K. and Eskay, R.L. (1984). Identification of Estrogen-Responsive LHRH Neurons in the Guinea Pig Hypothalamus. Brain Res. Bull., 12, 399-407.

Kubo, K., Gorski, R.A. and Kawakami, M. (1975). Effects of Estrogen on Neuronal Excitability in the Hippocampal-Septal-Hypothalamic Systems. Neuroendocrinol., 18, 176-191.

Landau, T.I. (1976). Effect of Subcutaneous vs Intraperitoneal Administration of An Anti-Estrogen, CI-628, on Estradiol and Estradiol Benzoate-Stimulated Lordosis in the Ovariectomized Rat. Pharmacol. Biochem. Behavior, 5, 473-476.

Landau, T.I. (1977). Relationship Between the Effects of the Anti-Estrogen CI-628, on Sexual Behavior, Uterine Growth, and Cell Nuclear Estrogen Retention After Estradiol-17β Benzoate Administration in the Ovariectomized Rat. Brain Res., 135, 119-138.

Lincoln, D.W. (1967). Unit Activity in the Hypothalamus, Septum, and Preoptic Area of the Rat: Characteristics of Spontaneous Activity and the Effect of Oestrogen. J. Endocrinol., 37, 177-189.

Lincoln, D.W. and Cross, B.A. (1967). The Effects of Oestrogen on the Responsiveness of Neurons in the Hypothalamus, Septum and Preoptic Area of Rats with Light-Induced Persistent Oestrus. J. Endocrinol., 37, 191-203.

Lofstrom, A., Eneroth, P. and Gustafsson, J. (1977). Effect of Estradiol Benzoate on Catecholamine Levels and Turnover in Discrete Areas of the Median Eminence and Limbic Forebrain and on Serum Luteinizing Hormone, Follicle Stimulating Hormone and Prolactin Concentrations in the Ovariectomized Female Rat. Endocrinol., 101, 1559-1569.

McEwen, B.S., Biegon, A., Davis, P.G., Krey, L.C., Luine, V.N., McGinnis, M.Y., Paden, C.M., Parsons, B. and Rainbow, T.C. (1982). Steroid Hormones: Humoral Signals Which Alter Brain Cell Properties and Function. Rec. Prog. Horm. Res., 38, 41-92.

McIlwain, H. and Bachelard, H.S. (1971). Biochemistry and the Central Nervous System. Churchill, Livingston, Edinburgh.

McIlwain, H. (1975). Practical Neurochemistry. Churchill, Livingston, Edinburgh.

Moss, R.L. and Dudley, C.A. (1984). Molecular Aspects of the Interaction Between Estrogen and the Membrane Excitability of Hypothalamic Nerve Cells. Prog. Brain Res., 61, 3-22.

Morrell, J.I., Schwanzel-Fukuda, M., Fahrbach, S.E. and D.W. Pfaff. (1984). Axonal Projections and Peptide Content of Steroid Hormone Concentrating Neurons. Peptides, 5, 227-239.

Nabekura, J., Oomura, Y., Minami, T., Mizuno, Y. and Fukuda, A. (1986). Mechanism of the Rapid Effect of 17β-Estradiol on Medial Amygdala Neurons. Science, 233, 226-228.

Oomura, Y., Minami, T. and Nabekura, J. (1986). Effect of Estradiol on the Amygdala and Ventromedial Hypothalamic Neurons. 16th Annual Neuroscience Meeting, Washington, D.C.

Orrego, F., (1979). Criteria for the Identification of Central Neurotransmitters, and Their Application to Studies with Some Nerve Tissue Preparations in vitro. Neuroscience 4, 1037-1057.

Pfaff, D.W. (1980). Estrogens and Brain Function. Springer-Verlag, New York.

Teyler, J.L., Vardaris, R.M., Lewis, D. and Rawitch, A.B. (1980). Gonadal Steroids: Effect on Excitability of Hippocampal Pyramidal Cells. Science, 209, 1017-1019.

Towle, A.C. and Sze, P.Y. (1983). Steroid Binding to Synaptic Plasma Membrane: Differential Binding of Glucocorticoids and Gonadal Steroids. J. Steroid Biochem., 18, 135-143.

Van Hartesvedt, C. and Joyce, J.N. (1986). Effects of Estrogen on the Basal Ganglia. Neuroscience and Biobehavioral Reviews, 10, 1-14.

Wade, G.M. and Blaustein, J.D. (1978). Effects of an Antiestrogen on Neural Estradiol Binding and on Behaviors in Female Rats. Endocrinol., 102, 245-251.

Whelan, R.E., Martin, J.V. and Olsen, K.L. (1975). Effect of Estrogen Antagonist on Hypothalamic Estrogen Receptors. Nature (Lond.), 258, 742-743.

Yagi, K. (1970). Effects of Estrogen on Unit Activity of Rat

Hypothalamus. J. Physiol. Soc. Jap., 32, 692-693.

Yagi, K. and Sawaki, Y. (1973). Feedback of Estrogen in the Hypothalamic Control of Gonadotropin Secretion. In: Neuroendocrine Control. (eds. K. Yagi and S. Yoshida). University of Tokyo Press, Tokyo.

11
Effects of Estrogens on the Electrophysiological Properties of Neurons and Adenohypophyseal Cells: Interactions with Dopamine

J.D. Vincent, J. Demotes-Mainard, E. Arnauld and J.M. Israel

INTRODUCTION

Apart from their classical action through the genetic protein synthesizing system, estrogens have been shown to influence the electrophysiological properties of both neuronal (Dufy et al., 1976 ; Moss & Dudley, 1984) and non neuronal cell types (Dufy & Vincent, 1980). Direct effects of 17 β -estradiol on membrane excitability have been demonstrated by intracellular recordings in rat medial amygdala brain slices (Nabekura et al., 1986). These rapid estrogen effects on membrane potential had already been described on GH3/B6 cells, a tumoral cell line of pituitary cells which releases prolactin (PRL) in response to estrogen stimulation (Dufy et al., 1979). Since they exhibit membrane excitability comparable to that observed in neurons, PRL secreting cells in culture offer us an accessible model to study the effect of estrogen on electrophysiological properties of cells. Some of our observations from tumoral and normal rat PRL cells will be given in this report.

Another way by which estrogen may influence membrane properties of excitable cells is to interact with the action of classical neurotransmitters. In this respect we shall consider the interactions between dopamine (DA) and estrogen in both neurones (i.e. caudate nucleus) and non neuronal cells (i.e. PRL cells).

EFFECT OF ESTROGENS ON PROLACTIN CELLS

Two types of material are used : adenohypophysial cells from tumoral origin and purified population of PRL secreting cells from normal rats.

The tumoral cells consist of a clonal strain of rat pituitary tumor cells (GH3/B6) which have been shown to secrete both growth hormone (GH) and PRL (Tixier-Vidal et al., 1975). These cells are stimulated by thyrotropin-releasing hormone (TRH) and 17β -

estradiol. Conversely and in contrast with normal PRL cells, they are poorly inhibited by DA.

The normal PRL cells are removed from Wistar rats (14 day-old female rats and lactating rats) ; the method of cell dispersion is adapted from Hopkins and Farquhar (1973) and is described in detail in Israel et al. (1983). Briefly, after a trypsine treatment and mechanical dispersion, the cells are allowed to sediment on a continuous gradient of bovine serum albumine. PRL cells identified by immunoassay, are located in the light fractions of the gradient. They are then grown in DMEM supplemented with 10% new-born calf serum. Perifusion experiments and electrophysiological recordings are made on cells cultured for 8-20 days. The cells secrete large amount of PRL during continuous perifusion. TRH (10^{-10} to 10^{-7} M) stimulates basal release in a dose dependent manner while DA (10^{-10} to 10^{-7} M) inhibits the release of PRL.

Early Membrane Responses to Estradiol

Fifty per cent of GH3/B6 cells displayed spontaneous or evoked calcium-dependent action potentials (Dufy et al., 1979). Ejection of 17 β -estradiol (10^{-10} M) in the close vicinity of the cells elicited a slight depolarization of the membrane resting potential leading to a prolonged bursts of action potentials (Fig. 1).

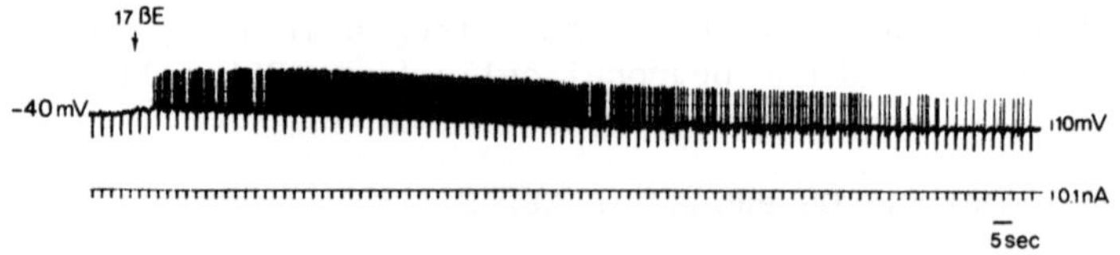

FIGURE 1

Estrogen (17 β -E ; 10^{-9}M, 300 msec) is ejected from a micropipette close to the GH3/B6 cell tested. After a slight depolarizing response, the cell begins to fire.

Seventeen α-estradiol had a much weaker effect at higher doses but prevented the effect of 17 β-estradiol. Progesterone and testosterone were not able to induce spiking activity. This indicate a high degree of stereo-specificity for the effects of estrogen on the electrical activity of the GH3 cells and suggest that estrogen may act directly on the electrical properties of the membrane of PRL cells, at least for PRL cells of tumoral origin. Indeed, normal PRL cells were not directly responsive to local application of 17 β-E, even ejected at high concentration (up to 10^{-6} M).

Modulation by Estrogen of the Effects of Substances Controlling PRL Secretion : DA and TRH

After 8 days in culture, the totality of normal rat PRL cells are excitable and display spontaneous or evoked calcium dependent action potentials.

DA (10^{-7} M) applied by a micropipette in the vicinity of a cell induced in 80% of the recorded population a rapid hyperpolarizing response lasting for several seconds (Fig. 2). When the cell was spontaneously firing, spiking was suppressed by DA (Fig. 2A). This response to DA involves a decrease in membrane input resistance (Fig. 2B). Since reversal potential is found to be near -100 mV, it is suggested that DA may act on PRL cell membrane by increasing a potassium conductance.

The DA response was mimicked by specific agonists of DA receptors such as bromocriptine or RU 24926. Conversely, the response was inhibited by DA antagonists such as haloperidol (Fig. 3A) and blocked more specifically by the D2 antagonist spiperone (Fig. 3B).

TRH applied at various concentrations (10^{-8} to 10^{-6} M) was unable to modify the electrical properties of the PRL cells which responded to DA. On the other hand, PRL cells which were unresponsive to DA exhibited a clear-cut electrophysiological response to TRH (Fig. 4). This response consisted of a depolarization with a decrease in the input resistance and a reversal potential near -20 mV (Fig. 5).

When the PRL cells were treated with 17 β-E (10^{-11} M in the medium) for at least 5 days, we observed a dramatic change in the proportion of the cells which responded to DA and TRH respectively. The majority of the cells (more than 80%) now responded to TRH and were unresponsive to DA.

On the one hand, estrogen are known to increase the number of TRH receptors on PRL cells (Brunet et al., 1977, 1980 ; Gershengorn et al., 1979). On the other hand, DA binding sites are decreased by 17 β-E treatment (Heiman & Ben-Jonathan, 1982). Bression et al. (1983) have also described the conversion of high affinity to low affinity DA binding sites in chronic estrogenized rats.

Our results therefore confirm that estrogens are able to modify the way by which the electrical activity of the membrane of PRL cell responds to factors which control the release of their secretory product.

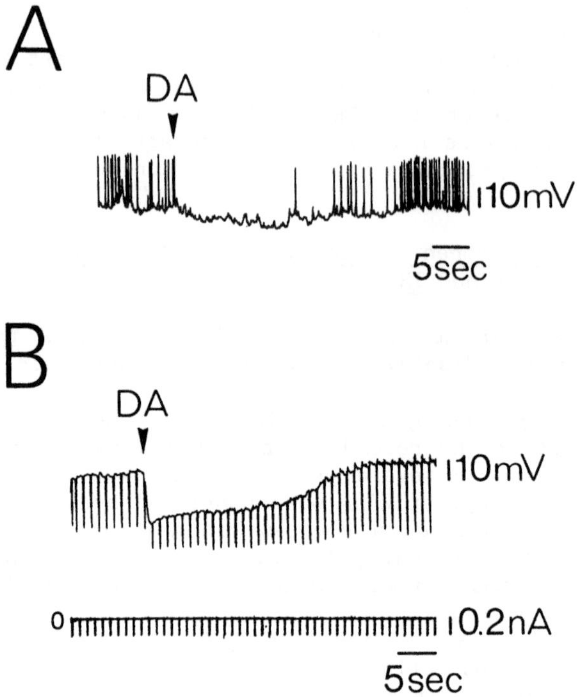

FIGURE 2

Effect of dopamine on normal prolactin cells
A : The ejection of dopamine (DA : 10^{-8}M ; ejection by gravity, 0,5 sec) on a spontaneous firing cell lead to a hyperpolarizing response which induces the cessation of the electrical activity. This effect lasts for several seconds.
B : Small hyperpolarizing currents (0,2 nA, 100 msec, 1 Hz) induce rapid voltage deflexions. During the dopamine response (10^{-7}M, 100 msec) the amplitude of the deflexions is reduced, indicating a decrease of the membrane resistance (i.e. an increase of the membrane conductance).

INTERACTIONS BETWEEN ESTROGEN AND DOPAMINE ON CAUDATE NEURONS

Evidence for interaction between estrogens and striatal DA receptors is still indirect. Clinically, estrogen has been reported to improve symptoms of tardive dyskinesia following chronic treatment with neuroleptics (Bedard et al., 1977), while elevated concentration of estrogens during pregnancy or oral contraceptive use can induce chorea (Nausieda et al., 1979). Experimentally, it has been observed that estrogens can influence the circling

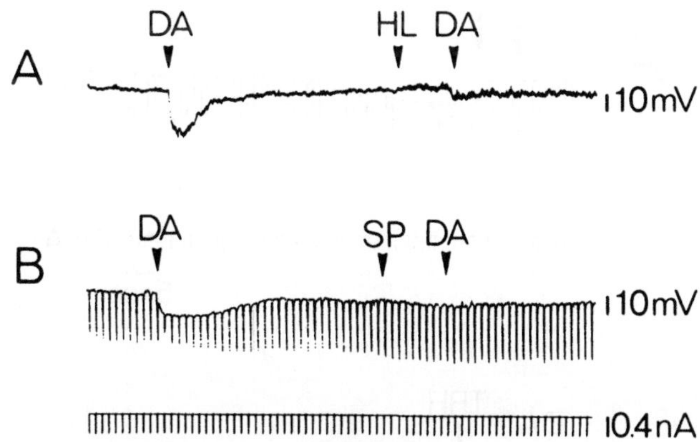

FIGURE 3

Pharmacological characterization of the dopamine induced response

A : A second ejection of DA is inhibited by an ejection of haloperidol (HL).

B : The DA response is blocked by spiperone (SP). HL and SP do not induce modification of the electrical activity of the cell.

behavior induced by amphetamine and apomorphine in rats with unilateral lesion of the nigro-striatal pathway (Review in Van Hartesveldt & Joyce, 1986). Estrogens are also known to interact with other aminergic transmission : i.e. serotoninergic (Biegon & McEwen, 1982) and noradrenergic (Wagner et al., 1979) receptors.

We have previously reported that estrogens are able to influence the global activity of caudate neurons and to reverse their sensitivity to DA (Arnauld et al., 1981).

The electrical activity of caudate neurons and their response to iontophoretically applied dopamine were tested at different time intervals before and after an i.m. injection of estradiol benzoate (EB) in rats ovariectomized two weeks previously. One hundred and three neurons were recorded before estradiol injection and, of these, 25% were spontaneously active and 75% were driven by glutamate. The firing activity of the majority (86%) of the neurons was inhibited on application of DA. At 2-6 hours after the EB injections, the proportion of spontaneously active neurons was elevated to 87% and their firing rate was dramatically increased. Concomitantly, the proportion of cells affected by DA decreased to

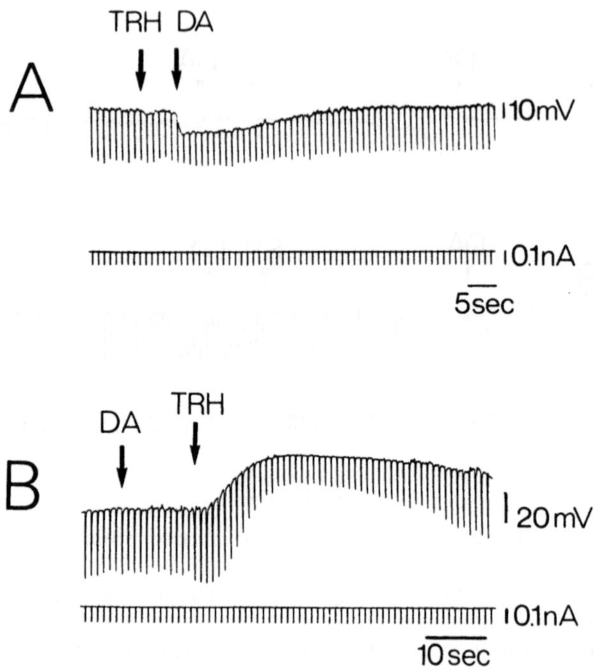

FIGURE 4

Effect of TRH on normal prolactin cells
A : An ejection of TRH (10^{-7} M, 100 msec) on a cell which responds to dopamine does not modify the electrical activity of the cell.
B : A cell which does not respond to DA respond to TRH.

39% ; moreover, the majority of cells responsive to DA was now excited following it (Table).

In an attempt to further characterize the receptors involved in in the changes in responsiveness of caudate neurons to DA, we have performed iontophoretic applications of D1 and D2 agonists and antagonists after a single dose of EB (20 μg i.m.) in ovariectomized rats.

Neurons were recorded from the head of the left caudate nucleus according to the coordinates from König and Klippel (1963): A, 7.2 to 8.5 mm ; L, 2.5 to 3.5 mm ; V, +2 to -2 mm. Of 22 neurons recorded before or immediately after EB injection 3 were spon-

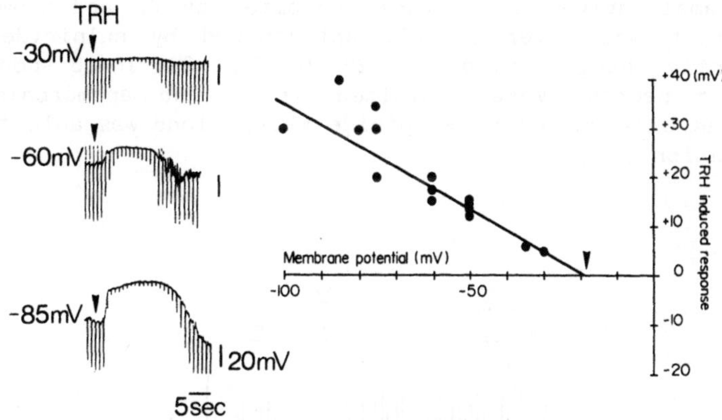

FIGURE 5

Analysis of the TRH-induced response
- Left : TRH is ejected on three cells at different resting potentials (-30, -60 and -85 mV). The amplitude of the responses varies with the potential of the cell.
- Right : The amplitude of the TRH-induced response recorded from 15 cells is plotted against the membrane potential. The extrapollated relation crosses the axe at -20 mV indicating the reversal potential of the response.

	Number of neurons	Electrical activity: Spontaneously active neurons (%)	Electrical activity: Glutamate-driven neurons (%)	Response to DA: Effect (%)	Response to DA: No effect (%)
Control	103	25	75	86 (i:98 ; e:2)	14
After EB: 0– 2 h	25	20	80	88 (i:100; e:0)	12
2– 4 h	23	48	52	39 (i:57 ; e:43)	61
4– 6 h	21	76	24	57 (i:11 ; e:89)	43
6– 8 h	15	87	13	73 (i:73 ; e:27)	27
8–12 h	18	83	17	83 (i:75 ; e:25)	17

TABLE : Electrical activity and response to DA of caudate neurons in ovariectomized rats before and at various time intervals after estradiol benzoate (EB) i.m. injection (20 μg in sesame oil).
i = neurons inhibited by DA in % of cells affected by DA ; e = neurons excited by DA in % of cells affected by DA.

taneously active (14%) and 19 were silent, driven by glutamate. The glutamate-driven cells were inhibited by DA and apomorphine; this effect was never clearly antagonized by sulpiride or was inconstantly antagonized by SCH 23390 (Fig. 6). The 3 spontaneously active neurons were inhibited by DA and apomorphine. This effect was antagonized by sulpiride which alone was able to induce an excitation (Fig. 7).

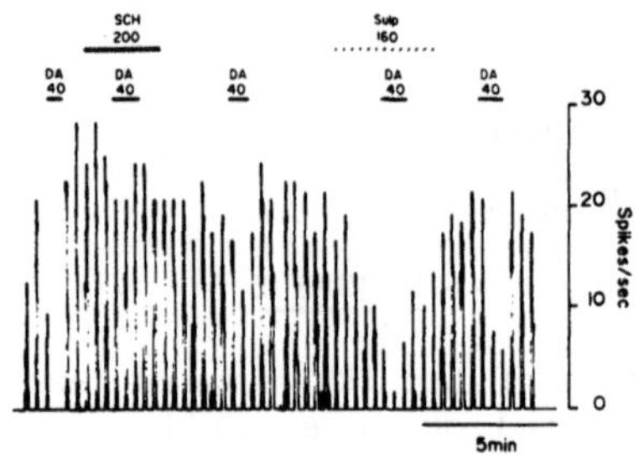

FIGURE 6

2 hours after EB, silent neuron recorded under glutamate activation (80 nA ; pulse duration : 10 sec ; intervals : 10 sec). The DA-induced inhibition is antagonized by SCH 23390 but not by sulpiride which has an inhibitory action (Intensity in nA).

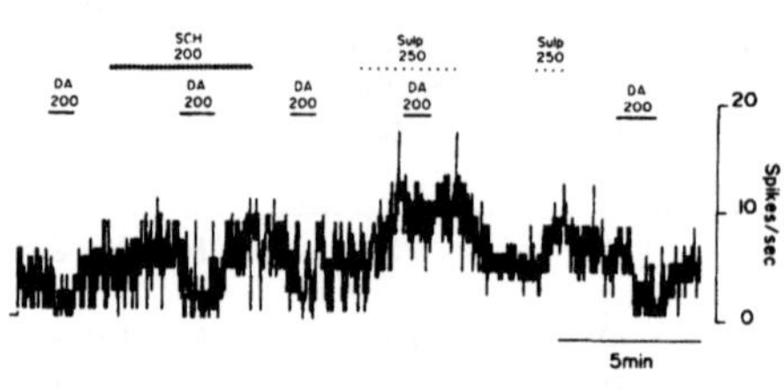

FIGURE 7

Spontaneously active neuron recorded 3 hours after EB ; DA induces an inhibition not antagonized by SCH 23390 but by sulpiride which has an excitatory action (intensity in nA).

These findings suggest that DA inhibition of firing observed in ovariectomized animals are mainly a D2 receptor-mediated action. Nevertheless, glutamate activation of neurons does not allow for correct DA receptor characterization since neither sulpiride nor SCH 23390 give a clear-cut effect on the DA inhibition.

Of the 29 neurons recorded 5 hours after EB injection 12 were spontaneously active (42%) and 17 silent, glutamate-driven. DA inhibited the firing of the majority of glutamate-driven neurons (15) but no clear-cut effects of antagonists could be observed. In the spontaneously active neurons only one neuron showed an inhibition by DA and apomorphine, antagonized by sulpiride, while 11 neurons were excited by DA ; 5 of these neurons were also inhibited by apomorphine (Fig. 8). The excitatory effects of DA was antagonized by SCH 23390 while sulpiride was without effect (Fig. 9) or excited the neuron (Fig. 10).

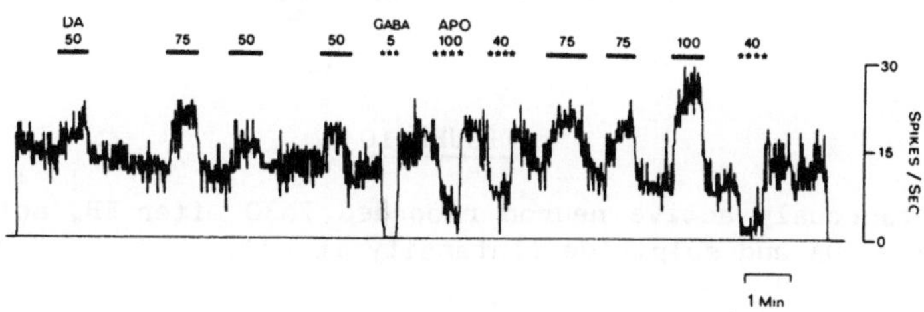

FIGURE 8

Spontaneously active neuron recorded 7 hours after EB, activated by DA and inhibited by apomorphine (APO) (Intensity in nA).

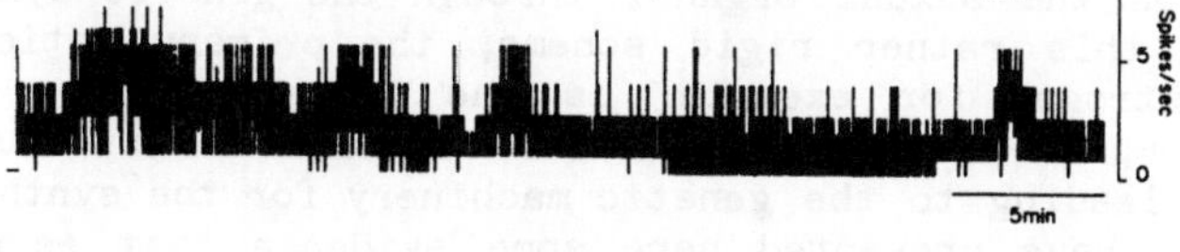

FIGURE 9

Spontaneously active neuron recorded 6 hours after EB. DA-induced activation is antagonized by SCH but not by sulpiride (Intensity in nA).

We conclude from this study that spontaneously firing caudate neurons usually inhibited by iontophoretic application of dopamine become excited by this amine 5 hours after an i.m. EB injection.

This DA induced excitation, never observed in the absence of estrogen treatment, is mediated by a D1 receptor while D2 activation results in an inhibition of firing either before or after EB injection.

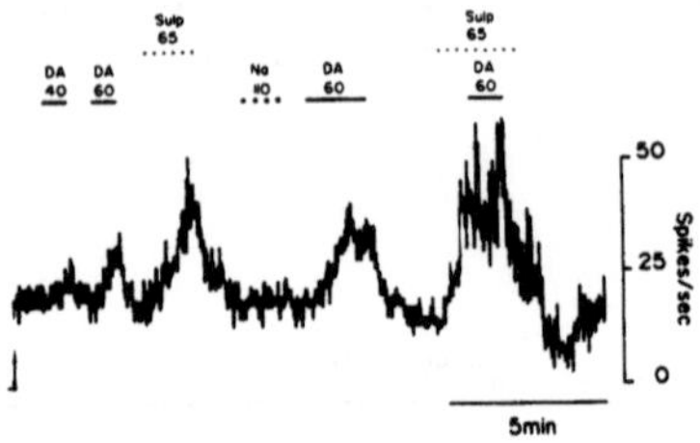

FIGURE 10

Spontaneously active neuron recorded 7h30 after EB, activated both by DA and sulpiride (Intensity in nA).

CONCLUSION

The functions of the brain depend largely upon the complex interactions of nervous and hormonal factors. Indeed, it is gradually emerging that the only way to account for brain mechanism which have a physiological relevance is to assume that they depend on the simultaneous action of multiple influences. Steroid hormones have long been considered as a separate entity even though it has been demonstrated that they act at the brain level. They have been thought to act exclusively, both on the brain and on the sexual organs, through the genetic synthesizing system. In this rather rigid scheme, the primary action of sex steroid, estrogen for exemple, is the formation of a cytosolic receptor complex which is translocated into the nucleus as the first step leading to the genetic machinery for the synthesis of a protein. We have presented here some evidence that estrogen may act on membrane excitability of anterior pituitary cells. This can provide a model for direct effects of estrogen on neuronal excitability. In addition we have also examined the possibility of interactions between estrogen and neurostransmitters at the level both of the anterior pituitary and of the caudate nucleus. These observations open a new field in the understanding of brain mechanisms concerning a regulatory role for steroid hormones in conventional neurotransmission.

ACKNOWLEDGMENTS

The authors thank D.A. Poulain and D.T. Theodosis for editorial assistance. This work was supported by grants from INSERM (U.176), CNRS (GRECO 85), Université de Bordeaux II and FRM.

REFERENCES

Arnauld, E., Dufy, B., Pestre, M. and Vincent, J.D. (1981). Effect of estrogens on the response of caudate neurons to microiontophoretically applied dopamine. Neurosci. Lett., 21, 325-331.

Bedard, P., Langelier, P., Villeneuve, A. (1977). Estrogens and extrapyramidal system. Lancet (ii), 1367-1368.

Biegon, A., McEwen, B.S. (1982). Modulation by estradiol of serotonine-1 receptors in brain. J. of Neurosci., 2, 199-205.

Bression, D., Brandi, A.M., Le Dafniet, M., Cesselin, F., Hamon, M., Martinet, M., Kerdelhue, B. and Peillon, F. (1983). Modifications of the high and low affinity pituitary domperidone-binding site in chronic estrogenized rats. Endocrinology, 113, 1799-1805.

Brunet, N., Gourdji, D., Moreau, M.F., Grouselle, D., Bournaud, F. and Tixier-Vidal, A. (1977). Effect of 17 β-estradiol on prolactin secretion and thyroliberin responsiveness in two rat prolactin continuous cell lines. Definition of an experimental model. Ann. Biol. anim. Biol. Biophys., 17, 413-424.

Brunet, N., Gourdji, D. and Tixier-Vidal, A. (1980). Effect of 17β -estradiol on thyroliberin responsiveness in GH3/B6 rat prolactin cells. Mol. Cell. Endocr., 18, 123-136.

Dufy, B., Partouche, C., Poulain, D.A., Dufy-Barbe, L. and Vincent, J.D. (1976). Effect of estrogens on the electrical activity of identified and unindentified hypothalamic units. Neuroendocrinology, 22, 38-47.

Dufy, B., Vincent, J.D., Fleury, H., Du Pasquier, P., Gourdji, D. and Tixier-Vidal, A. (1979). Membrane effects of thyrotropin-releasing hormone and estrogen shown by intracellular recording from pituitary cells. Science, 204, 509-511.

Dufy, B. and Vincent, J.D. (1980). Effects of sex steroids on cell membrane excitability : a new concept for the action of steroids on the brain. In "Hormones and the Brain" (D.E. Wied & Van Keep P.A., eds.) pp 29-42, MTP Press, Lancaster.

Gershengorn, M.C., Marcus-Samuels, B.E. and Geras, E. (1979). Estrogens increase the number of thyrotropin-releasing hormone receptors on mammotropic cells in culture. Endocrinology, 105, 171-176.

Heiman, M.L., Ben-Jonathan, N. (1982). Rat anterior pituitary dopaminergic receptors are regulated by estradiol and during lactation. Endocrinology, 111, 1057-1060.

Hopkins, C.R. and Farquhar, M.G. (1973). Hormone secretion by cells dissociated from rat anterior pituitaries. J. Cell. Biol., 59, 276-303.

Israel, J.M., Denef, C. and Vincent, J.D. (1983). Electrophysiological properties of normal somatotrophs in culture : an intracellular study. Neuroendocrinology, 37, 193-199.

König, J.F.R. and Klippel, R.A. (1963). The rat brain : a stereotaxic atlas. Williams and Wilkins, Baltimore.

Moss, R.L. and Dudley, C.A. (1984). Molecular aspects of the interaction between estrogen and the membrane excitability of hypothalamic nerve cells. Prog. Brain Res., 61, 3-19.

Nabekura, J., Oomura, Y., Minami, T., Mizuno, Y. and Fukuda, A. (1986). Mechanisms of the rapid effect of 17 β-estradiol on medial amygdala neurons. Science, 223, 226-227.

Nausieda, P.A., Koller, W.C., Weiner, W.J. and Klawans, H.L. (1979). Chorea induced by oral contraceptives. Neurology, 29, 1605-1609.

Tixier-Vidal, A., Gourdji, D. and Tougard, C. (1975). A cell culture approach to the study of anterior pituitary cells. Int. Rev. Cytol., 41, 173-239.

Van Hartesveldt, C. and Joyce, J.N. (1986). Effect of estrogens on basal ganglia. Neurosci. Biobehav. Rev., 10, 1-14.

Wagner, H.R., Crutcher, K.A. and Davis, J.N. (1979). Chronic estrogen treatment decreases β-adrenergic response in rat cerebral cortex. Brain Res., 171, 147-151.

Session II
Interactions between Receptors in the Membrane

Part A

Possible Interactions between Molecular and Supramolecular Entities in the Membrane

Chairman: J. Axelrod

12
Receptor Modification and Receptor Disposition in Membranes

Deirdre Cooney, Andreas Holzhöfer, Fritz Boege, Christian Dees, Rolf Jürss, Alan K. Keenan, Mirko Hekman and Ernst J.M. Helmreich

ABSTRACT

This report deals with the functional significance of covalent modifications of the β_1-adrenoceptor from turkey erythrocytes. Two types of modification are being considered: phosphorylation and glycosylation. The effects of these modifications were tested using a reconstitution assay with pure receptor and GTP-binding proteins incorporated into lipid vesicles.

INTRODUCTION

Recently, much attention has been directed to the role of covalent modifications of receptors during desensitization. In principle, covalent modification could provide a signal either for termination of hormonal activation or initiation of sequestration. The first possibility can be readily studied in a reconstituted system, because it should result in functional uncoupling of receptor and G-proteins. Uncoupling of signal transmission is the main event in heterologous desensitization as it occurs in turkey erythrocytes. In contrast to heterologous desensitization of β_1-adrenergic signal transmission, homologous desensitization involves subsequent removal of receptors from its original location in the plasma membrane at the cell surface. This implies that in homologous desensitization the ß-adrenoceptor must be mobilized. Therefore, it has been speculated that covalent modification might be instrumental in removing constraints due to association of the receptor with a cytoskeletal element as a prerequisite for mobilization and removal of the receptor from the membrane (Helmreich and Elson, 1984). Therefore, we shall address problems related to quantitative measurements of mobility of hormone receptors which couple to G-proteins and control adenylate cyclase activity in intact cells.

So far as covalent modifications are concerned, the main interest has been focussed on receptor phosphorylation. The discovery of a specific ß-receptor kinase in S49 kin-mutant cells by Lefkowitz and colleagues (Benovic et al., 1986) which phosphorylates the activated agonist-occupied receptor has highlighted this development. Unlike phosphorylation, other covalent modifications such as glycosylation of ß-adrenoceptors have received much less interest. Therefore, the Würzburg laboratory has made efforts directed towards an understanding of the role of glycosylation of ß-adrenoceptors. The progress made up to now is reported.

RESULTS AND DISCUSSION

Phosphorylation of Turkey Erythrocyte $ß_1$-Adrenoceptors

Isoprenaline-induced desensitization of turkey erythrocyte adenylate cyclase is associated with a decrease in the mobility of photolabeled $ß_1$-adrenoceptor on SDS-polyacrylamide gels and with an increase in phosphate incorporation into the receptor (Stadel et al., 1983). It has been proposed that the phosphorylated, desensitized receptor can no longer couple effectively with the G-proteins of the signal transmission chain, since purified desensitized turkey $ß_1$-receptors showed reduced ability to couple with the ß-receptor deficient adenylate cyclase system of <u>Xenopus laevis</u> erythrocytes (Strulovici et al., 1984). It was our objective to compare the functional state of purified desensitized receptor with receptor purified from untreated cells by testing its ability to couple with pure G_s in a defined reconstituted milieu. The behavior of the purified reconstituted system was also compared with that of membranes prepared from desensitized and control cells. In the reconstituted system any reduction in coupling can be attributed unequivocally to changes in the receptor, while changes in the coupling efficiency of the components of the signal transmission chain in intact membranes might be due to changes in coupling partners other than the receptor, for example the G-proteins.

Turkey erythrocyte β_1-receptors were desensitized by treatment of intact cells with 10^{-5} M isoprenaline for 3 h at 37°C. Membranes were prepared from desensitized cells and receptors purified according to previously published procedures (Hekman et al., 1984). Photoaffinity labeling of purified desensitized receptor indicated an apparent molecular weight shift on SDS gels in accordance with Stadel et al. (1983). The purified receptors were reconstituted with pure turkey erythrocyte G_s in phospholipid vesicles as described by Feder et al. (1986). Hormone- and time-dependent G_s-activation was induced by addition of isoprenaline and GTPγS. The adenylate cyclase and GTPase activities were followed in native membranes and in reconstituted $R \cdot G_s$ vesicles. Both hormone-dependent adenylate cyclase and GTPase activities were significantly reduced in native membranes prepared from desensitized cells (Fig. 1 A and B). However, the receptors purified from desensitized cells and from control (non-desensitized) cells behaved alike with respect to coupling efficiency to G_s (Fig. 2A). Moreover, GTPase activity elicited by purified receptor in reconstituted lipid vesicles was not altered by desensitization (Fig. 2B).

Altered mobility of the photoaffinity-labeled β_1-receptor from desensitized erythrocytes on SDS gels indicative of receptor phosphorylation agrees with the data of Stadel et al. (1983). In addition, we have observed two receptor bands in the 40-45 kDa Mr range

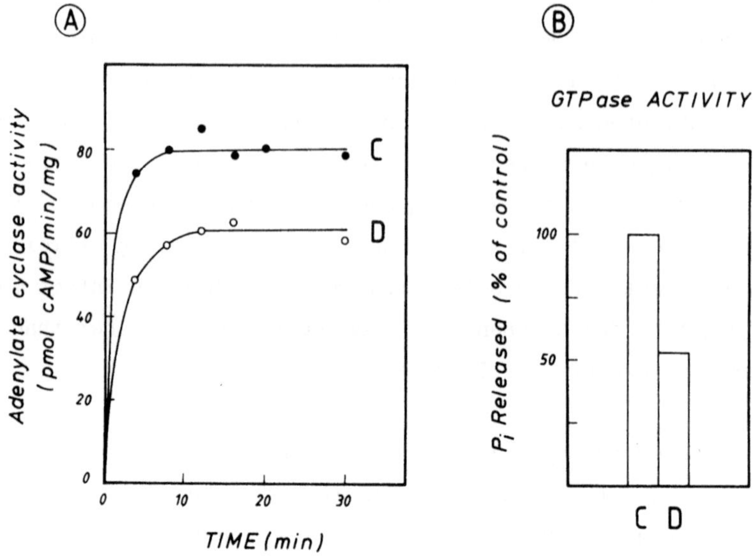

Fig. 1. Comparison of $R \cdot G_s$ coupling efficiency in turkey erythrocyte membranes prepared from control (C) and desensitized (D) cells. (A): Turkey erythrocyte membranes (1 mg/ml) were activated with 60 nM GTPγS and 10 µM l-isoproterenol at 30°C. At indicated times, 50 µl aliquots were withdrawn and quenched with 200 µM d,l-propranolol. After activation the cyclase activity was assayed as described in Hekman et al. (1984). (B): GTPase activity in control (C) and desensitized (D) membranes (200 µg/ml) was measured in the presence of 0.25 µM [γ-^{32}P]GTP, 500 µM AppNHp, 2 mM mercaptoethanol, 6 mM $MgCl_2$ and 10 µM l-isoproterenol at 30°C for 20 min according to Cassel and Selinger (1976).

by the silver staining method. This also lends support to the inference that the β_1-adrenoceptor polypeptide has become covalently modified by phosphorylation as a consequence of desensitization. However, contrary to the conclusions of Strulovici et al. (1984), we failed to detect any significant impairment in R•G coupling efficacy in reconstituted lipid vesicles. When the coupling between R and G_s was

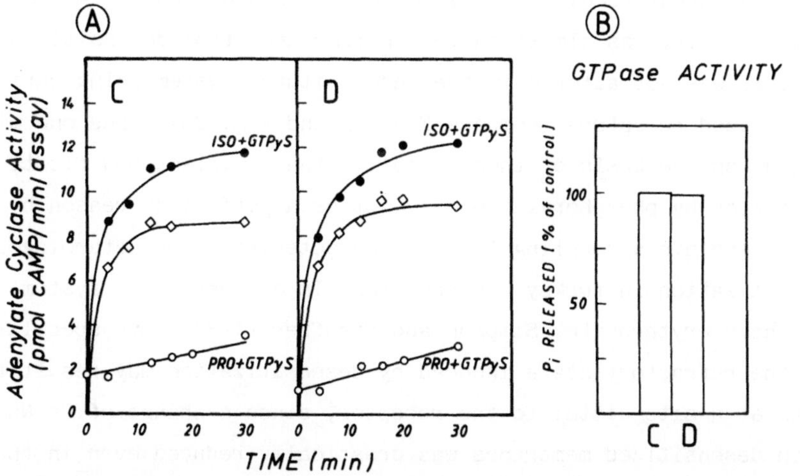

Fig. 2. Comparison of ß-receptor-G_s coupling using purified receptor prepared from control (C) and desensitized (D) turkey erythrocyte membranes. Vesicles containing purified ß-receptor and G_s-protein were prepared in the presence of 0.24% lauroyl sucrose by the sephadex G-50 method using a lipid mixture containing PE/PS/cholesterol hemisuccinate at a ratio of 12:8:5. The molar ratios of G_s to ß-receptor after incorporation into the vesicles were 2.92 and 3.14 for control (C) and desensitized (D) ß-receptor, respectively. Hormone- and time-dependent activation of G_s (A) was carried out at 30°C with 60 nM GTPγS and 10 μM l-isoproterenol (•) or 60 nM GTPγS and 10 μM d,l-propranolol (o). The net specific activation is also shown (◇). Hormone-induced GTPase activity (B) was measured using the same R•G_s vesicles as in (A) in the presence of 0.25 μM [γ-^{32}P]GTP, 100 μM AppNHp, 100 μM ascorbic acid, 0.9 mM $MgCl_2$ and 10 μM l-isoproterenol. The vesicles were incubated at 30°C for 30 min.

measured in turkey erythrocyte membranes prepared from desensitized cells, the expected reduction (about 25-30% compared to controls) of hormone-dependent cyclase activation (in the presence of GTPγS) was clearly apparent (see: Fig. 1A). Moreover, whereas hormone-dependent GTPase activity was almost halved in membranes from desensitized cells, it was not altered in the reconstituted system using purified desensitized receptors (compare Fig. 1B and Fig. 2B). Thus one might conclude on the basis of these results, that covalent modification of receptor by phosphorylation may not be a sufficient reason by itself to explain uncoupling that is characteristic for heterologous desensitization in turkey erythrocytes. In another avian system, the pigeon erythrocyte, Simpson and Pfeuffer (1980) have speculated that the refractory state induced by desensitization may result from events at a site distal to the receptor, since exchange of GppNHp by GTP in desensitized membranes was drastically reduced even in the absence of hormone. It appears quite likely therefore that phosphorylation at the level of the transducer via an as yet unidentified protein kinase alone or in conjunction with receptor modification is responsible for $R{\cdot}G_s$ uncoupling during desensitization. An involvement of multiple protein kinases in control of hormonally activated adenylate cyclase has already been considered, however, only for phosphorylation of turkey erythrocyte $ß_1$-receptor during desensitization (Nambi et al., 1985).

Glycosylation of Turkey Erythrocyte $ß_1$-Adrenoceptors

It has been known for some time that in turkey erythrocytes there are two $ß_1$-adrenoceptor forms, one with an apparent Mr weight of 50 kDa and one with a Mr weight of 40 kDa (cf. e.g. Burgermeister et al., 1982). But the structural and functional consequences of the receptor microheterogeneity in turkey erythrocytes have not been clarified. We have recently described a specific protease in turkey erythrocytes which converts the larger, 50 kDa form of the $ß_1$-adrenoceptor (P50) to the smaller 40 kDa form (P40) which retains the specific ligand binding site (Jürss et al., 1985). The specific receptor protease can be inhibited by glutathione, dithiothreitol and ophenanthroline, but not by EDTA. This proteolytic conversion is attenuated in intact erythrocytes, presumably because of inhibition of the protease by endogenous glutathione. This and the occurence of both forms in the intact native cell assures us that processing of the β_1-adrenoceptor is not an artefact of preparation of ghosts or membranes. Recently we have obtained evidence to indicate that the P40 receptor has lost in the course of proteolytic conversion most, if not all, N-glycosidically linked sugars (Jürss et al., 1985; Boege et al., 1986). Moreover, we have been able to separate affinity-purified P50 and P40 receptors by chromatography on wheat germ agglutinin. This made it possible to compare the purified P50 and P40 β_1-adrenoceptors from turkey erythrocytes with respect to their relative capabilities to couple in lipid vesicles with pure G_s-proteins prepared from turkey erythrocytes or rabbit liver. It is interesting to note that the N-deglycosylated P40 receptor couples to pure G_s-proteins in lipid vesicles to a larger extent and with a faster rate in response to activation with l-isoproterenol and GTPγS (Fig. 3). For example, the k_{on} values for the pseudo first order rate of activation of G_s by P40 were $0.47 \pm 0.045\ min^{-1}$ in the case of liver G_s and $0.22 \pm 0.02\ min^{-1}$ in the case of turkey erythrocyte G_s, whereas the corresponding values for P50 were $0.34 \pm 0.006\ min^{-1}$ and $0.14 \pm 0.02\ min^{-1}$, respectively.

We do not know whether these functional differences between glycosylated and N-deglycosylated β_1-adrenoceptor forms are biologically significant. One possibility one might consider is that the disposition of the glycosylated receptor in the lipid bilayer on vesiculization might be more restricted and strained than in the case of the deglycosylated receptor. Glycosylation might effect the position of the receptor inserted in the lipid vesicle, perhaps making it less accessible for its coupling partner. Another possibility is a conformational change in the receptor polypeptide due to glycosylation which would make the glycosylated P50 form less active for interaction with and activation of G_s.

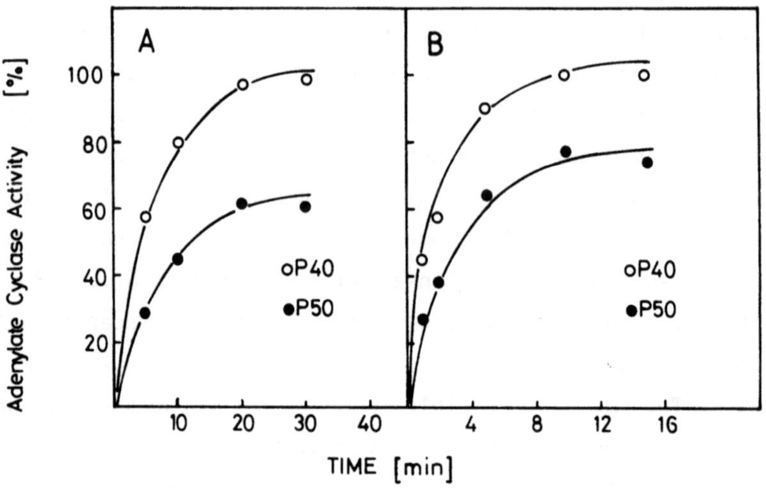

Fig. 3. Coupling efficiency of purified 50 kDa ß-receptor (•) and 40 kDa ß-receptor (o) with G_s-proteins purified from turkey erythrocytes (A) and rabbit liver (B). This figure is reproduced from Boege et al. (1986).

Receptor Mobility

Until recently, no information was available on the mobility of a hormone receptor which activates adenylate cyclase via G-proteins in general, and of ß-adrenoceptors in particular. We thought it worthwhile therefore to initiate studies of the lateral mobility of the ß-adrenoceptor. The ß-adrenergic system offers unique advantages for this kind of study because of the different biological consequences depending on whether the same site on the receptor is occupied by an agonist or an antagonist. In contrast to an agonist-occupied ß-receptor, receptors occupied by antagonists are incapable of functional coupling with G-proteins, activation and desensitization. We were able to show that the majority of the ß-adrenoceptors when occupied by an antagonist were immobilized in the membranes of cultured liver cells over a temperature range of 4°C to 37°C. A patched appearance and a nonrandom distribution of the specifically attached fluorescence on the surface of the cells were apparent in video-amplified fluorescence microscopy pictures (see in: Henis et al., 1982). Thus the immobility of the ß-receptors correlated with their inhomogenous distribution. This finding allowed us to conclude that the ß-receptors on these cultured liver cells are prepatched. Preincubation with isoproterenol had two effects: (i) It decreased the patchy appearance of the bound antagonist fluorescence, and (ii) it gradually and relatively slowly mobilized the ß-receptor suggesting that dispersal of receptors may be related to desensitization. This raises questions about the mechanism of adenylate cyclase activation via ß-receptors. In view of the lateral immobility of the major portion of the ß-receptors, several possibilities can be considered: (a) The coupling partners, the G-proteins, may be freely mobile in the membrane and thus can meet and be activated by the agonist-occupied ß-receptor and can shuttle between receptor and adenylate cyclase. (b) The G-protein and the catalytic units may be sequestered in the same local regions on the cell surface as the ß-receptors. Then lateral diffusion over micrometer distances measured by fluorescence recovery after photobleaching is not required, and lateral or rotational diffusion over submicrometer distances of free or agonist-occupied receptors may suffice for activation. (c) We cannot rule out the possibility that a small fraction of the receptors undetectable in our measurements is laterally mobile and that this fraction is responsible for adenylate cyclase activation. Accordingly, the following questions should be considered: (a) What are the mobility and distribution properties of the G-proteins and the catalytic units? (b) What are the forces and structures which cause clustering and immobilization of the ß-receptors in the absence of agonist?

Perhaps, the most serious shortcoming of our previous work was the use of a fluorescent antagonist, NBD-alprenolol, rather than a fluorescent agonist. Since the use of a fluorescent ß-adrenergic agonist is precluded mainly because of the too low affinity of available ß-agonists, we have synthesized several fluorescent glucagon derivatives. Glucagon is a polypeptide hormone which, like ß-adrenoceptors, stimulates adenylate cyclase via G-proteins (Iyengar and Birnbaumer, 1982). Moreover, the availability of a suitable fluorescent glucagon derivative would make it possible to study lateral mobility of hormone receptors on intact primary hepatocytes with the microphotolysis technique, a further attraction in the use of such a derivative. We have therefore synthesized fluorescent glucagon derivatives which appear to be suitable probes for the glucagon receptor in intact cells utilizing the previously reported modification of Trp-25 to 2-thiol-tryptophan (Wright and Rodbell, 1980) (Fig. 4). These derivatives retain the ability to maximally stimulate adenylate cyclase and moreover possess sufficiently high binding affinities. Among the fluorescent glucagon derivatives tested, NBD-glucagon (4-nitrobenzo-

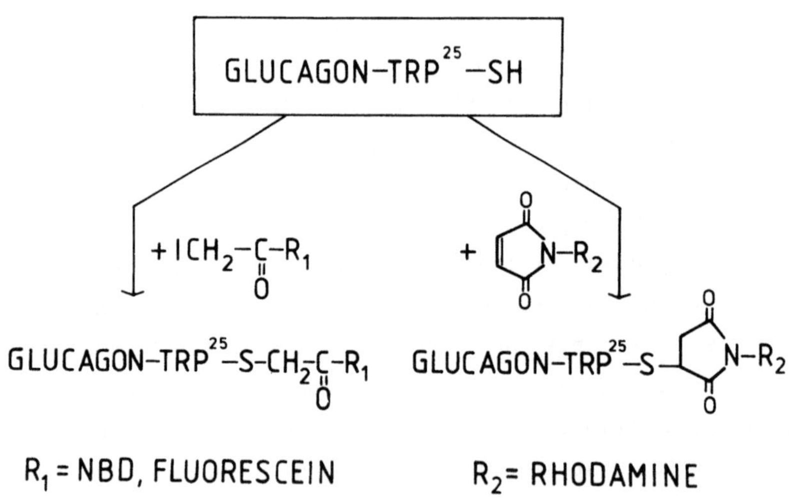

Fig. 4. Fluorescent glucagon derivatives. Schematic representation of the chemical reaction scheme used for labeling glucagon with fluorescent groups. Glucagon was modified to 2-thiol-tryptophan-glucagon by the method of Wright and Rodbell (1980) before reaction with the appropriate sulfhydryl specific reagent.

2-oxa-1,3-diazolyl-glucagon) bound with only a slightly reduced affinity as compared to native glucagon whereas the fluorescein and rhodamine glucagon derivatives had a 9 fold reduced affinity which still corresponds to K_D values of 10 nM. However, the higher quantum yield and relative solvent insensitivity of rhodamine and fluorescein glucagons as compared to NBD-glucagon may make the former derivatives more suitable. The photobleaching recovery and video intensification fluorescence microscopy experiments using these glucagon derivatives are presently being carried out. These experiments will hopefully provide insights into the role of location and mobility of hormone receptors on the cell surface in adenylate cyclase activation, receptor desensitization and sequestration.

ACKNOWLEDGMENTS

The fluorescent glucagon derivatives were synthesized by Drs. Mie-Jae Im and Helmut W. Klein. Miss Gaby Pollak participated in this work in partial fulfillment of the requirements for a Ph.D. degree of the Faculty for Chemistry and Pharmacy of the University of Würzburg, F.R.G. The glucagon derivatives were functionally tested on primary hepatocytes by Miss Gaby Pollak and Miss Barbara Freeman. The preliminary photobleaching experiments were carried out in the laboratory of Dr. Reiner Peters at the Max Planck Institute of Biophysics in Frankfurt by Dr. Helmuth Heithier and Dr. Larry Ward. L.W. wishes to acknowledge an award of a A. v. Humboldt fellowship. A report of this work is being prepared.

REFERENCES

Benovic, J.L., Strasser, R.H., Caron, M.G. and Lefkowitz, R.J. (1986). ß-Adrenergic receptor kinase: Identification of a novel protein kinase that phosphorylates the agonist-occupied form of the receptor. Proc. Natl. Acad. Sci. USA 83, 2797-2801.

Boege, F., Jürss, R., Cooney, D., Hekman, M., Keenan, A.K. and Helmreich, E.J.M. (1986). Functional and structural characterization of the two $ß_1$-adrenoceptor forms in turkey erythrocytes with Mr weights of 50 kDa and 40 kDa. Biochemistry, in press.

Burgermeister, W., Hekman, M. and Helmreich, E.J.M. (1982). Photoaffinity labeling of the ß-adrenergic receptor with azide derivatives of iodocyanopindolol. J. Biol. Chem. 257, 5306-5311.

Cassel, D. and Selinger, Z. (1976). Catecholamine-stimulated GTPase activity in turkey erythrocyte membranes. Biochim. Biophys. Acta 452, 538-551.

Feder, D., Im, M.J., Klein, H.W., Hekman, M., Holzhöfer, A., Dees, C., Levitzki, A., Helmreich, E.J.M. and Pfeuffer, T. (1986). Reconstitution of $ß_1$-adrenoceptor-dependent adenylate cyclase from purified components. EMBO J. 5, 1509-1514.

Hekman, M., Feder, D., Keenan, A.K., Gal, A., Klein, H.W., Pfeuffer, T., Levitzki, A. and Helmreich, E.J.M. (1984). Reconstitution of ß-adrenergic receptor with components of adenylate cyclase. EMBO J. 3, 3339-3345.

Helmreich, E.J.M. and Elson, E.L. (1984). Mobility of proteins and lipids in membranes. Adv. Cyclic Nucleotide and Protein Phosphorylation Res. 18, 1-62.

Henis, Y.I., Hekman, M., Elson, E.L. and Helmreich, E.J.M. (1982). Lateral motion of ß-receptors in membranes of cultured liver cells. Proc. Natl. Acad. Sci. USA 79, 2907-2911.

Iyengar, R. and Birnbaumer, L. (1982). Hormone receptor modulates the regulatory component of adenylyl cyclase by reducing its requirement for Mg^{2+} and enhancing its extent of activation by guanine nucleotides. Proc. Natl. Acad. Sci. USA 79, 5179-5183.

Jürss, R., Hekman, M. and Helmreich, E.J.M. (1985). Proteolysis-associated deglycosylation of $ß_1$-adrenergic receptor in turkey erythrocytes and membranes. Biochemistry 24, 3349-3354.

Nambi, P., Peters, J.R., Sibley, D.R. and Lefkowitz, R.J. (1985). Desensitization of the turkey erythrocyte ß-adrenergic receptor in a cell-free system. J. Biol. Chem. 260, 2165-2171.

Simpson, I.A. and Pfeuffer, T. (1980). Functional desensitization of ß-adrenergic receptors of avian erythrocytes by catecholamines and adenosine 3',5'-phosphate. Eur. J. Biochem. 111, 111-116.

Stadel, J.M., Nambi, P., Shorr, R.G.L., Sawyer, D.F., Caron, M.G. and Lefkowitz, R.J. (1983). Catecholamine-induced desensitization of turkey erythrocyte adenylate cyclase is associated with phosphorylation of the ß-adrenergic receptor. Proc. Natl. Acad. Sci. USA 80, 3173-3177.

Strulovici, B., Cerione, R.A., Kilpatrick, B.F., Caron, M.G. and Lefkowitz, R.J. (1984). Direct demonstration of impaired functionality of a purified desensitized ß-adrenergic receptor in a reconstituted system. Science 225, 837-840.

Wright, D.E. and Rodbell, M. (1980). Preparation of 2-thioltryptophan-glucagon and $(tryptophan\text{-}S\text{-}glucagon)_2$. J. Biol. Chem. 255, 10884-10887.

The work of this laboratory is supported by SFB 176 Würzburg, Projects A1 and A4, and by grants from the Fritz Thyssen Foundation and the Fonds der Chemischen Industrie e.V.

13
Membrane Phospholipid Metabolism and Transmitters

M.J. Berridge

INTRODUCTION

Many neurotransmitters are now known to act by using a ubiquitous signal transduction mechanism based on the hydrolysis of a unique lipid (Downes, 1983; Nahorski et al, 1986). The lipid in question is phosphatidylinositol 4,5-bisphosphate ($PtdIns4,5P_2$) located within the inner leaflet of the plasma membrane where it is stored as a precursor to be used by the receptor mechanism to generate second messengers (Downes & Michell, 1985; Berridge, 1984; Berridge & Irvine, 1984). The transmitters known to operate through this inositol lipid mechanism include acetylcholine (muscarinic), norepinephrine (α_1), histamine (H_1), 5-hydroxytryptamine ($5\text{-}HT_2$), vasopressin (V_1), substance P, bradykinin, neurotensin and glutamate. Upon binding its appropriate transmitter, the receptor operates through a GTP-binding protein (Gp) which activates a phosphoinositidase to cleave $PtdIns4,5P_2$ at its phosphodiester bond to release inositol 1,4,5-trisphosphate ($Ins1,4,5P_3$) which diffuses into the cytosol leaving diacylglycerol (DG) behind in the membrane. This hydrolysis of $PtdIns4,5P_2$ is a key event in signal transduction because both products give rise to second messengers and thus represents a bifurcation in the signal pathway. The DG functions by stimulating protein kinase C (C-kinase) (Nishizuka, 1986) whereas $Ins1,4,5P_3$ mobilizes calcium (Berridge, 1984; Berridge & Irvine, 1984). In this article I shall attempt to describe how the dual signalling system based on the DG/C-kinase and $Ins1,4,5P_3$/calcium pathways might participate in the control of neural activity. Particular attention will be paid to the role of the DG/C-kinase pathway in carrying our receptor-receptor interactions. Such transmodulation of receptor activity represents one of the major functions of protein kinase C.

FORMATION AND METABOLISM OF SECOND MESSENGERS

A unique feature of phosphatidylinositol (PtdIns), which sets it apart from all the other membrane phospholipids, is that it can be further phosphorylated to form the polyphosphoinositides. Addition of a phosphate at the 4-position of the inositol headgroup produces phosphatidylinositol 4-phosphate which is then phosphorylated to the $PtdIns4,5P_2$ used by the receptor. As for other signal transduction mechanisms, a GTP-binding protein called Gp plays a role in linking receptors to the phosphoinositidase (Taylor & Merritt, 1986). Evidence that Gp functions in signal transduction is based on the observation that the breakdown of $PtdIns4,5_2$ to form $Ins1,4,5P_3$ and DG can be stimulated by the addition of non-hydrolyzable analogues of GTP to either permeabilized cells or isolated membranes. The identity of Gp is unknown but there are intriguing suggestions that it may be encoded for by the <u>ras</u> oncogene (Fleischman et al., 1986; Wakelam et al., 1986).

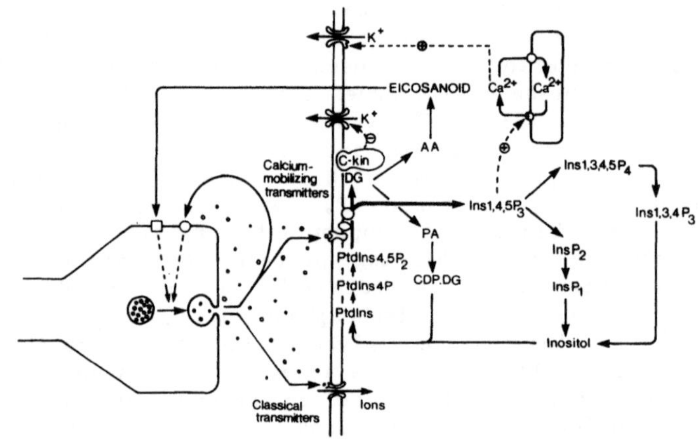

Fig.1 Proposed role of inositol lipid hydrolysis in neurotransmission. The model depicts what might happen when a classical transmitter, which opens ion channels, is released together with a calcium-mobilizing transmitter. The latter stimulates the hydrolysis of $PtdIns4,5P_2$ to form $Ins1,4,5P_3$ and DG. The neutral DG can either be phosphorylated to phosphatidic acid (PA) or it can be hydrolyzed to release arachidonic acid (AA). Synaptic transmission may be modulated either through autoreceptors, sensitive to the released transmitters, or through the eicosanoids originating from the post-synaptic receptors. See text for further details.

The subsequent metabolism of the two second messengers is complicated because each can be inactivated by two separate pathways (Fig.1). The neutral DG which remains within the plasma membrane can either be phosphorylated to phosphatidic acid by a DG-kinase or it can be hydrolyzed by a DG lipase. In the former mechanism, DG will be recycled back to PtdIns and could be reused for signalling. In the latter mechanism, DG is used as a precursor of the eicosanoids (prostaglandins, thromboxanes and leukotrienes) because the DG lipase forms monoacylglycerol which is further hydrolyzed to release arachidonic acid. The bifurcating signal pathway may also contribute to arachidonic acid release by stimulating phospholipase A_2 activity which is controlled by lipocortin. There is some evidence that the latter is phosphorylated and hence inactivated by the DG/C-kinase limb of the bifurcating signal pathway (Touqui et al., 1986). In blood platelets, about 15% of the arachidonic acid comes from the DG lipase pathway whereas the remainder is derived from other phospholipids through the actions of phospholipase A_2 (Mahadevappa & Holub, 1986). Since eicosanoids function as local hormones and are produced in large quantities during stimulation of neural tissue (Wolfe, 1982) they offer a unique mechanism for receptor-receptor interactions especially for the transfer of information between pre- and post-synaptic sites (Fig.1).

The model outlined in figure 1 illustrates a possible role for eicosanoids which could also throw some light on why transmitters are frequently colocalized and released together to activate separate receptor mechanisms. Release of a classical transmitter may be responsible for rapid transmission by directly opening ion channels whereas the co-release of a peptide, like substance P, may act through the inositol lipid mechanism to modulate neurotransmission as will be described in more detail later. Part of this modulation may be due to the release of eicosanoids which feedback to influence pre-synaptic events. Electrical stimulation of preganglionic nerve fibres leads to an increased release of prostaglandins in guinea-pig superior cervical ganglion (Trevisani et al., 1982). Since the effect was blocked by decreasing the Ca^{2+}/Mg^{2+} ratio or by adding atropine, it was argued that prostaglandins were being produced in response to the release of acetylcholine. There also is evidence that such endogenously produced prostaglandins play a physiological role by feeding back to reduce acetylcholine release by nerve stimulation (Inoue et al., 1984). By using this local hormonal system, the post-synaptic membrane is able to 'inform' the presynaptic cell that it has received a meaningful signal and thus provides a mechanism for coupling pre-and post-synaptic events (Fig.1).

The Ins1,4,5P_3 which is released to the cytosol can also be metabolized through two separate pathways. In one pathway it is dephosphorylated sequentially to free inositol (Storey et al., 1984) which can be re-used to synthesize PtdIns. The final dephosphorylation step is inhibited by lithium (Hallcher & Sherman, 1980) which could explain the therapeutic action of this ion in controlling manic-depressive illness (Berridge, 1986). The alternative pathway begins with a soluble kinase which phosphorylates Ins1,4,5P_3 to form Ins1,3,4,5P_4 (Irvine et al., 1986; Hansen et al., 1986; Stewart, et al., 1986) which was recently recognized as the precursor of Ins1,3,4P_3 (Batty et al., 1985).

Stimulation of cerebral cortex resulted in a rapid increase in the formation of both Ins1,4,5P_3 and Ins1,3,4,5P_4 (Batty et al., 1985). The latter is dephosphorylated to Ins1,3,4P_3 which characteristically increases in cells very much more slowly than Ins1,4,5P_3 (Irvine et al., 1985: Burgess et al., 1985: Hansen et al., 1986). Ins1,3,4P_3 is then dephosphorylated back to free inositol through a pathway yet to be described.

DUAL SIGNAL HYPOTHESIS

The receptor-mediated hydrolysis of PtdIns4,5P_2 yields two second messengers which control two separate but interacting signal pathways. The water-soluble Ins1,4,5P_3 released to the cytoplasm functions as a second messenger to release calcium from the endoplasmic reticulum (ER) (Fig.2) (Berridge & Irvine, 1984). The function of the other inositol polyphosphates is uncertain but it has been suggested that both Ins1,3,4P_3 (Irvine et al., 1984) and Ins1,3,45P_4 (Batty et al., 1985) might have second messenger functions. One suggestion is that Ins1,3,4,5P_4 might function to control calcium entry across the plasma membrane (Hansen et al., 1986). The neutral DG, which remains within the plane of the plasma membrane, activates protein kinase C which phosphorylates a specific set of cellular proteins (Nishizuka, 1984, 1986). Phorbol esters mimic the action of DG by activating protein kinase C and have provided an effective mechanism of stimulating this limb of the signal pathway independently of the Ins1,4,5P_3/calcium limb. Conversely, the latter can be activated by calcium ionophores so bypassing the action of Ins1,4,5P_3. By combining an ionophore to raise calcium and a phorbol ester to activate protein kinase C, Kaibuchi et al., (1983) were able to mimic the stimulatory effect of thrombin on blood platelets. The action of many other hormones and neurotransmitters can also be reproduced in this way thus emphasizing the dual role of the two signal pathways in controlling a whole range of cellular responses.

As we learn more about this bifurcating signal pathway it becomes apparent that the two limbs have qualitatively different roles to play in the overall cellular response to external agonists. By mobilizing intracellular calcium, Ins1,4,5P_3 plays a primary role in initiating a whole range of calcium-dependent responses. The importance of the Ins1,4,5P_3/calcium limb of the signal pathway is evident from the ability of Ins1,4,5P_3 to induce a whole range of physiological responses when added to permeabilized cells. It can stimulate contraction in skinned smooth muscle (Somlyo et al., 1985) or skeletal muscle (Vergara et al., 1985; Volpe et al., 1985). In slime mold it can stimulate both cyclic GMP formation (Europe-Finner & Newell, 1985) and actin polymerization (Europe-Finner & Newell, 1986). Addition of Ins1,4,5P_3 to permeabilized blood platelets initiates the whole panoply of responses normally found in intact cells including protein phosphorylation (Lapetina et al., 1984; Israels et al., 1985; Watson et al., 1986), shape change (Israels et al., 1985; Authi et al., 1986), aggregation (Israels et al., 1985;

Authi et al., 1986; Watson et al., 1986), secretion of 5-HT (Brass & Joseph, 1985; Israels et al., 1985; Watson et al., 1986) and the formation of thromboxane A_2 (Israels et al., 1985; Authi et al., 1986; Watson et al., 1986). Injection of $Ins1,4,5P_3$ into intact cells is also able to trigger a complex array of physiological responses which include phototransduction and adaptation in *Limulus* photoreceptors (Brown et al., 1984; Fein et al., 1984), membrane depolarization and calcium mobilization in *Xenopus* oocytes (Oron et al., 1985), early events of fertilization including membrane depolarization and the cortical reaction of sea urchin eggs (Whitaker & Irvine, 1984; Slack et al., 1986; Turner et al., 1986) and *Xenopus* eggs (Busa et al., 1986), membrane hyperpolarization due to increased potassium currents in NG108-15 cells (Higashida et al., 1986), smooth muscle (Klockner & Isenberg, 1985) and lacrimal gland (Evans & Marty, 1986). All these effects observed in permeabilized or intact cells can be ascribed to the ability of $Ins1,4,5P_3$ to mobilize calcium.

As we learn more about the DG/C-kinase pathway it becomes apparent that it has two main functions. Firstly, it works together with the $Ins1,4,5P_3$/calcium pathway to bring about the final response often in a synergistic manner as described earlier. Secondly, it functions as a modulator of either the calcium signalling pathway (homologous interactions) or of other signal pathways (heterologous interactions).

HOMOLOGOUS INTERACTIONS

A number of interactions have been identified within the inositol lipid pathway all centred around protein kinase C (Fig. 2). Stimulation of protein kinase C by DG requires calcium and an increase in the intracellular level of calcium may contribute to the activation of the enzyme perhaps by increasing its binding affinity for DG (Dougherty & Niedel, 1986). Brocklehurst et al. (1985) have suggested that protein kinase C in the adrenal medulla may be activated by calcium, entering via voltage-dependent calcium channels, independently of DG. In most cells, however, where both limbs of the signal pathway are activated, DG and calcium probably act synergistically with each other to stimulate the enzyme. Most of the other homologous interactions concern the ability of the DG/C-kinase pathway to modulate various aspects of the signalling system in particular to reduce intracellular calcium (Fig. 2). Pretreatment of cells with a phorbol ester reduces subsequent agonist-induced increases in intracellular calcium in rat basophilic leukaemia cells (Sagi-Eisenberg et al., 1984), blood platelets (Drummond, 1985), liver (Cooper et al., 1985), Swiss 3T3 cells (Hesketh et al., 1985), neutrophils (Naccache et al., 1985), vascular smooth muscle (Brock et al., 1985), adrenal glomerulosa cells (Kojima et al., 1986), WRK-1 cells (Monaco & Mufson, 1986) and human A431 cells (Moolenaar et al., 1986). This reduction in calcium signalling is achieved through a negative feedback effect whereby the DG/C-kinase pathway inhibits further hydrolysis of

PtdIns4,5P_2 (Orellana et al., 1985; Labarca et al., 1984; Brock et al., 1985; Monaco & Mufson, 1986; Kojima et al., 1986) or through a positive feedback effect to stimulate calcium pumping (Drummond, 1985; Rickard & Sheterline, 1985) (Fig. 2). Another important action of the DG/C-kinase pathway appears to be the activation of inositol trisphosphatase which will accelerate the hydrolysis of Ins1,4,5P_3 so reducing its calcium-mobilizing action (Connolly et al., 1986; Molina y Vedia & Lapetina, 1986).

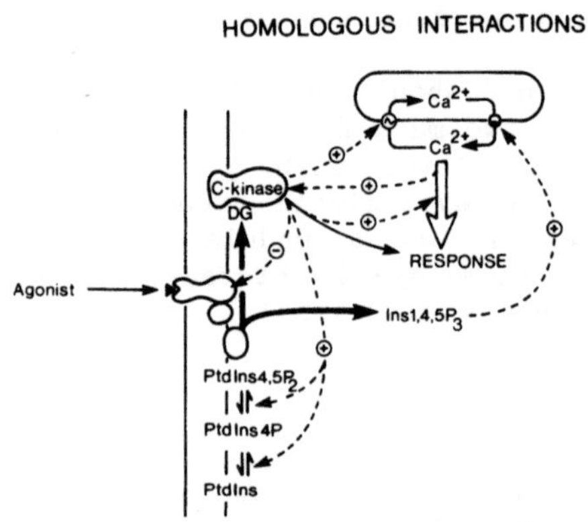

Fig. 2. Summary of the major interactions operating within the inositol lipid pathway. Protein kinase C (C-kinase) is a focal point for many of these interactions.

There are a number of potential targets (receptor, Gp, phosphoinositidase) where the DG/C-kinase pathway could act to reduce the agonist-dependent hydrolysis of PtdIns4,5P_2. In liver, which has a number of receptors working through the inositol lipid pathway, phorbol esters selectively inhibit the α_1-receptors while the effects of vasopressin and angiotensin II were relatively unimpaired (Cooper et al., 1985). A similar situation exists in Swiss 3T3 cells where inositol phosphate formation induced by bombesin was inhibited by phorbol esters whereas the action of PDGF was unaffected (Sturanic et al., 1986). Such selective actions suggest that the DG/C-kinase pathway exerts its negative feedback effect by phosphorylating receptors rather than Gp or the phosphoinositidase. The phorbol ester-induced inhibitory effect in cultured smooth muscle cells is associated with a rapid phosphorylation of the α_1-adrenergic receptor (Leeb-Lundberg et al., 1985).

The DG/C-kinase pathway may also influence calcium signalling by altering the rate in which PtdIns is converted to PtdIns4,5P_2 (De Chaffoy de Courelles et al., 1984; Halenda & Feinstein, 1984; Taylor et al., 1984). The increase in PtdIns4,5P_2 labelling in thymocytes brought about by phorbol esters resulted in an increased formation of Ins1,4,5P_3 when cells were subsequently challenged with concanavalin A (Taylor et al., 1984). This feedback mechanism is likely to be particularly important in the nervous system which has a neurone-specific protein called B50 which appears to regulate PtdIns4P kinase (Jolles et al., 1981). This regulatory function of B50 is controlled by protein kinase C (Aloyo et al., 1983).

HETEROLOGOUS INTERACTIONS

The DG/C-kinase pathway plays a significant role in receptor-receptor interactions by modulating the activity of other second messenger pathways (Fig.3), particularly the cyclic nucleotides (Nishizuka, 1986). There are numerous examples where the stimulation of receptors which hydrolyse inositol lipids potentiates those receptors which form cyclic AMP. For example, the production of cyclic AMP in the pineal by β-receptors is enhanced by activation of α_1-receptors (Sugden et al., 1985). Similarly, adenosine- or VIP-induced cyclic AMP formation in brain is increased by the simultaneous activation of α_1, 5-HT_2 or H_1 receptors all of which stimulate inositol lipid hydrolysis (Hollingsworth & Daly, 1985). This enhancement of cyclic AMP formation is mediated through the DG/C-kinase pathway because phorbol esters augment the accumulation of cyclic AMP by α_1-receptors in pineal gland (Sugden et al., 1985), frog erythrocytes (Sibley et al., 1986) and S49 lymphoma cells (Bell et al., 1985). Phorbol esters also enhance the adenosine-induced increase in cyclic AMP in brain (Hollingsworth et al., 1985). There are indications that protein kinase C acts to facilitate the interaction between G_s and adenylate cyclase (Bell et al., 1985). Protein kinase C phosphorylates G_i in blood platelets so reducing the activity of the pathway which inhibits adenylate cyclase (Katada et al., 1985).

The reciprocal relationship whereby cyclic nucleotides influence inositol lipid metabolism has not received as much attention except in a few cell types such as blood platelets. Thrombin-induced platelet responses such as the increase in intracellular calcium, phosphorylation of the 40K protein, aggregation and the release of 5-HT are all inhibited by agents which act through cyclic AMP (Yamanishi et al., 1983; Rink & Sanchez, 1984; Zavoico & Feinstein, 1984). Cyclic AMP acts by inhibiting the hydrolysis of PtdIns4,5P_2 so reducing the formation of both DG (Imai et al., 1983) and Ins1,4,5P_3 (Watson et al., 1984; Shukla, 1985; Nakashima et al., 1986). Cyclic GMP plays a similar role in inhibiting agonist-induced hydrolysis of PtdIns4,5P_2 in blood platelets (Takai et al., 1981; Nakashima et al., 1986) and in

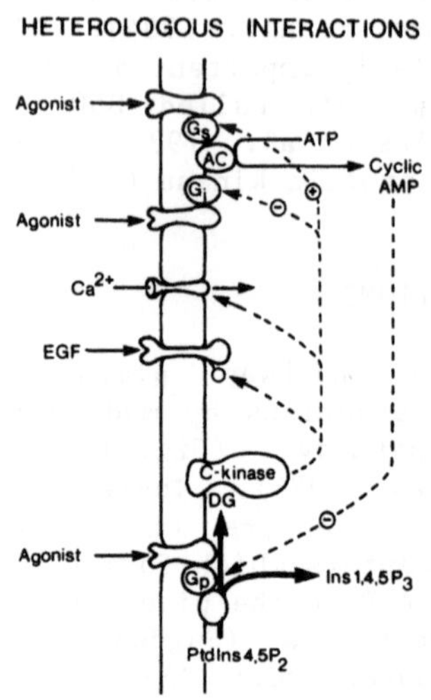

Fig.3. Summary of the heterologous interactions operating between the inositol lipid pathway and other signalling mechanisms.

smooth muscle (Rapoport, 1986). Cyclic nucleotdies may act by interfering with the enzymes which supply the receptor mechanism with $PtdIns4,5P_2$ (Dechaffoy de Courcelles et al., 1986; Lapetina, 1986).

Many excitable cells gate calcium across the plasma membrane using voltage-dependent calcium channels which represent another target for the modulatory effect of protein kinase C. In *Aplysia* neurones, the entry of calcium is enhanced (DeRiemer et al., 1984) whereas in two cell-lines derived from mammalian cells (PC12 & RINm4F) it was reduced (Di Virgilio et al., 1986; Harris et al., 1986). This reduction in the activity of voltage-sensitive calcium channels by protein kinase C might explain the well-known inactivation phenomenon which sets in during continuous depolarization.

Receptors which have tyrosine kinase activi'y are also targets for protein kinase C as exemplified by the EGF receptor. Phosphorylation of the EGF receptor on threonine 654 by protein kinase C (Davis & Czech, 1985) results in a decrease both in the affinity of the receptor for EGF (Brown et al., 1979) as well as its tyrosine kinase activity (Davis & Czech, 1985). Such transmodulation of other receptor mechanisms represents one of the major functions of the DG/C-kinase signal pathway.

INOSITOL LIPID METABOLISM AND NEUROTRANSMISSION

The precise function of those transmitters which use inositol lipids is still somewhat of a mystery but they are likely to operate at both pre- and post-synaptic sites. A pre-synaptic action in modulating the release of transmitter is an obvious function. For example, certain transmitters may use $Ins1,4,5P_3$ to adjust the resting level of calcium and so influence the amount of transmitter released during each action potential. Such a phenomenon was recently described in the adrenal which has both muscarinic and nicotinic receptors. Activation of the former gives $Ins1,4,5P_3$ which releases calcium which, although insufficient to stimulate the release of adrenaline is capable of greatly potentiating the secretory response resulting from stimulating the nicotinic receptor (Forsberg et al., 1986). A role for the DG/C-kinase pathway has also been suggested based on the fact that the phorbol esters can act synergistically with a calcium ionophore to stimulate the release of transmitter from PC 12 cells (Pozzan et al., 1984) and from neurones found in both the peripheral (Tanaka et al., 1984) and central nervous system (Tanaka et al., 1986).

The inositol lipid receptor mechanism also has an important role to play in mediating post-synaptic events. The accommodation which sets in following stimulation of pyramidal neurones in the hippocampus can be prevented by treating the cells with a phorbol ester (Baraban et al., 1985). The mechanism is thought to depend upon the activation of protein kinase C which then acts to close the potassium channels responsible for hyperpolorizing the membrane. Closure of potassium channels may turn out to be one of the main functions of protein kinase C in neural tissue. The photoreceptors of Hermissenda also have potassium channels which are closed by protein kinase C (Farley & Auerbach, 1986). Likewise, the closure of potassium channels which results in increased excitability of neurones in the dorsal raphe (Aghajanian, 1985) and globus pallidus (Stanfield et al., 1985) may also be due to the activation of the DG/C-kinase limb of the signal pathway. In the case of such cells situated in situ, it is difficult to establish precisely the nature of the second messenger system responsible for regulating neuronal activity. One way of approaching this problem is to study cells in culture. An interesting example is the study on NG108-15 cells which respond to bradykinin with a characteristic electrophysiological response, an early hyperpolarizing phase is followed by a longer-lasting depolarizing response (Higashida et al., 1986; Higashida & Brown, 1986). Previous biochemical studies had already established that bradykinin stimulated the hydrolysis of $PtdIns4,5P_2$ to form $Ins1,4,5P_3$ (Yano et al., 1984, 1985) and an increase in the intracellular level of calcium (Reiser & Hamprecht, 1985). This $Ins1,4,5P_3$/calcium pathway appears to be responsible for the early phase of the electrophysiological response because the injection of either $Ins1,4,5P_3$ or calcium induced a membrane hyperpolarization identical to that observed by adding bradykinin (Higashida et al., 1986; Higashida & Brown, 1986). On the other hand, the DG/C-kinase limb of the signal pathway seems to regulate the later depolarizing phase since it can be induced by adding a phorbol ester. As this depolarizing phase results from the closure of a potassium channel, we see yet another example of protein kinase C having an effect on neuronal excitability by modulating a potassium conductance.

CONCLUSION

There is increasing evidence that the inositol lipid signal pathway plays a central role in mediating the action of neurotransmitters and particularly those concerned with neuromodulation. It is argued that the second messengers $Ins1,4,5P_3$ and DG may function in modulating the amount of transmitter released during each action potential. At post-synaptic sites these second messengers seem to function primarily to modulate potassium channels which are opened by the $Ins1,4,5P_3$/calcium pathway but closed following activation of protein kinase C. Another important modulatory function concerns the ability of this receptor mechanism to interact and to adjust the activity of other receptors and particularly those functioning through adenylate cyclase.

The functional significance of why many transmitters are packaged and released together is still very much of a mystery. An examination of those transmitters which are co-localized reveals that one is a classical transmitter which functions to open ion channels whereas the other is often a typical calcium-mobilizing transmitter capable of stimulating the hydrolysis of inositol lipids leading to the neuromodulatory events described above. In addition, the activation of such receptors invariably gives rise to the formation of eicosanoids which thus provide an elegant mechanism for transferring information back from post- to the pre-synaptic endings.

References

Aghajanian, G.K. (1985). Modulation of a transient outward current in serotonergic neurones by α_1-adrenoreceptors. Nature (Lond.) 315, 501-503.

Aloyo, V.J., Zwiers, H. & Gispen, W.H. (1983). Phosphorylation of B-50 protein by calcium-activated, phospholipid-dependent protein kinase and B-50 protein kinase. J. Neurochem. 41, 649-653.

Authi, K.S., Evenden, B.J. & Crawford, N. (1986). Metabolic and functional consequences of introducing inositol 1,4,5-trisphosphate into saponin-permeabilized human platelets. Biochem. J. 233, 709-718.

Baraban, J.M., Snyder, S.H. & Alger, B.E. (1985). Protein kinase C regulates ionic conductance in hippocampal pyramidal neurones: Electrophysiological effects of phorbol esters. Proc. Natl. Acad, Sci. U.S.A. 82, 2538-2542.

Batty, I.R., Nahorski, S.R. & Irvine, R.F. (1985). Rapid formation of inositol (1,3,4,5) tetrakisphosphate following muscarinic stimulation of rat cerebral cortical slices. Bioch. J. 232, 211-215.

Bell, J.D., Buxton, I.L.O. & Brunton, L.L. (1985). Enhancement of adenylate cyclase activity in S49 lymphoma cells by phorbol esters. J. Biol. Chem. 260, 2625-2628.

Berridge, M.J. (1986). An alternative hypothesis of Li^+ mode of action based on intervention of the inositol lipid signal pathway. In "Drug Receptors and Dynamic Processes in Cells", Alfred Benzon Symposium 22, 334-337. (Eds. J.S. Skou, A. Giesler & S. Norn), Munksgaard, Copenhagen.

Berridge, M.J. & Irvine, R.F. (1984). Inositol trisphosphate, a novel second messenger in cellular signal transduction. Nature (Lond.) 312, 315-321.

Brass, L.F. & Joseph, S.K. (1985). A role for inositol trisphosphate in intracellular Ca^{2+} mobilization and granule secretion in platelets. J. Biol. Chem. 260, 15172-15179.

Brock, T.A., Rittenhouse, S.E.,Powers, C.W., Ekstein, L.S., Gimbrone, M.A. & Alexander, R.W. (). Phorbol ester and 1-oleoyl-2-acetylycerol inhibit angiotensin activation of phospholipase C in cultured vascular smooth muscle cells. J. Biol. Chem. 260, 14158-14162.

Brocklehurst, K.W., Morita, K. & Pollard, H.B. (1985). Characterization of protein kinase C and its role in catecholamine secretion from bovine adrenal-medullary cells. Biochem. J. 228, 35-42.

Brown, J.E., Rubin, L.J., Ghalayini, A.J., Tarver, A.P., Irvine, R.F., Berridge, M.J. & Anderson, R.E. (1984). Myo-inositol polyphosphate may be a messenger for visual excitation in Limulus photoreceptors. Nature (Lond.) 311, 160-163.

Brown, K.D., Dicker, P. & Rozengurt, E. (1979). Inhibition of epidermal growth factors binding to surface receptors by tumor promotors. Biochem. Biophys. Res. Commun. 86, 1037-1043.

Burgess, G.M., McKinney, J.S., Irvine, R.F. & Putney, J.W. (1985). Inositol (1,3,4) trisphosphate and inositol (1,4,5) trisphosphate formation in Ca-mobilizing hormone activated cells. Biochem. J. 232, 237-248.

Busa, W.B., Ferguson, J.E., Joseph, S.K., Williamson, J.R. & Nuccitelli, R. (1985). Activation of frog (Xenopus laevis) eggs by inositol trisphosphate. I. Characterization of Ca^{2+} release from intracellular stores. J. Cell Biol. 101, 677-682.

Connolly, T.M., Lawing, W.J. Jnr. & Majerus, P.W. (1986). Protein kinase C phosphorylates human platelet inositol trisphosphate 5'-phosphomonoesterase, increasing phosphatase activity Cell 46, 951-958.

Cooper, R.H., Coll, K.E. & Williamson, J.R. (1985). Differential effects of phorbol ester on phenylephrine and vasopressin-induced Ca^{2+} mobilization in isolated hepatocytes J. Biol. Chem. 260, 3281-3288.

Davis, R.J. & Czech, M.P. (1985). Platelet-derived growth factor mimics phorbol diester action on epidermal growth factor receptor phosphorylation at threonine-654. Proc. Natl. Acad. Sci. U.S.A. 82, 4080-4084.

De Chaffoy de Courcelles, D., Roevens, P. & Van Belle, H. (1984). 12-O-Tetradecanoylphorbol 13-acetate stimulates inositol lipid phosphorylation in intact human platelets. FEBS 173, 389-393.

De Chaffoy de Courcelles, D., Roevens, P. & van Belle, H. (1986). Agents that elevate platelet cAMP stimulate the formation of phosphatidylinositol 4-phosphate in intact human platelets. FEBS Letters 195, 115-118.

De Riemer, S.A., Strong, J.A., Albert, K.A., Greengard, P. & Kaczmarek, L.K. (1985). Enhancement of calcium current in Aplysia neurones by phorbol ester and protein kinase C. Nature (Lond.) 313, 313-316.

Di Virgilio, F., Pozzan, T., Wollheim, C.B., Vicentini, L.M. & Meldolesi, J. (1986). Tumor promotor phorbol myristate acetate inhibits Ca^{2+} influx through voltage-gated Ca^{2+} channels in two secretory cell lines, PC12 and RINmSF. J. Biol. Chem. 261, 32-35.

Dougherty, R.W. & Niedel, J.E. (1986). Cytosolic calcium regulates phorbol diester binding affinity in intact phagocytes J. Biol. Chem. 261, 4097-4100.

Downes, C.P. (1985) Receptor-dependent generation of intracellular signals from inositol phospholipids in parotid gland and brain. Biochem. Soc. Trans. 13, 1107-1110

Downes, C.P. & Michell, R.H. (1985). Inositol phospholipid breakdown as a receptor-controlled generator of second messengers. In Molecular Mechanisms of Transmembrane Signalling P. Cohen & M. Houslay eds pp3-56, Elsevier Science Publishers.

Drummond, A.H. (1985). Bidirectional control of cytosolic free calcium by thyrotropin-releasing hormone in pituitary cells. Nature (Lond.) 315, 752-755.

Europe-Finner, G.N. & Newell, P.C. (1985). Inositol 1,4,5-trisphosphate induces cyclic GMP formation in Dictyostelium discoideum. Biochem. Biophys. Res. Commun. 130, 1115-1122.

Europe-Finner, G.N. & Newell, P.C. (1986). Inositol 1,4,5-trisphosphate and calcium stimulate actin polymerization in Dictyostelium discoidium. J. Cell Sci. 82, 41-51.

Evans, M.G. & Marty, A. (1986). Potentiation of muscarinic and α-adrenergic responses by an analogue of guanosine 5'-trisphosphate. Proc. Natl. Acad. Sci. U.S.A. 83, 4099-4103.

Fein, A., Payne, R., Corson, D.W., Berridge, M.J. & Irvine, R.F. (1984). Photoreceptor excitation and adaptation by inositol 1,4,5-trisphosphate. Nature (Lond.) 311, 157-160.

Fleischman, L.F., Chahwala, S.B. & Cantley, L. (1986). Ras-transformed cells: altered levels of phosphatidylinositol-4,5-bisphosphate and catabolites. Science 231, 407-410.

Forsberg, E.J., Rojas, E. & Pollard, H.B. (1986). Muscarinic receptor enhancement of nicotinic-induced catecholamine secretion may be mediated by phosphoinositide metabolism in bovine adrenal chromaffin cells. J. Biol. Chem. 261, 4915-4920.

Halenda, S.P. & Feinstein, M.B. (1984). Phorbol myristate acetate stimulates formation of phosphatidylinositol 4-phosphate and phosphatidylinositol 4,5-bisphosphate in human platelets. Biochem. Biophys, Res. Commun. 124, 507-513.

Hallcher, L.M. & Sherman, W.R. (1980). The effects of lithium ion and other agents on the activity of myo-Inositol-1-phosphatase from bovine brain J. Biol. Chem. 255, 1089-1090.

Hansen, C.A., Mah, S. & Williamson, J.R. (1986). Formation and metabolism of inositol 1,3,4,5-tetrakisphosphate in liver. J. Biol. Chem. 261, 8100-8103.

Harris, K.M., Kongsamut, S. & Miller, R.J. (1986). Protein kinase C mediated regulation of calcium channels in PC-12 pheochromocytoma cells. Biochem. Biophys. Res. Commun. 134, 1298-1305.

Hesketh, T.R., Moore, J.P., Morris, J.D.H., Taylor, M.V., Rogers, J., Smith, G.A. & Metcalfe, J.C. (1985). A common sequence of calcium and pH signals in the mitogenic stimulation of eukaryotic cells. Nature (Lond) 313, 481-484.

Higashida, H. & Brown, D.A. (1986). Two polyphosphoinositide metabolites control two K^+-currents in a neuronal cell. Nature (Lond.) 323, 333-335.

Higashida, H., Streaty, R.A., Klee, W. & Nirenberg, M. (1986). Bradykinin-activated transmembrane signals are coupled via N_o or N_i to production of inositol 1,4,5-trisphosphate, a second messenger in N105-15 neuroblastoma-glioma hybrid cells. Proc. Natl. Acad. Sci. U.S.A. 83, 942-946.

Hollingsworth, E.B., Sears, E.B. & Daly, J.W. (1985). FEBS Letters 184, 339-342.

Hollingworth, E.B. & Daly, J.W. (1985). Accumulation of inositol phosphates and cyclic AMP in guinea-pig cerebral cortical preparations. Biochim. Biophys. Acta. 847, 207-216.

Imai, A., Hattori, H., Takahashi, M. & Nozawa, Y. (1983). Evidence that cyclic AMP may regulate Ca^{2+} mobilization and phospholipases in thrombin-stimulated human platelets. Biochem. Biophys. Res. Commun. 112, 693-700.

Inoue, T. & Takeda, K. (1984). Prostaglandin-induced inhibition of acetylcholine release from neuronal elements of dog tracheal tissue. J. Physiol 349, 553-570.

Irvine, R.F., Anggard, E.A., Letcher, A.J. & Downes, C.P. (1985). Metabolism of inositol (1,4,5) trisphosphate and inositol (1,3,4) trisphosphate in rat parotid glands. Biochem. J. 229, 505-511.

Irvine, R.F., Letcher, A.J., Heslop, J.P. & Berridge, M.J. (1986). The inositol tris/tetrakisphosphate pathway-demonstration of $Ins(1,4,5)P_3$-3-kinase activity in animal tissue. Nature (Lond.) 320, 631-634.

Irvine, R.F., Letcher, A.J., Lander, D.J. & Downes, C.P. (1984). Inositol trisphosphates in carbachol-stimulated rat parotid glands. Biochem. J. 223, 237-243.

Israels, S.J., Robinson, P., Docherty, J.C. & Gerrard, J.M. (1985). Activation of permeabilized platelets by inositol-1,4,5-trisphosphate. Thromb. Res. 40, 499-509.

Jolles, J., Zwiers, H., Dekker, A., Wirtz, K.W.A. & Gispen, W.H. (1981). Corticotropin-(1-24)-tetracosapeptide affects protein phosphorylation and polyphosphoinositide metabolism in rat brain. Bioch. J. 194, 283-291.

Kaibuchi, K., Takai, Y.,Sawamura, M., Hoshijima, M., Fujikura, T. & Nishizuka, Y. (1983). Synergistic functions of protein phosphorylation and calcium mobilization in platelet activation. J. Biol. Chem. 258, 6701-6704.

Katada, T., Gilman, A., Watanabe, Y., Bauer, S. & Jacobs, K.H. (1985). Eur. J. Biochem. 151, 431-437.

Klockner, U. & Isenberg, G. (2985). Calcium activated potassium currents as an indicator for intracellular (i.c.) Ca-transients. Single smooth muscle cells from trachea and urinary bladder. Pflug. Archiv. 405, R61.

Kojima, I., Shibata, H. & Ogata, E. (1986). Phorbol ester inhibits angiotensin-induced activation of phospholipase C in adrenal glomerulosa cells. Biochem. J. 237, 253-258.

Labarca, R., Janowsky, A., Patel, J. & Paul, S.M. (1984). Biochem. Biophys. Res. Commun. 123, 703-709.

Lapetina, E.G. (1986). Incorporation of synthetic 1,2-diacylglycerol into platelet phosphatidylinositol is increased by cyclic AMP. FEBS Letters 195, 111-114.

Lapetina, E.G., Watson, S.P. & Cuatrecasas, P. (1984). myo-Inositol 1,4,5-trisphosphate stimulates protein phosphorylation in saponin-permeabilized human platelets. Proc. Natl. Acad. Sci. U.S.A. 81, 7431-7435.

Leeb-Lundberg, L.M.F., Cotecchia, S.,Lomasney, J.M., DeBernardis, J.F., Lefkowitz, R.J. & Caron, M.G. (1985). Phorbol esters promote α_1-adrenergic receptor phosphorylation and receptor uncoupling from inositol phospholipid metabolism. Proc. Natl. Acad. Sci. U.S.A. 82, 5651-5655.

Molina y Vedia, L.M. & Lapetina, E.G. (1986) Phorbol 12,13-dibutyrate and 1-oleyl-2-acetyldiacylglycerol stimulates inositol trisphosphate dephosphorylation in human platelets. J. Biol. Chem. 261, 10493-10495

Monaco, M.E. & Mufson, R.A. (1986). Phorbol ester inhibition of the hormone-stimulated phosphoinositide cycle in WRK-1 cells. Biochem. J. 236, 171-175.

Moolenaar, W.H., Kruijer, W.,Tilly, B.C., Verlaan, I., Bierman, A.J. & de Laat, S.W. (1986). Growth factor like actions of phosphatidic acid. Nature (Lond.) 323, 171-173.

Naccache, P.H., Molski, T.F.P., Borgeat, P., White, J.R. & Sha'afi, R.I. (1985). Phorbol esters inhibit the fMet-leu-phe- and leukotriene B_4 stimulated calcium mobilization and enzyme secretion in rabbit neurotrophils. J. Biol. Chem. 260, 2125-2131.

Nahorski, S.R., Kendall, D.A., & Batty, I. (1986). Receptors and phosphoinositide metabolism in the central nervous system. Biochem. Pharm. 35, 2447-2454.

Nakashima, S., Tohmatsu, T., Hattori, H., Okano, Y. & Nozawa, Y. (1986). Inhibitory action of cyclic GMP on secretion, polyphosphoinositide hydrolysis and calcium mobilization in thrombin-stimulated human platelets. Biochem. Biophys. Res. Commun. 135, 1099-1104.

Nishizuka, Y. (1984). The role of protein kinase C in cell surface signal transduction and tumor promotion. Nature (Lond.) 308, 693-697.

Nishizuka, Y. (1986). Studies and perspectives of protein kinase C. Science 233, 305-312.

Orellana, S.A., Solski, P.A. & Brown, J.H. (1985). Phorbol ester inhibits phosphoinositide hydrolysis and calcium mobilization in cultured astrocytoma cells. J. Biol. Chem. 260, 5236-5239.

Oron, Y., Dascal, N., Nadler, E. & Lupu, M. (1985). Inositol 1,4,5-trisphosphate mimics muscarinic response in Xenopus oocytes. Nature (Lond.) 313, 141-143.

Pozzan, T., Gatti, G., Dozio, N., Vicentini, L.M. & Meldolesi, J. (1984). Ca^{2+}-dependent and -independent release of neurotransmitters from PC12 cells: a role for protein kinase C activation? J. Cell Biol. 99, 628-638.

Rapoport, R.M. (1986). Cyclic guanine monophosphate inhibition of contraction may be mediated through inhibition of phosphatidylinositol hydrolysis in rat aorta. Circulation Res. 58, 407-410.

Reiser, G. & Hamprecht, B. (1985). Bradykinin causes a transient rise of intracellular Ca^{2+}-activity in cultured neural cells. Pflugers Archiv. 405, 260-264.

Rickard, J.E. & Sheterline, P. (1985). Evidence that phorbol ester interferes with stimulated Ca^{2+} redistribution by activating Ca^{2+} efflux in neutrophil leucocytes. Biochem. J. 231, 623-628.

Rink, T.J. & Sanchez, A. (1984). Effects of prostaglandins I_2 and forskolin on the secretion from platelets evoked at basal concentrations of cytoplasmic free calcium by thrombin, collagen, phorbol ester and exogenous diacylglycerol. Biochem. J. 222, 833-836.

Sagi-Eisenberg, R., Lieman, H. & Pecht, I. (1985). Protein kinase C regulation of the receptor-coupled calcium signal in histamine secreting rat basophilic leukaemia cells. Nature (Lond.) 313, 59-60.

Shukla, S.D. (1985). Platelet activating factor-stimulated formation of inositol trisphosphate in platelets and its regulation by various agents including Ca^{2+}, indomethacin, CV-3988, and forskolin. Arch. Bioch. Biophys. 240, 674-681.

Sibley, D.R., Jeffs, R.A., Daniel, K., Nambi, P. & Lefkowitz, R.J. (1986). Phorbol diester treatment promotes enhanced adenylate cyclase activity in frog erythrocytes. Archiv. Biochem. Biophys. 244, 373-381.

Slack, B.E., Bell, J.E. & Benos, D.J. (1986). Inositol-1,4,5-trisphosphate injection mimics fertilization potentials in sea urchin eggs. Am. J. Physiol. 250, C340-C344.

Stanfield, P.R., Nakajima, Y. & Yamaguchi, K. (1985). Substance P raises neuronal membrane excitability by reducing inward rectification. Nature (Lond.). 315, 498-501.

Stewart, S.J., Prpic, V.,Powers, F.S., Bocckino, S.B., Isaaks, R.E. & Exton, J.H. (1986). Perturbation of the human T cell antigen receptor-T3 complex leads to the production of inositol tetrakisphosphate: Evidence for conversion from inositol trisphosphate. Proc. Natl. Acad. Sci. U.S.A. in press.

Storey, D.J., Shears, S.B., Kirk, C.J. & Michell, R.H. (1984). Stepwise enzymatic dephosphorylation of inositol 1,4,5-trisphosphate to inositol in liver. Nature (Lond.) 312, 374-376.

Sturani, E., Vicentini, L.M., Zippel, R., Toschi, L., Pandiella-Alonso, A., Comoglio, P.M. & Meldolesi, J. (1986). PDGF-induced receptor phosphorylation and phosphoinositide hydrolysis are unaffected by protein kinase C activation in mouse Swiss 3T3 and human skin fibroblasts. Biochim. Biophys. Res. Commun. 137, 343-350.

Sugden, D., Vanecek, J., Klein, D.C., Thomas, T.P. & Anderson, W.B. (1985). Activation of protein kinase C potentiates isoprenaline-induced cyclic AMP accumulation in rat pinealocytes. Nature (Lond.) 314, 359-361.

Tanaka, C., Taniyama, K. and Kusunoki, M.(1984). A phorbol ester and A23187 act synergistically to release acetylcholine from the guinea pig ileum. FEBS Letters. 175, 165-169

Tanaka, C., Fujiwara, H. & Fujii, Y. (1986). Acetylcholine release from guinea pig caudate slices evoked by phorbol ester and calcium, FEBS Letters. 195, 129-134

Takai, Y., Kaibuchi, K., Matsubara, T. & Nishizuka, Y. (1981). Inhibitory action of guanine 3',5'-monophosphate on thrombin-induced phosphatidylinositol turnover and protein phosphorylation in human platelets Bioch. Biophys. Res. Commun. 101, 61-67.

Taylor, C.W. & Merritt, J.E. (1986). Receptor coupling to polyphosphoinsitide turnover: A parallel with the adenylate cyclase systems. Trends in Pharmacol. Sci. 7, 238-242.

Taylor, M.V., Metcalfe, J.C., Hesketh, T.R., Smith, G.A. & Moore, J.P. (1984). Mitogens increase phosphorylation of phosphoinositdes in thymocytes. Nature (Lond.) 312, 462-465.

Touqui, L., Rothhut, B.,Shaw, A.M., Fradin, A., Vargaftig, B.B. & Russo-Marie, F. (1986). Platelet-activation - a role for a 40K anti-phospholipase A_2 protein indistinguishable from lipocortin Nature (Lond.) 321, 177-180.

Trevisani, A., Biondi, C., Bulluzzi, O., Borasio, P.G., Cappuzzo, A., Ferretti, M.E. & Perri, V. (1982). Evidence for increased release of prostaglandins of E-type in response to orthodromic stimulation in the guinea-pig superior cervical ganglion. Brain Res. 236, 375-381.

Turner, P.R., Jaffe, L.A. & Fein, A. (1985). Regulation of cortical vesicle exocytosis in sea urchin eggs by inositol 1,4,5-trisphosphate and GTP-binding protein. J. Cell Biol. 102, 70-76.

Vergara, J., Tsien, R.Y. & Delay, M. (1985). Inositol 1,4,5-trisphosphate: A possible chemical link in excitation-contraction coupling in muscle. Proc. Natl. Acad. Sci. U.S.A. 82, 6352-6356.

Volpe, P., Salviati, G., Di Virgilio, F. & Pozzan, T. (1985). Inositol 1,4,5-trisphosphate induces calcium release from sarcoplasmic reticulum of skeletal muscle. Nature (Lond.) 316, 347-349.

Watson, S.P., McConnell, R.T. & Lapetina, E.G. (1984). The rapid formation of inositol phosphates in human platelets by thrombin is inhibited by prostacyclin. J. Biol. Chem. 259, 13199-13203.

Watson, S.P., Ruggiero, M., Abrahams, S.L. & Lapetina, E.G. (1986). Inositol 1,4,5-trisphosphate induces aggregation and release of 5-hydroxytrypamine from saponin-permeabilized human platelets. J. Biol. Chem. 261, 5368-5372.

Whitaker, M. & Irvine, R.F. (1984). Inositol 1,4,5-trisphosphate microinjection activates sea urchin eggs. Nature (Lond.) 312, 636-639.

Yamanishi, J., Takai, Y.,Kaibuchi, K., Sano, K., Castagna, M. & Nishizuka, Y. (1983). Synergistic functions of phorbol ester and calcium in serotonin release from human platelets. Biochem. Biophys. Res. Commun. 112, 778-786.

Yano, K., Higashida, H., Hattori, H. & Zozawa,Y. (1985). Bradykinin-induced transient accumulation of inositol trisphosphate in neuron-like cell line NG 108-15 cells. FEBS Letters 181, 403-406.

Yano, K., Higashida, H., Inoue, R. & Nozawa, Y. (1984). Bradykinin-induced rapid breakdown of phosphatidylinositol 4,5-bisphosphate in neuroblastoma x glioma hybrid NG 108-15 cells. J. Biol. Chem. 259, 10201-10207.

Zavoico, G.B. & Feinstein, M.B. (1984). Cytoplasmic Ca^{2+} in platelets is controlled by cyclic AMP: Antagonism between stimulators and inhibitors of adenylate cyclase. Biochem. Biophys. Res. Commun. 120, 579-585.

14
Allosteric Sites and Conformational Transitions of the Acetylcholine Receptor: Models for Short-term Regulation of Receptor Response

J.P. Changeux, J. Giraudat, M. Dennis, M. Goeldner, C. Hirth, C. Mulle, F. Révah, A. Devillers-Thiéry and T. Heidmann

INTRODUCTION

The word "receptor" is commonly used with two separate significations : for the site complementary to the structure of the chemical messenger and for the protein which carries the site. From a functional point of view, receptor-receptor interactions (Fuxe et al., 1981, 1983, 1986, Agnati et al., 1983a,b, 1984) may be considered as representing interactions between distinct sites for specific ligands without any preconception about the protein components involved. One may thus legitimately distinguish at the cell, or synaptic, level : 1) intramolecular allosteric interactions between topologically distinct sites mediated by a unique protein assembly via conformational transitions (see Monod et al., 1963, 1965, Changeux 1981, Changeux et al., 1976, 1983, 1984) and 2) indirect intermolecular coupling between distinct protein entities transmitted by close contact within the membrane, by electric fields and/or intracellular second messengers. However, even in these last instances at least one allosteric protein integrates the interaction between one primary messenger and the coupling signals produced by another receptor (see Changeux 1986).

The acetylcholine nicotinic receptor (AChR) from vertebrate neuromuscular junction and fish electric organ is a ligand regulated ion channel involved in fast synaptic transmission which can be taken as an example of membrane bound allosteric protein able to integrate intramolecular interactions between distinct categories of pharmacologically active ligands (Changeux 1981). Eventhough, several of the endogenous counterparts of these ligands have not been identified, it appears legitimate to briefly present and discuss some of these properties, at least as plausible elementary mechanisms involved in receptor-receptor interactions at the sub-molecular level.

I. The acetylcholine receptor : a *quasi* symmetrical transmembrane protein.

The AChR light form is a glycoprotein of approx. 300 KD Mol. W. (268 KD of protein) made up of four different polypeptides chains of exact molecular weight (in *Torpedo*) α (50,116), β (53, 681), γ (56,279) and δ (57,565) (Noda *et al.*, 1982, 1983ab) assembled in a rather uncommon 2α.1β.1γ.1δ stoichiometry (Reynolds and Karlin 1978, Lindstrom *et al.*, 1979, Raftery *et al.*, 1980, Saitoh *et al.*, 1980).

The purified and membrane-bound AChR appears by electron microscopy on "en face" views as rosettes, 80 Å in diameter, with a central depression of about 25 Å in width (Cartaud *et al.*, 1973) and five peaks of density, regularly disposed around a 5-fold axis of symmetry (Bon *et al.*, 1984, Brisson and Unwin 1985). When viewed on the side the light form appears roughly as a cylinder (110 Å long) which extends above the lipid bilayer by 55 Å into the synaptic cleft and by about 15 Å into the cytoplasm. Selective proteolysis, antibody binding and radioisotope iodination of the memebrane-bound AChR show that all five subunits traverse the membrane. The AChR thus appears as a roughly cylindrical bundle with a "transverse polarity" (Changeux and Thiéry 1968) and a rotational axis of 5-fold symmetry perpendicular to the plane of the membrane (Changeux 1981, Kistler and Stroud 1981, Wise *et al.*, 1981, McCarthy *et al.*, 1986, Hucho, 1986).

The cDNAs coding for the four chains have been cloned and sequenced in *Torpedo californica* (Ballivet *et al.*, 1982, Noda *et al.*, 1982, 1983 a,b, Claudio *et al.*, 1983) in *T. marmorata* (for the α chain : Giraudat *et al.*, 1982, Sumikawa *et al.*, 1982, Devillers-Thiéry *et al.*, 1983) and in several vertebrate species, including human (see Noda *et al.*, 1983c, Kubo *et al.*, 1985 and Boulter *et al.*, 1986). As initially noticed from amino terminal sequence analysis (Raftery *et al.*, 1980), the sequences show striking homologies (from 10 to 60 per cent identity with an average of 40 per cent) (Noda *et al.*, 1983b) which support a *quasi*-symmetrical organisation of the receptor oligomer.

Reconstitution experiments with the purified protein incorporated into lipid vesicles and planar lipid bilayers show that the $\alpha_2\beta\gamma\delta$ oligomer contains the ion channel (selective for Na+, K+, Ca^{++}) and *all* the structural elements required for its opening by nicotinic agonists and for its regulation by allosteric effectors (review in Popot *et al.*, 1981, Montal *et al.*, 1986). Supporting this conclusion, injection of the four mRNAS coding for the receptor subunits into *Xenopus* oocytes yields functional acetylcholine-gated ion channels and in *Torpedo* the presence of the four subunits is required for a fully functional receptor (Mishina *et al.*, 1984). Yet, with mouse muscle receptor, three subunits (including the α-subunit) suffice to give the ionic response (Boulter et al., 1986) and even two with neuronal receptor (Boulter et al, unpublished).

The AChR is thus a compact transmembrane channel protein to which regulatory signals have access from both the outside and the inside of the cell.

II. Models of transmembrane topology of the subunits.

The aligned sequences of the four AChR subunits show a similar hydrophobicity profile which justifies a common subdivision of the homologous chains into : 1) a large hydrophilic amino-terminal domain of 210-224 aminoacids ; 2) a compact hydrophobic region of 68 residues subdivided into three segments of 19-27 uncharged amino acids (numbered I, II and III) ; 3) a small hydrophilic domain of 109-146 amino acids ; 4) a carboxy-terminal segment of 20 hydrophobic residues (numbered IV). On the basis of these primary sequence data models of transmembrane organisation common to all homologous subunits have been proposed (Claudio et al., 1983, Devillers-Thiéry et al., 1983, Noda et al., 1983b, Guy 1984, Finer-Moore and Stroud 1984, McCarthy et al., 1986, Criado et al., 1985, Ratnam et al., 1986). Their common features are : 1) the orientation of the large hydrophilic domain towards the synaptic cleft ; 2) the orientation of the small hydrophilic domain towards the cytoplasm ; 3) the assignment of the four hydrophobic segments I to IV (only I to III for Ratnam et al., 1986) to transmembrane α-helices. All assume that the acetylcholine binding site is located in the large hydrophilic domain on the α subunit. Differences concern the number and orientation of the transmembrane segment(s) which would contribute to the walls of the ionic channel, in all instances, assumed to lie in the axis of quasi-symmetry of the receptor molecule. In one group of models, the chain traverses the lipid bilayer four times and the carboxy terminal end faces the synaptic cleft, helix I (Noda et al., 1983b) or III (Devillers-Thiéry et al., 1983) being the transmembrane components of an uncharged channel. Another model postulates an additional fifth "amphipathic" helix, named A, formed at the expenses of the cytoplasmic domain thus reorienting helix IV to expose the carboxy terminal to the cytoplasmic face. Helix A would then delineate a charged channel (Guy 1984, Finer-Moore and Stroud 1984, Mishina et al., 1985). A recent model proposes that helix IV is not transmembrane but cytoplasmic and that a transmembrane "hair pin" loop forms from the large hydrophilic domain (from amino acid 142 to 192 on the α subunit) ; the segment 159-192 would then make an amphipathic helix appropriate for lining a charged channel (Ratnam et al., 1986).

Tests for these models have been developed in several laboratories using different experimental approaches which include : 1) site directed mutagenesis and expression of cloned cDNAs for the α subunit (Mishina et al., 1985) ; 2) immunochemical studies using antibodies directed against peptides of defined sequences (review Lindstrom 1986) ; 3) covalent labeling of specific sites of the AChR with radioactive affinity (or photoaffinity) reagents and identification of the labeled peptides. The results obtained in our laboratory by this last method are described in the next two paragraphs.

III. Localisation of the acetylcholine binding site in the large NH_2-terminal hydrophilic domain.

The $\alpha_2\beta\gamma\delta$ oligomer carries two primary acetylcholine binding sites to which nicotinic agonists, antagonists and snake venom α toxins bind in a reversible and mutually exclusive manner. The two α-subunits contribute to these sites which exhibit positive cooperative interactions but nevertheless possess quite different binding properties (review Karlin 1980, 1983, Changeux et al., 1984). Southern blot hybridization of genomic DNA with specific cDNA probes supports the existence of a single gene for the α-subunit in Torpedo (Klarsfeld et al., 1984) and in other species (Merlie 1984, review McCarthy et al., 1986). The different properties of the two acetylcholine binding sites may result from differences of subunit environment in the receptor oligomer (eventhough differences of subunit posttranslational modification cannot be excluded).

Attempts to identify the amino acids which contribute to the acetylcholine binding site have been made on the basis of the observation that disulfide bond reduction perturbs the pharmacological response to agonists and permits the labeling of the α-subunit by cholinergic affinity ligands specific for sulfhydryl groups (such as the maleimide reagents MBTA or MPTA) (review Karlin 1969, 1983). Kao et al. (1984) have demonstrated by peptide mapping and sequencing that the residues cystein 192 and possibly 193, a tandem unique to the α-subunit, represent the sites of incorporation of MBTA, yet, exclusively on the reduced receptor. Owing to the restricted reactivity of these reagents limited to free sulfhydryls, it was however unclear to what extent cysteins 192 and 193 and the surrounding regions actually participate in the binding of acetylcholine to the native receptor.

In order to probe in greater detail the structure of the acetylcholine binding sites on the native AChR, we used p N,N-dimethylaminobenzenediazonium fluoroborate (DDF) (Goeldner and Hirth 1980), a derivative of TDF, a compound described as the first affinity labeling reagent of electric organ AChR (Changeux et al., 1967a, Weiland et al., 1979). In the dark DDF behaves as a competitive antagonist but irreversibly blocks the binding of cholinergic ligands and snake venom α toxins following irradiation at 295 nm. Under appropriate conditions [^{3}H] DDF predominantly labels the α subunit of the native AChR and this labeling is prevented by cholinergic agonists and antagonists (Lagenbuch-Cachat et al., 1987). Comparative peptide mapping of the α subunit labeled by [^{3}H] MPTA and [^{3}H] DDF revealed differences in the distribution of the two markers. [^{3}H] DDF was incorporated into three distinct CNBr fragments in a carbamylcholine-sensitive manner. The predominant one corresponded to segment α 179-207, and was the only one labeled by [^{3}H] MPTA at the level of cysteins 192 and 193. In this fragment, [^{3}H] DDF labeled tyrosine 190, cysteins 192 and 193 and possibly subsequent amino acids (Dennis et al., 1986 and unpublished). This region is also involved in α-bungarotoxin binding (Wilson et al., 1985, Neumann et al., 1986). Another fragment specifically labeled by [^{3}H] DDF was derived from a more aminoterminal region of the large hydrophilic domain of the α subunit (unpublished). Several segments of the large hydrophilic domain may thus participate in the binding of cholinergic ligands (see also Mulac-Jericevic and Atassi 1986, McCormick and Atassi 1984 but Criado et al., 1986). These results obtained with the native AChR are, thus, consistent with the above mentioned models which postulate that the large hydrophilic domain of the α subunit is at least in part exposed to the synaptic cleft and carries the acetylcholine binding site.

IV. The high affinity sites for "channel blockers".

In both the native postsynaptic membranes (Steinbach 1968, Magleby and Pallotta 1978, Adams 1981, Adams and Galvan 1986) and reconstituted purified receptor (Popot et al., 1981), the ionic response to acetylcholine is inhibited by a class of particularly potent pharmacological agents, the noncompetitive blockers, which are postulated to interfere directly or indirectly (or both) with the ion channel (references in Neher and Steinbach 1978). The binding site(s) at which these noncompetitive blockers exert their effect are different from the acetylcholine binding site (Weber and Changeux 1974) but strongly interact with them under equilibrium conditions (Cohen et al., 1974, Grünhagen and Changeux 1976, Eldefrawi et al., 1977, Krodel et al., 1979, review Heidmann et al., 1983). These sites have been subdivided into two main categories (Heidmann et al., 1983) : 1) high-affinity sites, sensitive to histrionicotoxin (Eldefrawi et al., 1977) and present as a unique copy per receptor pentamer ; and 2) low-affinity sites, insensitive to histrionicotoxin, much more numerous (10 to 20 times the number of α-bungarotoxin sites) and lipid dependent, that is, most likely located at the interface of the receptor with membrane lipids.

Early attempts to identify the chain (or chains) of the AChR involved in the high-affinity site were done with T. marmorata receptor and an azido derivative of the potent local anesthetic trimethisoquin. Only the δ chain incorporated the label in an agonist-dependent manner (Oswald *et al.*, 1980, Oswald and Changeux 1981a). Both [^{3}H]-labeled perhydrohistrionicotoxin and [^{3}H] phencyclidine, under U.V. irradiation (Oswald and Changeux 1981b) gave similar results. However, with *T. californica* and quinacrine mustard (Kaldany and Karlin 1983), or triphenylmethylphosphonium (Muhn and Hucho 1983) or with *T. ocellata* and azidophencyclidine (Haring *et al.*, 1984), labeling occurred primarily at the level of the α or the β chains, respectively. With *T. marmorata*, [^{3}H]-chlorpromazine labeled all four chains after U.V. irradiation (Oswald and Changeux 1981b).

Careful measurements confirmed a stoichiometry of one chlorpromazine high-affinity binding site per two α-toxin sites. Since, when bound to this site, chlorpromazine attaches covalently to any of the four chains of the AChR, it was proposed (Heidmann *et al.*, 1983) that this site lies in the *axis of quasi-symmetry* of the $\alpha_2\beta\gamma\delta$ oligomer where the distances to all five chains are minimum. Minor sequence differences between species of *Torpedo* and differences in ligand structure would then account for the various labeling patterns mentioned above. A rather similar location for an allosteric site has been described for diphosphoglycerate (Arnone 1972) (and other ligands, Perutz *et al.*, 1986) in hemoglobin.

Amino acids labeled by [^{3}H]-chlorpromazine were identified in this laboratory (Giraudat *et al.*, 1986, 1987). Photolabeling of the membrane bound receptor with [^{3}H]-chlorpromazine was carried out under equilibrium in the presence of carbamylcholine and in the absence, or presence, of phencyclidine, a selective ligand for the high affinity site for noncompetitive blockers. The δ-subunit was purified, digested with trypsin and the resulting fragments fractionated by reversed-phase HPLC. The labeled peptide(s) could not be purified to homogeneity because of its (their) marked hydrophobic character. Sequence analysis of the fraction of tryptic peptides containing the specifically labeled material revealed the presence of three amino-terminal sequences and produced a clear release of radioactivity on cycle 6. In order to identify which fragment carried a labeled residue in position 6, the pool of tryptic peptides was subjected to CNBr subcleavage and the specifically labeled material repurified by HPLC and sequenced. A mixture of two amino terminal sequences was found and the release of radioactivity was now observed on cycle 5. One of the sequences present in the CNBr derived fraction (Ser-Thr-Ala-Ile-Ser-Val ...) corresponds to a δ-subunit tryptic peptide shortened by its amino terminal methionine residue (Met 257) upon CNBr subcleavage. The concomitant shift of Ser 262 and of radioactivity release from cycle 6 to cycle 5 demonstrates that this residue is labeled by [^{3}H]-chlorpromazine. Sequence analysis of analogous material derived from control batches demonstrated that labeling of residue Ser 262 was prevented by phencyclidine. Residue Ser 262 which lies within the hydrophobic segment II of the δ-subunit is thus part (or in close vicinity) of the high-affinity site for noncompetitive blockers (Giraudat et al., 1986). Similar experiments were repeated with the β-chain and sequence analysis resulted in the identification of Ser 254 and Leu 257 as residues labeled by [^{3}H]-chlorpromazine in a phencyclidine protectable manner (Giraudat et al., 1987). These residues are also located in the hydrophobic and potentially transmembrane segment M II of the β chain, a region homologous to that containing chlorpromazine-labeled Ser 262 in the δ chain.

Recently, using a different noncompetitive blocker [^{3}H] triphenylmethylphosphonium, Oberthür et al. (1986) have identified Ser 262 at the labeled amino acid on the δ subunit a finding consistent with our original observation with [^{3}H]-chlorpromazine. Preliminary report from the same group (Hucho et al., 1986) also provide suggestive evidence for a labeling on the α and β subunits at homologous positions.

Homologous regions of different receptor subunits thus contribute to the unique high affinity site for noncompetitive blockers, a finding consistent with the proposal (Heidmann et al., 1983) that this site is located in the axis of quasi symmetry of the AChR molecule. Moreover, the segments M II are non homologous of the regions labeled by affinity reagents of the acetylcholine binding site on the α subunits. These structural data are thus consonant with the concept that the interactions between the acetylcholine binding site and the high affinity site for noncompetitive blockers are indirect or allosteric.

V. Correlation of ion channel opening with rapid [^{3}H]-chlorpromazine labeling.

An attractive method for identifying the structural elements involved in the ion channel gated by acetylcholine is the combination of site directed mutagenesis on subunit cDNA, expression of the mutated mRNA in *Xenopus* oocytes and electrophysiological recordings from the oocyte membrane (Mishina et al., 1985, 1986, Sakmann et al., 1985). The data published up to now are however disappointing since multiple deletions of important segments of the α subunit (Mishina et al., 1985) in the putative transmembrane segments I, II, III, IV and in the amphipathic helix A lead to a decrease (to less than 1 %) of the membrane potential response to acetylcholine (Mishina et al., 1985).

Our approach has been to compare the rapid kinetic properties of ion transport through the channel and [^{3}H]-chlorpromazine labeling. First, the effect of chlorpromazine and phencyclidine, the compounds that we used in the structural studies, was investigated on single channel currents recorded by the patch clamp method (Neher and Sakmann 1976, Hamill et al., 1981) on

C2 mouse myotubes (Changeux et al., 1986a). Under conditions where both compounds are positively charged (pH 7.2), 10-200 nM chlorpromazine or phencyclidine lead to shortened mean burst times as expected for high-affinity channel blockers (slow channel block, Neher and Steinbach 1978). Yet, this shortening appeared voltage independent (at variance with the effect of many, though not all, channel blockers) and, in addition, did not vary in a simple linear manner with concentration. Eventhough the electrophysiological data obtained by this method are accounted for by a variety of models, the effects observed at low concentration are consistent with the notion that, in this range of concentrations, chlorpromazine and phencyclidine block the ion channel by steric hindrance and that their high-affinity site is located in (or near) the ion channel.

The subsequent analysis of the interaction of [^{3}H]-chlorpromazine with the AChR is based on the observation that purified membrane preparations from both E. electricus (Kasai and Changeux 1971) and Torpedo (Hazelbauer and Changeux 1974) preserve the regulation of ion channel opening by acetylcholine as measured by the translocation of tracer cations (review Heidmann et al., 1983a, Cash et al., 1985). The development of methods for rapid measurement, in parallel, of agonist binding (Heidmann and Changeux 1979, Boyd and Cohen 1980) and ion fluxes (review Cash et al., 1985) with the same membrane preparation (Heidmann et al., 1983a, Neubig and Cohen 1980) lead to the conclusion that the binding of agonists triggers the opening of the ion channel in a relatively high range of concentrations [(45-80 µM acetylcholine or Dansyl-C6-choline (a fluorescent agonist) and 300-800 µM carbamylcholine] consistent with the estimates of the local concentration of acetylcholine in the synaptic cleft during transmission (see Katz and Miledi 1977).

The kinetics of covalent attachment of chlorpromazine to the AChR was resolved in a rapid mixing photolabeling apparatus (Heidmann and Changeux 1984, 1986). Rapid addition of [^{3}H]-chlorpromazine to AChR-rich membrane fragments followed by brief (∿5msec) U.V irradiation of the mixture results in the simultaneous covalent labeling of all four chains of the AChR, in a time-dependent manner. The rate of labeling increases (10^2-10^3 fold) under conditions of simultaneous addition of chlorpromazine and acetylcholine (or agonist) and varies with chlorpromazine concentration in the explored concentration range (up to 10µM) as expected for a diffusion-controled bimolecular binding reaction with an on rate constant close to 10^7 M^{-1} sec^{-1}. It increases with agonist concentration with apparent dissociation constants close to 30 µM for acetylcholine and 400 µM for carbamylcholine (Heidmann and Changeux 1984, 1986). It is blocked by snake venom α toxins and does not occur with competitive antagonists. Moreover, under conditions where rapid desensitization takes place with a significant amplitude (see chapter VI), the rates of chlorpromazine incorporation decline with the time course (half life 6 s^{-1} for 100 μM acetylcholine, maximum 15 s^{-1}) and concentration dependence reported for the rapid inactivation (desensitization) of the ion flux response measured by chemical kinetics (Cash et al., 1985). Thus, the enhanced rate of chlorpromazine incorporation elicited by acetylcholine binding to the AChR takes place at the level of a site which becomes accessible by diffusion when the channel opens. The selective labeling of helix II by [^{3}H]-chlorpromazine (Giraudat et al., 1986) was done in the presence of cholinergic agonists but under equilibrium conditions. It is not known yet to what extent the same aminoacid is labeled after rapid mixing. If it appears to be the same, then, the putative transmembrane helix II from each subunit might plausibly contribute to the walls of the ion channel. At the membrane level the transport of ions through the AChR would then take place in an uncharged environment (Devillers-Thiéry et al., 1983, Noda et al., 1983, see Furois-Corbin and Pullman 1986). Following this scheme, the charged groups distributed at both ends of the helix II segments could participate to the cationic selectivity of the channel. Such organisation differs from those postulated by models which give to an amphipathic charged helix a crucial role in ion transport (Guy 1984, Finer Moore and Stroud 1984, Stroud and Finer Moore 1985, Mishina et al., 1985, Ratnam et al., 1986).

VI. Structural evidence for the regulation of AChR response by desensitization.

In addition to the fast all-or-none opening or activation of the ion channel, AChR mediates a "high order" regulation of the ionic response. Prolonged exposure to exogenous acetylcholine causes within seconds or minutes a reversible decline of the conductance response to cholinergic agonists (Katz and Thesleff, 1957). This phenomenon of desensitization comprises two main kinetic processes : a fast one at the rate of 2-7 per second and a slow one at the rate of 0.01-0.1 per second (Feltz and Trautmann 1980, Sakmann et al., 1980, Magleby and Pallotta 1981). It occurs on neuromuscular preparations without addition of exogenous acetylcholine as long as the firing rates of the motor nerve is sufficiently high, in the presence of an acetylcholinesterase blocking agent, or, even, in its absence, if the interval between stimuli is shorter than 20 msec (Magleby and Pallotta 1981).

Desensitization persists after purification of postsynaptic membrane fragments and even after purification of the AChR and reconstitution into lipid bilayers (review Popot et al., 1981, Montal et al., 1986). It is thus an intrinsic property of the AChR protein and does not require any obligatory covalent modification. The molecular transitions of the AChR engaged in desensitization have been resolved by a variety of rapid mixing techniques with fluorescent agonists (Heidmann and Changeux 1979, Prinz and

Maelicke 1983) or radioactive ligands (Boyd and Cohen 1980) and by following intrinsic (Grünhagen and Changeux 1976, Bonner et al., 1976) or extrinsic (Grünhagen and Changeux 1976, Grünhagen et al., 1977, Dunn et al., 1980) changes of fluorescence.

Rapid mixing studies with the fluorescent agonist Dansyl-C6-choline (Heidmann and Changeux 1979) and [^{3}H] acetylcholine (Boyd and Cohen 1980) show that, in the absence of preincubation with agonist, 20 percent of the receptor spontaneously exists in a state of high affinity (3nM for acetylcholine and 2 nM for Dansyl-C6-choline), the remaining 80 percent exhibiting a low affinity for agonists. Beyond the details of the analysis, which are often linked to a particular kinetic model (see Changeux et al., 1984, Hucho 1986) two major conformational transitions have been resolved between these states : a "slow" one, with an apparent rate constant of about 0.01 per second ; and an "intermediate" one, with an apparent rate constant of 2 per second (Dunn et al., 1980) to 50 per second (Heidmann and Changeux 1980). The state stabilized after equilibration binds acetylcholine and Dansyl-C6-choline with the same high affinity as the 20 per cent of the population in the membrane present before mixing with the agonist.

An exceptional advantage of the in vitro preparations of functional AChR is that they make possible a comparison of the kinetics of both agonist binding and permeability response in the same time scale and with the same membrane preparation. Such parallel measurements show that the slow transition coincides with the slow desensitization phase of the ionic response (Neubig et al., 1982, Heidmann et al., 1983) and that the intermediate transition monitored with Dansyl-C6-choline fits with the fast desensitization (Heidmann et al., 1983a). In addition, they demonstrate the conjecture of Katz and Thesleff (1957) that the high affinity state for agonists of the AChR is a desensitized state of the AChR (the slowly one) where the ion channel is shut. The data are accounted for by a minimal four-state model (Neubig and Cohen 1980, Heidmann and Changeux 1980) which has already been extensively discussed (see Changeux et al ., 1984, Changeux and Heidmann 1987)

where A corresponds to the active state with the channel open, and I and D correspond to the rapidly and slowly desensitized states, respectively. These states are discrete and interconvertible. Their respective dissociation constants would be at least 50 to 100 μM (R), less than 1 μM (I), and 3 to 5 nM (D) for acetylcholine (Boyd and Cohen 1980) and Dansyl-C6-choline (Heidmann and Changeux 1979, 1980). In agreement with the "concerted" model of allosteric transition (Monod, Wyman and

Changeux 1965) several of these states [A (Jackson 1984) and D (Heidmann and Changeux 1979)] have been shown to occur spontaneously prior to ligand binding.

VII. Regulation of the "efficacy" of AChR response by allosteric effectors acting on desensitization.

Among other consequences, the model predicts that the "efficacy" of the receptor response to acetylcholine pulses might be regulated by effectors which differentially stabilize the I and D state or change the transition rates towards these states. (Changeux et al., 1976, Heidmann and Changeux, 1979, 1982, Changeux and Heidmann 1987). The simplest situation is that where the effector stabilizes the desensitized conformation of the AChR when it binds to the acetylcholine site. Under physiological conditions, acetylcholine may play this role (Magleby and Pallotta 1981, see chapter VI) leading to an "homosynaptic" regulation of postsynaptic response efficacy. Interestingly, the dissociation constant for the desensitized state fits with the nonquantal leak concentration of acetylcholine measured at the level of the postsynaptic membrane in the presence of esterase inhibitor (Katz and Miledi 1977). Accordingly, the non-quantal release of neurotransmitter may exert a regulatory role in pre-setting the $R \rightleftharpoons D$ equilibrium to a finite steady-state value.

Regulation of desensitization may also take place via allosteric sites distinct from the agonist binding site as long as their affinity differ in at least some of the multiple conformational states of the AChR. The best characterized allosteric sites are the high and low affinity sites for noncompetitive blockers. The high affinity site has been related to the ion channel (see chapters IV, and V), yet it might still be available for regulatory ligands in its close channel conformation where, despite a reduced accessibility, its affinity might still be higher than in the active conformation. The low affinity sites possess a different pharmacological specificity (Heidmann et al., 1978, 1983b) and are readily available to regulate desensitization.

At equilibrium, these noncompetitive blockers indeed reversibly enhance (to different extent) the binding of cholinergic ligands to the acetylcholine receptor site, converting the shape of their binding curve from a sigmoid to a hyperbola (Cohen et al., 1974). Conversely, cholinergic ligands potentiate their binding (Grünhagen and Changeux 1976, Krodel et al., 1979, Heidmann et al., 1983b). Under conditions of rapid mixing they enhance the rate of fast desensitization of the electrophysiological (Magazanik and Vyskocil 1975, Magleby and Pallotta 1981) and ion flux response (references in Maleque et al., 1982, Karpen et al., 1982) and, in parallel, accelerate the intermediate and slow transitions of the menbrane-bound AChR recorded by diverse methods (Grünhagen and Changeux 1976, Weiland et al., 1977, Heidmann and Changeux 1979, Heidmann et al., 1983b, Cohen and Boyd 1979, Young and Sigman 1981, 1983, Sine and Taylor 1982).

The endogenous counterparts of the noncompetitive blockers are not known but physiologically significant effectors such as Ca^{++} ions and electrical potential regulate these transitions. Ca^{++} ions accelerate desensitization and/or stabilize the high affinity state (Magazanik and Vyskocil 1970, Cohen et al., 1974, Heidmann and Changeux 1979, Oswald 1983) in particular when applied to the cytoplasmic side of the membrane (Miledi 1980). Hyperpolarisation also accelerates rapid desensitization while depolarization has the opposite effect (references in Takeyasu et al., 1983).

In a systematic investigation on the effects of peptides on the nicotinic peripheral receptor carried out in this laboratory, a polypeptide hormone of 49 amino acids produced by the thymus, thymopoietin (references in Audhya et al., 1981), was found to affect desensitization of electric organ and muscle AChR. Its effect was recorded both by patch clamp methods on C2 mouse muscle cells in culture and by fluorescence spectroscopy with Dansyl-C6-Choline and *T. marmorata* AChR-rich membranes (Revah et al., 1987). Previous work indicated that thymopoietin causes a decrease of muscle action potential recorded by electromyography in rat (Goldstein 1974) and binds to *Torpedo* AChR with a high affinity (K_D $4x10^{-10}$ M) and in an α-bungarotoxin sensitive manner (Venkatasubramanian et al., 1986). Thymopoietin, in addition, accelerates desensitization via an original mechanism : its effect on desensitization becomes manifest only in the presence of Ca^{++} ions while its binding to the agonist site does not require Ca^{++}. Recordings with outside-out patches further show that Ca^{++} ions have to be applied on the cytoplasmic face of the AChR to exert their effect. In other words, the enhanced desensitization of the AChR to acetylcholine is elicited by the conjunction of two distinct signals, Ca^{++} ions and thymopoietin, on both sides of the membrane.

In humans, *myasthenia gravis* is often associated with thymic abnormalities and thymectomy frequently leads to an improvement of the patients. This effect might then result from a release of a direct and/or allosteric blocking action of thymopoietin on motor end plate AChR. An implication of thymopoietin in the pathology of *myasthenia gravis* has thus been suggested in addition to, or as alternative of, an autoimmune reaction to the AChR (Goldstein 1974). It may account for the often observed discrepancy between the levels of circulating anti-AChR antibodies and the severity of the myasthenic symptoms.

Covalent modifications such as phosphorylation have been shown, in the past, to regulate the allosteric transition of classical regulatory enzymes such as phosphorylase b (review Krebs and Beavo 1979) and it was suggested that such effect might occur in the case of the acetylcholine receptor (Changeux et al., 1983) and of the ion channels (review Nestler and Greengard 1984). Interestingly, recent evidence indicates that phosphorylation of the AChR accelerates desensitization of the ion flux response (Albuquerque et al., 1986, Middleton et al., 1986, Huganir et al., 1986). Covalent modifications may thus regulate the response efficacy of the AChR by modulating its desensitization transitions in addition to, or as alternative of, the reversible binding of allosteric ligands.

The allosteric transitions of the AChR between a resting "activable" and desensitized "inactive" conformations thus appear as built-in molecular mechanisms available for the regulation of synapse efficacy at the postsynaptic level by extra and intra-cellular communication signals

VIII. Conclusions : from intramolecular transitions to intermolecular interactions.

AChR carries several categories of topologically distinct binding sites and the interactions between these sites are mediated by several distinct allosteric transitions. Similar allosteric sites and conformational transitions have been recognized with other ionic channels, including the voltage sensitive Na^+ channel (Table 1). One example is the GABA-benzodiazepine receptor (Fischer and Olsen 1986) which is regulated by a wide variety of allosteric effectors. Among the negative effectors are the β-carbolines (Hunkeler et al., 1981, Braestrup et al., 1982) and some barbiturates which can be considered as homologous to the local anesthetics and other noncompetitive blockers of the AChR (Cohen et al., 1974, Cohen 1978, Heidmann and Changeux 1979). Positive ones are the benzodiazepines (Tallman et al., 1978, Supavilai and Karobath 1980, Guidotti et al., 1978, Skerritt et al., 1982), which, yet, have no equivalent with nicotinic AChR. Such ligands may lead to a "potentiation" or to a "depression" of the response depending of the initial balance between the R and D states and of the relative affinities of these ligands for the R and D states respectively (Monod et al., 1965, Rubin and Changeux 1966).

AChR and other ligand-regulated ion channels, however, exhibit distinctive properties which are not shared with classical cytoplasmic allosteric proteins (Table 1). In particular, their transmembrane disposition creates geometrical constraints which reduce the symmetry properties accessible (see Changeux 1969) to rotational axes perpendicular to the plane of the membrane, an organisation which favors the transport of permeant ions through the membrane (Table 1). As a consequence the molecule possesses a transverse polarity which makes possible the binding of different categories of ligands to both sides of the membrane and thus permits the integration of multiple extracellular and intracellular signals by "concerted" transitions of its transmembrane structure.

Such allosteric transitions constitute, at least, one category of elementary mechanism able to mediate intramolecular receptor-receptor interactions. In addition, they may be involved in intermolecular coupling between distinct receptor molecules. Within the cytoplasmic membrane, nearest-neighbor interactions may take place between receptor and/or channel proteins which may form two-dimensional lattices and thus exhibit positive cooperative interactions (see Changeux et al., 1967b, Changeux 1969, Mazet et al., 1987). In the case of Torpedo AChR, side-by-side dimers have also been identified (review Popot and Changeux 1984) . Yet, the physiological significance of these supra-molecular organisations remains uncertain. Intermolecular interactions between topologically distant receptors and/or ionic channels require the propagation of signals at the cellular level. These signals might be internal second messengers emitted by receptors distributed in different domains of the neuronal surface. These second messengers may interact with the considered receptor either directly (e.g. Ca^{++} ion in the case of the AChR) or indirectly via covalent modifications (e.g. as a consequence of the activation of a protein kinase [see Nestler and Greengard 1984]). The electrical potential of the postsynaptic cell, in particular, though not necessarily, its firing (Hebb, 1949) may serve as "global" signal regulating the allosteric transitions of ligand-regulated and voltage-sensitive ion channels, again, either directly, if a difference of dipole moment exists between their activable and inactivable conformation or indirectly, via second messengers and covalent modifications.

Timing relationships between converging signals produced by distinct communication pathways might simply result from the "concerted" character of the allosteric transitions of at least one allosteric receptor or ion channel, thus making possible the short-term association of these pathways (Heidmann and Changeux 1982, Changeux and Heidmann 1987, Changeux et al., 1986b).

Significant differences may exist between these ligand-regulated (or voltage-sensitive) ion channels and the enzyme-linked receptors such as the β-adrenergic receptor (Dixon et al., 1986), or the acetylcholine muscarinic receptor (Kubo et al., 1986) which appear as more complex supra-molecular assemblies. Yet, it is timely to test the validity of these molecular models with the adequate preparation under physiologically relevant situations.

CLASSICAL ALLOSTERIC PROTEIN	**ACETYLCHOLINE RECEPTOR** (Na^+, K^+ channel)
SYMMETRICAL OLIGOMER	300 KD $\alpha_2\beta\gamma\delta$ Sequence homologies quasi-5-fold Sy
ALLOSTERIC SITES	Active sites : (+) Ach (-) d-tubocurarine
	Allosteric sites : (-) Non competitive blockers (-) Thymopoietin (-) Ca^{++}, Hyperpolarisation (-) Phosphorylation
POSITIVE COOPERATIVITY	Between non identical Ach binding sites
ALLOSTERIC TRANSITIONS between few, discrete, pre-existing, states	Activation (low affinity Ach) **R ⇄ D** Desensitization (high affinity Ach) **I, D**

TABLE 1 : Comparison of the properties of several ligand-regulated ion channels including the acetylcholine nicotinic receptor from muscle and electric organ with those of classical allosteric proteins. References for classical allosteric proteins can be found in Monod et al. (1963, 1965), for the GABA receptor in Fisher and Olsen (1986), for the voltage-sensitive Na^+ channel in Noda et al. (1984) and for the gap junction protein in Kumar and Gilula (1986).

GABA RECEPTOR (Cl^- channel)	**VOLTAGE-SENSITIVE Na^+ channel**	**GAP JUNCTION PROTEIN** (connexon)
~200 KD (?) $\alpha_2\beta_2$ (?) 2-fold Sy (?)	260 KD single polypeptide **4x[~275 AA] repeats** quasi 4-fold Sy	300 KD α_6 6-fold Sy
Active sites : (+) GABA (-) Bicuculline	Active sites : (+) Depolarisation (-) Tetrodotoxin	
Allosteric sites : (+) Diazepam (-) β-Carbolines (±) Barbiturates (-) Picrotoxin	Allosteric sites : (+) Veratridine (+) Scorpion toxins (+) Sea anemon toxins	Allosteric sites : (+) Phosphorylation (-) Ca^{++}, H^+
?	Positive cooperativity for Na^+ transport	Positive cooperativity between connexons
Activation (low affinity, GABA) **R ⇄ A** Desensitization (high affinity, GABA) **D**	Activation : **R ⇄ A** Inactivation : **I**	

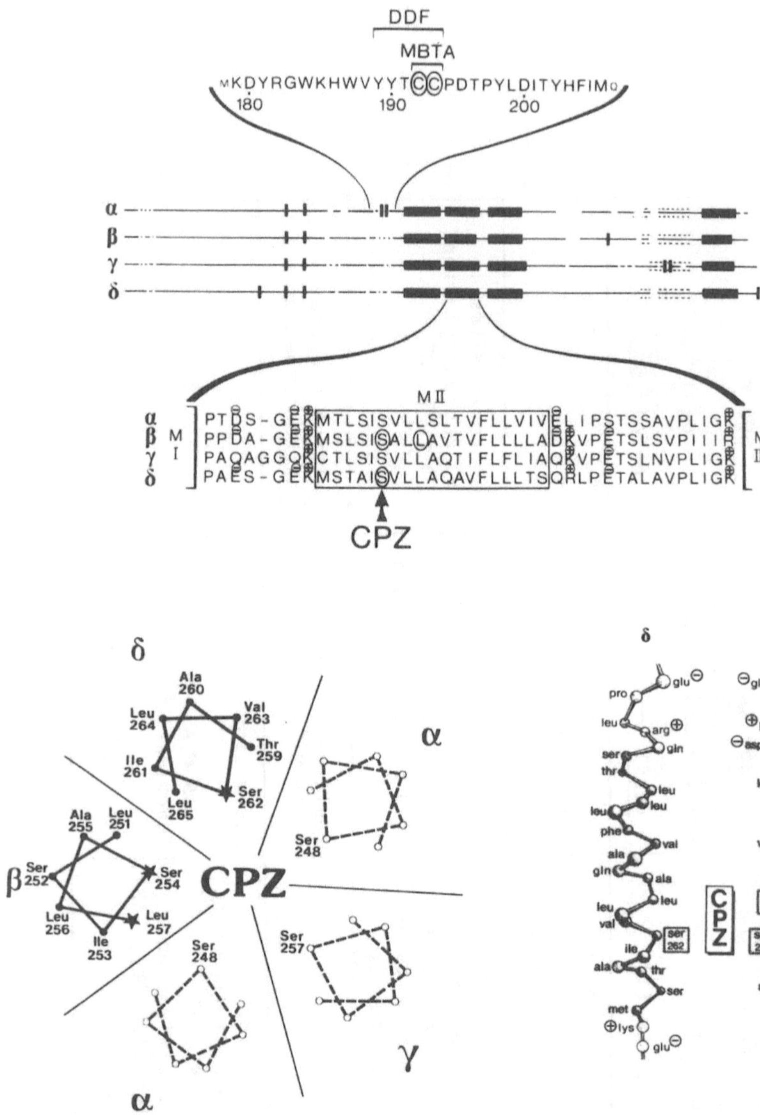

FIGURE : Location of amino acids belonging to the acetylcholine binding site (DDF, MBTA) and to the high affinity site for the noncompetitive "channel" blocker chlorpromazine (CPZ) on the amino acid sequence of the four subunits of the AChR from Torpedo marmorata. Top : aligned sequences of the four subunits. Bottom : hypothetical model of the front and side-views of the high affinity site for chlorpromazine (from Giraudat et al., 1987).

References :
- Adams, P.R. (1981). Acetylcholine receptor kinetics. J. Membrane Biol., 58, 161-174.
- Adams, P.R. and Galvan, M. (1986). Voltage-dependent currents of vertebrate neurons and their role in membrane excitability. Advances in Neurology, 44, 137-170.
- Agnati, L., Fuxe, K., Benfenati, F., Calza, L., Battistini, N. and Ogren, S. O. (1983a). Receptor-receptor interactions : possible new mechanisms for the reaction of some antidepressant drugs. In : Frontiers in Neuropsychiatric research (E. Usdin, M. Goldstein, A.T. Friedhoff and Gergotas, A., eds), p.p 301-318, Mcmillan Press London.
- Agnati, L.F., Fuxe, K., Benfenati, F., Battistini, N., Härfstrand, A., Hökfelt, T., Tatemoto, K. and Mutt, V. (1983b). Neuropeptide Y in vitro selectively increases the number of α_2-adrenergic binding sites in membranes of the medulla oblongata of the rat. Acta Physiol. Scand., 118, 293-295.
- Agnati, L.F., Fuxe, K., Benfenati, F., Battistini, N., Zini, I., Camurri, M. and Hökfelt, T. (1984). Postsynaptic effects of neuropeptide comodulators at central monoamine synapses. Neurology and Neurobiology, 8B,191-198.
- Alburquerque, E.X., Deshpande, S.S., Aracava, Y., Alkondon, M. and Daly, J.W. (1986). A possible involvement of cyclic AMP in the expression of desensitization of the nicotinic acetylcholine receptor. FEBS Lett. 199, 113-120.
- Arnone, A. (1972). X-ray diffraction study of binding of 2,3-diphosphoglycerate to human deoxyhaemoglobin. Nature 231 : 146-149.
- Audhya, T., Schlesinger, D.H. and Goldstein, G. (1981). Complete amino acid sequences of bovine thymopoietins I, II, and II : closely homologous polypeptides. Biochemistry 20, 6195-6200.
- Ballivet, M., Patrick, J., Lee, J. and Heinemann, S. (1982). Molecular cloning of cDNA coding for the gamma-Subunit of Torpedo acetylcholine receptor. Proc. Natl. Acad. Sci. USA 79, 4466-4470.
- Bon, F., Lebrun, E., Gomel, J., Van Rappenbusch, R., Cartaud, J., Popot, J. L. and Changeux, J. P. (1984). Image analysis of the heavy form of the acetylcholine receptor from Torpedo marmorata. J. Mol. Biol., 176, 205-237.
- Boulter J., Evans, K.L., Martin, G., Gardner, P.D., Connolly, J., Heinemann, S. and Patrick, J. (1986). Mouse muscle acetylcholine receptor molecular cloning of α, β, γ and δ-subunit cDNA's and expression in Xenopus Laevis oocytes. Abstr. Soc. for Neurosci. 12, 40.2, 146.
- Boyd, N. D. and Cohen, J. B. (1980). Kinetics of binding of [^{3}H] acetylcholine and [^{3}H] carbamoylcholine to Torpedo postsynaptic membranes : slow conformational transitions of the cholinergic receptor. Biochemistry 19, 5344 - 5358.
- Braestrup, C., Schmiechen, R., Neff, G., Nielsen, M., Petersen,, E.N. (1982). Interaction of convulsive ligands with benzodiazepine receptors. Science 216, 1241-1243.

- Brisson, A. and Unwin, P.N.T (1985). Quaternary structure of the acetylcholine receptor. Nature 315, 474-477
- Cartaud, J., Benedetti, L., Cohen, J. B., Meunier, J. C. and Changeux, J. P. (1973). Presence of a lattice structure in membrane fragments rich in nicotinic receptor protein from the electric organ of Torpedo marmorata. FEBS Lett., 33, 109-113.
- Cash, D., Aoshima, H., Pasquale, E. and Hess, G. (1985). Acetylcholine receptor-mediated ion fluxes in Electrophorus electricus and Torpedo californica membrane vesicles. Rev. Physiol. Biochem. Pharmacol., 102, 74-117.
- Changeux, J.P. (1969). Remarks on the symmetry and cooperative properties of biological membranes. In "Symmetry and function in biological systems at the macromolecular level". A Engström and B. Strandberg edit. 235-256, Wiley, New-York.
- Changeux, J.P. (1981). The acetylcholine receptor : An "allosteric" membrane protein. Harvey Lectures, 75, 85-254.
- Changeux, J.P. (1986). Coexistence of neuronal messengers and molecular selection. Progr. Brain Res. 68, 373-403.
- Changeux, J. P. and Thiéry, J.P. (1968). On the excitability and cooperativity of biological membranes. In "Regulatory Functions of Biological Membranes". Ed. J. Jarnefelt, 116-138, Elsevier Amsterdam.
- Changeux, J.P. and Heidmann, T. (1987). Allosteric receptors and molecular models of learning. "New Insights into Synaptic Function", Salk Institute G. Edelman, W.E. Gall and W.M. Cowan Wiley, New York (in press).
- Changeux, J.P, Podleski, T. and Wofsy, L. (1967a). Affinity labeling of the acetylcholine receptor. Proc. Natl. Acad. Sci. Wash. 58, 2063-2070.
- Changeux, J.P., Thiéry, J.P., Tung, Y. and Kiettel, C. (1967b). On the cooperativity of biological membranes. Proc. Natl. Acad. Sci. USA 57, 335-341.
- Changeux, J. P., Benedetti, L., Bourgeois, J. P., Brisson, A., Cartaud, J., Devaux, P. F., Grünhagen, H. H., Moreau, M., Popot, J. L., Sobel, A. and Weber, M. (1976). Some structural properties of the cholinergic protein in its membrane environment relevant to its function as a pharmacological receptor. Cold Spring Harb. Symp. Quant. Biol. 40, 203-210
- Changeux, J. P., Bon, F., Cartaud, J., Devillers-Thiery, A., Giraudat, J., Heidmann, T., Holton, B., Nghiem, H. O., Popot, J. L, Van Rapenbusch, R. and Tzartos, S. (1983). Allosteric properties of the acetylcholine receptor protein from Torpedo marmorata. Cold Spring Harbor Symp. Quant. Biol. 48, 35-52
- Changeux, J.P., Devillers-Thiéry, A. and Chemouilli, P. (1984). Acetylcholine receptor : an allosteric protein. Science 225, 1335-1345.
- Changeux, J.P., Pinset, C. and Ribera, A.B. (1986a). Single channel analysis of the effects of low concentrations of chlorpromazine and phencyclidine on the acetylcholine receptor in C2 mouse myotubes. J. Physiol. (London) 378, 495-513.

- Changeux, J.P., Klarsfeld, A. and Heidmann, T. (1986b). The acetylcholine receptor and molecular models for short and long term learning. Dahlem Konferenzen. In "The cellular and molecular bases of learning". J.P. Changeux and M. Konishi, ed. Springer-Verlag, Berlin (in press).
- Claudio, T., Ballivet, M., Patrick, J. and Heinemann, S. (1983). Nucleotide and deduced amino acid sequences of Torpedo californica acetylcholine receptor gamma-subunit. Proc. Natl. Acad. Sci., USA 80, 1111-1115.
- Cohen, J.B. and Boyd, N.D. (1979). Conformational transitions of the membrane bound cholinergic receptor. In "Catalysis in Chemistry and Biochemistry", B. Pulman and O. Ginsburg, Eds (D. Reidel, Publ.).
- Cohen, J.B. (1978). Ligand binding properties of membrane-bound cholinergic receptor of Torpedo marmorata. In "Membrane Function". Solomon, A. K., Karnovsky, M., ed. Harvard Univ. Press. (Cambridge), 99-127.
- Cohen, J. B., Weber, M. and Changeux, J.P. (1974). Effects of local anesthetics and calcium on the interaction of cholinergic ligands with the nicotinic receptor protein from Torpedo marmorata. Mol. Pharmacol., 10, 904-932.
- Criado, M., Hochschwender, S., Sarin, V., Fox, J. L. and Lindstrom, J. (1985). Evidence for unpredicted transmembrane domains in acetylcholine receptor subunits. Proc. Natl. Acad. Sci., USA 82, 2004-2008.
- Criado, M., Sarin, V., Fox, J.L. and Lindstrom, J. (1986). Evidence that the acetylcholine binding site is not formed by the sequence α 127-143 of the acetylcholine receptor. Biochemistry 25, 2839-2846.
- Dennis, M., Giraudat, J., Kotzyba-Hibert, F., Goeldner, M., Hirth, C., Chang, J.Y. and Changeux, J.P. (1986). A photo-affinity ligand of the acetylcholine binding site predominantly labels the region 179-207 of the alpha subunit on native acetylcholine receptor from Torpedo marmorata. FEBS Lett., 207, 243-247.
- Devillers-Thiéry, A., Giraudat, J., Bentaboulet, M. and Changeux, J.P. (1983). Complete mRNA coding sequence of the acetylcholine binding alpha subunit of Torpedo marmorata acetylcholine receptor : A model for the transmembrane organization of the polypeptide chain. Proc. Natl. Acad. Sci., USA 80, 2067-2071.
- Dixon, R.A.F., Kobilka, B.K., Strader, D.J., Benovic, J.L., Dohlman, H.G., Frielle, T., Bolanowski, M.A., Bennett, C.D., Rands, E., Diehl, R.E., Mumford, R.A., Slater, E.E., Sigal, I.S., Caron, M.G., Lefkowitz, R.J. and Strader, C.D. (1986). Cloning of the gene and cDNA for mammalian β-adrenergic receptor and homology with rhodopsin. Nature 321, 75-79.
- Dunn, S. M. J., Blanchard, S. G. and Raftery, M.A. (1980). Kinetics of carbamylcholine binding to membrane-bound acetylcholine receptor monitored by fluorescence changes of a covalently-bound probe. Biochemistry 19,

- Eldefrawi, A. T., Eldefrawi, M. E., Albuquerque, E. X., Oliveira, A.C., . Mansour, N., Adler, M., Daly, J. W., Brown, G. B., Burgermeister, W. and Witkop, B. (1977). Perhydrohistrionicotoxin : a potential ligand for the ion conductance modulator of the acetylcholine receptor. Proc. Natl. Acad.Sci., USA 74, 2172-2176.
- Feltz, A. and Trautmann, A. (1980). Interaction between nerve-released acetylcholine and bath applied agonists at the frog end plate. J. Physiol., (London) 299, 533-552.
- Finer-Moore, J. and Stroud, R.M. (1984). Amphipathic analysis and possible formation of the ion channel in an acetylcholine receptor. Proc. Natl. Acad. Sci., USA 81, 155-159.
- Fischer, J.B and Olsen, R.W. (1986). Biochemical aspects of GABA/Benzodiazepine receptor function. Benzodiazepine/GABA Receptors and Chloride Channels : Structural and Functional properties, p.p 241-259, Alan R. Liss, Inc.
- Furois-Corbin, S. and Pullman, A. (1986). Theoretical studies of the packing of α-helices of poly (L-alanine) into transmembrane bundles. Possible significance for ion transfer. Biochim. Biophys. Acta. 860, 165-177.
- Fuxe, K., Agnati, L.F, Benfenati, F., Cimmino, M., Algeri, S., Hökfelt, T, and Mutt V. (1981). Modulation by cholecystokinins of ^{3}H-spiroperidol binding in rat striatum : evidence for increased affinity and reduction in the number of binding sites. Acta Physiol. Scand. 113, 567-569.
- Fuxe, K., Aganti, L., Härfstrand, A., Janson, A., Anderson, K., Ruggeri, M., Zoli, M. and Goldstein, M. (1986). Morphofunctional studies on the neuropeptide Y/adrenaline costoring nerve terminal systems in the dorsal cardiovascular region of the medulla oblongata. Focus on receptor-receptor interactions in co-transmission. Progr. Brain Res. 68, 303-320.
- Fuxe, K., Agnati, L., Benfenati, F., Celani, M., Zini, I., Zoli, M. and Mutt., V. (1983). Evidence for the existence of receptor-receptor interaction in the central nervous system. Studies on the regulation of monoamine receptors by neuroleptics. J. Neuronal Trans. 18, 165-179
- Giraudat, J., Dennis, M., Heidmann, T., Chang, J.Y. and Changeux, J.P. (1986). Structure of the high affinity binding site for noncompetitive blockers of the acetylcholine receptor : Serine-262 of the δ subunit is labeled by ^{3}H chlorpromazine. Proc. Natl. Acad. Sci., U.S.A. 83, 2719-2723.
- Giraudat, J., Dennis, M., Heidmann, T., Haumont, P.T, Lederer, F. and Changeux, J.P. (1987). Structure of the high-affinity binding site for noncompetitive blockers of the acetylcholine receptor : ^{3}H chlorpromazine labels homologous residues in the beta and delta chains. (submitted).
- Giraudat, J., Devillers-Thiéry, A., Auffray, C., Rougeon, F. and Changeux, J.P. (1982). Identification of a cDNA clone coding for the acetylcholine binding subunit of Torpedo marmorata acetylcholine receptor. EMBO J. 1, 713-717.

- Goeldner, M.P. and Hirth, C.G. (1980). Specific photoaffinity labeling induced by energy transfer : application to irreversible inhibition of acetylcholinesterase. Proc. Natl. Acad. Sci. USA 77, 6439-6442.
- Goldstein, G. (1974). Isolation of bovine thymin : a polypeptide hormone of the thymus. Nature 247, 11-14.
- Grünhagen, H. H. and Changeux, J.P. (1976). Studies on the electrogenic action of acetylcholine with Torpedo marmorata electric organ. Quinacrine : a fluorescent probe for the conformational transitions of the cholinergic receptor protein in its membrane bound state. J. Mol. Biol., 106, 497-516.
- Grünhagen, H.H., Iwatsubo, M. and Changeux, J.P. (1977). Fast kinetic studies on the interaction of cholinergic agonists with the membrane-bound acetylcholine receptor from Torpedo marmorata as revealed by quinacrine fluorescence. Eur. J. Biochem. 80, 225-242.
- Guidotti, A., Toffano, G. and Costa, E. (1978). An endogenous protein modulates the affinity of GABA and benzodiazepine receptors in rat brain. Nature 275 : 553-555.
- Guy, H.R. (1984). Structural model of the acetylcholine receptor channel based on partition energy and helix packing calculations. Biophys. J., 45, 249-261.
- Hamill, O.P., Marty, A., Neher, E., Sakmann, B. and Sigworth, F.J. (1981). Improved patch clamp techniques for high resolution current recording from cells and cell-free patches. Pfugers Archiv. 391, 85-100.
- Haring, R., Kloog, Y. and Sokolovsky, M. (1984). Localization of phencyclidine binding sites on α and β subunits of the nicotinic acetylcholine receptor from Torpedo ocellata electric organ using azido phencyclidine. J. Neuroscience 4, 627-637.
- Hazelbauer, G. and Changeux, J.P. (1974). Reconstitution of a chemically excitable membrane. Proc. Natl. Acad. Sci., USA 71, 1479-1483.
- Hebb, D.O. (1949). The organization of behavior : a neuro-psychological theory. Wiley, New York.
- Heidmann, T. and Changeux, J.P. (1979). Fast kinetics studies on the interaction of a fluorescent agonist with the membrane-bound acetylcholine receptor from Torpedo marmorata. Eur. J. Biochem., 94, 281-296.
- Heidmann, T. and Changeux, J. P. (1980). Interaction of a fluorescent agonist with the membrane-bound acetylcholine receptor from Torpedo marmorata in the millisecond time range : resolution of an "intermediate" conformational transition and evidence for positive cooperative effects. Biochem. Biophys. Res. Comm., 97, 889-896.
- Heidmann, T. and Changeux, J.P. (1982). Un modèle moléculaire de régulation d'efficacité d'un synapse chimique au niveau postsynaptique. C. R. Acad. Sci., 295, 665-670.

- Heidmann, T. and Changeux, J. P. (1984). Time-resolved photolabeling by the noncompetitive blocker chlorpromazine of the acetylcholine receptor in its transiently open and closes ion channel conformations. Proc. Natl. Acad. Sci., USA 81, 1897-1901.
- Heidmann, T. and Changeux, J.P. (1986). Characterization of the transient agonist-triggered state of the acetylcholine receptor rapidly labeled by the non-competitive blocker [^{3}H] chlorpromazine : additional evidence for the open channel conformation. Biochemistry 25, 6109-6113.
- Heidmann, T., Sobel, A. and Changeux, J.P. (1978). Recovery of allosteric interactions between a fluorescent cholinergic agonist and local anesthetics after removal of the detergent from cholate-solubilized membrane fragments rich in acetylcholine receptor. FEBS. Lett. 94, 397 - 404.
- Heidmann, T., Bernhardt, J., Neumann, E. and Changeux, J.P. (1983a). Rapid kinetics of agonist binding and permeability response analysed in parallel on acetylcholine receptor-rich membranes from Torpedo marmorata. Biochemistry 22, 5452-5459.
- Heidmann, T., Oswald, R.E. and Changeux, J.P. (1983b). Multiple sites of action for non competitive blockers on acetylcholine receptor-rich membrane fragments from Torpedo marmorata. Biochemistry 22, 3112-3127.
- Hucho, F.L, Oberthür, W. and Lottspeich, F. (1986). The ion channel of the nicotinic acetylcholine receptor is formed by the homologous helices M II of the receptor subunits. FEBS Lett. 205, 137-142.
- Hucho, F. (1986). The nicotinic acetyclholine receptor and its ion channel. Eur. J. Biochem., 158, 211-226.
- Huganir, R.L., Delcour, A.H., Greengard, P. and Hess, G.P. (1986). Phosphorylation of the nicotinic acetylcholine receptor regulates its rate of desensitization. Nature 321, 744-776.
- Hunkeler, W., Möhler, H., Pieri, L., Polc, P., Bonetti, E.P., Cumin, R., Schaffner, R. and Haefely, W. (1981). Selective antagonists of benzodiazepines. Nature 290, 514-516.
- Jackson, M.B. (1984). Spontaneous openings of the acetylcholine receptor channel. Proc. Natl. Acad. Sci. USA, 81, 3901-3904.
- Kaldany, R.R.J. and Karlin, A. (1983). Reaction of quinacrine mustard with the acetylcholine receptor from Torpedo californica : Functional consequences and sites of labeling. J. Biol. Chem., 258, 6232-6242.
- Kao, P., Dwork, A., Kaldany, R., Silver, M., Wideman, J., Stein, S., and Karlin A. (1984). Identification of the α-subunit half-cystine specifically labeled by an affinity reagent for the acetylcholine receptor binding site. J. Biol. Chem., 259, 11662-11665.
- Karlin, A. (1980). Molecular properties of nicotinic acetylcholine receptors. In "Cell Surface reviews" (Poste G., Nicolson, G. L., Cotman, C. W., Eds) 6, 191 - 260. New York

- Karlin, A. (1983). Anatomy of a receptor. Neuroscience Comment., 1, 111-123.
- Karlin, A. (1969). Chemical modification of the active site of the acetylcholine receptor. J. Gen. Physiol., 54, 245-264.
- Karpen, J. W., Aoshima, H., Abood, L. G. and Hess G.P. (1982). Cocaine and phencyclidine inhibition of the acetylcholine receptor : Analysis of the mechanisms of action based on measurements of ion flux in the millisecond-to-minute time region. Proc. Natl. Acad. Sci., USA 79, 2509-2513.
- Kasai, M. and Changeux, J.P. (1971). In vitro excitation of purified membrane fragments by cholinergic agonists. I. Pharmacological properties of the excitable membrane fragments. II. The permeability change caused by cholinergic agonists. III. Comparison of the dose response curves to decamethonium with the corresponding binding curves of decamethonium to the cholinergic receptor. J. Memb. Biology., 6, 1-80.
- Katz, B. and Miledi, R. (1977). Transmitter leakage from motor nerve endings. Proc. R. Soc. Lond. B., 196, 59-72.
- Katz, B. and Thesleff, S. (1957). A study of the "desensitization" produced by acetylcholine at the motor end-plate. J. Physiol., (London) 138, 63-80.
- Kistler, J. and Stroud, R.M. (1981). Crystalline arrays of membrane-bound acetylcholine receptor. Proc. Natl. Acad. Sci., USA 78, 3678-3682.
- Klarsfeld, A., Devillers-Thiéry, A., Giraudat, J. and Changeux, J.P. (1984). A single gene codes for the nicotinic acetylcholine receptor alpha-subunit in Torpedo marmorata : Structural and developmental implications. EMBO J., 3, 35-41.
- Kloog, Y., Flynn, D., Hoffman, A. R. and Axelrod, J. (1980). Enzymatic carboxymethylation of the acetylcholine receptor. Biochem. Biophys. Res. Comm., 97, 1426-1480.
- Krebs, G. and Beavo, J. (1979). Phosphorylation, dephosphorylation of enzymes. Annu. Rev. Biochem. 48, 923.
- Krodel, E.K., Beckman, R.A. and Cohen, J.B. (1979). Identification of a local anesthetic binding site in nicotinic post-synaptic membranes isolated from Torpedo marmorata electric tissue. Mol. Pharmacol., 15, 294-312.
- Kubo, T., Noda, M., Takai, T., Tanabe, T., Kayano, T., Shimizu, S., Tanaka, K., Takahashi, H., Hirose, T., Inayama, S., Kikuno, R., Miyata, T. and Numa, S. (1985). Primary structure of delta subunit precursor of calf muscle acetylcholine receptor deduced from cDNA sequence. Eur. J. Biochem., 149, 5-13.
- Kubo, T., Fukuda, K., Mikami, A., Maeda, A., Takahashi, H., Mishina, M., Haga, T., Haga, K., Ichiyama, A., Kangawa, K., Kojima, M., Matsuo, H., Hirose, T. and Numa, S. (1986). Cloning sequencing and expression of complementary DNA encoding the muscarinic acetylcholine receptor. Nature 323, 411-416.
- Kumar, N.M. and Gilula N.B. (1986). Cloning and characterization of human and rat liver cDNAs coding for a gap junction protein. J. Cell Biol. 103, 767-776.

- Lagenbuch-Cachat, J., Bon, C., Goeldner, M., Hirth, C. and Changeux, J.P. (1987). Photoaffinity labelling by aryldiazonium derivatives of Torpedo marmorata acetylcholine receptor. Submitted.
- Lindstrom, J. (1986). Probing nicotinic acetylcholine receptors with monoclonal antibodies. Trends in Neurosci. 9, 401.
- Lindstrom, J., Merlie, J. P., Yogeeswaran, G. (1979). Biochemical properties of acetylcholine receptor subunits from Torpedo californica. Biochemistry 18, 4465-4470.
- McCarthy, M.P., Earnest, J.P., Young, E.F., Choe, S. and Stroud, R.M. (1986). The molecular neurobiology of the acetylcholine receptor. Ann. Rev. Neurosci., 383-413.
- McCormick, D.J. and Atassi, Z. (1984). Localization and synthesis of the acetylcholine-binding site in the alpha-chain of the Torpedo californica acetylcholine receptor. Biochem. J., 224, 995-1000.
- Magazanik, L.G., and F. Vyskocil (1970). Dependence of acetylcholine desensitization on the membrane potential of frog muscle fibre and on the ionic changes in the medium. J. Physiol., (Lond.) 210, 507-518.
- Magazanik, L.G. and Vyskocil (1975). The effect of temperature on desensitization kinetics at the post-synaptic membrane of the frog muscle fibre. J. Physiol., (London), 249, 285-300.
- Magleby, K.L. and Pallotta, B.S. (1981). A study of desensitization of acetylcholine receptors using nerve-released transmitter in the frog. J. Physiol., (London) 316, 225-250.
- Maleque, M. A., Souccar, C., Cohen, J. B. and Albuquerque, E.X. (1982). Meproadifen reaction with the ionic channel of the acetylcholine receptor : potentiation of agonist-induced desensitization at the frog neuromuscular junction. Mol. Pharmacol., 22, 636-647.
- Mazet, J.L., Schindler, H. and Changeux, J.P. (1987). Aggregation and coperativity induced by long range lipid effects on the acetylcholine receptor protein reconstituted in planar lipid bilayers. In preparation.
- Merlie, J. P. (1984). Biogenesis of the acetylcholine receptor, a multisubunit integral membrane protein. Cell 36, 573-575.
- Middleton, P., Jaramillo, F. and Schuetze, M. (1986). Forskolin increases the rate of acetylcholine receptor desensitization at rat soleus endplates. Proc. Natl. Acad. Sci. U.S.A, 83, 4967-4971.
- Miledi, R. (1980). Intracellular calcium and desensitization of acetylcholine receptors. Proc. R. Soc., London (Biol.) 209, 447-452.
- Mishina, M., Kurosaki, T., Tobimatsu, T., Morimoto, Y., Noda, M., Yamamoto, T., Terao, M., Lindstrom, J., Takahashi, T., Kuno, M. and Numa, S. (1984). Expression of functional acetylcholine receptor from cloned cDNAs. Nature 307, 604-608.
- Mishina, M., Tobimatsu, T., Imoto, K., Tanaka, K., Fujita, Y., Fukuda, K., Kurasaki, M., Takahashi, H., Morimoto, Y., Hirose, T., Inayama, S., Takahashi, T., Kuno, M. and Numa, S. (1985). Location of functional regions of acetylcholine receptor alpha-subunit by site-directed mutagenesis. Nature 313, 364-368.

- Mishina, M., Takai, T., Imoto, K., Noda, M., Takahashi, T., Numa, S., Methfessel, C. and Sakmann, B. (1986). Molecular distinction between fetal and adult forms of muscle acetylcholine receptor. Nature 321, 406.
- Monod, J., J.P. Changeux, and F. Jacob (1963). Allosteric proteins and cellular control systems. J. Mol. biol., 6, 306-328.
- Monod, J., Wyman, J. and Changeux, J. P. (1965). On the nature of allosteric transitions : a plausible model. J. Mol. Biol. 12, 88-118.
- Montal, M., Anholt, R. and Labarca, P. (1986). The reconstituted receptor. In "Ion channel reconstitution". C. Miller, ed., Plenum, 157-204.
- Muhn, P. and Hucho, F. (1983). Covalent labeling of the acetylcholine receptor from Torpedo electric tissue with the channel blocker [^{3}H] triphenylmethylphophonium by ultraviolet irradiation. Biochemistry 22, 421-425.
- Mulac-Jericevic, B. and Atassi, M.Z. (1986). Segment α-182-198 of Torpedo californica acetylcholine receptor contains a second toxin-binding region and binds anti-receptor antibodies. FEBS Lett. 199, 68-74.
- Neher, E. and Sakmann, B. (1976). Single channel currents recorded from membrane of denervated frog muscle fibers. Nature 260, 799-802.
- Neher, E. and Steinbach, J.H. (1978). Local anaesthetics transiently block currents through single acetylcholine-receptor channels. J. Physiol., (London) 277, 153-176.
- Nestler, E. and Greengard, P. (1984). Protein phosphorylation in the nervous system. Wiley New York.
- Neubig, R. R. and Cohen, J. B. (1980). Permeability control by cholinergic receptors in Torpedo post synasptic membranes : Agonist dose response relations measured at second and millisecond times. Biochemistry 19, 2770-2779.
- Neubig, R.R., Boyd, N.D. and Cohen J.B. (1982). Conformations of Torpedo acetylcholine receptor associated with ion transport and desensitization. Biochemistry 21, 3460-3467.
- Neumann, D., Barchan, D., Safran, A., Gershoni, J.M. and Fuchs, S. (1986). Mapping of the α-bungarotoxin binding site within the α-subunit of the acetylcholine receptor. Proc. Natl. Acad. Sci. USA 83, 3008-3011.
- Noda, M., Takahashi, H., Tanabe, T., Toyosato, M., Furutani, Y., Hirose, T., Asai, M., Inayama, S., Miyata, T. and Numa, S. (1982). Primary structure of alpha-subunit precursor of Torpedo californica acetylcholine receptor deduced from cDNA sequence. Nature 299, 793-797.
- Noda, M., Takahashi, H., Tanabe, T., Toyosato, M., Kikyotani, S., Hirose, T., Asai, M., Takashima, H., Inayama, S., Miyata, T. and Numa, S. (1983a). Primary structures of beta and delta-subunit precursors of Torpedo californica acetylcholine receptor deduced from cDNA sequences. Nature 301, 251-255

- Noda, M., Takahashi, H., Tanabe, T., Toyosato, M., Kikyotani, S., Furutani, Y., Hirose, T., Takashima, H., Inayama, S., Miyata, T. and Numa, S. (1983b). Structural homology of Torpedo californica acetylcholine receptor subunits. Nature 302, 528-532.
- Noda, M., Furutani, Y., Takahashi, H., Toyosato, M., Tanabe, T., Shimizu, S., Kikyotani, S., Kayano, T., Hirose, T., Inayama, S. and Numa, S. (1983c). Cloning and sequence analysis of calf cDNA and human genomic DNA encoding α-subunit precursor of muscle acetylcholine receptor. Nature 305, 818-823.
- Noda, M., Shimizu, S., Tanabe, T., Takai, T., Kayano, T., Ikeda, T., Takahashi, H., Nakayama, H., Kanaoka, Y., Minamino, N., Kangawa, K., Matsuo, H., Raftery, M.A., Hirose, T., Inayama, S., Hayashida, H., Miyata, T. and Numa, S. (1984). Primary structure of Electrophorus electricus sodium channel deduced from cDNA sequence. Nature 312, 121-127
- Oberthür, W., Muhn, P., Baumann, H., Lottspeich, F., Wittmann-Liebold, B. and Hucho, F. (1986). The reaction site of a noncompetitive antagonist in the delta-subunit of the nicotinic acetylcholine receptor. EMBO J. 5, 1815-1819.
- Oswald, R.E. (1983). Effects of calcium on the binding of phencyclidine to acetylcholine receptor-rich membrane fragments from Torpedo californica electroplaque. J. Neurochem. 41, 1077.
- Oswald, R., Sobel, A., Waksman, G., Roques, B. and Changeux, J. P. (1980). Selective labeling by [^{3}H-]trimethisoquin azide of polypeptide chains present in acetylcholine receptor rich membranes from Torpedo marmorata. FEBS Lett., 111, 29-34.
- Oswald, R.E. and Changeux, J.P. (1981a). Selective labeling of the delta-subunit of the acetylcholine receptor by a covalent local anesthetic. Biochemistry 20, 7166-7174.
- Oswald, R. and Changeux, J.P. (1981b). Ultraviolet light-induced labeling by noncompetitive blockers of the acetylcholine receptor from Torpedo marmorata. Proc. Natl. Acad. Sci., USA 78, 3925-3929.
- Perutz, M.F., Fermi, G., Abraham, D.J., Poyart, C. and Bursaux, E. (1986). Hemoglobin as a receptor of drugs and peptides : X-ray studies of the stereochemistry of binding. J. Am. Chem. Soc. 108, 1064-1078.
- Popot, J.L. and Changeux, J.P. (1984). The nicotinic acetylcholine receptor : structure of an oligomeric integral membrane protein. Physiol. Rev. 64, 1162-1184.
- Popot, J. L., Cartaud, J. and Changeux, J. P. (1981). Reconstitution of a functional acetylcholine receptor : incorporation into artificial lipid vesicles and pharmacology of the agonist-controlled permeability changes. Eur. J. Biochem. 118, 203-214.
- Prinz, H. and Maelicke, A. (1983). Interaction of cholinergic ligands with the purified acetylcholine receptor protein : Equilibrium binding studies. J. Biol. Chem. 258, 10263-10271.
- Raftery, M. A., Hunkapiller, M., Strader, C. D. and Hood, L. E. (1980). Acetylcholine receptor : complex of homologous subunits. Science 208, 1454-1457.

- Ratnam, M., Le Nguyen, D., Rivier, J., Sargent, P.B. and Lindstrom, J. (1986). Transmembrane topography of nicotinic acetylcholine receptor: Immunochemical tests contradict theoretical predictions based on hydrophobicity profiles. Biochemistry 25, 2633-2643.
- Reynolds, J. A. and Karlin, A. (1978). Molecular weight in detergent solution of acetylcholine receptor from Torpedo californica. Biochemistry 17, 2035-2038.
- Revah, F., Mulle, C., Tapan, A., Goldstein, G. and Changeux, J.P. (in preparation).
- Rubin, M.M. and Changeux, J.P. (1966). On the nature of allosteric transitions ; implications of non exclusive ligand binding. J. Mol. Biol. 21, 265-274.
- Saitoh, T., Oswald, R., Wennogle, L.P. and Changeux, J.P. (1980). Conditions for the selective labelling of the 66000 dalton chain of the acetylcholine receptor by the covalent non-competitive blocker 5-azido-^{3}H- trimethisoquin. FEBS Lett., 116, 30-36.
- Sakmann, B. , Patlak, J. and Neher, E. (1980). Single acetylcholine activated channels show burst-kinetics in presence of desensitizing concentrations of agonist. Nature 286, 71-73.
- Sakmann, B., Methfessel, C., Mishina, M., Takahashi, T., Takai, T., Kurasaki, M., Fukuda, K. and Numa, S. (1985). Role of acetylcholine receptor subunits in gating of the channel. Nature, 318, 538-543.
- Schoffeniels, E. and Nachmansohn, D. (1957). An isolated single electroplax preparation. I. New data on the effect of acetylcholine and related compounds. Biochim. Biophys. Acta. 26, 1-15.
- Sine, S.M. and Taylor, P. (1982). Local anesthetics and histrionicotoxin are allosteric inhibitors of the acetylcholine receptor. Studies of clonal muscle cells. J. Biol. Chem. 257, 8106-8114.
- Skerritt, J.H., Willow, M., Johnston, G.A.R. (1982). Diazepam enhancement of low affinity GABA binding to rat brain membrane. Neurosci. Lett. 29, 63-66.
- Stallcup, W. B. and Patrick, J. (1980). Substance P enhances cholinergic receptor desensitization in a clonal nerve cell line. Proc. Natl. Acad. Sci., USA 77, 634-638.
- Steinbach, A.B. (1968). Alteration by xylocaine (lidocaine) and its derivatives of the time course of the end-plate potential. J. Gen. Physiol. 52, 144-161.
- Strader, C.D. and Raftery, M.A. (1980). Topographic studies of Torpedo acetylcholine receptor subunits as a transmembrane complex. Proc. Natl. Acad. Sci., USA 77, 5807-5811.
- Stroud, R.M. and Moore, J.F. (1985). Acetylcholine receptor structure, function, and evolution. Ann. Rev. Cell Biol. 1, 317-51.
- Sumikawa, K., Houghton, M., Smith, J. C., Bell, L., Richards, B. M. and Barnard, E.A. (1982). The molecular cloning and characterization of cDNA coding for the alpha subunit of the acetylcholine receptor. Nucleic Acid. Res., 10, 5809-5822.

- Supavilai, P. and Karobath, M. (1980). The effect of temperature and chloride ions on the stimulation of [^{3}H] flunitrazepam binding by the muscimol analogues THIP and piperidine-4 sulfonic acid. Neurosci. Lett. 19, 337-341.
- Takeyasu, K., Udgaonkar, J. B. and Hess, G.P. (1983). Acetylcholine receptor : Evidence for a voltage-dependent regulatory site for acetylcholine. Chemical kinetic measurements in membrane vesicles using a voltage clamp. Biochemistry 22, 5973-5978.
- Tallman, J.F., Thomas, J.W. and Gallager, D.W. (1978). GABAergic modulation of benzodiazepine binding site sensitivity. Nature 274, 383-385.
- Venkatasubramanian, K., Audhya, T. and Goldstein, G. (1986). Binding of thymopoietin to acetylcholine receptor. Proc. Natl. Acad. Sci., USA,83, 3171-3174.
- Weber, M. and Changeux, J.P. (1974). Binding of Naja nigricollis 3H-alpha-toxin to membrane fragments from Electrophorus and Torpedo electric organs. 1. Binding of the tritiated alpha-neurotoxin in the absence of effector. Mol. Pharmacol. 10, 1-14.
- Weber, M. and Changeux, J.P. (1974). id. 2. Effect of the cholinergic agonists and antagonists on the binding of the tritiated α-neurotoxin. Mol. Pharmacol. 10, 15-34.
- Weber, M. and Changeux, J.P. (1974). id. 3. Effects of local anaesthetics on the binding of the tritiated α-neurotoxin. Mol. Pharmacol. 10, 35-40.
- Weiland, G., Frisman, D. and Taylor, P. (1979). Affinity labeling of the subunits of the membrane associated cholinergic receptor. Mol. Pharmacol. 15, 213-226.
- Weiland, G., Georgia, B, Lappi, S., Chignell, C.F. and Taylor, P. (1977). Kinetics of agonist-mediated transitions in state of the cholinergic receptor. J. Biol. Chem. 252, 7648-7656.
- Wilson, P.T., Lentz, T.L., Hawrot, E. (1985). Determination of the primary amino acid sequence specifying the alpha-bungarotoxin binding site on the alpha subunit of the acetylcholine receptor from Torpedo californica. Proc. Natl. Acad. Sci. USA, 82, 8790-8794.
- Wise, D.S., Schoenborn, B.P. and Karlin, A. (1981). Structure of acetylcholine receptor dimer determined by neutron scattering and electron microscopy. J. Biol. Chem. 256, 4124-4126.
- Young, A. P. and Sigman, D. S. (1981). Allosteric effects of volatile anesthetics on the membrane-bound acetylcholine receptor protein. I. stabilization of the high-affinity state. Mol. Pharmacol. 20, 498-505.
- Young, A.P. and Sigman, D.S. (1983). Conformational effects of volatile anesthetics on the membrane-bound acetylcholine receptor protein : facilitation of the agonist-induced affinity conversion. Biochemistry 22, 2154-2161.

15
Chloride Channel Modulation at GABAergic Synapses in Primary Cultures of Cortical Neurons

Stefano Vicini, Jean Marc Mienville and Erminio Costa

INTRODUCTION

The opening of Cl^- channels mediated by γ-aminobutyric acid (GABA) can be modulated by benzodiazepine (BZ) or beta-carboline (BC) recognition site ligands (Choi et al., 1977) which per se do not act on Cl^- channels. We studied these ligands at GABAergic synapses in primary cultures of rat cortical neurons and selected for this study flunitrazepam (a positive allosteric modulator) DMCM* (a negative allosteric modulator) and the imidazobenzodiazepine RO-151788** (a high affinity ligand with low intrinsic activity) (Hunkeler et al., 1981). Chan et al. (1985) have shown that flunitrazepam facilitates the Cl^- fluxes elicited by exogenous GABA in primary cultures of CNS neurons and DMCM decreases these ionic fluxes; moreover RO-151788 blocks the action of both compounds showing little or no intrinsic action. In the present study, the modulation of GABA mediated Cl^- channel opening was studied during the release of GABA from presynaptic endings at GABAergic synapses formed among rat cortical neurons in primary culture. The direct action of these three compounds on Cl^- channel was also studied in membrane patches excised from these neurons. This study was intended to further understanding of the physiological role of the allosteric modulation of GABA recognition sites by exogenous and endogenous ligands of BZ-BC recognition sites.

METHODS

Primary cultures of cortical neurons were prepared from 1-day-old rat cerebral cortices following the procedure described

*DMCM = methyl-6,7-dimethoxy-4-ethyl-β-carboline-3-carboxylate.
**RO-151788 = ethyl-8-fluoro-5,6-dihydro-5-methyl-6-oxo-4H-imidazo [1,5][1,4]benzodiazepine-3-carboxylate.

elsewhere (Vicini et al., 1986). The cells were studied one to three weeks after plating at room temperature (21-23 C). The culture medium was replaced with a mM solution of NaCl 145, KCl 5, $CaCl_2$ 2, HEPES-NaOH 5 at pH 7.4. The whole cell recording configuration (Hamill et al. 1981)) was achieved after disruption of the tight seal membrane patch obtained between a glass micropipette and the neuronal surface. This pipette was filled with a mM solution of K-Gluconate 145, EGTA 1.1, $MgCl_2$ 2 and HEPES-KOH 10, pH 7.2, in order to obtain a Cl^- reversal potential close to the cell resting membrane potential (rmp) and to discriminate between outward going, Cl^--mediated inhibitory synaptic currents (IPSCs) and inward going excitatory currents (EPSCs) when the post-synaptic cell was voltage-clamped at values slightly lower than the rmp. Synaptic currents were triggered either by generating an action potential in a pre-synaptic neuron by current injection through another patch-clamp amplifier set in current-clamp mode (LIST-electronics EPC-7) (Fig. 1) or by pressure application of 15 uM L-Glutamate. Drugs to be tested for this action on GABA mediated Cl^- currents were locally applied for 4 to 5 sec by pressure during the interval between two successive presynaptic stimulations. In another set of experiments, the patch pipette after establishing a tight seal with the membrane was carefully brought away from the cell leading to the formation of an outside-out membrane patch. In this case, we used mM solutions containing symetrical chloride concentrations: Tris-Cl 145, NaCl 3, $MgCl_2$ 1, $CaCl_2$ 1, glucose 5 and tetrodotoxin 0.0005, pH 7.4 in the bath and CsCl 125, TEA-Cl 20, $MgCl_2$ 1, EGTA 11 and HEPES 20 pH 7.2 in the pipette, yielding a chloride equilibrium potential close to 0 mV. The patch was then brought close to micropipettes containing solutions of various drugs to be tested. Samples of

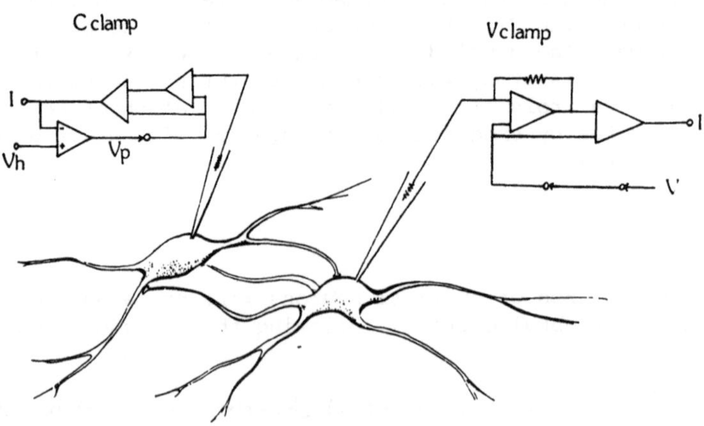

Figure 1. Double recording configuration. Presynaptic neuron in current clamp and postsynaptic neuron in voltage clamp.

channel activity were filtered at 2 KHz, digitized at 20 KHz and analyzed on a LSI-11/73 computer (INDEC-SYSTEM). A semiautomated procedure (Middleton et al., 1986) generated distribution of amplitudes, open times and durations of bursts assuming as a burst a group of openings separated by channel closures lasting less than 5 msec.

RESULTS

Inhibitory Synaptic Currents from Cortical Neurons

In primary cultures of neurons prepared from cortices of neonatal rats, the activity of excitatory and inhibitory synapses can be recorded using whole cell configuration recording. Using double patch recordings, action potentials could be generated in pre-synaptic neurons eliciting IPSCs or EPSCs in the post-synaptic cells (Fig. 2). After termination of the recording sessions, the presence of glutamic acid decarboxylase (GAD) could be detected immunocytochemically in the pre-synaptic cell of the paired neurons (Fig. 1). The inhibition of the IPSCs by bicuculline gave additional evidence that GABA released from presynaptic terminals was operative in the generation of the IPSCs. The IPSCs decay

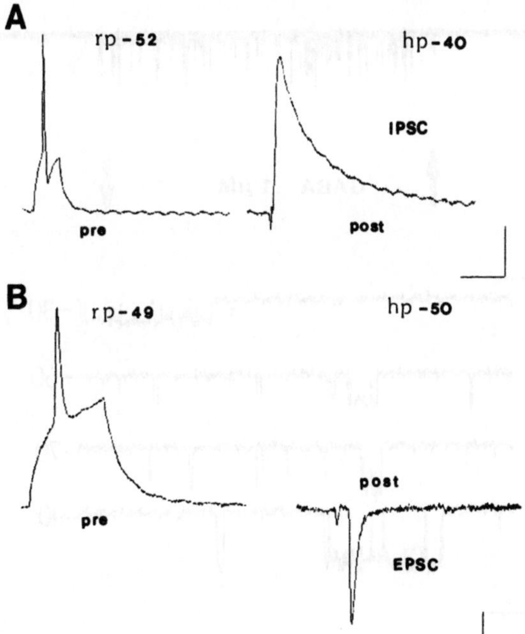

Figure 2. Inhibitory postsynaptic outward current (A) and excitatory postsynaptic inward current (B) triggered by presynaptic action potential. Calibration 15 mV and 50 pA (A,B) 30 msec A 10 msec B.

curve was biphasic with a voltage dependent fast component. The time constant for this decay was 15 ± 4 msec (mean ± SD; N=10) and was calculated by measuring the inward IPSC in post-synaptic cells held at -130 mV, to obtain a potential 70 mV hyperpolarized with respect to the synaptic current reversal (-60 mV) potential. This condition was comparable to the one chosen to study Cl^- channels from outside-out patches held at -70 mV, where reversal potential was 0 mV.

GABA-Activated Single Cl^- Channels from Membrane Patches Excised from cortical Neurons

A pipette filled with 1 uM GABA was kept close to the membrane patch; the GABA leaking from the pipette activated the Cl^- channels present in the membrane patch as shown in Fig. 3a. Often spontaneous openings were observed even without adding external GABA. The conductance levels of these channels were identical to those responding to GABA with a main conductance level of 30 pS, sublevels at 19 and 10 pS and a mean open time of about 1 msec.

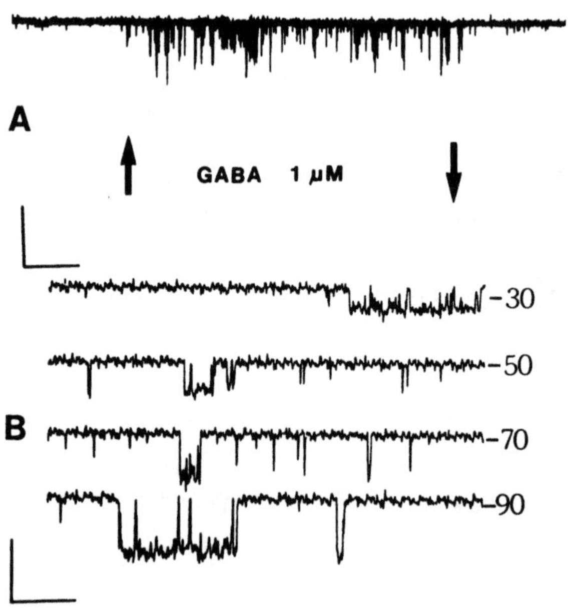

Figure 3. GABA-activated Cl^- channel in an outside-out patch excised from a cultured cortical neuron. Calibrations 2 pA (A,B) 1 sec A, 10 msec B. In A, holding potential -70 mV. From the voltage related changes in channel amplitude reported in B conductance can be calculated to be 31 pS.

When the spontaneous activity was very low, GABA increased the number of open and bursting channels. In patches with high spontaneous activity, the bursting time distribution was similar to that elicited by the transmitter leaking from pipettes filled with 1 uM GABA. This activity could be fitted by double exponential probability density curves with time constants 0.4 and 8 msec. Very likely active concentrations of GABA are present close to the membrane patch due to a continuous release of transmitter from the great number of synaptic terminals present in the surroundings or even on the excised membrane patch.

Both GABA-activated currents and spontaneous events were completely abolished by the application of bicuculline methiodide 20 uM. The fast time constant in the burst duration distribution curve reflects the great number of fast openings probably related to the monoliganded state of GABA receptor (Bormann et al., 1985). We observed an average number of 4 openings per burst of GABA activated Cl^- currents and an average frequency of 24 $\pm$ 6 (mean $\pm$ SD; N=20) openings per second. Probably these values were overestimated since it was impossible to obtain patches without superimposed openings. For this reason the closing time parameters were not evaluated. Up to five channels could be seen simultaneously active in a patch, but only records including no more than two superimposed openings were analyzed. Always, after a certain time, the channel activity present in a membrane patch tended to decrease. The reason for this decline of activity is not clear, perhaps uncontrollable metabolic factors could be responsible for this decrease in channel activity which becomes resistent to the presence of very high concentrations of exogenous GABA.

Modulation of Inhibitory Synaptic Activity.

Applications of 4 uM flunitrazepam increased the amplitude of the synaptic currents and the decay time as shown in Fig. 4a. DMCM 10 uM decreased both the amplitude and the decay time of these currents (Fig. 4a), the effect being more prominent on the amplitude. RO-151788 10 uM blocked the effects of both compounds but it also had an intrinsic effect (Fig. 4a) expressed by a decrease in the decay time constant of the IPSC without changing the amplitude. Data are summarized in Table 1.

Modulation of Chloride Channels Activity.

When the excised patch was close to the pipette containing flunitrazepam 1 uM (Fig. 4b) a dramatic increase in channel activity occurred. The numbers of single openings and bursting events were increased. The conductance of the channel, the mean open time and the number of the openings per burst were unchanged. Since due to the high frequency of occurring events a detailed kinetic analysis could not be carried out, the channel activity

TABLE 1

Action of Flunitrazepam, DMCM and RO-151788 on Amplitude and Decay Time Constant of Transynaptically Generated Cl^- Currents in Cultured Cortical Neurons

Compound	uM	Amplitude in %	Decay Time Constant in %
Flunitrazepam	4	116 (5)	174 (5)
DMCM	10	65 (3)	82 (3)
RO-151788	10	91 (7)	78 (7)

DMCM = methyl-6,7-dimethoxy-6-4-ethyl-β-carboline-3-carboxylate; RO-151788 = ethyl-8-fluoro-5,6-dihydro-5-methyl-6-oxo-4H-imidazo [1,5][1,4] benzodiazepine-3-carboxylate. Amplitude and decay time constant are reported as % of variation from controls (=100). Currents were recorded from cortical neurons held at -40 mV. In parenthesis N of observations.

present in the membrane patches was reduced by applying 2 uM bicuculline methiodide. When flunitrazepam was applied to bicuculline treated membrane patches, the activity was brought back to control levels, and with this rate of channel activity a kinetic analysis became possible. It was found that flunitrazepam only slightly increased the bursting time while it caused a two-fold increase in the frequency of openings and a small increase in the percentage of bursting channels (Table 2). DMCM applied at 5 uM concentrations decreased the frequency of channel openings in the membrane patches (Fig. 4b) without modification of the step amplitude. Again, rather than a decrease in the mean bursting time, the total number of openings was decreased with a significative decrease in the percentage of bursting channel (Table 2). RO-151788 10 uM affected neither the channel activity (Fig. 4b) nor the mean bursting time and conductance (Table 2).

DISCUSSION

This study shows that flunitrazepam and DMCM, two ligands for the allosteric modulatory site of GABA receptor change the characteristics of the Cl^- currents generated by GABAergic transmission. When these ligands are applied to membrane patches excised from GABAergic innervated cells, they increase or decrease the probability of GABA mediated channel openings. In contrast, the imidazobenzodiazepine RO-151788 reduces the IPSC decay time but fails to change the GABA mediated Cl^- channel activity in membrane patches. During GABAergic synaptic currents many channels open at the same time while in an outside-out patch the openings observed can be attributed to the activity of few channels. When comparing the probability density fitting of the bursting time

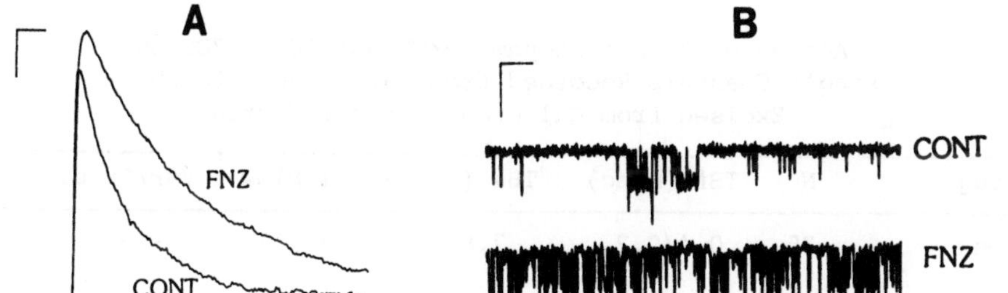

Figure 4. GABA-ergic synaptic currents A and GABA activated Cl^- channels B in control condition and after different drugs. Calibrations 50 pA and 15 msec in A, 2 pA 50 msec in B.

distribution in membrane patches held at -70 mV and the decay of synaptic currents produced under the same driving force conditions it seems that the time constant of the decay is longer than the mean bursting time. This could indicate that during synaptic transmission multiple binding of GABA to specific recognition sites takes place, causing repetitive bursts of openings. Distorsion of the decay phase of the current due to a space-clamp phenomenon (Johnston et al., 1983) or kinetic changes due to excision of the membrane patch might also be considered. The dramatic increase in the number of openings, both fast and bursting, seen with flunitrazepam application has a reflection in the increased decay time and amplitude of the GABA mediated synaptic currents.

In the presence of flunitrazepam, the increase in channel openings and bursts at the onset of synaptic neurotransmission produced the observed effects on the pattern of the synaptic currents. However, the occurence of multiple openings

Table 2

Action of Flunitrazepam, DMCM and RO-151788 on Single Channels Recorded from Outside-Out Patches Excised from Cultured Cortical Neurons

Drug	N	TBf (msec)	TBs (msec)	% Slow	Var/Freq.
None	25	0.4/0.2	7.8/3	37/17	---
FNZ	4	0.4/0.3	8.9/4	55/30	2.4
DMCM	8	0.4/0.1	7.8/4	25/9	0.4
RO-151788	5	0.4/0.1	7.4/3	40/16	1.0

Values are mean ± SD. FNZ=flunitrazepam (1 uM); DMCM (5 uM) and RO-151788 (10 uM). N=Number of membrane patches included in this study; TBf and TBs= fast and slow time constants from burst duration distribution fitting; % Slow=area under slow component, Var/Freq=channel opening frequency variation. Patches were held at -70 mV. The step amplitude and the mean channel opening time were not changed by the drugs under investigation.

during synaptic transmission makes it difficult to compare synaptic currents and membrane patches channel activity accurately. DMCM has just the opposite effect, on one hand it decreases the total number of active channels and thereby decreases the amplitude of the synaptic current and on the other hand it reduces the relative number of bursting channels thereby accelerating the decay time. Of extreme interest is the effect of RO-151788 on the two experimental models. It accelerates the synaptic current decay time constant without any significative variation of the amplitude. Thus, the effect of RO-151788 differs from that of DMCM which lowers the amplitude and accelerates the decay time constant. Furthermore, RO-151788 fails to change single channel activity when applied to the excised membrane patches indicating that its effect strictly depends on synaptic transmission. Since RO-151788 does not change the effects of exogenous GABA acting on the excised membrane patch it is surmized that this compound affects the actions of an unknown compound coreleased with GABA during synaptic activity. An interaction of RO-151788 with calcium influx during presynaptic depolarization seems unlikely because variation of the extracellular calcium concentration during IPSC experiments failed to modify the time constant of the decay and affected only the amplitude (Vicini et al., 1986). Hence, the possibility exists that RO-151788 displaces an endogenous ligand of the BZ/BC recognition site that is operative during synaptic transmission. From the profile of the RO-151788 action one might infer that the unknown compound might be acting as an endogenous positive allosteric modulator of GABA mediated Cl^- channel activation.

REFERENCES

Bormann, J. and Clapham, D.E. (1985). γ-Aminobutyric acid receptor channels in adrenal chromaffin cells: A patch-clamp study. Proc. Natl. Acad. Sci. USA 82, 2168-2172.

Hamill, O.P., Marty, A., Neher, E., Sakmann, B. and Sigworth, F.J. (1981). Improved patch-clamp techniques for high resolution current recording from cells and cell-free membrane patches. Pfluger Arch. 391, 85-100.

Chan, C.Y. and Farb, D.H. (1985). Modulation of neurotransmitter action: Control of the γ-aminobutyric acid response through the benzodiazepine receptor. J. Neurosci. 5, 2365-2373.

Choi, D.W., Farb, D.H. and Fishbach, G.D. (1977). Chlordiazepoxide selectively augments GABA action in spinal cord cell cultures. Nature 269, 342-344.

Hunkeler, W., Mohler, H., Pieri, L., Polc, P., Bonetti, E.P., Cumin, R., Schaffner, R. and Haefely, W. (1981). Selective antagonists of benzodiazepines. Nature 290, 514-516.

Johnston, D. and Brown, T.H. (1983). Interpretation of voltage-clamp measurements in hippocampal neurons. J. Neurophysiol. 50, 464-486.

Middleton, P., Jaramillo, F. and Schuetze S.M. (1986). Forskolin increase the rate of acetylcholine receptor desensitization at rat soleus end plates. Proc. Natl. Acad. Sci. USA, 83, 4967-4971.

Vicini, S., Alho, H., Costa, E., Mienville, J.M., Santi, M.R. and Vaccarino, F.M. (1986). Modulation of γ-aminobutyric acid mediated inhibitory synaptic currents in dissociated cortical cell cultures. Proc. Natl. Acad. Sci. USA, in press.

16
Mechanisms of Receptor-coupled Signal Transduction

Richard A. Cerione

INTRODUCTION

The mechanisms by which specific cell surface receptor proteins transmit signals to enzymes, transport proteins or other biological effectors has been the subject of extensive investigation. Among the most studied of these receptor-coupled signal transduction systems have been those involved in the adrenergic regulation of adenylate cyclase activity. Most of the primary components of the catecholamine-mediated stimulatory and inhibitory pathways of adenylate cyclase have now been purified. Both stimulation and inhibition of the enzyme activity is initiated by the binding of hormones to specific receptors; the β-adrenergic receptor being involved in stimulation while the α_2-adrenergic receptor is responsible for triggering inhibition of the enzyme activity. Each of these receptors is known to be comprised of a single type of polypeptide with an apparent Mr ≅ 65,000 daltons (Benovic et al., 1984; Regan et al., 1986). Following their interaction with hormones, these receptors go on to interact with, and promote the activation of, distinct GTP binding proteins which are typically designated as N_s or G_s (the stimulatory GTP binding protein) and N_i or G_i (the inhibitory GTP binding protein). Both N_s and N_i are heterotrimeric in structure. They contain distinct GTP binding subunits, designated as α_s (Mr ≅ 45,000) and α_i (Mr ≅ 41,000), but seemingly identical β (Mr ≅ 36,000) and γ (Mr ≅ 5-10,000) subunits (Northup et al., 1983; Bokoch et al., 1984; Codina et al., 1984). The activation of these "N" proteins is directly reflected by the binding of the "active" guanine nucleotide, GTP. This, in turn, primes the N proteins for stimulating (or inhibiting) the effector enzyme until a GTPase activity deactivates these proteins and ultimately allows for a recycling of the components of the system.

An analogous type of three component (receptor, GTP binding protein, effector enzyme) system appears to be operating in vertebrate visual transduction. Here, again, the primary components of the system have been purified and extensively characterized. In visual transduction it is the absorption of light by the photoreceptor, rhodopsin, which initiates the signalling by enabling rhodopsin to interact with a heterotrimeric GTP binding protein, transducin (where α_t = 39,000; β = 36,000; γ_T = 5,000) (Stryer et al., 1981; Fung, 1983). The activated transducing·GTP complex can then go on to interact with, and stimulate, the cyclic GMP-dependent phosphodiesterase. The latter enzyme, by hydrolyzing cyclic GMP, effects a reduction in the Na^+ flux across retinal rod membranes and thus a hyperpolarization of the membrane which is ultimately communicated to the optic nerve.

Over the past few years a great deal of progress has been made in the development of methods for functionally reconstituting the interactions between pure receptors, GTP binding proteins, and effector enzymes in phospholipid vesicle systems (Brandt et al., 1983; Cerione et al., 1985a; 1986a). In the first section presented below we will consider the use of these systems in probing the mechanisms by which receptors promote the activation of GTP binding proteins, using both the adenylate cyclase system and the phototransduction system as models for such studies. In the second section, we will describe the use of reconstitution approaches in monitoring the potential interactions between receptors for growth factors and polypeptide hormones (such as insulin) and specific GTP binding proteins.

Reconstitution of Receptor-Coupled Signal Transduction

The interactions of receptors with GTP binding proteins in lipid vesicles results in the stimulation of two types of activities in the GTP binding protein; these being a high affinity guanine nucleotide binding and a GTPase activity. The high affinity binding of guanine nucleotides, such as [^{35}S] GTPγS, appears to be a direct reflection of the activation of the GTP binding protein (Asano et al., 1984). The receptor-stimulated GTPase activity is a natural outcome of the receptor-stimulated GTP binding and is a direct reflection of the deactivation of the GTP binding protein. Since each turnover of the GTPase activity represents one turn of the activation-deactivation cycle of the GTP binding protein, GTPase assays provide a convenient means for monitoring the effects which receptors have on the complete cycle. Figure 1 (Left Panel) shows the effects of the pure β-adrenergic receptor on the GTPase activities of different pure GTP binding proteins in various two component phospholipid vesicle systems. The results of these studies illustrate that the pure β-adrenergic receptor (BAR) is able to effectively couple with the pure N_s protein. This interaction is promoted by agonist binding to the receptor (Cerione et al., 1984) and effects about an 8 fold stimulation of the GTPase activity in N_s. The agonist stimulation of the GTPase activity can be fully accounted for by the agonist-

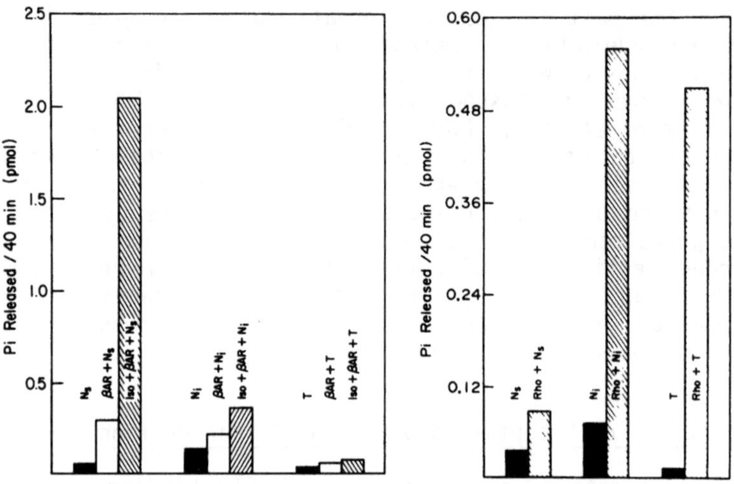

Figure 1. (Left): GTPase activity in phospholipid vesicles containing pure preparations of BAR and N_s, N_i or transducin (T). Reconstitution incubations were initiated with 10 pmol of the different protein components. Phospholipid vesicles containing the reconstituted proteins were obtained and assayed for GTPase activity as described in Cerione et al. (1985a). (-) Isoproterenol = 10^{-5}M. (Right): GTPase activity in phospholipid vesicles containing pure preparations of rhodopsin (RHO) and N_s, rhodopsin and N_i, or rhodopsin and transducin (T). Reconstitution incubations were initiated with 36 pmol of rhodopsin and 10 pmol of the different GTP binding proteins. Phospholipid vesicles containing the reconstituted proteins were obtained and assayed for GTPase activity as described in Cerione et al. (1985a). Rhodopsin was light-activated by exposure to room light.

induced increases in the rate and extent of [^{35}S]GTPγS binding under these experimental conditions (Cerione at al., 1985a). While the β-adrenergic receptor is also able to interact with N_i, as evidenced by the agonist-stimulated GTPase in this reconstituted system, the effectiveness of this interaction is significantly reduced relative to that observed with N_s. In the case of transducin, there is almost no interaction between this GTP binding protein and the β-adrenergic receptor. Overall, these results are keeping with the postulated physiological role of the β-adrenergic receptor (i.e., it couples with the stimulatory GTP binding protein) and thus provides support for the fidelity of these reconstituted systems.

The results presented in Figure 1 (right panel) illustrate that the photoreceptor, rhodopsin, is able to stimulate the GTPase activity of N_i and transducin while having almost no effect on the GTPase activity of N_s. The ability of rhodopsin to interact with N_i and not with N_s is consistent with other biochemical data suggesting that N_i may be structurally more similar to transducin than to N_s (Manning & Gilman, 1983). Interestingly, when similar reconstitution studies are performed using the purified α_2-adrenergic receptor, transducin shows little ability to substitute for N_i in interacting with the inhibitory receptor (Cerione et al., 1986b). Thus there are apparently some important differences in the coupling domains of the N_i and transducin proteins. Under optimal (receptor-coupled) conditions the turnover numbers for the GTPase activities of N_s, N_i and transducin all approach the same maximum values (1-2 min^{-1} per mol of GTP binding protein) (Cerione et al., 1985a).

The availability of each of the primary components of the phototransduction system in mg quantities (as well as the individual α and βγ subunit species) has enabled us to probe in more detail the mechanisms of the receptor-promoted activation-deactivation cycle of this GTP binding protein. In agreement with the initial work of Fung (1983), the presence of both the α and βγ subunit components of transducin are necessary to observe optimal coupling between this protein and light-activated rhodopsin in these reconstituted phospholipid vesicle systems. The promotion by the βγ species of the rhodopsin-stimulated GTPase activity in α_T can be completely accounted for by the promotion by βγ of GTP binding to the α_T subunit (Cerione et al., 1986c). Thus, each turn of the receptor-promoted activation-deactivation cycle of the GTP binding protein must involve an interaction between the receptor and the holo-GTP binding protein.

Interestingly, the addition of the βγ subunit complex to phospholipid vesicles containing rhodopsin and holo-transducin will actually stimulate the GTPase cycle of the transducin molecules. This stimulation is dependent on both [$MgCl_2$] and [$\beta\gamma_T$] and appears to reflect the promotion by $\beta\gamma_T$ of the rhodopsin-α_T interactions which occur at the outset of each new activation-deactivation cycle (Cerione et al., 1986c). These results directly suggest that the association-dissociation equilibrium of the transducin molecule is an important component of this cycle.

The importance of subunit dissociation in the activation of heterotrimeric GTP binding proteins has also been suggested from studies using more complex systems in which we have reconstituted a receptor-coupled effector enzyme activity. The inclusion of pure N_i, or pure transducin, in phospholipid vesicles containing BAR, N_s, and partially purified preparations of adenylate cyclase (C) results in a guanine nucleotide-dependent inhibition of the N_s GppNHp-, or hormonal-, stimulated adenylate cyclase activity (Cerione et al., 1985b, 1986a). In these cases, the $\beta\gamma$ complexes of N_i, or transducin, appear to be responsible for mediating the inhibitory effect. An example of this is shown in Figure 2, where the reconstituted inhibition of the isoproterenol-stimulated adenylate cyclase activity by holo-transducin can be fully accounted for by the $\beta\gamma_T$ complex. One possible explanation for this inhibition was proposed by Gilman and colleagues (Katada et al., 1984). Specifically, it was suggested that a re-association of $\beta\gamma$ with α_s would convert the active α_s species to an inactive holoprotein and thereby effect an inhibition of N_s-stimulated adenylate cyclase activity.

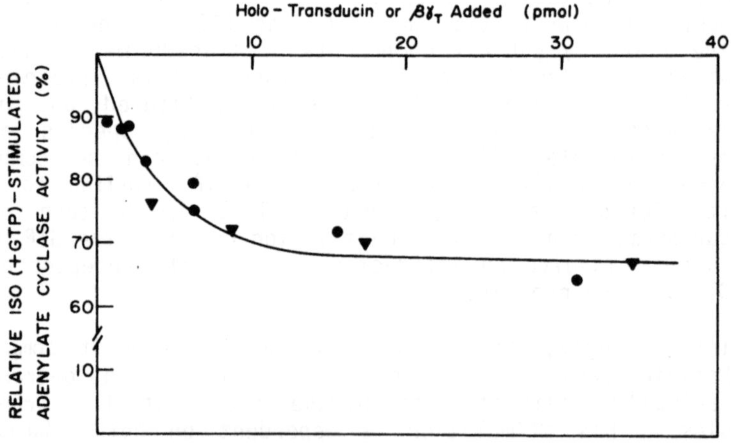

Figure 2. Dose response relationships for the $\beta\gamma_T$, or holo-transducin mediated inhibition of isoproterenol (plus GTP)-stimulated adenylate cyclase activity. The resolved C preparation,

10 pmol of N_s, and 10 pmol of BAR were added to reconstitution incubations as described in Cerione et al. (1986a). Aliquots of the isolated vesicles were then incubated with $\beta\gamma_T$ (●) or holo-transducin (▼) in the amounts shown, prior to assaying their isoproterenol (plus GTP)-stimulated adenylate cyclase activity. Isoproterenol (ISO) = GTP = 100 μM; $MgCl_2$ = 2 mM. Data was obtained from Cerione et al. (1986a).

The following scheme (Figure 3) can be used to summarize the results obtained from the different reconstitution studies. The first step is the interaction of the agonist occupied-, or light-activated-, receptor (R*) with the inactive GTP binding protein (N). In most cases, the inactive N protein will be tightly complexed with GDP (Ferguson, et al., 1986). The main kinetic advantage provided by the receptor is likely to be the stimulation of the dissociation of GDP from the N protein. This in turn frees the guanine nucleotide binding site for a subsequent interaction

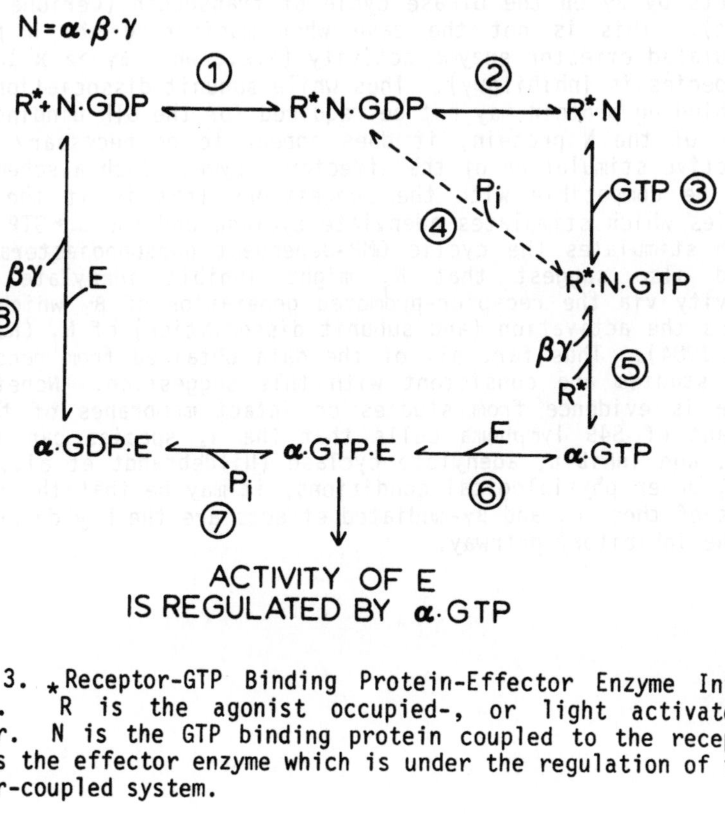

Figure 3. Receptor-GTP Binding Protein-Effector Enzyme Interactions. R* is the agonist occupied-, or light activated-, receptor. N is the GTP binding protein coupled to the receptor and E is the effector enzyme which is under the regulation of this receptor-coupled system.

with GTP. It is this interaction with GTP (step 3) that is suggested to directly result in the dissociation of the subunits of the holo-GTP binding protein (step 5). The dissociation of βγ from this complex should facilitate the dissociation of R^* from the α subunit, since an intact holo-GTP binding protein appears to be necessary for the effective interaction of this protein with the receptor. The free receptor can return to step 1, interact with new holo-GTP binding protein molecules, and in this way promote the activation of multiple N proteins. Similarly, the βγ species generated at step 5, can reassociate with α•GDP species which are generated at the end of each cycle (step 7), and promote further interactions with the receptor. This would be consistent with the observations that βγ can stimulate the GTPase cycle in holo-transducin.

Under conditions where subunit dissociation of holo-N should be blocked, i.e., βγ >> 10 x N, we still observe stimulatory effects by βγ on the GTPase cycle of transducin (Cerione et al., 1986c). This is not the case when monitoring the N protein-stimulated effector enzyme activity (i.e., when βγ >> x 10 N, the βγ species is inhibitory). Thus while subunit dissociation of the GTP binding protein may not be required for the GTP binding-GTPase cycle of the N protein, it does appear to be necessary for the effective stimulation of the effector enzyme. Such a scheme would then be compatible with the suggestions that is it the α_s•GTP species which stimulates adenylate cyclase and the α_T•GTP complex which stimulates the cyclic GMP-dependent phosphodiesterase. It would also suggest that N_i might inhibit adenylate cyclase activity via the receptor-promoted generation of βγ which accompanies the activation (and subunit dissociation) of N_i (Katada et al., 1984). Thus far, all of the data obtained from reconstitution studies are consistent with this suggestion. Nonetheless, there is evidence from studies on intact membranes of the cyc^- variant of S49 lymphoma cells that the α_i species can interact with, and inhibit, adenylate cyclase (Hildebrandt et al., 1984). Thus, under physiological conditions, it may be that the relative rates of these α_i and βγ-mediated effects are the key determinants in the inhibitory pathway.

The Use of Reconstitution Approaches to Probe the Role of GTP Binding Proteins in Growth Factor Action.

Various lines of evidence have suggested that GTP binding proteins might be important components of polypeptide hormone-, or growth factor-coupled signal transduction pathways (Kamata & Feramisco, 1984; Johnson et al., 1986). We are using reconstitution approaches similar to those described in the preceding section to examine the possible functional interactions between these types of receptors and different purified GTP binding proteins. The purified insulin receptor was chosen as one model for these studies since receptor preparations suitable for reconstitution studies can be obtained rather easily from human placental membranes (Petruzelli et al., 1984; Yu & Czech, 1984, Pike et al., 1986). In addition a great deal of structural information about this receptor has been obtained and it is infact known to share some important basic structural and functional features with other members of the growth factor receptor/tyrosine kinase family (which includes the receptors for the epidermal growth factor, the platelet derived growth factor, and the insulin-like growth factor I). The insulin receptor is heterotetrameric in structure and is made up of two types of polypeptides (Ullrich et al., 1985). One of these, designated α (Mr 135,000 daltons), is exposed to the extracellular space and appears to contain the insulin binding site. The other, designated β (Mr 95,000 daltons), has been suggested to span the membrane and to contain the tyrosine kinase active site and acceptor sites for an insulin-promoted autophosphorylation of the receptor (with both of these sites being exposed to the cytoplasm (Ullrich et al., 1985).

We have found that the insulin receptor/tyrosine kinase can be functionally inserted into phospholipid vesicles using methods similar to those developed for the components of the adenylate cyclase and vertebrate phototransduction systems (Krupinski et al., 1986). Different GTP binding proteins can be co-inserted into these vesicles with the insulin receptor preparations, enabling a direct examination of the potential interactions between this receptor and different GTP binding proteins in a lipid milieu. One sensitive monitor for such interactions has been to examine the effects of the various GTP binding proteins on insulin-promoted phosphorylation reactions. Thus far, little or no effect has been observed in studies where the insulin receptor has been co-inserted with the stimulatory GTP binding protein (N_s), transducin, or various variants of the ras p21 protein (as prepared by Gibbs et al., 1984) into vesicles. However, some interesting results have been obtained when the receptor is co-inserted into vesicles with a mixture of the bovine brain GTP binding proteins N_i and N_o (N_o:N_i = 70:30). Specifically, an insulin-stimulated phosphorylation of the α_i and α_o subunits is observed which is about equal in intensity to the insulin-

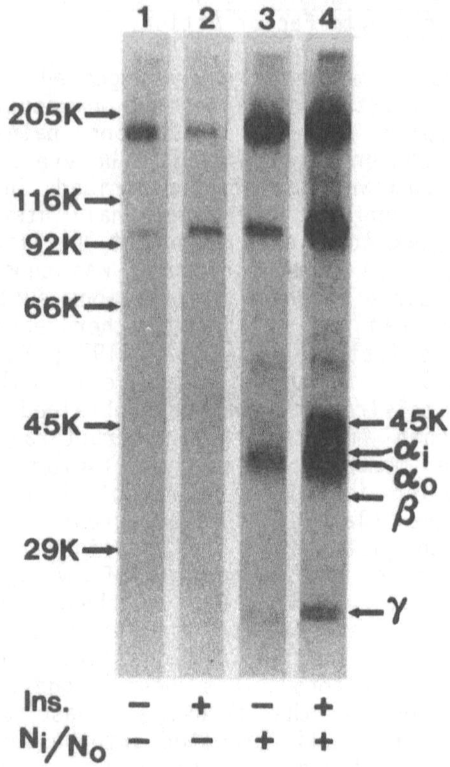

Figure 4. Insulin-promoted phosphorylation of a tyrosine-agarose-purified insulin receptor and a mixture of N_i and N_o in lipid vesicles. The insulin receptor was solubilized from human placental membranes as described by Pike et al. and then partially purified by a combination of wheat germ lectin chromatography and tyrosine agarose chromatography. The mixture of N_o and N_i was solubilized and purified from bovine brain membranes using the procedures of Robishaw and Sternweis (1984) essentially as described in Cerione et al. (1986b). The purified insulin receptor (5 pmol) and GTP binding proteins (100 pmol) were added to reconstitution incubations and phospholipid vesicles were isolated from Extracti-gel resin (Cerione et al., 1985a). Aliquots of the reconstituted vesicles were added to phosphorylation incubations for 10 min. at 30°C. The arrows on the left indicate the positions for molecular weight standards. The arrows on the right indicate the positions of the different subunits of the GTP binding proteins as determined by protein staining of the reconstituted vesicles.

stimulated autophosphorylation of the β subunit of the receptor (see Figure 4, lanes 3 and 4). This phosphorylation appears to be essentially equally distributed between these two alpha subunits. No phosphorylation of the β subunits of the GTP binding proteins has been observed, although an insulin-promoted phosphorylation of a polypeptide which corresponds to the mobility of the γ subunit also occurs. There is also an insulin-promoted phosphorylation of a 45,000 dalton polypeptide which is the primary contaminant in the bovine brain N_o/N_i preparation (representing ≅ 10% of the total protein). We have not yet identified this polypeptide (it is not the α_s subunit), however it can be removed from the N_o/N_i preparation by hydroxyapatite chromatography without any effect on the insulin-stimulated phosphorylation of the GTP binding proteins.

The co-reconstitution of the insulin receptor with the GTP binding proteins are essential for an optimal insulin-promoted phosphorylation of the alpha subunits. The direct addition of the purified GTP binding proteins (in cholate solution) to the tyrosine agarose-purified insulin receptor results in little or no insulin-stimulated phosphorylation of the GTP binding proteins. The increased phosphorylations of the 170,000 dalton polypeptide in the receptor preparation (this contains the EGF receptor which essentially co-purifies with the insulin receptor through tyrosine agarose chromatography), as well as the increased phosphorylation of the insulin receptor, in the presence of N_o/N_i are also the outcome of the co-reconstitution of these proteins (for example compare lanes 2 and 4 in Figure 4). Specifically, these effects appear to result from an increased efficiency of insertion of the EGF and insulin receptors into the lipid vesicles which is induced by the GTP binding proteins. This increase in efficiency has varied from < 1.5 x to several fold in different experiments and appears to be relatively specific for N_o and N_i.

At the present time these reconstituted systems are being used to determine the functional implication of the insulin-promoted phosphorylations of these GTP binding proteins. The development of these systems should allow correlations between phosphorylation and the ability of growth factor receptors to promote the activation-deactivation cycles of GTP binding proteins, and to begin to determine the different protein components operating in these receptor-coupled signal pathways.

REFERENCES

1. Asano, T., Pedersen, S.E., Scott, C.W., and Ross, E.M. (1984). Biochemistry 23, 5460-5467.
2. Benovic, J.L., Shorr, R.G.L., Caron, M.G., and Lefkowitz, R.J. (1984). Bicohemistry 23, 4510-4518.

3. Bokoch, G.M., Katada, T., Northup, J.K., Ui, M., and Gilman, A.G. (1984). J. Biol. Chem. 259, 3560-3567.
4. Brandt, D.R., Asano, T., Pedersen, E.M., Ross, E. (1983). Biochemistry 22, 4357-4362.
5. Cerione, R.A., Staniszewski, C., Benovic, J.L., Lefkowitz, R.J., Caron, M.G., Gierschik, P., Somers, R., Spiegel, A.M., Codina, J., and Birnbaumer, L. (1985). J. Biol. Chem. 260, 1483-1500.
6. Cerione, R.A., Staniszewski, C., Gierschik, P., Codina, J., Somers, R.L., Birnbaumer, L., Spiegel, A.M., Caron, M.G., and Lefkowitz, R.J. (1986a). J. Biol. Chem., 261, 9514-9520.
7. Cerione, R.A., Codina, J., Benovic, J.L., Lefkowitz, R.J., Birnbaumer, L., and Caron, M.G. (1984). Biochemistry 23, 4519-4525.
8. Cerione, R.A., Regan, J.W., Nakata, H., Codina, J., Benovic, J.L., Gierschik, P., Somers, R.L., Spiegel, A.M., Birnbaumer, L., Lefkowitz, R.J., and Caron, M.G. (1986b). J. Biol. Chem., Vol, 3901-3909.
9. Cerione, R.A., Lakonishok, M., Somers, R.L., and Spiegel, A.M., (1986c). submitted.
10. Cerione, R.A., Codina, J., Kilpatrick, B.F., Staniszewski, C.F., Gierschik, P., Somers, R.L., Spiegel, A.M., Birnbaumer, L., Caron, M.G., Lefkowitz, R.J. (1985b). Biochemistry 24, 4499-4503.
11. Codina, J., Hildebrandt, J.D., Sekura, R.D., Birnbaumer, M., Bryan, J., Manclark, C.R., Iyengar, R., and Birnbaumer, L. (1984). J. Biol. Chem. 259, 5871-5886.
12. Ferguson, K.M., Higashijima, T., Smigel, M.D., and Gilman, A.G. (1986). J. Biol. Chem. 261, 7393-7399.
13. Fung, B.,.-K. (1983). J. Biol. Chem. 258, 10495-10502.
14. Gibbs, J.B., Sigel, I.S., Poe, M., and Scolnick, E.M. (1984). PNAS 81, 5704-5708.
15. Hildebrandt, J.D., Codina, J., and Birnbaumer, L. (1984). J. Biol. Che. 259, 13178-13185.
16. Johnson, R.M., Connelly, P.A., Sisk, R.B., Pobiner, B.F., Hewlett, E.L., and Garrison, J.C. (1986). PNAS, 83, 2032-2036.
17. Kamata, T., and Feramisco, J.R. (1984). Nature 310, 147-149.
18. Katada, T., Bokoch, G.M., Smigel, M., Ui, M., and Gilman, A.G. (1984). J. Biol. Chem., 259, 3586-3595.
19. Krupinski, J., Lakonishok, M., Benovic, J.L., and Cerione, R.A. (1986). submitted.
20. Manning, D.R., and Gilman, A.G. (1983). J. Biol. Chem., 258, 7059-7063.
21. Northrup, J.K., Smigel, M.D., Sternweis, P.C., and Gilman, A.G. (1983). J. Biol. Chem., 258, 11369-11376.
22. Petruzzelli, L., Herrera, R., and Rosen, O.M. (1984). PNAS 81, 3327-3331.
23. Pike, L.J., Eakes, A.T., and Krebs, E.G. (1986). J. Biol. Chem., 261, 3782-3789.

24. Regan, J.W., Nakata, H., DeMarinis, R.M., Caron, M.G., and Lefkowitz, R.J. (1986). J. Biol. Chem. (in press).
25. Sternweis, P.C., and Robishaw, J.D. (1984). J. Biol. Chem. 259, 13806-13813.
26. Stryer, L., Hurley, J.B., and Fung, B.K.-K. (1981). Curr. Top. Membr. Transp. 15, 93-108.
27. Ullrich, A., Bell, J.R., Chen, E.Y., Herrera, R., Petruzzelli, L.M., Dull, T.J., Gray, A., Coussens, L., Liao, Y.-C., Tsubokawa, M., Mason, A., Seeburg, P.H., Grunfeld, C., Rosen, O.M., and Ramanchandran, J. (1985). Nature 313, 756-761.
28. Yu and Czech (1984). J. Biol. Chem., 259, 5277-5286.

17
The G_i-protein as a Target for Receptor–Receptor Interactions

Karl H. Jakobs, Peter Gierschik, Rüdiger Grandt and Rainer Marquetant

INTRODUCTION

The hormone-sensitive adenylate cyclase system, which serves as a transmembrane signal transduction system for a wide variety of hormones and neurotransmitters, is a dually regulated signalling system. On the one hand, many hormonal agents, after interaction with their specific receptors, cause an increase in cyclic AMP formation by the adenylate cyclase. This action is mediated by the stimulatory guanine nucleotide-binding regulatory component G_s, which transfers the information from the hormone-activated receptor to the adenylate cyclase. On the other hand, a similarly large number of hormonal agents, after interaction with their specific receptors, induce inhibition of cyclic AMP formation by the adenylate cyclase. This action involves the coupling function of the inhibitory guanine nucleotide-binding regulatory component G_i. Thus, the hormone-sensitive adenylate cyclase by itself is a system exhibiting receptor-receptor interaction, at least at the level of the adenylate cyclase, in that the action of one receptor (stimulatory or inhibitory to the adenylate cyclase) on cyclic AMP formation can be modulated by activation of another set of receptors (inhibitory or stimulatory to the adenylate cyclase).

Evidence has been accumulated that there may also be a receptor-receptor interaction at the level of the G-proteins. The G-proteins of the adenylate cyclase system are apparently heterotrimeric proteins with dissimilar α-subunits (α_s and α_i) but very similar, if not identical, β- and γ-subunits (Gilman, 1984; Birnbaumer et al., 1985). Since activation of the G-proteins by the respective receptors hypothetically involves dissociation of the G-proteins into the activated, GTP-liganded α-subunits and the βγ-dimers, it is reasonable to assume that the βγ-dimers released from one activated G-protein can affect the activity of the other G-protein by binding to its α-subunit. This hypothesis, although not yet proven for the native membrane situation, is one of the most favored models to explain inhibition of adenylate cyclase by the activated G_i protein. Is is hypothesized that the $\beta\gamma$-dimer of G_i released during inhibitory receptor activation interacts with free α_s and, thereby, inactivates the G_s protein by forming its holoprotein conformation. The decrease in free active α_s, then, will cause a fall in adenylate cyclase activity. Although this model cannot fully explain the mechanisms of G_i-induced inhibition of adenylate cyclase, it points to the very attractive hypothesis that a receptor-receptor interaction is in principle not only possible at the level of the adenylate cyclase but also at the level of the G-proteins. In this chapter, two other types of receptor-receptor interactions will briefly be discussed in which G-proteins, particularly the G_i-protein, are apparently involved.

ACTIVATION OF THE G_i PROTEIN BY STIMULATORY HORMONE RECEPTORS

It is generally assumed that stimulatory hormone receptors only cause stimulation of adenylate cyclase, while inhibitory receptors only induce inhibition of the enzyme. However, both indirect and direct evidence has been accumulated that stimulatory hormone receptors in addition to their primary stimulatory action on the adenylate cyclase can also somehow induce inhibition of adenylate cyclase. This action most probably involves the G_i-protein. First, it is well known that the G_i-protein and its action on the adenylate cyclase can be inhibited by various agents and treatments such as by pertussis toxin, which ADP-ribosylates $G_{i\alpha}$, by N-ethylmaleimide, most probably also affecting $G_{i\alpha}$, by activated protein kinase C, apparently phosphorylating $G_{i\alpha}$ (see below) and by treatment with various proteolytic enzymes (Ui et al., 1984; Jakobs et al., 1984). All of these treatments can completely abolish inhibitory hormone receptor-induced and G_i-mediated inhibition of adenylate cyclase. One would assume that these treatments have no effect on stimulation of adenylate cyclase. Indeed, these treatments have usually no or only a small effect on stimulation of adenylate cyclase by forskolin, which apparently stimulates cyclic AMP formation by a direct interaction with the adenylate cyclase catalytic moiety. Similarly, basal adenylate cyclase activity and basal cyclic AMP levels are usually not enhanced by these G_i-inactivating treatments. However, as shown with various cells and tissues and with various stimulatory hormones, these treatments can cause a large increase in stimulation of adenylate cyclase by stimulatory hormones, an action primarily mediated by the G_s protein (Ui et al., 1984; Jakobs et al., 1984). These data, thus, suggest that in the course of adenylate cyclase activation by stimulatory hormones not only the G_s-protein but possibly also the G_i-protein is activated.

The second evidence for an activation of G_i by stimulatory hormones comes from data obtained in membrane preparations, by studying the action of stimulatory hormones on the adenylate cyclase activity and its regulation by guanine nucleotides. One of the best studied examples in this regard is the human platelet adenylate cyclase system. In these membranes, the stimulatory hormone, prostaglandin E_1 (PGE_1), acting via its own receptor, and forskolin, activating the adenylate cyclase directly, can increase adenylate cyclase activity by maximally about 30- and 50-fold, respectively. The forskolin-stimulated adenylate is only minimally affected by the addition of GTP, whereas the stimulation by PGE_1 is potentiated by GTP (Aktories and Jakobs, 1985). However, this increase in activity by GTP is only observed at low GTP concentrations (< 1 µM), whereas at higher concentrations GTP has an inhibitory action in the presence of the stimulatory hormone PGE_1, but not in the presence of forskolin. Thus, it appears that the stimulatory hormone can promote the adenylate cyclase inhibition by GTP. Furthermore, after treatment of the membranes with agents inactivating G_i, the inhibitory phase of GTP's action is abolished, suggesting that the GTP-induced inhibition observed in the presence of the stimulatory hormone somehow involves an activated G_i-protein. It has been reported that GTP-dependent inhibition of adenylate cyclase by α_2-adrenergic agonists in human platelet membranes is markedly less pronounced in the presence of PGE_1 than in presence of forskolin (Katada et al., 1984; Bockaert et al., 1985). Based on the above mentioned hypothesis of G_i-mediated adenylate cyclase inhibition, it has been argued that the $\beta\gamma$-dimer hypothetically released during activation of G_i is less activate against the PGE_1-activated α_s than the α_s found in the presence of forskolin. However, a much simpler explanation is that PGE_1 by itself partially activates G_i and that, therefore, the action of the inhbitory hormone is less pronounced than in the presence of forskolin, which by itself does not activate G_i.

An additional argument for an activation of G_i by stimulatory hormones has been obtained by studying the adenylate cyclase inhibition by stable GTP analogs and its modulation by hormones. Inhibition of adenylate cyclase by stable GTP analogs such as guanosine 5'-γ-thiotriphosphate (GTPγS) is preceded by a lag phase. As shown e.g. in membranes of human platelets (Jakobs and Aktories, 1983), this lag phase is shortened in the presence of adenylate cyclase inhibitory hormones such as epinephrine. Interestingly, not only adenylate cyclase inhibitory hormones but also PGE_1, which stimulates human platelet adenylate cyclase, causes an acceleration of the inhibitory action of stable GTP analogs in membranes of human platelets (unpublished observations).

Additional evidence for a possible activation of G_i by stimulatory hormones has been obtained in reconstitution experiments. Using purified $_1$-adrenoceptors and purified G-proteins, Asano et al. (1984) reported that activation of $_1$-adrenoceptors by the agonist isoproterenol increases GTPase activity and GTPγS binding of G_i, although the affinity of the β-adrenoceptor for G_i was less than that for G_s.

The strongest evidence for an activation of G_i and subsequent inhibition of adenylate cyclase by adenylate cyclase stimulatory hormones was obtained by studying the action of these hormones on an otherwise maximally stimulated adenylate cyclase. Since an inhibitory action of these hormones cannot be demonstrated on basal adenylate cyclase activity (here they are stimulatory by definition), we studied the action of these hormones on adenylate cyclase activity activated by exogenous purified, GTP analog-preactivated G_s protein (Jakobs et al., 1985). Under these conditions, we observed that in S49 lymphoma cyc$^-$ membranes, which are defective in the endogenous $G_{s\alpha}$-subunit, the β-adrenergic agonist isoproterenol causes a reduction in stimulated enzyme activity. This reduction was observed at the same concentrations at which isoproterenol stimulates basal adenylate cyclase activity in membranes of wild-type S49 lymphoma cells and it was blocked by β-adrenoceptor antagonists. Furthermore, the inhibition was not dependent on the presence of GTP, indicating that it was probably not due to exchange of G_s-protein-bound GTP analog (active) by GTP, which would form an inactive G_s species. Additionally, the inhibition of the G_s-protein-activated adenylate cyclase by isoproterenol was not additive with that caused by activation of the endogenous G_i-protein. Finally, agents and treatments which reduce or abolish G_i-induced inhibition of adenylate cyclase such as Mn^{2+}, proteases and pertussis toxin treatment reduced or abolished the isoproterenol-induced adenylate cyclase inhibition in membranes of cyc$^-$ S49 lymphoma cells. Similar data have been obtained in membranes of human platelets with an adenylate cyclase maximally stimulated by exogenous preactivated G_s-protein and using PGE_1 and PGI_2 as primarily stimulatory hormones (unpublished observations). Interestingly, under none of the many conditions studied, isoproterenol inhibited cyc$^-$ adenylate cyclase stimulated only by forskolin. These data suggest that the primary action of the stimulatory hormoes is via an interaction with their receptors on the G_s-protein, which is defective in cyc$^-$ cells, and that this G_s-protein activation may somehow lead to an altered activity of the G_i-protein, resulting in inhibition of adenylate cyclase.

In summary, there is apparently an additional receptor-receptor interaction in the hormone-sensitive adenylate cyclase system, in that stimulatory hormones not only via G_s activate adenylate cyclase but apparently can also cause simultaneously inhibition of adenylate cyclase by an activation of G_i. Although the mechanism of this interaction is far from being resolved, it appears that the concomitant activation of G_s and G_i by adenylate cyclase stimulatory hormones can serve as a rapidly acting regulatory mechanism to protect the cell from overstimulation by one specific intracellular messenger.

THE G_i PROTEIN AS TARGET FOR INTRACELLULAR EFFECTORS

The G-proteins apparently not only act as coupling components between hormone-activated receptors and effector moieties in the signal transduction cascade, recent evidence suggests that these proteins may also serve as targets for intracellular effector enzymes, particularly protein kinases activated by intracellular second messengers. The so far best characterized example of G-protein modification by intracellular effectors is the phosphorylation of G_i by the calcium-activated and phospholipid-dependent protein kinase (protein kinase C). Based on apparent discrepancies in the actions of thrombin on cyclic AMP formation in human platelet membranes and intact human platelets, we were prompted to study the possible regulation of the hormone-sensitive adenylate cyclase system by intracellular effectors of the phosphoinositide system. In platelet membranes prepared in Ca^{2+}-free buffers, thrombin potently inhibits adenylate cyclase, whereas in intact platelets thrombin potently activates the phosphoinositide metabolism but has no or only a small inhibitory effect at high concentrations on cyclic AMP accumulation (Aktories and Jakobs, 1984). Therefore, we studied whether protein kinase C, which is activated by thrombin in intact human platelets by the stimulated breakdown of phosphoinositides (Takai et al., 1984), can interfere with the inhibitory adenylate cyclase system involving the G_i-protein. Several approaches were used to study this possible regulation. First, we treated intact cells (human platelets and S49 lymphoma wild-type and cyc^- cells) for a short period of time with phorbol esters, which activate protein kinase C in a receptor-independent manner, and studied the regulation of adenylate cyclase activity in membranes prepared from these cells. Second, similar studies were performed using membranes prepared in the absence of Ca^{2+} and, thereby, in the virtual absence of protein kinase C, to which purified activated protein kinase C was added. Finally, the possible phosphorylation of components of the adenylate cyclase inhibitory pathway by protein kinase C was studied.

In the studies using membranes prepared from phorbol ester-treated cells, we observed that the treatment had no effect on the stimulatory signal transduction to the adenylate cyclase, involving stimulatory receptors, the G_s protein and the adenylate cyclase itself (Jakobs et al., 1985a). There was one exception, namely in the presence of stimulatory hormone (PGE_1) and high GTP concentrations (≧1 μM). Under this condition, the phorbol ester treatment caused an abolition of the inhibitory action of GTP, which resulted in a 2- to 3-fold activation of the adenylate cyclase by phorbol esters compared to controls. Inhibition of adenylate cyclase by agents acting directly on the adenylate cyclase such as calcium ions and dideoxyadenosine was also not affected by the phorbol ester treatment. Furthermore, the short time phorbol ester treatment did not change the number of inhibitory hormone receptors, as studied on platelet α_2-adrenoceptors. However, the treatment largely decreased or abolished the G_i-protein-mediated and agonist-induced inhibition of adenylate cyclase. This was observed for epinephrine and thrombin in membranes of phorbol ester-treated human platelets and for somatostatin in membranes of S49 lymphoma wild-type and cyc$^-$ cells, the latter being defective in the α-subunit of the G_s-protein. The impairment of G_i-mediated inhibition of adenylate cyclase by the phorbol ester treatment of intact cells was independent of the presence of adenylate cyclase activators, i.e. it was observed on basal adenylate cyclase activity as well as on enzyme activity stimulated by forskolin or a stimulatory hormone.

In the second approach to study the modulation of hormone-sensitive adenylate cyclase by activated protein kinase C, we added purified protein kinase C to kinase C-depleted human platelet membranes. In these studies, almost identical data were obtained as with the phorbol ester treatment of intact cells (Watanabe et al., 1985). The forskolin-activated and hormone-stimulated adenylate cyclase activities were not altered by the addition of protein kinase C, except when GTP at a high concentration was combined with the stimulatory hormone PGE_1. In accordance with the data obtained by phorbol ester treatment of intact cells, addition of protein kinase C abolished the inhibitory hormone-induced and G_i-mediated inhibition of adenylate cyclase. When the activity of G_i was maintained at a low level, i.e., at a low GTP concentration (0.1 μM), we even observed that the inhibitory hormone was able to cause an increase in adenylate cyclase activity in membranes treated with activated purified protein kinase C. In contrast to the data observed with inhibitory hormones in the presence of GTP, the activated protein kinase C apparently had no effect on direct activation of G_i by stable GTP analogs such as GTPγS (Bauer and Jakobs, 1986).

All of these functional data were in principle very similar to those observed after treatment of intact cells or membranes with pertussis toxin, which inactivates the G_i-protein by ADP-ribosylating its α-subunit (Ui et al., 1984). Therefore, it was studied whether activated protein kinase C can phosphorylate the G_i-protein and, if so, which of its subunits is phosphorylated. Using holo-G_i-protein purified from rabbit liver plasma membranes as well as its resolved subunits (α_i and βγ), it was found that partially purified protein kinase was able to phosphorylate the resolved α_i-subunit. Maximal phosphorylation was up to about 0.6 Mol phosphate incorporated per Mol α_i-subunit, which value corresponded well to the maximal GTPγS binding capacity of the same resolved protein. The phosphorylation of $G_{i\alpha}$ was inhibited in the presence of the βγ-dimer, which was able to suppress the phosphorylation almost completely. In contrast to free α_i ("inactive"), the GTPγS-liganded α_i was not a good substrate for the protein kinase C-induced phosphorylation). Partially similar data were obtained using intact and detergent-solubilized membranes of human platelets. Addition of protein kinase C phosphorylated several membrane proteins, also one with the apparent molecular weight of the α_i-subunit. Addition of excess of βγ-dimer to the solubilized membranes decreased the phosphorylation of only this one polypeptide, which comigrated with the ADP-ribosylated substrate of pertussis toxin. These data, thus, suggest that in the membranes at least part of the G_i-protein is present in its dissociated form, since apparently only this form is phosphorylated by protein kinase C (Katada et al., 1985).

All of these functional and structural data provide strong evidence that the G_i-protein may be a natural substrate of activated protein kinase C. Thus, the G_i-protein apparently not only acts as coupling component in the hormone-sensitive adenylate cyclase system, but also serves as target for the interaction of receptors causing activation of the phosphoinositide metabolism with those receptors affecting the adenylate cyclase activity. The phosphorylation of G_i by activated protein kinase C and its subsequent inactivation may also explain why several hormones and neurotransmitters, which activate the phosphoinositide metabolism and, thereby, activate protein kinase C, can cause a large increase in the stimualtory actions of hormones on cyclic AMP formation. This potentiation was particularly observed in neuronal cells and tissues, which are exceptionally rich in protein kinase C (Takai et al., 1984). In order to prove this hypothesis, it will be necessary to show that activation of protein kinase C by phosphoinositide metabolism-stimulating hormones and neurotransmitters indeed causes phosphorylation of G_i in intact cells and tissues.

CONCLUSIONS

The hormone-sensitive adenylate cyclase system as a dually regulated enzyme by itself provides a basis for receptor-receptor interactions. In addition, it appears that at least one component of the system, namely the G_i-protein, can serve as a target for additional receptor-receptor interactions, in that this G-protein may also be activated by adenylate cyclase stimulatory hormone receptors and may additionally be a target for interactions of the hormone-sensitive adenylate cyclase system with the hormone-sensitive phosphoinositide system. It is feasible that such types of receptor-receptor interactions may not only occur at the level of the G_i-protein, but that other components of transmembrane signalling systems may be involved as well in receptor-receptor interactions.

ACKNOWLEDGEMENTS

The authors studies reported herein were supported by the Deutsche Forschungsgemeinschaft.

REFERENCES

Aktories, K. and Jakobs, K.H. (1984). N_i-mediated inhibition of human platelet adenylate cyclase by thrombin. Eur. J. Biochem., 145, 333-338.

Aktories, K. and Jakobs, K.H. (1985). Regulation of platelet cyclic AMP formation. In The Platelets: Physiology and Pharmacology. (ed. G.L. Longenecker). Academic Press, Orlando.

Asano, T., Katada, T., Gilman, A.G. and Ross, E.M. (1984) Activation of the inhibitory GTP-binding protein of adenylate cyclase, G_i, by β-adrenergic receptors in reconstituted phospholipid vesicles. J. Biol. Chem., 259, 9351-9354.

Bauer, S. and Jakobs, K.H. (1986). Phorbol ester treatment impairs hormone- but not stable GTP analog-induced inhibition of adenylate cyclase. FEBS Lett., 198, 43-46.

Birnbaumer, L., Codina, J., Mattera, R., Cerione, R.A., Hildebrandt, J.D., Sunyer, T., Rojas, F.J., Caron, M.G., Lefkowitz, R.J. and Iyengar, R. (1985). Regulation of hormone receptors and adenylyl cyclases by guanine nucleotide binding N proteins. Rec. Progr. Hormone Res., 41, 41-94.

Bockaert, J., Deterre, P., Pfister, C., Guillon, G. and Chabre, M. (1985). Inhibition of hormonally regulated adenylate cyclase by the βγ subunit of transducin. EMBO J., 4, 1413-1417.

Gilman, A.G. (1984). G Proteins and dual control of adenylate cyclase. Cell, 36, 577-579.

Jakobs, K.H. and Aktories, K. (1983). Synergistic inhibition of human platelet adenylate cyclase by stable GTP analogs and epinephrine. Biochim. Biophys. Acta, 732, 352-358.

Jakobs, K.H., Aktories, K. and Schultz, G. (1984). Mechanisms and components involved in adenylate cyclase inhibition by hormones. Adv. Cyclic Nucleotide Protein Phosphoryl. Res., 17, 135-143.

Jakobs, K.H., Aktories, K., Minuth, M. and Schultz, G. (1985). Inhibition of adenylate cyclase. Adv. Cyclic Nucleotide Protein Phosphoryl. Res., 19, 137-150.

Jakobs, K.H., Bauer, S. and Watanabe, Y. (1985a). Modulation of adenylate cyclase of human platelets by phorbol ester. Impairment of the hormone-sensitive inhibitory pathway. Eur. J. Biochem., 151, 425-430.

Katada, T., Bokoch, G.M., Northup, J.K., Ui, M. and Gilman, A.G. (1984). The inhibitory guanine nucleotide-binding regulatory component of adenylate cyclase. Properties and function of the purified protein. J. Biol. Chem., 259, 3568-3577.

Katada, T., Gilman, A.G., Watanabe, Y., Bauer, S. and Jakobs, K.H. (1985). Protein kinase C phosphorylates the inhibitory guanine-nucleotide-binding regulatory component and apparently suppresses its function in hormonal inhibition of adenylate cyclase. Eur. J. Biochem., 151, 431-437.

Takai, Y., Kikkawa, U., Kaibuchi, K. and Nishizuka, Y. (1984). Membrane phospholipid metabolism and signal transduction for protein phosphorylation. Adv. Cyclic Nucleotide Protein Phosphoryl. Res., 18, 119-158.

Ui, M., Katada, T., Murayama, T., Kurose, H., Yajima, M., Tamura, M., Nakamura, T. and Nogimori, K. (1984). Islet-activating protein, pertussis toxin: A specific uncoupler of receptor-mediated inhibition of adenylate cyclase. Adv. Cyclic Nucleotide Protein Phosphoryl. Res., 17, 145-151.

Watanabe, Y., Horn, F., Bauer, S. and Jakobs, K.H. (1985). Protein kinase C interferes with N_i-mediated inhibition of human platelet adenylate cyclase. FEBS Lett., 192, 23-27-

Part B
Interactions between Receptors in Synaptic Membranes in the Central Nervous System

Chairman: J.P. Changeux

18

On the Role of Receptor–Receptor Interactions in Synaptic Transmission: Biochemical and Autoradiographical Studies on the Interactions between α_2-adrenergic and Neuropeptide Y Receptors in the Nucleus Tractus Solitarius

Kjell Fuxe, Luigi F. Agnati, Anders Härfstrand, Bertil B. Fredholm, Madhu Kalia and Menek Goldstein

INTRODUCTION

In 1980 we postulated that receptors can modulate the binding characteristics of oneanother in membrane preparations under equilibrium conditions, and we introduced the concept of receptor-receptor interaction, an event which may represent an important intramembrane integrative mechanism in synaptic transmission (see Agnati et al., 1983a,b,c,d, 1986, Fuxe and Agnati 1985, Fuxe et al., 1981, 1983, 1984a,b, 1986). In the present paper we will present our recent work on the interactions between neuropeptide Y (NPY) and α_2-adrenergic receptors in the nucleus tractus solitarius (NTS) using biochemical binding techniques and quantitative receptor autoradiography.

MATERIAL AND METHODS

Male specific pathogen free male Sprague-Dawley rats (200g, ALAB, Stockholm, Sweden) have been used. In the biochemical binding experiments the α_2 agonist ^{3}H-paraminoclonidine (^{3}H-PAC, NEN USA, spec. act. 42.0 Ci/mmol) and the α_2 antagonist ^{3}H-idazoxane (Amersham, spec. act. 40.0 Ci/mmol) have been used as radioligands for the α_2-adrenergic recognition sites. For details on the binding procedures, see Agnati et al. (1983a) and Fuxe et al. (1984b).

^{3}H-PAC was used as a radioligand for the α_2-adrenergic recognition sites and 125I-NPY (Amersham U.K. spec. act. 2000 Ci/mmol) as radioligand for the NPY recognition sites. The binding procedures for 125I-NPY was based on the procedure of Goldstein et al. (1986). For details on the methods for quantitative receptor autoradiography, see Benfenati et al., 1986; and Agnati et al., 1984 and for quantitation of 125I-NPY binding in the NTS, see Härfstrand et al., 1986a.

In the experiments involving irreversible α_2-adrenergic receptor blockade by means of EEDQ, the substance was given into the cisterna magna (ic) in a dose of 10 µg, 24 hours before sacrifice. In some rats the α_2-adrenergic receptors were selectively protected from irreversible blockade by EEDQ by pretreatment of the animals 15 minutes earlier with the selective α_2 adrenoreceptor blocking compound idazoxan (Doxey et al., 1983) (30 µg, i.c.).

Tyrosine hydroxylase (TH) and NPY immunoreactivities (IR) within the NTS area were demonstrated according to the biotin avidin immunoperoxidase procedure (see Härfstrand et al., 1987). For details on the relationship of the various catecholaminergic neural groups with the classical neuroanatomical description of the NTS area, see Kalia et al., (1985) and Kalia and Fuxe (1985).

In the statistical analysis Student's t-test as well as multiple comparison procedures (Dunn test) were used (Snedecor and Cochran, 1980; Hollander and Wolfe, 1972).

RESULTS

Pre- and postsynaptic features of adrenergic and NPY transmission within the nucleus tractus solitarius

High densities of α_2-adrenergic agonist binding sites are present where there are high densities of TH immunoreactive nerve terminals medial to the tractus solitarius. A high density of NPY IR nerve terminals is predominantly found in the ventral part of the NTS and within the dorsal motor nucleus of the vagus (dmnX), while the highest density of 125I-NPY binding sites exists within the dorsal and dorsomedial part of the NTS (fig. 1). Thus, the pre- and the postsynaptic features of the NPY IR nerve terminals of the NTS are not present in the same locations. Such observations have lead us to postulate the existence of a "volume transmission" within the central nervous system (Agnati et al., this volume). Fig. 2 schematically illustrates the distribution of the α_2-adrenergic and NPY binding sites within the NTS. The overlap area is found to be located within the dorsal parts of the nucleus tractus solitarius (see Fuxe et al., 1986) where *inter alia* the baroreceptor afferents terminate (Kalia et al., 1981).

Modulations of the interactions between NPY and clonidine binding sites in membrane preparations from the medulla oblongata

The role of calcium and magnesium ions. NPY (10 nM) in the presence of calcium (5mM) and magnesium (5mM) ions reduces the affinity of the ^{3}H-PAC binding sites, while leaving the B-max values unaffected. However, in the absence of calcium and magne-

fig 1

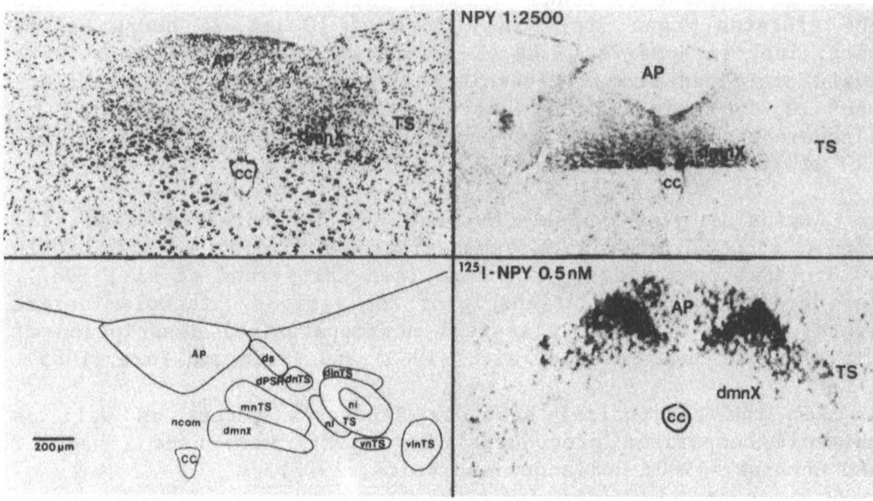

Fig. 1. Distribution of NPY immunoreactive nerve terminals and 125I-NPY binding sites in the NTS and the dorsal motor nucleus of the vagus (dmnX) at the area postrema (AP) level. The various subnuclei are illustrated and the Nissl staining is shown in the upper left part. Abbreviations used: ds= dorsal strip; dPSR= dorsal parasolitary region; mNTS= medial subnucleus of the NTS; ni= interstitial subnucleus of the NTS; dNTS= dorsal subnucleus; dlNTS= dorsolateral subnucleus; nI= intermediate nucleus; vNTS= ventral subnucleus; vlNTS= ventrolateral subnucleus.

fig 2

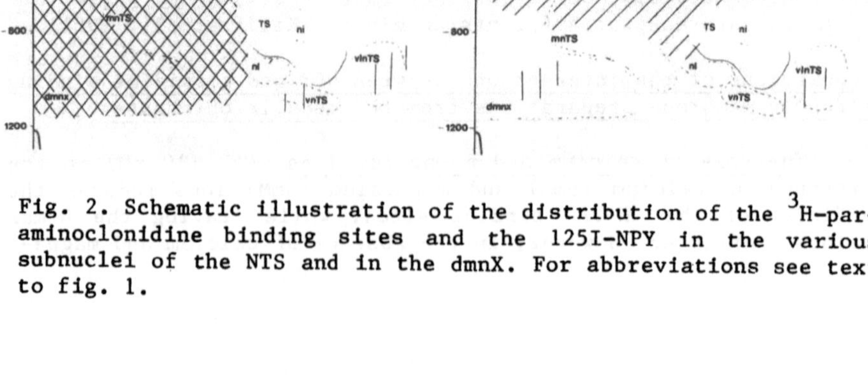

Fig. 2. Schematic illustration of the distribution of the ^{3}H-paraminoclonidine binding sites and the 125I-NPY in the various subnuclei of the NTS and in the dmnX. For abbreviations see text to fig. 1.

sium ions no modulation is observed in the presence of NPY (fig. 3).

The role of GTP. GTP (100 µM) markedly increases the K_D-value and reduces the B-max value of the ^{3}H-PAC binding sites (see fig. 4). This action of GTP on the high affinity α_2 agonist binding sites is not counteracted by NPY (10 nM). In agreement with these results it is shown in fig. 5 that GTP induces a loss of ^{3}H-PAC binding which is not modulated by NPY (10 nM). Unexpectedly, however, the islet activating protein (IAP; pertussis toxin) was ineffective in modulating the GTP induced displacement of ^{3}H-PAC binding, even though such an effect has been reported earlier (see Nomura et al., 1985). The failure of IAP to affect binding could not be accounted for by a failure of IAP to ADP-ribosylate 39 and 41kDa proteins in the medulla oblongata membranes (for incubation conditions, see fig. 5). These results indicate that the N protein involved with the α_2-adrenergic receptor in the medulla oblongata is not sensitive to IAP.

The effects of NPY on the GTP induced reduction in the affinity of clonidine for the ^{3}H-idazoxan binding sites in medulla oblongata membranes are shown in fig. 6. A ^{3}H-idazoxan (α_2-adrenergic antagonist) binding prodecure very similar to that for ^{3}H-PAC was used. The main difference was a 20 min incubation time instead of a 10 min incubation time as used for ^{3}H-PAC (see Fuxe et al., 1984b). As seen in fig. 6 GTP (100 µM) significantly increased the IC_{50} value of clonidine to displace ^{3}H-idazoxan binding. The GTP induced shift in the clonidine displacement curve is shown to be counteracted by NPY (10 nM). NPY (10 nM) alone has no significant effect.

Studies on the ability of α_2-adrenergic receptors to modulate the binding characteristics of 125I-NPY binding sites within the nucleus tractus solitarius

Effects of clonidine. As seen in fig. 7 clonidine (20 nM) can *in vitro* substantially reduced the binding of NPY in the NTS and also produced a small but significant reduction in the NPY binding in the area postrema. When clonidine was instead given into the lateral ventricle (3.75 nmol in 30 µl artificial CSF, 15 min) of awake unrestrained rats, an even larger reduction of NPY binding was observed within the NTS, while NPY binding in the area postrema was not significantly affected (fig. 8).

EEDQ experiments. As seen in fig. 9 pretreatment with EEDQ (10 µg, intracisternally (ic)) 24 hours earlier resulted in a marked disappearance of the ^{3}H-PAC binding within the NTS. This disappearance was found to be partly counteracted by pretreatment of the animals with idazoxan (50 µg, ic) 15 min prior to the EEDQ injection (ic). Idazoxan given alone (30 µg, ic, 24 hours before killing) did not affect the ^{3}H-PAC binding.

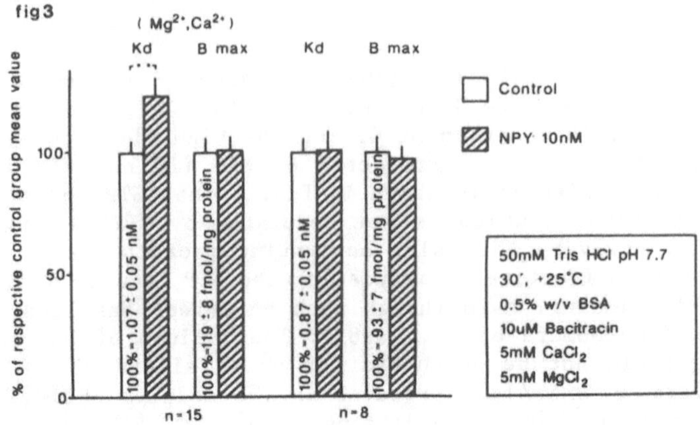

Fig. 3. Effects of NPY (10 nM) in the presence or absence of ions (Mg^{2+}, Ca^{2+}) on ^{3}H-PAC binding parameters in membrane preparations from the rat medulla oblongata (without substantia gelatinosa). Means ± s.e.m. Student's paired t test. ** p<0.01.

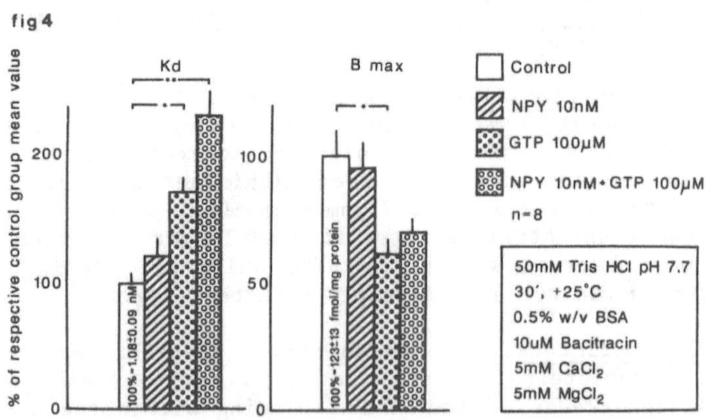

Fig. 4. Effects of NPY (10 nM) in the presence or absence of GTP (100 μM) on ^{3}H-PAC binding parameters in membrane preparations from the rat medulla oblongata (without substantia gelatinosa). Means ± s.e.m. are shown. Dunn test. * p< 0.05; ** p<0.01.

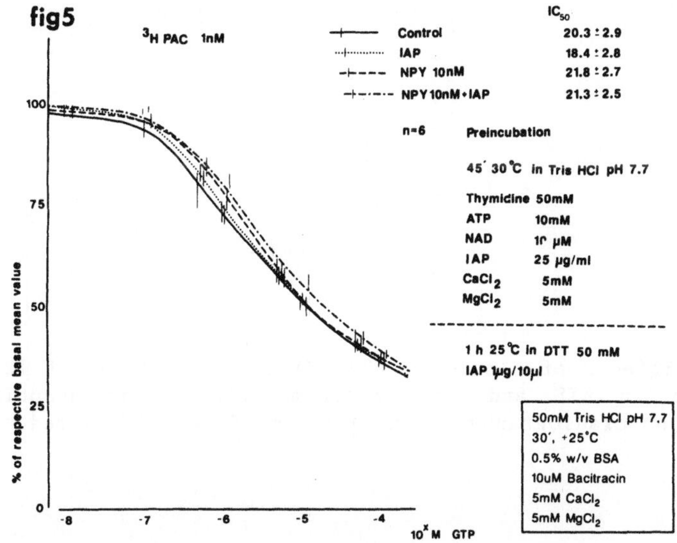

Fig. 5. Effects of GTP on ^{3}H-PAC binding in medulla oblongata membrane preparations after NPY and/or IAP pretreatment. Means ± s.e.m. are shown. The IC_{50} values are given as nM. The activation conditions of IAP and the preincubation conditions are also given. In the square the composition of the incubation buffer is shown.

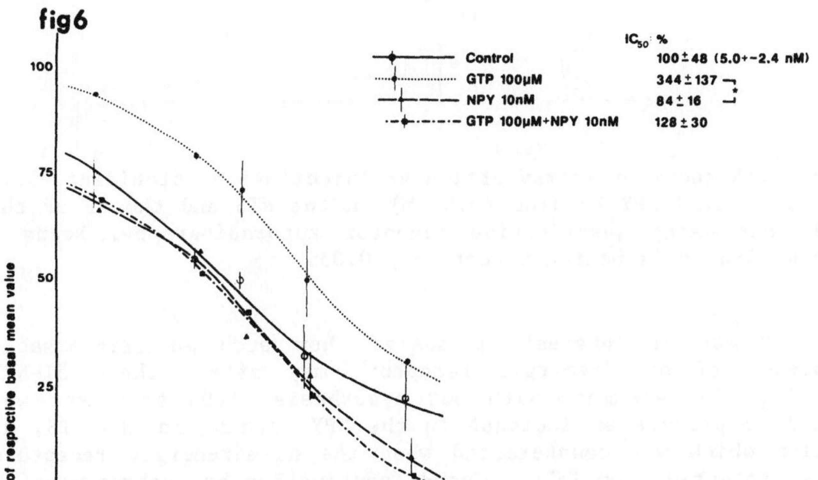

Fig. 6. Counteraction by NPY of the GTP induced reduction in the affinity of clonidine for the ^{3}H-idazoxan binding sites in medulla oblongata membranes. Means ± s.e.m. are shown for the IC_{50} values (nM). Dunn test * $p<0.05$; n= 6.

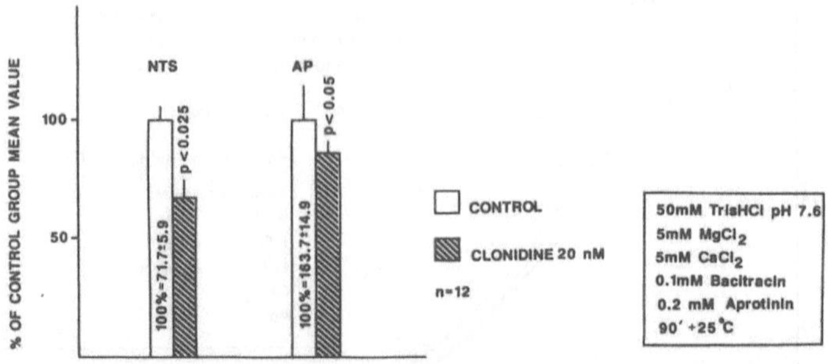

Fig. 7. Effects of clonidine (20 nM) on the binding of 125I-NPY (0.5 nM) in NTS and area postrema (AP) using quantitative receptor autoradiography. Means ± s.e.m. Student's paired t test.

fig 8

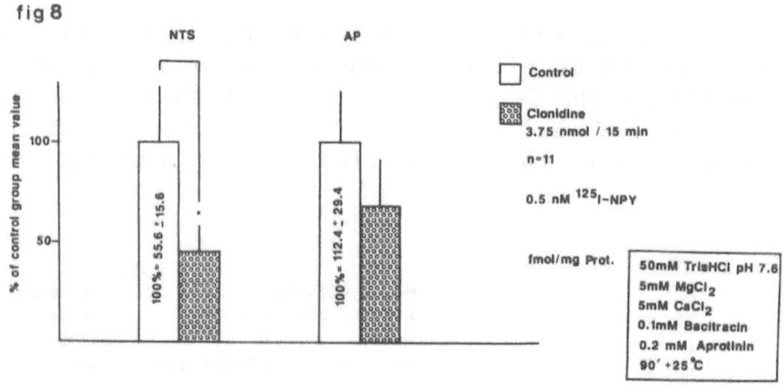

Fig. 8. Effects of intraventricular injections of clonidine (3.75 nmol) on 125I-NPY binding (0.5 nM) in the NTS and the AP of the male rat using quantitative receptor autoradiography. Means ± s.e.m. Student's paired t test. * $p<0.05$.

It was of interest to analyze how such an irreversable blockade of α_2-adrenergic receptors may affect the 125I-NPY binding. In agreement with our hypothesis, EEDQ treatment was found to produce an increase in the NPY binding in the NTS, an action which was counteracted when the α_2-adrenergic receptors were protected from EEDQ induced inactivation by pretreatment of idazoxan (see in fig. 10). EEDQ treatment did not significantly influence the NPY binding in the area postrema.

On the role of membrane fluidity. Following the procedures of Heron et al. (1980) we have modulated the fluidity of the membranes present in sections of the medulla oblongata by incubating the sections for 1 hour at room temperature with the unsatu-

fig 9

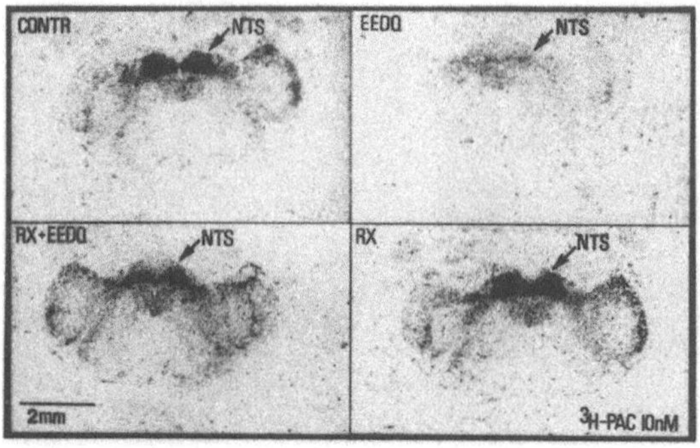

Fig. 9. Receptor autoradiography showing the effects of EEDQ treatment on ^{3}H-PAC binding (10 nM) in the NTS of the medulla oblongata with or without prior pretreatment with an α_2-adrenergic antagonist RX 781094 (Idazoxan). For further details, see text.

rated fatty acid linoleic acid (6 mg/ml) or with the saturated fatty acid cholesteryl-hemisuccinate (40 mg/ml). As shown in fig. 11, linoleic acid alone and clonidine (20 nM) alone reduced the 125I-NPY binding in the NTS, but no further reduction was observed with the combined treatment. Similar results were obtained with the saturated fatty acid.

DISCUSSION

Based on the present and previous results we have introduced the model shown in fig. 12 to explain the available information on the interactions between NPY receptors and α_2-adrenergic receptors in the membrane. This model is based on the observations that both receptors are capable of inhibiting the adenylase cyclase activity in slices of the NTS area (see Härfstrand et al., 1986b). The GTP binding site on the Ni protein to the α_2-agonist binding sites is not interfered with by an activation of the NPY receptor. Instead, the present experiments have shown that the α_2 adrenergic agonist promoted coupling of the Ni protein to the adenylate cyclase is substantially counteracted by activation of the NPY receptor. These results favour the existence of an antagonistic interaction between the two receptors. This view is also favoured by the fact that the major result of the NPY modulation of the binding characteristics of the α_2 adrenergic agonist site is an increase in the K_D-value. The reduction of affinity induced by NPY may also be produced by a direct effect on the protein containing the α_2-agonist site. Although the increase of the K_D-value is only in the order of

fig 10

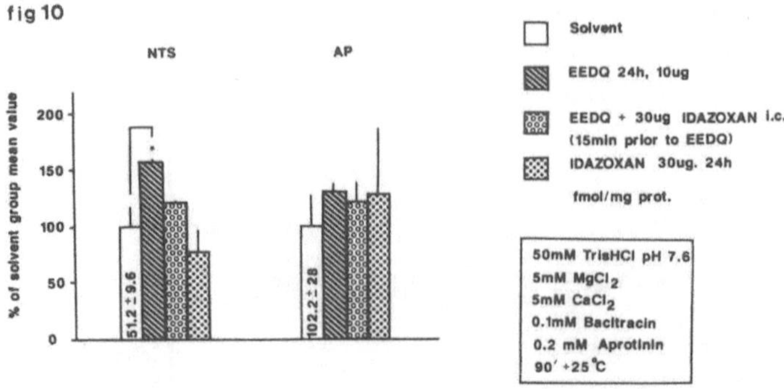

Fig. 10. Prevention of the EEDQ (ic) induced increase in 125I-NPY binding in the NTS by ic pretreatment with idazoxan. Means $\pm$ s.e.m. Dunn test. * $p<0.05$. n= 4.

20-40% it could reflect a major change in the binding kinetics (see Changeux, this symposium). Finally, our physiological experiments on central cardiovascular control (see Härfstrand, this symposium) indicate that the modulation observed in the NPY and α_2-adrenergic transmission lines as a result of putative receptor-receptor interactions may have an important physiological consequence. Thus, both adrenaline and NPY when given centrally, reduce the arterial blood pressure and heart rate and when given together they antagonize the actions of oneanother (see paper by Härfstrand et al., this volume).

The present results with the irreversable monoamine blocking agent EEDQ, has given additional evidence that α_2 adrenergic and NPY receptors may interact <u>in vivo</u> . Thus, EEDQ increased 125I-NPY binding probably as a result of its ability to irreversibly block the α_2-adrenergic receptor in low doses.

Further evidence for membrane interactions between the two receptors was obtained in the experiments aiming to change membrane fluidity by incubating the sections in the presence of suspensions in linoleic acid or cholesteryl hemisuccinate. Thus, both the unsaturated and saturated fatty acid reduced the 125I-NPY binding. When clonidine was subsequently added it could not induce any further lowering of the 125I-NPY binding. Thus, the results indicate that clonidine in order to modulate the NPY receptor requires optimal membrane fluidity. Taken together the results support the existence of intramembrane interactions between receptors.

In conclusion, the present results are compatible with the existence of an intramembrane antagonistic interaction between α_2-adrenergic and NPY receptors within the nucleus tractus solitarius. This interaction appears to play an important role in

fig 11

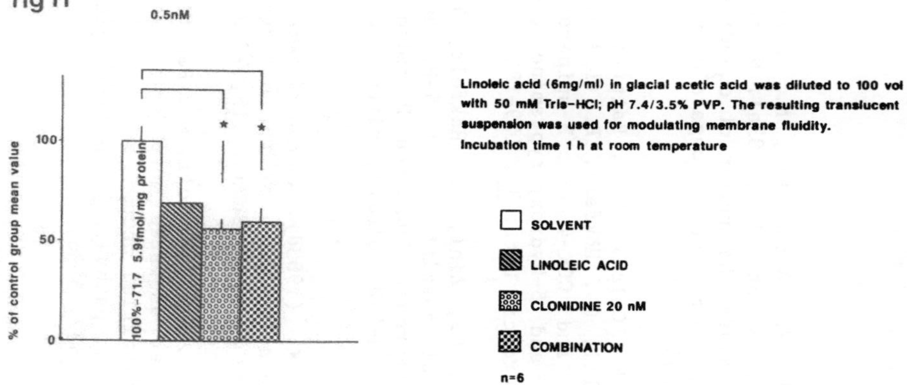

fig 12

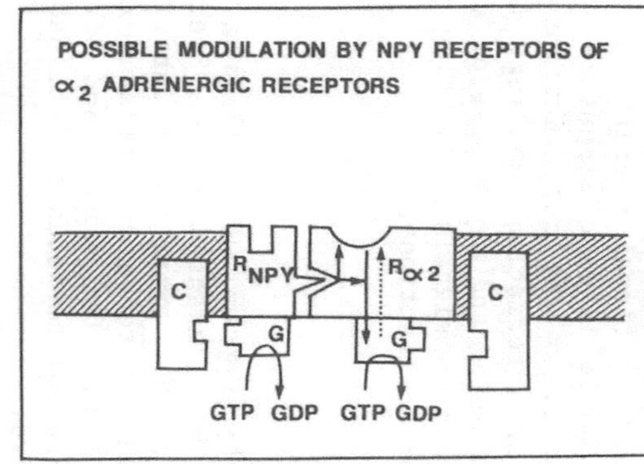

fig 13

The functional meaning of the rec-rec interactions

a) site of interaction among the multiple transmission line of the n-dimensional synapse
b) Filtration of signals to the intracellular device
c) integration of signals to the intracellular device
d) construction of ultromicrocircuits for information handing
e) storage of information (permanent or semipermanent interactions)
f) possible new targets for paracrine and endocrine signals which besides modulating the receptors could modulate the probability of receptor - receptor interactions

Fig. 11. Studies on the interactions between linoleic acid and clonidine in their modulation of 125I-NPY binding in nucleus tractus solitarius as evaluated by quantitative receptor autoradiography. Means $\pm$ s.e.m. Dunn test. * $p<0.05$.

Fig. 12. Possible mechanisms for the modulation by NPY receptors of α_2-adrenergic receptors in the medulla oblongata. C= adenylate cyclase; G= coupling protein (GTP binding protein); R_{NPY}= NPY-receptor; $R\alpha_2$= α_2-adrenergic receptor.

Fig. 13. Possible functional roles of receptor-receptor interactions in synaptic transmission.

central cardiovascular regulation (see Fuxe et al., 1986). In fig. 13 we summarize the possible functional meaning of the receptor-receptor interactions. We would like to emphasize that receptor-receptor interactions may allow the constructions of ultramicrocircuits of importance for information handling. They can also allow the storage of information and may represent also new targets for paracrine and endocrine signals modulating the probability of the receptor-receptor interactions.

ACKNOWLEDGEMENTS

This work has been supported by a grant (04X-715) from the Swedish Medical Research Council, by a grant from the Swedish Environmental Work Health Fund, by a CNR-I grant and by Ostermans fond. We are grateful for the excellent technical assistance of Kerstin Nilsson, Birgitta Nyberg, Ulla-Britt Finnman and Barbro Tinner, and for the excellent secretarial assistance of Elisabeth Sandqvist, Anne Edgren and Monia Särd.

REFERENCES

Agnati, L.F., Fuxe, K., Benfenati, F., Battistini, N., Härfstrand, A., Tatemoto, K., Hökfelt, T, & Mutt, V. (1983a). Neuropeptide Y in vitro selectively increases the number of α_2-adrenergic binding sites in membranes of the medulla oblongata of the rat. Acta Physiol. Scand. 118, 293-295.

Agnati, L.F., Fuxe, K., Benfenati, F., Celani, M.F., Battistini, N,. Mutt, V., Cavicchioli, L., Galli, G., Hökfelt, T. (1983b). Differential modulation by CCK-8 and CCK-4 of ^{3}H-spiperone binding sites linked to dopamine and 5-hydroxytryptamine receptors in the brain of the rat. Neurosci. Lett. 35, 179-183.

Agnati, L.F., Fuxe, K., Benfenati, F., Zini, I., Hökfelt, T. (1983c). On the functional role of coexistence of 5-HT and substance P in bulbospinal 5-HT neurons. Substance P reduces affinity and increases density of ^{3}H-5-HT binding sites. Acta Physiol. Scand. 117, 299-301.

Agnati, L.F., Celani, M.F., Fuxe, K. (1983d). Cholecystokinin peptides in vitro modulate the characteristics of the striatal ^{3}H-N-propylnorapomorphine sites. Acta Physiol. Scand. 118, 79-81.

Agnati, L.F., Fuxe, K., Benfenati, F., Zini, I, Zoli, M., Fabbri, L. & Härfstrand, A. (1984). Computer assisted morphometry and microdensitometry of transmitter identified neurons with special reference to the mesostriatal dopamine pathway. I. Methodologica aspects. Acta Physiol. Scand. Suppl. 532, 5-36.

Agnati, L.F., Fuxe, K., Zoli, M., Merlo Pich, E., Benfenati, F., Zini, I. and Goldstein, M. (1986). Aspects on the information handling by the central nervous system: focus on cotransmission in the aged rat brain. In: Progress in Brain Research, Vol 68. (T. Hökfelt, K. Fuxe and B. Pernow eds.), pp. 291-301.

Benfenati, F., Cimino, M., Agnati, L.F. & Fuxe, K. (1986). Quantitative autoradiography of central neurotransmitter receptors: methodological and statistical aspects with special reference to computer-assisted image analysis. Acta Physiol. Scand. 128, 129-146.

Doxey, J.C., Roach, A.G. & Smith, C.F.C. (1983). Studies on RX 781094: a selective, potent and specific antagonist of α_2-adrenoceptors. Br. J. Pharmac. 78, 489-505.

Fuxe, K., Agnati, L.F., Benfenati, F., Celani, M.F., Zini, I., Zoli, M. Mutt, V. (1983). Evidence for the existence of receptor-receptor interactions in the central nervous system. Studies on the regulation of monoamine receptors by neuroleptics. J. Neuronal Trans. 18, 165-179.

Fuxe, K., Agnati, L.F., Andersson, K., Martire, M., Ögren, S-O., Giardino, L., Battistini, N., Grimaldi, R., Farabegoli, C., Härfstrand, A. & Toffano, G. (1984a). Receptor-receptor interactions in the central nervous system. Evidence for the existence of heterostatic synaptic mechanisms. In: Regulation of transmitter function: Basic and Clinical Aspects, Developments in Neuroscience Series, Vol. 17 (E.S. Vizi & K. Magyar, eds.), pp. 129-140.

Fuxe, K., Agnati, L.F., Härfstrand, A., Martire, M., Goldstein, M., Grimaldi, R., Bernardi, P., Zini, I., Tatemoto, K. & Mutt, V. (1984b). Evidence for a modulation by neuropeptide Y of the α-2 adrenergic transmission line in central adrenaline synapses. New possibilities for treatment of hypertensive disorders. Clin. and Exp. Hypertension, Theory and Practice, A6(10&11). 1951-1956.

Fuxe, K. and Agnati, L.F. (1985). Receptor-receptor interactions in the central nervous system. A new integrative mechanism in synapses. Med. Res. Rev. vol.5, No.4, 441-482.

Fuxe, K., Agnati, L.F., Härfstrand, A., Janson, A.M., Neumeyer, A., Andersson, K., Ruggeri, M., Zoli, M. and Goldstein (1986). Morphofunctional studies on the neuropeptide Y/adrenaline costoring terminal systems in the dorsal cardiovascular region of the medulla oblongata. Focus on receptor-receptor interactions in cotransmission. In: Progress in Brain Research, Vol. 68. (T. Hökfelt, K. Fuxe and B. Pernow eds.), Elsevier Science Publishers, Amsterdam, pp. 303-320.

Goldstein, M., Kusano, N., Adler, C. and Meller, E. (1986). Characterization of central neuropeptide Y receptor binding sites and possible interactions with α_2-adrenoceptors. In: Progress in Brain Research. Vol. 68. (T. Hökfelt, K. Fuxe and B. Pernow, eds.). Elsevier Science Publishers B.V., pp. 331-336.

Heron, D.S., Shinitzky, M., Hershkowitz, M. and Samuel, D. (1980). Lipit fluidity markedly modulates the binding of serotonin to mouse brain membranes. Proc. Natl. Acad. Sci. USA, Vol. 77(2), 7463-7467.

Hollander, M. and Wolfe, D.A. (1973). Nonparametric statistical methods. John Wiley & Sons, New York, U.S.A.

Härfstrand, A., Fuxe, K., Agnati, L.F., Kitayama, I., Cintra, Janson, A.M., Kalia, M., Vanderhaegen, J-J., Goldstein, M. and Terenius, L. (1987a). Intracisternal administration of cholecystokinin-8 counteracts the central cardiovascular effects of adrenaline and NPY. A study based on the existence of cholecystokinin, phenylethanolamine N-methyltransferase and neuropeptide Y immunoreactivity in the same neurons of the nucleus tractus solitarius. Neurochem. Int., in press.

Härfstrand, A., Fuxe, K., Agnati, L.F., Benfenati, F. and Goldstein, M. (1986a). Receptor autoradiographical evidence for high densities of 125I-neuropeptide Y binding sites in the nucleus tractus solitarius of the normal male rat. Acta Physiol. Scand. 128, 195-200.

Härfstrand, A., Fredholm, B. and Fuxe, K. (1987b). Inihibitory effects of neuropeptide Y on cyclic AMP accumulation in slices of the nucleus tractus solitarius region of the rat. Neurosci. Lett. in press.

Hökfelt, T, Holets, V.R., Staines, W., Meister, B., Melander, T., Schalling, M., Schultzberg, Freedman, J., Björklund, H., Olson, L., Lindh, B., Elfvin, L-G., Lundberg, J.M., Lindgren, J.Å., Samuelsson, B., Pernow, B., Terenius, L., Post, C., Everitt, B. and Goldstein, M. (1986). Coexistence of neuronal messengers - an overview. In: Progress in Brain Research. Vol. 68. (T. Hökfelt, K. Fuxe and B. Pernow eds.), pp. 33-70.

Kalia, M., Fuxe, K. and Goldstein, M. (1985). Rat medulla oblongata. II. Dopaminergic, noradrenergic (A1 and A2) and adrenergic neurons, nerve fibers, and presumptive terminal processes. J. Comp. Neurol. 233, 308-332.

Kalia, M. and Fuxe, K. (1985). Rat medulla oblongata. I. cytoarchitectonic considerations. J. Comp. Neurol. 233, 285-307.

Kalia, M.P. (1981). Localization of aortic and carotid baroreceptor and chemoreceptor primary afferents in the brain stem. In: Perspectives in Cardiovascular Research (J.P. Buckley & C.M. Ferrario, eds.), Raven Press, New York, pp. 9-21.

Meller, E., Bohmaker, K., Goldstein, M. and Friedhoff, A.J. (1985). Inactivation of D_1 and D_2 dopamine receptors by N-ethoxycarbonyl-2-ethoxy-1,2-dihydroquinoline in vivo: Selective protection by neuroleptics. J. Pharmacol. exp. 233, 656-662.

Nomura, Y., Kitamura, Y. and Segawa, T. (1985). Decrease of clonidine Binding affinity to α_2-adrenoceptor by ADP-ribosylation of 41.000-Dalton proteins in rat cerebral cortical membranes by islet-activating protein. J. Neurochemistry, 364-368.

Saria, A.,Theodorsson-Norheim, E. & Lundberg, J.M. (1984). Evidence for specific neuropeptide Y-binding sites in rat brain synaptosomes. Eur. J. Pharmacol. 107(1), 105-107.

Snedecor, G.W. and Cochran, W.G. (1980). Statistical methods. 7th Ed., The Iowa State University Press, Ames Iowa, U.S.A.

Undén, A., Tatemoto, K., Mutt, V. & Bartfai, T. (1984). Neuropeptide Y receptor in the rat brain. Eur. J. Biochem. 145, 525-530.

19

Aspects on the Integrative Capabilities of the Central Nervous System: Evidence for 'Volume Transmission' and its Possible Relevance for Receptor–Receptor Interactions

L.F. Agnati, K. Fuxe, E. Merlo Pich, M. Zoli, I. Zini, F. Benfenati, A. Härfstrand and M. Goldstein

INTRODUCTION

The hypothesis is introduced that the central nervous system (CNS) is endowed with the capability of handling information, not only based on the topological organization of its elements and the patterns of impulse flow along the neural networks (wiring transmission), but also based on the electrotonic currents and the spread of humoral signals in the extracellular fluid (volume transmission), which, in turn, man affect the computing characteristics of the neural networks themselves (Agnati et al. 1986b). Thus, in the CNS we can also recognize interactions, and therefore integrative aspects, at various hierarchical levels and at different degrees of miniaturization, and between the two types of information handlings. One example, at a high hierarchical level and a low degree of miniaturization, may be represented by the above mentioned interactions between the volume transmission and the wiring transmission at the network level, while another example, at a low hierarchical level and a high degree of miniaturization, may be represented by the receptor-receptor interactions, i.e., by the integration at membrane level of the inputs arriving through different transmission lines.

In the present paper data will be presented which can give indirect support for the existence of volume transmission and also data which favour a possible action of volume transmission on the receptor-receptor interactions. Our attention will mainly be focused on the cholecystokinin (CCK)/dopamine (DA) system projecting to the subcortical limbic structures (Hökfelt et al. 1980). Thus, by means of morphometrical and microdensitometrical procedures, we have characterized the DA and CCK ascending system from the ventral tegmental area (VTA) to the nucleus accumbens. We have studied tyrosine hydroxylase (TH)/CCK codistribution and coexistence in the cell body area of the ventral mesencephalon and in the terminals at the level of the caudal nucleus accumbens. Previous work has shown that coexistence of multiple neuro-

fig1

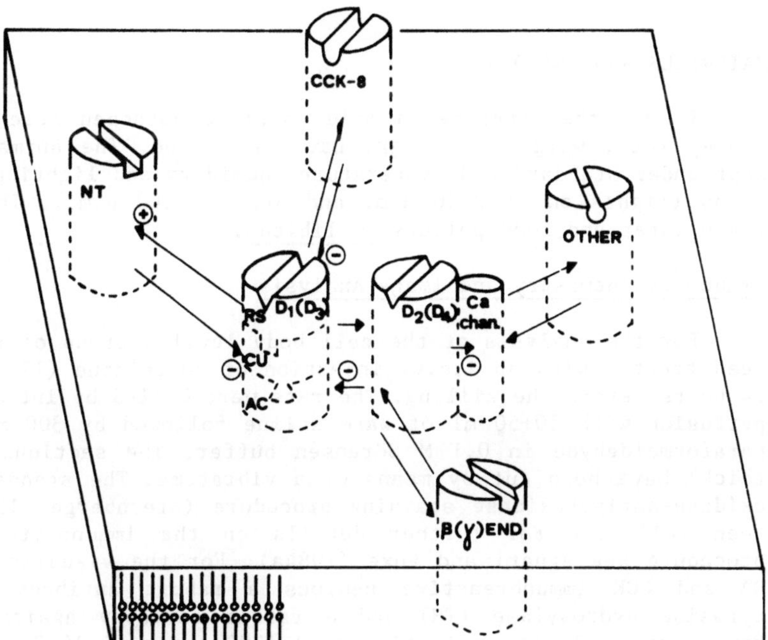

Fig. 1. Schematic representation of receptor-receptor interactions in limbic subcortical membranes, e.g., n. accumbens. Some of the interactions shown have been tested (see Fuxe et al. 1985). Abbreviations: RS= recognition site; CU= coupling unit; AC= adenylate cyclase.

active substances in nerve terminal populations may be related at the postsynaptic level, to an intramembrane mechanism: the receptor-receptor interaction (see fig. 1) (Agnati et al. 1980, 1983, 1984b, 1986a; Fuxe et al. 1983, 1986b; Fuxe and Agnati 1985; Rostene et al. 1983). However, this mechanism has been demonstrated in synapses, in which coexistence of multiple transmitters is present, and in local circuits, (which have been considered as electro-metabolic integrative units) and very likely under conditions, in which "volume transmission" (see Agnati et al. 1986a,b and below) is operative. The existence of an intramembrane interaction between DA and CCK receptors at the level of the n. accumbens has been evaluated <u>in vivo</u> by means of quantitative receptor autoradiography. The existence of an interaction between CCK-8 and DA at a post-recognition site level has been further investigated studying the effects of chronic central administration of CCK-8 on the receptor and behavioural recovery of the DA transmission line after irreversible blockade of the DA receptor.

MATERIALS AND METHODS

In all the experiments male specific pathogen free Sprague Dawley rats, weighing 300 gr, have been used. The animals were kept under standardized temperature, humidity and lighting conditions (lights on at 8.00 a.m. and off at 8.00 p.m.). They were given water and food pellets ad libitum.

Immunocytochemistry and Image Analysis

For the analysis at the cell body level a group of rats has been treated with an i.c.v. injection of colchicine (120 ug/rat) 24 hours before the killing. The rats were killed by intracardiac perfusion with 50+50 ml of warm saline followed by 300 ml of 4% paraformaldehyde in 0.1 M Sörensen buffer. The sections (40 um thick) have been cut by means of a vibratome. The standard peroxidase-antiperoxidase staining procedure (Sternberger 1979) has been followed. For further details on the immunocytochemical procedure see Agnati and Fuxe (1984a). For the visualization of DA and CCK immunoreactive neurons a rabbit antibody against tyrosine hydrosylase (TH) and a rabbit antibody against CCK-8 have been used, at a dilution of 1:1500 and of 1:1000, respectively. The antisera have previously been characterized (Markey et al. 1980; Vanderhaghen et al. 1980).

Four sections per animal have been analyzed at the level of the ventral mesencephalon (colchicine treated rats, n=4) and at the level of the caudal accumbens (non colchicine treated rats, n=4). For the morphometrical an microdensitometrical analysis, a MOP AMO II semiautomatic and an IBAS fully automatic image analyzer (both from Zeiss Kontron) have been utilized. The image analysis procedures have been described in previous papers (Agnati et al. 1984a, book by Agnati and Fuxe 1985; Zoli et al. 1986).

Receptor Autoradiography

The animals (n=4) were killed by intracardiac perfusion with 50+50 ml of saline followed by 300 ml of 0.1% paraformaldehyde in 0.1 M Sörensen buffer. The sections (14 μm thick) were cut with a cryostat (Bregma 1.2 and 2.2 mm levels according to the Paxinos and Watson Atlas) and left at $+4^{o}C$ for 1.30 h in a dryer. The sections were incubated with a fixed concentration close to the Kd value (1.4 nM) of the D2 receptor antagonist 3H-spiperone (16.7 Ci/mmol, NEN). The incubation time was 60 min at room temperature (incubation buffer: Tris HCl 85 mM, NaCl 120 mM, KCl 5 mM, $CaCl_2$ 2 mM, $MgCl_2$ 2 mM, pH 7.4) Unspecific binding was defined as the binding in the presence of (+)butaclamol (1 μM).

The sections were rinsed twice in ice cold Tris HCl buffer, 170 mM, pH 7.4 (5 min), and finally dipped in distilled water. The sections were then transferred to X-ray film cassettes together with commercial tritium standards (Amersham) and exposed to 3H-Ultrofilm (LKB) at -20°C for 45 days. The films were developed by a Kodak D19 developer (22°C, 4 min). The quantitative evaluation of the autoradiograms was carried out by means of a Tesak VDC 501 image analyzer according to previously described procedures (Benfenati et al. 1985, 1986).

Surgery and Drug Administration

The rats were anaesthetized using ketamine (Ketalar) in a dose of 100 mg/kg. A stainless steel cannula (0.65 mm diameter) connected to an Alzet mini-osmotic pump (Model 2001) filled with either CCK-8 sulfated (Bachem) or saline was placed in the lateral ventricle (A/P:1.0; M/L:1.4; V/D:4.0 mm with respect to bregma). Cannulae were secured to the skull using stainless-steel screws and dental acrylic.

In one series of experiments, rats (n=4) receiving a continuous intraventricular infusion of CCK 8 (2 ng/h) or saline were injected with n-ethoxycarbonyl-2-ethoxy-1,2-dihydroquinoline (EEDQ, 6.0 mg/kg i.p.) or its vehicle 24 h after onset of the i.v.t. infusion of CCK-8. EEDQ produces profound and irreversible in vivo inactivation of DA receptors in rats, with concomitant development of catalepsy and hypolocomotion (Hamblin and Creese 1983; Meller et al. 1985; Merlo Pich in preparation). The animals were killed for receptor autoradiography on the 1st, 2nd, 4th and 7th day after EEDQ administration.

A second group of animals (n=32) was also stereotaxically implanted with chronic bilateral guide cannulae (0.65 mm diameter) 2 mm over the posterior part of nucleus accumbens for the injection of DA (25 µg per 1 µl) (A/P:9.8 mm; M/L:1.5 mm: V/D:4.5 mm; relative to the interaural line; according to the Paxinos and Watson atlas (1982). The rats received a continuous intraventricular infusion of CCK 8 (2 ng/h) or saline and were injected with EEDQ (6.0 mg/kg i.p.) or its vehicle 24 h after onset of the i.v.t infusion of CCK-8. The rats were evaluated behaviourally during the recovery period.

Behavioural Procedures

Catalepsy was scored on the 1st, 2nd, 4th and 7th day after EEDQ injection according to the "bar" method (Erzin-Waters and Seeman 1982). On the 2nd day the rats were placed in novel rectangular plexiglass cages (20x40x30 cm) with two horizontal infrared photocell beams along the long axis 3 cm above the floor and exploratory locomotor activity was measured for 30 minutes. After habituation (1 hr), the rats were removed and microinjected with DA (25.0 µg/ul side) bilaterally in to the nucleus accumbens, using a Hamilton 10.0 µl syringe mounted on a high precision infusion pump (Carnegie Medicin) connected via PE-10 tubing to a stainless-steel 0.3 mm diameter wide injection needle. In situ, the injection needle extended 2 mm beyond the guide tips. Photocell beam interruption recording as well as experimental contingencies were controlled by means of an Apple II computer.

RESULTS

Morphometrical Studies on the TH/CCK Mesolimbic Pathway

TH/CCK codistribution and coexistance have been studied at the cell body level in the ventral mesencephalon and at the terminal level in the caudal part of the nucleus accumbens (fig. 2). We have found that the codistribution of TH and CCK immunoreactivities (IR) at the cell body level is present mainly in the A10 group and in the medial part of the substantia nigra (SN), while it is less pronounced in the lateral part of the SN. In agreement with these results coexistence is at the highest value (compatible with 100% coexistence) in the A10 group and is about 100% in a projection area of the A10 group, the nuc. accumbens caudalis (see fig. 2). Thus, the presynaptic analysis shows the possible existence of synapses endowed with multiple transmission lines (Agnati et al. 1984b), in this case probably TH, CCK and DA messengers at the level of the caudal part of the nuc. accumbens.

Effects of CCK-8 on DA-2 Recognition Sites in the Nuc. Accumbens

By means of quantitative receptor autoradiography it is shown that, in vivo, CCK-8 given i.c.v. (2 ng/h for 24 hrs) can upregulate the 3H-spiperone binding sites in the nucleus accumbens (fig. 3) (see also Dumbrille-Ross and Seeman 1985). This effect is present especially in the caudal but also in the rostral accumbens, i.e., in regions, where the entity of TH/CCK-costoring terminals is high and low, respectively. In the striatum, no significant effect of CCK-8 on 3H-spiperone biniding sites was observed, neither at the rostral nor at the caudal level.

Effects of CCK-8 on the DA Transmission Line after Irreversible Blockade with EEDQ: Autoradiographical and Behavioural Studies

To test whether chronic i.c.v. administration (2 ng/h forl, 2, 4, 7 days) of CCK-8 could modulate the DA transmission line at other levels beyond the recognition site, we have studied DA receptor recovery after EEDQ (6 mg/kg i.p.) treatment evaluating both receptor density and behavioural parameters. We have observed that until 4 days after the receptor blockade the recovery of the 3H-spiperone recognition sites in the presence and in the absence of CCK-8 was almost the same. In this period the synthetic capability for DA recognition sites seems to be maximally activated, so that no further enhancement caused by CCK-8 is

fig 2

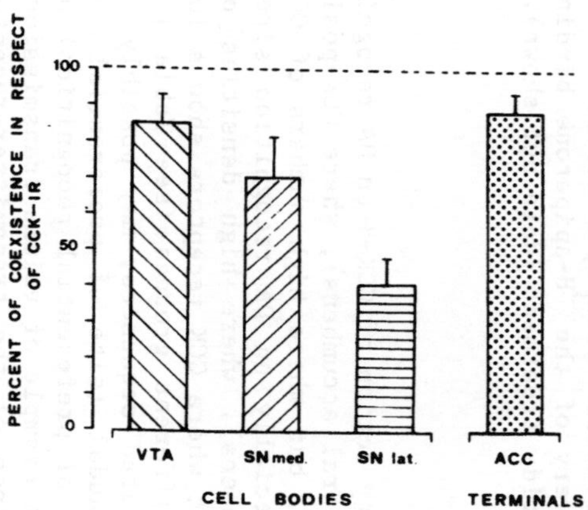

fig 3

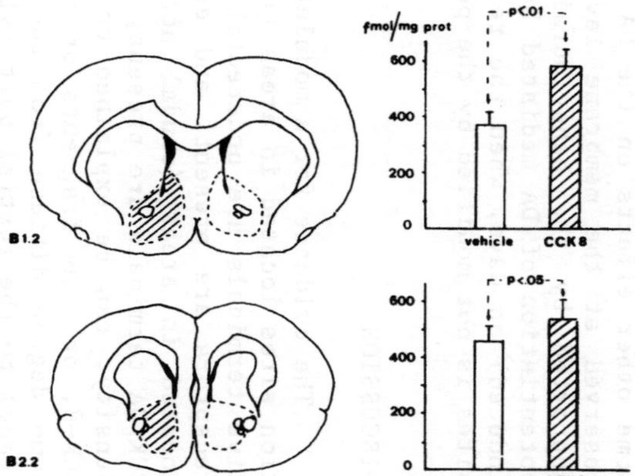

fig 4

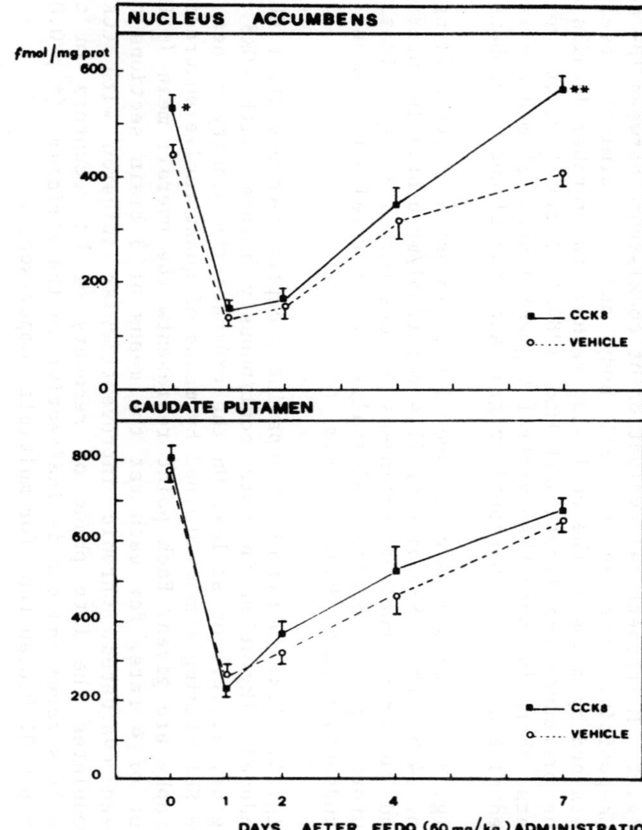

Fig. 2. Morphometrical evaluation of coexistence between TH- and CCK-positive profiles at cell body and at terminal levels, obtained by means of the occlusion method (for further details on the procedure see Agnati and Fuxe 1984a). Means ± s.e.m. n= 4 rats. Abbreviations: VTA= ventral tegmental area; SN med and lat= medial and lateral substantia nigra: Acc= caudal nuc. accumbens.

Fig. 3. Effects of a continuous i.v.t. infusion of CCK 8 (2 ng/h for 24 hr) on 3H-SPI binding (1.4 nM) (fmol/mg prot.) in anterior and posterior nucleus accumbens as determined by quantitative autoradiography. Statistical analysis was carried out by means of Student's t test. Means ± s.e.m. n= 6 rats.

Fig. 4. Time course of D2 dopamine receptor recovery after EEDQ induced alkylation in rats continuously infused with CCK8 (2 ng/hr, i.v.t.) or saline. On the ordinate the density values of 3H-SPI binding sites obtained by means of quantitative autoradiography are given. Each point represents the overall mean (±SEM) out of 4 rats, for each rat the means of 3 brain sections has been evaluated. Chronic intraventricular infusion with CCK-8 modulates the late phase of recovery of D2 receptors in the nucleus accumbens but is ineffective in the striatum. (* $p<0.05$; ** $p<0.01$ Duncan test for multiple comparisons).

possible. However, between 4 and 7 days there is an increase in the reappearance of 3H-spiperone binding sites for the CCK-8 treated animals, which reaches the high steady state level found after 24 hrs of CCK-8 treatment (see fig. 4). Hence, CCK-8 has some other effects on the DA transmission line besides the one observed at the membrane level. This view is also indirectly supported by the behavioural studies, since CCK-8 causes a potentiation of DA mediated behaviours (both spontaneous and DA induced) on days, when the recovery of the ^{3}H-spiperone binding sites is not modified by the peptide treatment (data not shown).

DISCUSSION

The evidence for a modulatory action of CCK-8 on DA recognition sites located in areas (rostral accumbens), where CCK-positive terminals are undetectable, but where high numbers of CCK receptors are present, and especially on DA recognition sites located in areas (caudal accumbens), where high densities of CCK/DA terminals are present, but where CCK receptors show a low density, can be explained on different grounds (see Table I). CCK-8, or some fragments or related sequences, may possibly to some degree diffuse from the caudal (site of preferential release) to the rostral part (site of preferential recognition) of the n. accumbens, representing an example of volume transmission. Thus, one of the possible meanings of the transmitter-receptor mismatch is that receptors exist for paracrine signals in the CNS

table 1

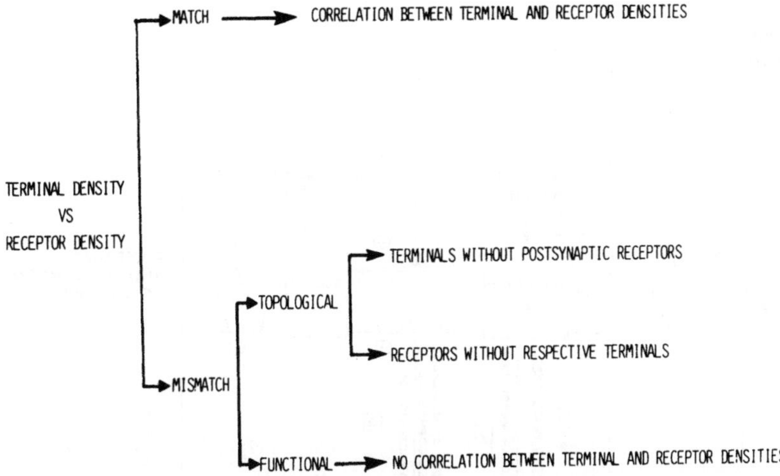

TRANSMITTER/RECEPTOR MISMATCH

- Volume transmission
- Low affinity receptors
- Receptor subtypes
- Different transmitter processing/ Receptor for transmitter fragments
- Isoreceptor interconversion
- Uncoupled receptors
- Intracytoplasmic receptors (receptor synthesis, transport, metabolism)
- Neuronal localization (presynaptic, postsynaptic, auto-receptors)

Table I. Upper panel: schematic representation of the possible match/mismatch between nerve terminal and receptor densities. Lower panel: possible explanation for the transmitter-receptor mismatch.

(volume transmission). Of course other possibilities must be considered as shown in Table I. Thus, in the present paper further evidence is given that a peptide can modulate the transmission lines for classical transmitters, and this phenomenon can take place not only at the synaptic level, but also at the local circuit level and even, and probably, for volume transmission signals. This view opens up the possibility that humoral and/or electrotonic signals, via the extracellular fluid, can affect

fig 5

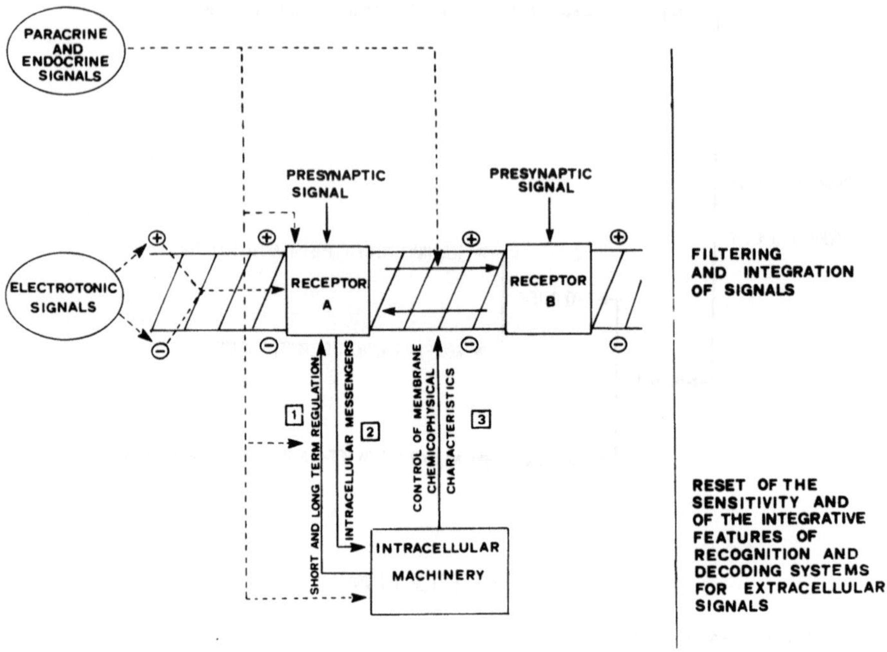

Fig. 5. Possible interactions between volume transmission and wiring transmission at membrane as well as at intracellular levels. At the intracellular level different targets on the modulatory action can be considered such as: 1. phosphorylation/dephosphorylation of the receptor complex. 2. cyclic nucleotide cascades. 3. lipid composition and polarization of the membrane.

recognition sites for transmitters (or for other humoral signals) acting on the filtering capabilities of the membrane and on the regulation of the sensitivity of the recognition sites and also on the entity of intramembrane macromolecular interactions (see fig. 5).

The present findings also offer important new evidence for a role of receptor-receptor interacions the in heterostatic regulation of costoring synapses (Fuxe et al. 1984, Fuxe and Agnati 1985, Fuxe et al. 1986b). Thus, a marked modulation of D2 receptors by CCK-8 is found in the caudal n. accumbens, where CCK-8 does not influence DA utilization (see Fuxe et al. 1984) nor DA release (Ruggeri et al. 1987), but where CCK peptides are exclusively costored in the DA terminals.

As put forward in previous papers (Agnati et al. 1986a,b) and already mentioned in the introduction, it has been suggested that in the CNS two main types of information transmission are present: the wiring transmission and the volume transmission (Table II). The thalamus may be a region where the volume transmission seems to be highly represented, since in this area there is a general mismatch between immunocytochemically visualized peptide terminals and their espective high affinity receptors. The volume transmission consists not only of humoral signals diffusing in the medium to reach the appropriate recognition sites, but also of electrotonic signals, which permeate the extracellular fluid of the brain. We can surmise that in the CNS the various types of information transmission listed in Table III are all operating. Therefore, at the postsynaptic level a very complex integration can take place like the one represented in fig. 5.

The information handling at membrane level has also to be considered together with a modulation of its integrative capabilities by the intracytoplasmic mechanisms. Of particular relevance seems to be the phosphorylation/dephosphorylation of membrane proteins (Greengard, this symposium) and the membrane viscosity (Fuxe, this symposium). The existence of an action of peptides on biochemical processes localized beyond the recognition site of a transmission line is also supported by the present data on CCK-8 effects on DA receptor turnover and on DA dependent behaviours. In fact, the receptor autoradiographical data may suggest that CCK-8, by acting on the set point (heterostatic regulation, Agnati et al. 1986a; Fuxe et al. 1986) of the DA recognition site density, can affect the load error of the negative feedback mechanism involved in the restoration of the density of DA recognition sites found prior to blockade of the DA receptors. The behavioural data may suggest that CCK-8 either can modulate the interplay between different networks involved in the control of the studied behaviours, or can affect the amplifying system associated with the DA recognition sites, increasing, in the presence of a reduction of the DA receptors, the gain of the DA receptor systems.

In conclusion, these findings as well as other data of our and other groups point to the possibility that the interneuronal communication can also occur according to non synaptic modalities (Vizi 1984), using both chemical and electrical signals. Thus, the integration of, in principle, two types of information transmission may occur in the CNS: one linked to the "wiring" of the neuronal network and the other one "open" to all the influences which can modulate the neural substrate. In shaping and modulating the working characteristics of these two types of transmission a role may be played not only by the neurons but also by the glia cells (see Table III). One site where integration between these two types of information transmission may take place is the membrane, where they may inter alia interact in the control of the receptor-receptor interactions.

table 2

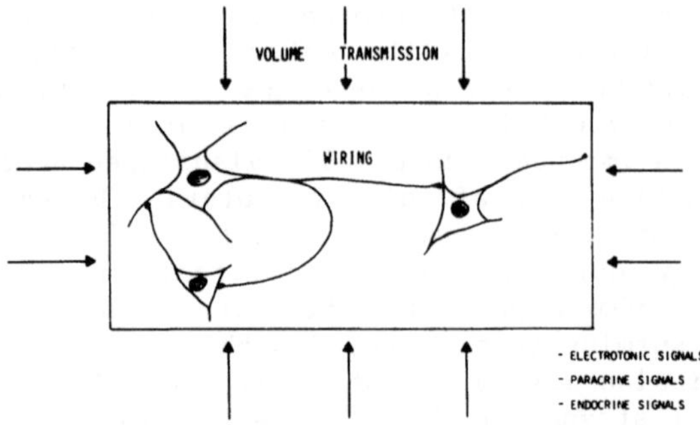

	SPEED OF TRANSMISSION	DEGREE OF DIVERGENCE	SEGREGATION ("SAFETY OF THE TRANSMISSION)	PLASTICITY	PREFERENTIAL INFORMATION PROCESSING
"WIRING TRANSMISSION" NEURON LINKED ELECTRO-CHEMICAL TRANSMISSION	HIGH	LOW TO MODERATE	HIGH	LOW TO MODERATE	ELEMENTARY ELABORATION SHORT TERM ACTION
"VOLUME TRANSMISSION" HUMORAL ("OPEN") ELECTRO-CHEMICAL TRANSMISSION	LOW	HIGH TO VERY HIGH	LOW	HIGH TO VERY HIGH	HOLISTIC ELABORATION LONG TERM ACTION

table 3

		GLIA FUNCTIONS		NEURON FUNCTIONS	
		PHYSICAL	BIOCHEMICAL	PHYSICAL	BIOCHEMICAL
WIRING TRANSMISSION	ELECTRICAL SYNAPSE	SYNAPSE SEGREGATION	MODULATION OF COUPLING VIA ECF	LOCATION OF SYNAPSE	MODULATION OF COUPLING VIA ICF
	CHEMICAL SYNAPSE	SYNAPSE SEGREGATION	MODULATION OF SYNAPTIC CLEFT MICROENVIRONMENT	LOCATION OF SYNAPSE	CONTROL OF THE EFFICACY OF TRANSMISSION LINES AND THEIR INTERPLAY
VOLUME TRANSMISSION	ELECTROTONIC SIGNALS	SHAPING OF ECF PATHWAYS FOR SIGNAL DIFFUSION	CONTROL OF ECF ION COMPOSITION	LOCATION OF SOURCES AND SINKS FOR SIGNALS	CONTROL OF SOURCES AND SINKS FOR SIGNALS
	HUMORAL SIGNALS	SHAPING OF ECF PATHWAYS FOR SIGNAL DIFFUSION	RELEASE , UPTAKE AND METABOLISM OF SIGNALS	LOCATION OF RELEASE AND RECOGNITION SITES FOR SIGNALS	RELEASE , UPTAKE AND METABOLISM OF SIGNALS

Table II. Schematic representation of the main features of the wiring versus the volume transmission.

Table III. Main features of the morphofunctional roles of neurons and glia cells in the wiring and in the volume transmission. (ECF= extracellular fluid; ICF= intracellular fluid).

ACKNOWLEDGEMENTS

This work has been supported by a Grant (04X-715) from the Swedish Medical Research Council and a CNR-I grant. For excellent secritarial assistance we are grateful to Monia Särd and Anne Edgren.

REFERENCES

Agnati, L.F., Fuxe, K., Zini, I., Lenzi, P., Hökfelt, T. (1980). Aspects on receptor regulation and isoreceptor identification. Med. Biol. 58, 182-187.

Agnati, L.F. Fuxe, K., Benfenati, F., Battistini, N. (1983). Neurotensin in vitro markedly reduces the affinity in subcortical limbic 3H-N-propylnorapomorphine binding sites. Acta Physiol. Scand., 119, 459-461.

Agnati, L.F., Fuxe, K. (1984a). Computer assisted morphometry and microdensitometry of transmitter identified neurons with special reference to the mesostriatal dopamine pathway. I. Methodological aspects. Acta Physiol. Scand., Suppl. 532, 5-32.

Agnati, L.F., Fuxe, K., Benfenati, F., Battistini, N., Zini, I., Camurri, M., Hökfelt, T. (1984b). Postsynaptic effects of neuropeptide comodulators at central monoamine synapses. In: Catecholamines: Neuropharmacology and Central Nervous System. Theoretical Aspects. (E. Usdin, A. Carlsson, A. Dahlström, J. Engel, eds.). Alan R. Liss Inc., New York, 191-198.

Agnati, L.F., Fuxe, K., (1985). Quantitative Neuroanatomy in Transmitter Research. McMillan Press, London.

Agnati, L.F., Fuxe, K., Zoli, M., Merlo Pich, E., Benfenati, F., Zini, I., Goldstein, M. (1986a). Aspects on the information handling by the central nervous system: focus on cotransmission in the aged rat brain. In: Coexistence of Neuronal Messengers. A new Principle in Chemical Transmission. (T. Hökfelt, K. Fuxe, B. Pernow, eds.). Prog. Brain Res. 68, Elsevier, Amsterdam, pp. 291-301.

Agnati, L.F., Fuxe, K., Zoli, M., Zini, I., Toffano, G., Ferraguti, F. (1986b). A correlation analysis of the regional distribution of central enkephalin and β-endorphin immunoreactive terminals and of opiate receptors in adult and old male rats. Evidence for the existence of two main types of communication in the central nervous system: the volume transmission and the wiring transmission. Acta Physiol. Scand. 128, 201-207.

Agnati, L.F., Fux, K., Zini, I., Härfstrand, A., Toffano, G., Goldstein, M. (1986c). Morphometrical and microdensitometrical studies on phenylethanolamine-N-methyltransferase and neuropeptide Y immunoreactive neurons in the rostral medulla oblongata of the adult and old rats. Neuroscience, submitted.

Benfenati, F., Agnati, L.F., Fuxe, K., Comino, M., Battistini, N., Merlo Pich, E., Farabegoli, C., Zini, I. (1985). Quantitative autoradiography as a tool to study receptors in neural tissue. Studies on 3H-ouabain binding sites and correlation with synaptic protein phosphorylation in different brain areas. In: Quantitative Neuroanatomy in Transmitter Research (L.F. Agnati, K, Fuxe eds.), McMillan Press, London, 381-396.

Benfenati, F., Cimino, M., Agnati, L.F., Fuxe, K. (1986). Quantitative autoradiography of central neurotransmitter receptors: methodological and statistical aspects with special reference to computer-assisted image analysis. Acta Physiol. Scand., 128, 129-146.

Dumbrill-Ross, A., Seeman, P. (1984). Dopamine receptor elevation by cholecystokinin. Peptides, 5, 1207-1212.

Erzin-Waters, C., Seeman, P. (1977). Tolerance to haloperidol catalepsy. Eur. J. Pharmacol., 41, 321-326.

Fuxe, K., Agnati, L.F., Benfenati, F., Celani, M., Zini,. Zoli, M., Mutt, V. (1983). Evidence for the existence of receptor-receptor interactions in the central nervous system. Studies on the regulation of monamine receptors by neuropeptides. J. Neural Transm., Suppl. 18, 165-179.

Fuxe, K., Agnati, L.F., Andersson, K., Martire, M., Ögren, S-O., Giardino, L., Battistini, N., Grimaldi, R., Farabergoli, C., Härfstrand, A. and Toffano, G. (1984a). Receptor-receptor interactions in the central nervous system. Evidence for the existence of heterostatic synaptic mechanisms. In: Regulation of Transmitter Function: Basic and Clinical Aspects: Developments in Neuroscience Series (E.S. Vizi and K. Magyar, eds.) 17, Elsevier Science Publ., pp. 129-140.

Fuxe, K., Agnati, L.F. (1985). Receptor-receptor interactions in the cental nervous system. A new integrative mechanism in synapses. Med. Res. Rev., 5, 441-482.

Fuxe, K., Meller, E., Goldstein, M., Benfenati, F., Agnati, L.F. (1986a). Analysis of 3H-spiperone binding sites in the rat striatum and frontoparietal cortex by means of quantitative receptor autoradiography after inactivation of dopamine receptors by N-ethoxycarbonyl-2-ethoxy-1,2-dihydroquinoline in vivo: selective protection by sulpiride in the striatum. Neurosci. Lett., 64, 163-168.

Fuxe, K., Agnati, L.F., Härfstrand, A., Janson, A.M., Neumeyer, A., Andersson, K., Ruggeri, M., Zoli, M., Goldstein, M. (1986b). Morphofunctional studies on the neuropeptide Y adrenaline costoring terminal systems in the dorsal cardiovascular region of the medulla oblongata. Focus on receptor-receptor interactions in cotransmission. In: Coexistence of Neuronal Messengers. A new Principle in Chemical Transmission (T. Hökfelt, K. Fuxe, B. Pernow eds.). Prog. Brain Res. 68, Elsevier, Amsterdam, pp. 303-320.

Hamblin, M.W., Creese, I. (1983). Behavioral and radioligand binding evidence for irreversible dopamine receptor blockade by N-ethoxycarbonyl-2-ethoxy-1,2-dihydroquinoline. Life Sci., 32, 2247-2255.

Hökfelt, T., Skirboll, L., Rehfeld, J.F., Goldstein, M., Markey, K., Dann, O. (1980). A subpopulation of mesencephalic dopamine neurons projecting to limbic areas contains a cholecystokinin-like peptide: evidence from immunohistochemistry combined with retrograde tracing. Neuroscience, 5, 2093-2124.

Markey, L., Kondo, S., Shenkmann, L., Goldstein, M. (1980). Purification and characterization of tyrosine hydroxylase from a clonal chromocytoma cell line. Mol. Pharmacol., 17, 79-85.

Meller, E., Bohmaker, K., Goldstein, M., Friedhoff, A.J. (1985). Inactivation of D1 and D2 dopamine receptors by N-ethoxycarbonyl-2-ethoxy-1,2-dihydroquinoline in vivo: selective protection by neuroleptics. J. Pharmacol. Exp. Ther., 233, 656-662.

Paxinos, G., Watson, C. (1982). The rat brain in stereotaxic coordinates. Academic Press, New York.

Rostene, W.H., Fischette, C.H., Mc Ewen, B.S. (1983). Modulation by vasoactive intestinal peptide (VIP) of serotonin 1 receptors in membranes from rat hippocampus, J. Neurosci., 3, 2414-2419.

Ruggeri, M., Ungerstedt, U., Agnati, L.F., Mutt, V., Härfstrand, A. and Fuxe, K. (198 7). Effects of cholecystokinin peptides and neurotensin on dopamine release and metabolism in the rostral and caudal part of the nucleus accumbens using intracerebral dialysis in the anaesthetized rat. Neurochem. Int.. in press.

Sternberger, L.A. (1979). Immunocytochemistry, 2nd ed., John Wiley, New York.

Vanderhaeghen, J.-J., Lotstra, F., De Mey, J., Gilles, C. (1980). Immunohistochemical localization of cholecystokinin and gastrin-like peptides in the brain and hypophysis of the rat. Proc. Natl. Acad. Sci., USA, 77, 1190-1194.

Vizi, E.S. (1984). Non-Synaptic Transmission between Neurons: Modulation of Chemical Transmission, John Wiley, New York.

Zoli, M., Fuxe, K., Agnati, L.F., Härfstrand, A., Terenius, L., Toni, R., Goldstein, M. (1986). Computer-assisted morphometry of transmitter-identified neurons: new openings for the understanding of peptide-monoamine interactions in the mediobasal hypothalamus. In: Neurohistochemistry: Modern Methods and Applications (P. Panula, Paivarainta, Soinila, eds.,), Alan R. Liss Inc., New York, pp. 137-172.

20

Radioautographic Approaches to the Study of Receptor–Receptor Interactions in the Central Nervous System

William Rostene, Jacqueline Besson, Emmanuel Moyse, Monique Dussaillant, Micheline Vial, Patrick Kitabgi and Alain Beaudet

INTRODUCTION

Communication between nerve cells involves far more than the direct excitation and inhibition of neuronal firing by chemical transmitters. Messengers liberated by neurons or glial cells, along with blood-born hormones, modulate the response properties of nerve cells. Such regulation involves changes in intracellular messengers and also, in some cases, modulation of genomic activity and expression of gene products implicated in nerve cell function. Chemically mediated regulations usually involve intracellular mechanisms but may also depend to a large degree on the responsiveness of membrane receptors. In this chapter we illustrate how monoamines, neuropeptides, and steroid hormones may interact in the central nervous system (CNS) both directly, and indirectly through receptor-receptor modulation. These studies were made possible by the recent development of receptor binding assay methodologies, which have provided us with new interpretations for the interactive mechanisms underlying neuronal communication.

Because the brain is so highly specialized anatomically and chemically, a high degree of spatial resolution is required to derive precise functional interpretations from binding studies. Radioautography has proven an ideal technique to assess both qualitatively and quantitatively the regional distribution of transmitter and peptide binding sites and to study the plasticity of these binding sites under different experimental conditions.

PEPTIDE - AMINE RECEPTOR MEDIATED INTERACTIONS.

In the past few years, a large number of neuropeptides, some of them colocalized with "classical" neurotransmitters (Hökfelt et al., 1980), were shown to directly interact with aminergic systems.

Vasoactive intestine peptide (VIP) - Amine Interactions

VIP was recently reported to stimulate in vitro the activity of serotonin-N-acetyltransferase, a rate-limiting enzyme in the synthesis of melatonin in rat pineal gland (Yuwiler, 1983), as well as that of cholineacetyltransferase, an enzyme involved in the synthesis of acetylcholine (Ach), in both the anterior hypothalamus and the hippocampus (Luine et al., 1984). Similarly, VIP was shown to stimulate the release of newly synthetized ^{3}H-serotonin (5-HT) from rat suprachiasmatic nucleus (SCN) (Hery et al., 1984).

All those effects are probably exerted through binding of the peptide to high affinity receptors, which have been shown to be present in high amounts in the pineal gland, the SCN and the hippocampal formation (for review see Rostène, 1984, and Figure 1).

Neurotensin (NT)-Amine Interactions

Pharmacological data suggest that NT may interact with nigro-striatal and mesolimbic cortical dopaminergic (DA) systems (Quirion, 1983; Kitabgi et al., 1986). In keeping with these data is the fact that the distribution of specific high affinity NT binding sites, as assessed by film or light microscopic radioautography following in vitro incubation of rat frozen brain sections with ^{3}H or ^{125}I-NT, shows a striking overlap with that of cell bodies of origin and terminal arbor of nigro-striatal and meso-corticoprefrontal-DA systems (Young and Kuhar, 1981; Quirion et al., 1982; Hervé et al., 1986; Moyse et al., 1987). The substantial decrease in NT binding observed in the ventral tegmental area (VTA) and in the substantia nigra, pars compacta (SN), after local injection of the specific neurotoxin, 6-OHDA, suggests that specific NT receptors in those areas might be associated with DA-containing somata (Palacios and Kuhar, 1981; Quirion et al., 1985; Hervé et al., 1986). The selective labeling of nerve cell bodies and proximal dendrites observed within the SN applying a high resolution radioautographic procedure is in keeping with this interpretation (Figure 2).

Similarly, in the human brain, high densities of NT binding sites are observed in the nucleus paranigralis and in the pars compacta of the SN, two structures which contain numerous DA cell bodies (Sadoul et al., 1984; Uhl et al., 1984; Sarrieau et al., 1985). Interestingly, a dramatic loss of NT binding is observed in Parkinson's disease, a degenerative disease essentially characterized by the destruction of the DA cells (Sadoul et al., 1984; Uhl et al., 1984; Rostène et al., 1986). These data strongly support the view that in the human and rat SN, DA cells are endowed with NT receptors and suggest a possible involvement of NT in the regulation of DA neurons.

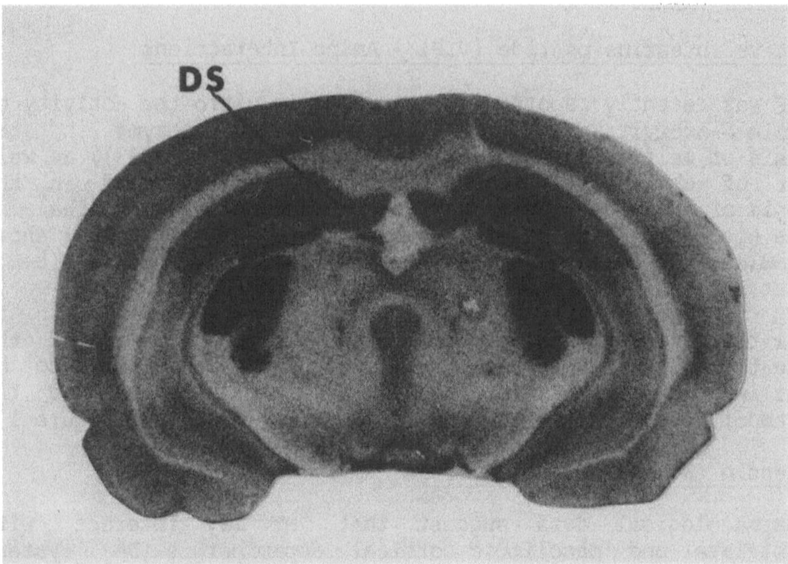

Figure 1: Film radioautography of ^{125}I-VIP binding sites at the level of the rostral midbrain. DS: dorsal subiculum of the hippocampal formation (from Besson et al., 1986)

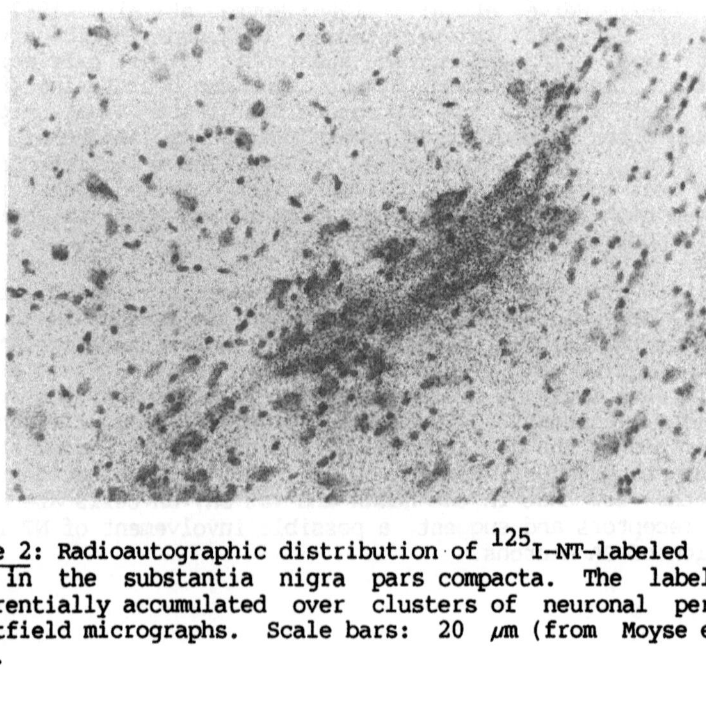

Figure 2: Radioautographic distribution of ^{125}I-NT-labeled binding sites in the substantia nigra pars compacta. The labeling is preferentially accumulated over clusters of neuronal perikarya. Brightfield micrographs. Scale bars: 20 μm (from Moyse et al., 1987).

The recent report by Moyse et al. (1987) on the presence of ^{125}I-NT binding sites in the DA cell- rich subdivision of the medial zona incerta and in the nucleus raphe linearis caudalis suggests that the coincidence between "neurotensinoceptive" and DA neurons might be more extensive than originally surmised. Conversely, not all DA cell groups seem to express NT binding, judging from the apparent sparing of the glomeruli of the olfactory bulb and the tuberoinfundibular system of the mediobasal hypothalamus (Moyse et al., 1987).

We recently proposed (Moyse et al., 1987) that NT receptors might also be associated with certain 5-HT and cholinergic neurons in the CNS. On the one hand, the presence of ^{125}I-NT binding sites in the medial and dorsal raphe nuclei, which contain the cell bodies of origin of ascending 5-HT pathways, might reflect an association of those binding sites with 5-HT-containing neurons. In agreement with this hypothesis is the observation that in certain 5-HT-innervated areas, such as the SCN, the pattern of distribution of NT binding closely resembles that of the 5-HT terminal arbor (Moyse et al., 1987). On the other hand, the presence of ^{125}I-NT labeled cells in the region of the basal forebrain cholineacetyl-transferase immunoreactive magnocellular neurons (Kimura et al., 1986) suggests that certain "neurotensinoceptive" cells might be cholinergic.

PEPTIDE - AMINE RECEPTOR - RECEPTOR INTERACTIONS

It was recently reported that the binding characteristics of transmitter receptors may be modulated by heterologous ligands, mainly at the post-synaptic level, without any change in presynaptic neurotransmission (Fuxe and Agnati, 1985). Such receptor-receptor interactions can take place between receptors located within the same synapse or in adjacent ones. This type of interaction may play an important role in short- and long-term regulation of synaptic plasticity involved for instance in learning and memory processes.

Cholecystokinin-like peptides were thus shown to modulate ^{3}H-spiroperidol binding in rat striatal membranes (Fuxe et al., 1981) and DA agonist binding in both subcortical limbic forebrain and striatal membranes (Agnati and Fuxe, 1983; Agnati et al., 1983a). NT was also found to reduce the affinity of DA agonist binding in the nucleus accumbens and olfactory tubercles (Agnati et al., 1983b) and to enhance the increase in ^{3}H-spiroperidol binding induced by haloperidol treatment in both the nucleus accumbens and the neostriatum (Simasko and Weiland, 1985). Moreover, blockade of DA receptors by chronic treatment with DA receptor antagonists such as neuroleptics may induce changes in NT binding in certain brain regions rich in either DA cell bodies such as the SN (Uhl and Kuhar, 1984), or in DA-terminal arbor such as the prefrontal cortex (Hervé et al., 1986). In the latter case, competition curves obtained by incubation of rostral coronal brain sections with increasing concentrations of unlabeled NT indicate that the change in ^{125}I-NT binding is due to an increased density of ^{125}I-NT binding sites without any modification of the affinity for the iodinated ligand. Moreover, the fact that a similar increase of NT binding sites can be obtained in the rat lateral prefrontal cortex following both lesion of DA neurons by 6-OHDA injection in the VTA, and DA receptor blockade with neuroleptic administration, suggests that DA transmission normally down-regulates NT binding sites in this area (Hervé et al., 1986). Finally, DA was reported to reduce the affinity and increase the number of ^{3}H-NT binding sites in subcortical limbic forebrain (Agnati et al., 1985) suggesting that NT-DA interactions may be exerted through mutual regulation of their respective receptors.

5-HT receptors were also shown to be modulated by neuropeptides. Substance P was reported to increase the density of ^{3}H-5-HT binding sites in the spinal cord (Agnati et al., 1982) and we reported that VIP, when applied directly in vitro on either broken membranes or tissue sections, was able to alter ^{3}H-5-HT binding to 5-HT_1 receptors (Rostène et al., 1983 a, b). We were able to document this effect by quantitative radioautography in two discrete brain structures, the SCN of the hypothalamus and the dorsal subiculum of the hippocampal formation.

In the SCN, VIP decreases the number of 5-HT_1 binding sites without affecting their affinity for ^{3}H-5-HT. Conversely, in the dorsal subiculum, VIP increases 5-HT_1 binding site density (Rostène et al., 1985). This effect is probably mediated through VIP receptors which were recently shown by quantitative radioautography to be highly concentrated in those two brain regions (Besson et al., 1986; Figure 1).

INFLUENCE OF PERIPHERAL HORMONES ON RECEPTOR - RECEPTOR INTERACTIONS

Steroid hormones may also affect peptidergic and aminergic receptor binding characteristics through activation of their respective CNS receptors. For instance, De Kloet et al. (1985) showed that estradiol (E_2) increased oxytocin binding site density in the ventromedial nucleus of the hypothalamus and in the central nucleus of the amygdala. Along the same lines, we recently observed a significant increase in ^{125}I-NT binding in the SCN of the female rat six weeks after castration (Moyse et al., 1986). This effet appears to be restricted to the rostral portion of the nucleus, which lies in the vicinity of E_2-concentrating neurons (Pfaff and Keiner, 1973). Finally, E_2 was found to significantly increase the density of 5-HT_1 receptors in the hypothalamus and lateral septum (Biegon et al., 1982).

The effets of steroid hormones are not restricted to E_2. Recent radioautographic data indeed showed that short-term adrenalectomy (ADX) increases 5-HT_1 binding site density in some regions of the rat brain, and that this effect is blocked by corticosterone (CORT) replacement (De Kloet et al., 1986).

Steroids may also affect peptide-amine receptor - receptor interactions. For instance, the VIP-induced increase in 5-HT_1 receptor density in the hippocampus reported above is no longer present following bilateral ADX and therefore appears to be dependent on the presence of adrenal steroids (Rostène et al., 1985; Table I). The mechanism by which corticosteroids exert this effect is not known, but might involve the regulation of enzymes or of other membrane-localized gene products which are required for the local interaction between VIP receptors and 5-HT_1 receptors. The pivotal role of adrenal steroids is also demonstrated by the fact that they regulate VIP levels and the developmental pattern of VIP in the hippocampus (Rotsztejn et al., 1980; Nobou et al., 1985); they also affect the primary effect of VIP on its receptor site, i.e. the stimulation of cyclic AMP formation (Harrelson et al., 1983).

Table I. In vitro effect of VIP (10^{-7}M) on ^{3}H-5-HT binding in various brain regions of normal and ADX rats. Results are expressesd as fmol/mg protein. $p<0.01$ vs respective control. (from Rostène et al., 1985).

	Normal		ADX	
	Control	VIP	Control	VIP
Dorsal subiculum	536±13	853±60*	620±40	761±60
CA_1 oriens	357±36	404±29	284±25	563±20*
CA_1 pyramidale	534±31	531±17	524±40	689±22*
CA_1 radiatum	638±27	608±24	681±61	776±47
CA_2 pyramidale	625±23	661±31	594±23	781±26*
CA_4 pyramidale	1101±45	1218±80	1173±68	1306±108
Hilus	664±27	714±28	688±48	816±38*
Superior colliculus	408±62	526±64	426±16	621±57*
Suprachiasmatic n.	666±14	574±23*	678±30	573±58
Lat. Septum dorsal	1303±88	1198±92	1054±94	710±71*

Finally, ADX can mask or unmask an effect that is not observed in normal animals (Rostène et al., 1985). For instance, VIP has no effect on ^{3}H-5-HT binding in the various subfields of the Ammon's horn or in the superior colliculus and septal regions in normal rats, but increases or decreases ^{3}H-5-HT binding in ADX animals (Rostène et al., 1985; Table I). It should be noted that the various effects of ADX take place mainly in the hippocampus and septum, brain structures known to contain the highest amount of binding sites for adrenal steroids (Mc Ewen, 1982; Sarrieau et al., 1986).

CONCLUSION

Receptor-receptor interactions represent an alternative way by which peptidergic innervation may modulate the activity of central "classical" neurotransmitter systems. The strategies described herein for studying binding sites for chemical messengers may help to develop new drugs working, for instance, through heterologous receptors, which could be used with a high degree of specificity in the treatment of several mental diseases. In the case of Parkinson's disease, it is possible that drugs which can mimick NT effects may be used as an alternative to the treatment with L-DOPA, in an attempt to overcome L-DOPA induced symptoms of tardive dyskinesia. Acting at the synaptic recognition site without changing presynaptic activity may be considered as the most important challenge in this new field of receptor interactions in the brain.

ACKNOWLEDGEMENTS

This work was supported by INSERM and by grants from the MRT (n° 85-C-1139), the Parkinson Foundation of Canada, and by a France - Quebec exchange program. We are indebted to Mrs D. Lhenry for typing the manuscript.

REFERENCES

- Agnati, L.F., Celani, M., and Fuxe K. (1983 a).Cholecystokinin peptides in vitro modulate the characteristics of the striatal ^{3}H-N-propylnorapomorphine sites. Acta Physiol. Scand., 118, 79-81.
- Agnati, L.F., and Fuxe, K. (1983). Subcortical limbic ^{3}H-N-propylnorapomorphine binding sites are markedly modulated by cholecystokinin-8 in vitro. Bioscience Reports 3, 1101-1105.
- Agnati, L.F., Fuxe, K., Battistini, N., Giardino, L., Benfenati, F., Martire, M., and Ruggeri, M. (1985). Further evidence for the existence of interactions between receptors for dopamine and neurotensin. Dopamine reduces the affinity and increases the number of ^{3}H-neurotensin binding sites in the subcortical limbic forebrain of the rat. Acta Physiol. Scand., 124, 125-128.
- Agnati, L.F., Fuxe, K., Benfenati, F., and Battistini, N. (1983 b). Neurotensin in vitro markedly reduces the affinity in subcortical limbic ^{3}H-N-propylnorapomorphine binding sites. Acta Physiol. Scand., 119, 459-461.
- Agnati, L.F., Fuxe, K., Benfenati, F., Zini, I., and Hökfelt, T. (1982). On the functional role of coexistence of 5-HT and substance P in bulbospinal 5-HT neurons. Substance P reduces affinity and increases density of ^{3}H-5-HT binding sites. Acta Physiol. Scand., 117, 299-301.
- Besson, J., Sarrieau, A., Vial, M., Marie, J.C., Rosselin, G., and Rostène, W. (1986). Characterization and autoradiographic distribution of vasoactive intestinal peptide binding sites in the

rat central nervous system. Brain Res., in press.
- Biegon, A., Fischette, C.T., Rainbow, T.C. and Mc Ewen, B.S. (1982) Serotonin receptor modulation by estrogen in discrete brain nuclei. Neuroendocrinology 35, 287-291
- De Kloet, E.R., Sybesma, H., and Reul H.M.H.M. (1986). Selective control by corticosterone of serotonin$_1$ receptor capacity in raphe-hippocampal system. Neuroendocrinology 42, 513-521.
- De Kloet, E.R., Voorhuis, T.A.M., and Elands, J. (1985). Estradiol induces oxytocin binding sites in rat hypothalamic ventromedial nucleus. Eur. J. Pharmacol., 118, 185-186.
- Fuxe, K., and Agnati, L.F. (1985). Receptor-receptor interactions in the central nervous system. A new integrative mechanism in synapses. Med. Res. Rev. 5, 441-482.
- Fuxe, K., Agnati, L.F., Benfenati, F., Cimmino, M., Algeri, S. Hökfelt, T., and Mutt, V. (1981). Modulation by cholecystokinins of ^{3}H-spiroperidol binding in rat striatum: evidence for increased affinity and reduction in the number of binding sites. Acta Physiol. Scand., 113, 567-569.
- Harrelson, A.L., Mc Ewen, B.S., and Rostène, W. (1983). Adrenalectomy modifies neurotransmitter-stimulated cyclic AMP accumulation in hippocampal slices. Abstr. Soc. Neurosc., 9, 26.13, p. 87.
- Hervé, D., Tassin, J.P., Studler, J.M., Dana, C., Kitabgi, P., Vincent, J.P., Glowinski, J., and Rostène, W. (1986). Dopaminergic control of ^{125}I-neurotensin binding site density in cortico-limbic structures of the rat brain. Proc. Natl. Acad. Sci. USA, 83, 6203-6207.
- Héry, M., Faudon, M., and Héry, F. (1984). Effect of VIP on serotonin release in the suprachiasmatic area of the rat: modulation by estradiol. Peptides 5, 313-318.
- Hökfelt, T., Lundberg, J.M., Schützberg, M., Johansson, O., Ljungdahl, A., and Rehfeld, J. (1980). Coexistence of peptides and putative transmitters in neurons. In Neural Peptides and Neuronal Communication (eds. E. Costa and M. Trabucchi), Raven Press, New York, pp. 1-23.
- Kimura, H., Mc Geer, P.L., and Peng, J.H. (1986). Choline-acetyltransferase-containing neurons in the rat brain. In Handbook of Chemical Neuroanatomy (eds. A. Björklund and T. Hökfelt), vol. 3, Elsevier, Amsterdam, pp. 51-67.
- Kitabgi, P., Checler, F., Mazella, J., and Vincent, J.P. (1986). Pharmacology and biochemistry of neurotensin receptors. Rev. Clin. Basic Pharmacol., in press
- Luine, V.N., Rostène, W.H., Rhodes, J., and McEwen, B.S. (1984). Activation of choline acetyltransferase by vasoactive intestinal peptide (VIP). J. Neurochem. 42, 1131-1134.
- Mc Ewen, B.S. (1982). Glucocorticoids and hippocampus: Receptors in search of a function. In Current Topics in Neuroendocrinology, (eds. D. Ganten and D.W. Pfaff), Springer-Verlag, Berlin, pp. 1-22.
- Moyse, E., Miller, M.M., Kitabgi, P., Rostène, W., and Beaudet, A. (1986). Effects of gonadal steroids on the binding of ^{125}I-neurotensin in rat suprachiasmatic nucleus. Abstr. Soc. Neurosc., in press.

- Moyse, E., Rostène, W., Vial, M., Léonard, K., Mazella, J., Kitabgi, P., Vincent, J.P., and Beaudet, A. (1987). Regional distribution of neurotensin binding sites in rat brain: A film and light microscopic radioautographic study using monoiodo ^{125}I-Tyr$_3$-neurotensin. Neuroscience, in press.
- Nobou, F., Besson, J., Rostène, W., and Rosselin, G. (1985). Ontogeny of vasoactive intestinal peptide and somatostatin in different structures of the rat brain: effects of hypo- and hypercorticism. Develop. Brain Res., 20, 296-301.
- Palacios, J.M., and Kuhar, M.J. (1981). Neurotensin receptors are located on dopamine-containing neurones in rat midbrain. Nature, 294, 587-589.
- Pfaff, D.W., and Keiner, M. (1973). Atlas of estradiol-containing cells in the central nervous system of the female rat. J. Comp. Neurol., 151, 121-158.
- Quirion, R. (1983). Interactions between neurotensin and dopamine in the brain: an overview. Peptides, 4, 609-615.
- Quirion, R., Chiueh, C.C., Everist, H.D., and Pert, A. (1985). Comparative localization of neurotensin receptors on nigrostriatal and mesolimbic dopaminergic terminals. Brain Res., 372, 385-389.
- Quirion, R., Gaudreau, P., St Pierre, S., Rioux, F., and Pert, C.B. (1982). Autoradiographic distribution of ^{3}H-neurotensin receptors in rat brain: visualization by tritium-sensitive film. Peptides, 3, 757-763.
- Rostène, W.H. (1984). Neurobiological and neuroendocrine functions of the vasoactive intestinal peptide (VIP). Prog. Neurobiology, 22, 103-129.
- Rostène, W.H., Fischette, C.T., Dussaillant, M., Mc Ewen, B.S. (1985). Adrenal steroid modulation of vasoactive intestinal peptide effect on serotonin$_1$ binding sites in the rat brain shown by in vitro quantitative autoradiography. Neuroendocrinology, 40, 129-134.
- Rostène, W.H., Fischette, C.T., Rainbow, T.C., and Mc Ewen, B.S. (1983 a). Modulation by vasoactive intestinal peptide of serotonin$_1$ receptors in the dorsal hippocampus of the rat brain: An autoradiographic study. Neuroscience Lett. 37, 143-148.
- Rostène, W.H., Fischette, C.T., and Mc Ewen, B.S. (1983 b).Modulation by VIP of serotonin$_1$ receptors in membranes from rat hippocampus. J. Neuroscience 3, 2414-2419.
- Rostène, W.H., Sarrieau, A., Moyse, E., Hervé, D., Kitabgi, P., Mc Ewen, B.S., Vial, M., Tassin, J.P., Vincent, J.P., and Beaudet, A. (1986). Imaging of neuropeptide-neurotransmitter interactions. In Progress Brain Research, (eds. E.R. De Kloet and D. De Wied), Elsevier, Amsterdam, in press.
- Rotsztejn, W.H., Besson, J., Briaud, B. , Gagnant, L., Rosselin, G., and Kordon, C. (1980). Effect of steroids on vasoactive intestinal peptide in discrete brain regions and peripheral tissues. Neuroendocrinology, 31, 287-291.

- Sadoul, J.L., Checler, F., Kitabgi, P., Rostène, W., Javoy-Agid, F., and Vincent, J.P. (1984). Loss of high affinity neurotensin receptors in substantia nigra from Parkinsonian subjects. Biochem. Biophys. Res. Commun., 125, 395-404.
- Sarrieau, A., Javoy-Agid, F., Kitabgi, P., Dussaillant, M., Vial, M., Vincent, J.P., Agid, Y., and Rostène, W.H. (1985). Characterization and autoradiographic distribution of neurotensin binding sites in the human brain. Brain Res., 348, 375-380.
- Sarrieau, A., Rostène, W.H., Moguilewsky, M., Philibert, D., and Mc Ewen, B.S. (1986). New routine methodology to study steroid hormone binding sites in heterogeneous tissues. In Receptor Binding in Drug Research, (ed. R.A. O'Brien), vol. 5, Marcel Dekker, New York, pp. 261-282.
- Simasko, S.M., and Weiland, G.A. (1985). Effect of neurotensin, substance P and TRH on the regulation of dopamine receptors in rat brain. Eur. J. Pharmacol., 106, 653-656.
- Uhl, G.R., and Kuhar, M.J. (1984). Chronic neuroleptic treatment ernhances neurotensin receptor binding in human and rat substantia nigra. Nature, 309, 350-352.
- Uhl, G.R., Whitehouse, P.J., Price, D.L., Tourtelotte, W.W., and Kuhar, M.J. (1984). Parkinson's disease: depletion of substantia nigra neurotensin receptors. Brain Res., 308, 186-190.
- Young, W.S., III, and Kuhar, M.J. (1981). Neurotensin receptor: localization by light microscopic autoradiography in rat brain. Brain Res., 206, 273-285.
- Yuwiler, A. (1983). Light and agonists alter pineal N-acetyltransferase induction by vasoactive intestinal polypeptide. Science 220, 1082-1083.

21

Long-term Receptor Regulation Induced by Denervation of Heterologous Afferent Fibers: Functional Significance

J.P. Tassin, D. Hervé, H. Simon, K. Taghzouti, M. Le Moal and J. Glowinski

INTRODUCTION

It is a classical feature that the denervation of the neuromuscular junction induces a supersensitivity of the cholinergic post-synaptic receptors (Cannon & Rosenblueth, 1949). In the central nervous system, the development of this "denervation supersensitivity" is more complex. In fact, in some cases, even the total destruction of an homogeneous population of afferent presynaptic fibers may not produce an hypersensitivity of the corresponding post-synaptic receptors (Von Voigtlander et al., 1973; Tassin et al., 1982). This suggests that the affinity or the number of postsynaptic receptor sites is not only regulated by chemical signals from presynaptic fibers but also by those delivered by some other neuronal inputs to the target cells.

By interacting with its own (homologous) receptor, a given neurotransmitter may modify the sensitivity of other (heterologous) receptors located on the same cell (Fig.1). For example, Woodward et al. have demonstrated that noradrenaline (NA) which is released from nerve terminals in the cerebellum facilitates the inhibitory effect of GABA on the activity of Purkinje cells (Woodward et al., 1979; Waterhouse et al., 1982). Conversely, it has also been shown that the inhibitory effect of GABA released from striato-nigral fibers on the activity of nigral efferent cells is reduced under the microiontophoretic application of DA on these cells (Waszczak & Walters, 1983). Electrophysiological studies have also allowed to reveal long term modifications in the sensitivity of receptors following denervation of heterologous fibers. Thus, in 1979, Wang et al. have shown that the serotonergic (5-HT) denervation of the amygdala increases the cells responsiveness not only to microiontophoretically applied 5-HT but also to NA and to a lesser extent to GABA. Moreover, in 1982, Brunello et al. (this book) have obtained biochemical evidence indicating that the absence of a certain type of innervation could prevent the modification in the density of an heterologous population of receptors induced by a pharmacological treatment. Indeed, these authors have shown that the down-regulation of cortical β-adrenergic receptors produced by a chronic treatment with antidepressants, requires the presence of 5-HT fibers (Brunello et al., 1982). Simultaneously we reported that the development of dopaminergic (DA) -D1- receptors denervation supersensitivity in the rat prefrontal cortex was dependent on the procedure (chemical or electrolytic) used in order to destroy the presynaptic DA neurons, and provided evidence indicating that non-DA fibers were intervening in the regulation of cortical D1 receptors (Tassin et al., 1982).

In the present review, first we will describe experiments in which attempts were made to identify the type(s) of neurons involved in the hetero-regulation of cortical D1 receptors. Then, we will show that the analysis of D1 receptors denervation supersensitivity in other brain structures innervated by DA neurons has revealed that this hetero-regulation also occurred in subcortical areas such as the nucleus accumbens. In this case, the non-DA fibers responsible for the modulation of D1 receptors sensivitity appeared to be part of a well-defined neuronal circuit, suggesting that this hetero-regulation had a functional significance. Indeed, in a third group of experiments, we have obtained evidence indicating that modifications of cortical and subcortical D1 receptors sensitivities induced by destruction of DA and non-DA neurons were associated with the development of a behavioural syndrom characteristic of cortico-limbic dysfunction (Le Moal et al., 1976).

I. Hetero-regulation of DA (D1) receptors denervation supersensitivity in the prefrontal cortex

Most of the DA innervation of cortical (prefrontal, cingular, entorhinal, rhinal) and subcortical (nucleus accumbens, septum, amygdala, striatum) structures arises from neuronal populations which are distributed in the ventral mesencephalic tegmentum (VMT) and the substantia nigra (SN) (Björklund & Lindvall, 1978). All projection areas of the different ascending DA systems contain post-synaptic DA receptors coupled positively to an adenylate cyclase (D1 receptors) (Bockaert et al., 1976; Kebabian et al., 1982; Krueger et al., 1976; Tassin et al., 1978) whose topographical distributions closely resemble those of DA nerve terminals (Tassin et al., 1976, 1978, 1984).

Chemical (6-OHDA) or electrolytic lesions of the VMT induced important decreases in the levels of DA in the prefrontal cortex (-93% and -76% respectively) indicating a massive destruction of the DA neurons. Six to ten weeks after bilateral electrolytic lesions of the VMT, the activity of the cortical DA-sensitive adenylate cyclase was markedly increased (+62%). This supersensitivity of D1 receptors was not associated with a change in the apparent affinity of DA for the receptor coupled to the enzyme and corresponded very likely to an enhanced number of receptor

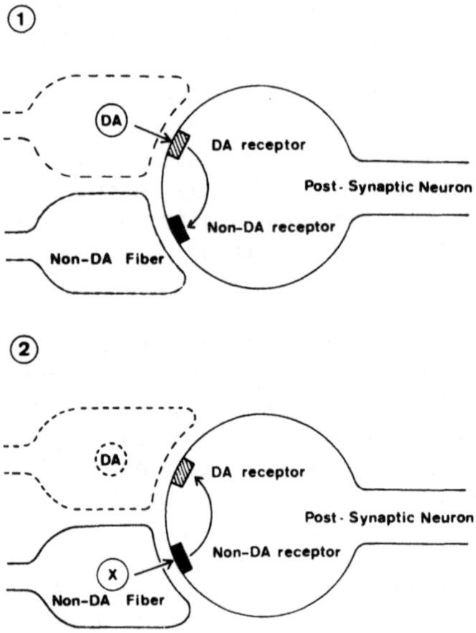

Fig.1: The two main types of long-term receptor hetero-regulation implicating DA systems. 1) DA fibers control the density of non-DA receptors. As an example, Neurotensin binding sites post-synaptic to DA neurons are regulated by DA afferences, particularly in cortico-limbic structures (Hervé et al., 1986). 2) Non-DA fibers modulate DA receptors. Two examples of this type of hetero-regulation are described in this paper.

sites (Tassin et al., 1982). Surprisingly, bilateral 6-OHDA lesions of the VMT, which induced an almost total destruction of the DA innervation in the prefrontal cortex, did not produce any D1 receptor supersensitivity. Two factors could contribute to the difference in the change of cortical D1 receptors properties seen following these two types of VMT lesions: 1) As revealed by biochemical data, NA as well as DA ascending pathways are destroyed following 6-OHDA injections into the VMT (Tassin et al., 1982). This was not the case after the electrolytic lesions since cortical levels of NA were not affected. 2) Besides DA neurons, other non-DA fibers or neurons can be destroyed following an electrolytic lesion of the VMT. This is the case for the 5-HT (Tassin et al., 1978b), substance-P (Sakanaka et al., 1983) or cholecystokinine (Studler et al., 1981) containing fibers which pass through or originate from the VMT.

According to these considerations, it can be postulated either that ascending NA neurons are necessary (first hypothesis) or that other non-DA fibers must be destroyed (second hypothesis) to obtain a denervation supersensitivity of the D1 receptors in the prefrontal cortex. Both hypothesis are not mutually exclusive and could be relevant.

In order to test the validity of the first hypothesis, namely to determine the influence of NA neurons, ascending NA pathways were destroyed selectively by bilateral injections of 6-OHDA made laterally to the pedunculus cerebellaris superior (P.C.S.) just before performing the electrolytic lesions of the VMT.

As a control, we first verified that the P.C.S. 6-OHDA lesions alone induced an almost complete degeneration of the ascending NA fibers, no change in the cortical levels of DA, an increase in β-adrenergic-sensitive adenylate cyclase activity (+104%) and no modification in DA-sensitive adenylate cyclase activity. As previously mentioned, the bilateral electrolytic lesions of the VMT alone did not affect the cortical levels of NA but decreased those of DA by 52%. In confirmation of this, the activity of the 3-adrenergic-sensitive adenylate cyclase was not affected by these lesions whereas the DA-sensitive adenylate cyclase activity was increased significantly (+38%) (Tassin et al., 1986).

When both types of lesions were performed simultaneously, cortical DA levels were reduced in a way similar to that observed in VMT lesioned rats while those of NA were decreased by 95%. In these conditions, although the β-adrenergic-sensitive adenylate cyclase activity was enhanced by +110%, a change in the DA-sensitive adenylate cyclase activity could not be seen any longer in the prefrontal cortex (Tassin et al., 1986).

It can therefore be proposed that the NA innervation of the prefrontal cortex plays a role in the expression of D1 receptor denervation supersensitivity by acting on cells rich in D1 receptors. In fact, a population of cells distributed in the deep cortical layers receiving the DA projection which are sensitive to the microiontophoretic application of DA are also sensitive to that of NA (Bunney & Aghajanian, 1976).

The demonstration of the influence of the NA innervation in the control of D1 receptors denervation supersensitivity does not exclude the possibility that non-DA fibers other than NA ones could also contribute to this regulation. Indeed, preliminary experiments have suggested that a supersensitivity of D1-receptors does occur in the prefrontal cortex, even after a complete destruction of the ascending NA fibers when extensive electrolytic lesions of the VMT are made. Therefore, a loss of non-catecholaminergic fibers may also affect the regulation of cortical D1 receptors sensitivity.

II <u>Hetero-regulation of D1 receptors denervation supersensitivity in a subcortical structure: functional significance</u>

As shown in electrophysiological studies, the stimulation of the mesocortico-prefrontal DA neurons leads to an inhibition of the activity of target cells in the prefrontal cortex (Ferron et al., 1984). Efferent prefronto-cortical neurons innervate some other ipsi- and contralateral cortical areas as well as several subcortical structures including those containing the cell bodies and the nerve terminals of the main ascending DA pathways (Beckstead, 1979). Thus, the mesocortico-prefrontal DA neurons, via their action on cortical efferent cells, may indirectly control DA transmission in areas such as the nucleus accumbens or the striatum, which are rich in DA nerve terminals, either by regulating the activity of corresponding DA cells or through presynaptic mechanisms.

Surprisingly, neither the chemical (6-OHDA) nor the electrolytic lesions of the VMT affected the sensitivity of D1 receptors in the nucleus accumbens, although both types of lesions reduced by 95% the levels of DA in this subcortical structure (Tassin et al., 1982). In fact, cells of the nucleus accumbens receive two types of information from DA neurons originating from the VMT. One is direct and mediated by the meso-nucleus accumbens DA fibers, the other is indirect and mediated by the mesocortico-prefrontal DA neurons which in turn control the prefronto-cortico-nucleus accumbens excitatory pathway (Fig.2). The absence of D1 denervation supersensitivity in the nucleus accumbens could therefore be due to the disinhibition of the cortico-nucleus accumbens pathway which may occur following the destruction of the mesocortical DA system. Bilateral ablations of the prefrontal cortex were then performed to prevent the effects of this disinhibition (Reibaud et al., 1984). These lesions induced a 55% decrease in the activity of the high affinity uptake of ^{3}H-glutamic acid in synaptosomes from the nucleus accumbens, confirming anatomical and biochemical observations indicating that numerous subcortical structures are innervated by glutamatergic neurons originating from the prefrontal cortex (Beckstead, 1979; Fonnum et al., 1981).

When the prefronto-cortical ablation was performed simultaneously with the bilateral injection of 6-OHDA into the VMT, a marked hypersensitivity of the D1 receptors was observed in the nucleus accumbens (+52% increase in Vmax and two-fold increase in Kapp for DA). The ablation of the prefrontal cortex alone induced only a small increase (14%, $p<0.01$) in D1 receptors sensitivity in this structure (Reibaud et al., 1984). These experiments indicate that the expression of the D1 receptors denervation supersensitivity in the nucleus accumbens is dependent on the activity of cortico-nucleus accumbens neurons. However, since the descending prefronto-cortical neurons innervate several subcortical structures in relation with the nucleus accumbens, it cannot be completely excluded that neurons other than the cortico-nucleus accumbens ones contribute to the observed effect (Kelley et al., 1982; Krettek & Price, 1978).

These results provide another example of an hetero-regulation of D1 receptors. As in the cerebral cortex, the hetero-regulation of D1 receptors demonstrated in the nucleus accumbens reveals the occurrence of specific interactions between identified neuronal systems which may have physiological significance. Indeed, the target cells of the VMT DA neurons in the nucleus accumbens integrate also information coming from the prefrontal cortex and changes in the activity of cortico-nucleus accumbens neurons can modulate the sensitivity of D1 receptors on these cells. It can be assumed that in each structure containing D1 receptors, specific hetero-regulations of these receptors occur depending on the nature and the origin of the afferent fibers and behavioral expressions of these hetero-regulations can be expected in certain conditions.

III Hetero-regulation of D1 receptors: possible behavioral consequences

Bilateral electrolytic lesions of the ventral mesencephalic tegmentum (VMT) in the rat induce behavioural deficits such as locomotor hyperactivity and disappearance of spontaneous alternations ("VMT syndrom") (Le Moal et al., 1969, 1976; Galey et al., 1979). As previously mentioned, these lesions destroy the DA cell bodies sending projections to the limbic and cortical areas including the nucleus accumbens and prefrontal cortex respectively. Correlation studies have indicated that the amplitude of the locomotor hyperactivity was proportional to the extent of destruction of the DA fibers innervating the prefrontal cortex and, more interestingly, to the development of a DA -D1- receptor supersensitivity in this area (Tassin et al., 1978, 1982b).

Since the destruction of the ascending NA pathways seemed to down-regulate the cortical D1 receptors denervation supersensitivity induced by the electrolytic lesion of the VMT (see § I), it was tempting to investigate whether chemical (6-OHDA) lesions of the NA innervation could affect the locomotor hyperactivity induced by the electrolytic lesion of the VMT.

As for the experiments made on D1 receptors, rats were lesioned either by bilateral electrocoagulations of the VMT, or by bilateral injections of 6-OHDA made laterally to the P.C.S. or simultaneously by both types of lesions (see § I). Three to four weeks later, the nocturnal (9 p.m.-7 a.m.) locomotor activity of each animal was recorded in a circular corridor (Le Moal et al., 1969).

As expected, the high frequency electrolytic lesions of the VMT (HF VMT) induced a mean increase of 56% ($p<0.01$) in the nocturnal locomotor activity of the rats. The chemical lesions (P.C.S., 6-OHDA) were without effect on the mean value of the nocturnal locomotor activity although the variability in this behavioral parameter from one animal to another seemed slightly higher than that observed for sham-operated animals. When both types of lesions were performed, the mean value of the rats locomotor activities was identical to that observed with the sham-operated animals (Taghzouti et al., submitted). Among the 16 double-lesioned rats, only one was hyperactive, i.e. presented a locomotor activity higher than 150% of the mean of the control values.

However, the absence of locomotor hyperactivity in the double-lesioned animals could be attributed to a non specific effect of the NA denervation on the locomotor behavior. Therefore, similar experiments were performed using as a test the spontaneous alternations, estimated in a Y-maze, another behavior disturbed by the electrolytic lesion of the VMT and involving cognitive processes (Galey et al., 1979; Taghzouti et al., 1986).

As observed for the locomotor activity, the NA denervation alone had no effect on the percentage of spontaneous alternations (76 $\pm$ 6%). At the opposite, all the VMT-lesioned animals which were hyperactive presented a profound perturbation of their spontaneous alternation behavior (59 $\pm$ 2%, $p<0.01$). Interestingly, when the destruction of the NA innervation was associated with the electrolytic lesion of the VMT, the mean value of the spontaneous alternations was not different from that found in control animals (78 $\pm$ 4%) (Taghzouti et al., submitted). These experiments strongly suggest that ascending NA fibers play a permissive role in the development of at least two of the characteristic behavioral deficits evoked by the electrolytic VMT lesion. They do not shed light, however, on the type of interaction which may occur between the NA fibers and the target cells of the VMT neurons affected by the lesion. Although the nature of the lesioned cells responsible for the behavioral syndrom has not been yet definitely identified, several data indicate that the destruction of the DA meso-cortico-limbic neurons could be the main factor implicated in the development of the "VMT syndrom" (Galey et al., 1977; Tassin et al., 1978).

Since a modulation of DA (D1) receptors by ascending NA neurons has been shown in the prefrontal cortex, this structure seems one of the priviledge sites where an interaction between the NA fibers and the target cells of the DA neurons is involved in the expression of the behavioral deficits examined. It cannot be excluded, however, that this type of interaction also occurs in other structures innervated by the DA VMT cells, such as the nucleus accumbens, the lateral septum, the amygdala or the head of the caudate nucleus (Björklund & Lindvall, 1978)). In any case, there is some indication that the behavioural recovery observed following NA denervation in VMT lesioned rats reflects the occurrence of antagonistic effects of NA and DA neurons on their common target cells.

CONCLUDING REMARKS

We have described two types of hetero-regulation of receptors which have been revealed by long term interruption of a given neurotransmission: 1) In the prefrontal cortex, D1 receptors are regulated by NA fibers; 2) In the nucleus accumbens, D1 receptors are modulated by non-DA, possibly glutamatergic prefronto-cortical efferent fibers.

These situations are of particular interest since they may represent two different modalities of hetero-regulation of messages mediated by corresponding afferent fibers with different functional significance.

In the first case, ie control of cortical D1 receptors by NA afferent fibers, an interaction between two types of monoaminergic systems is intervening. Although the identity of the NA receptor (α or β) involved is not established yet, this interaction implies two modulatory neuronal systems which participate to the control of the activity of cortical efferent neurons innervating sub-cortical structures (Fig.2). As already emphasized, the recovery from symptoms characteristic of a DA cortico-limbic dysfunction, such as locomotor hyperactivity, by a superimposed NA denervation suggests the existence of a necessary required equilibrium in the activity of these two types of systems for normal functions. However, a hierarchy exists between these systems since the NA denervation alone did not induce significant modifications in the locomotor behavior regulated by the DA neurons. Such interactions with similar hierarchies could also take place in other structures innervated both by DA and NA neurons. In fact, Taghzouti and Simon have shown recently that disruption of alternation behavior induced by 6-OHDA injections made into the lateral septum occurs only when NA afferences are preserved (Taghzouti & Simon, personal communication). In this context, it can be recalled that similar type of interactions have been seen in the basal ganglia between DA and 5-HT neurons since the catalepsy induced by the blockade of DA transmission is not seen any more following destruction of the

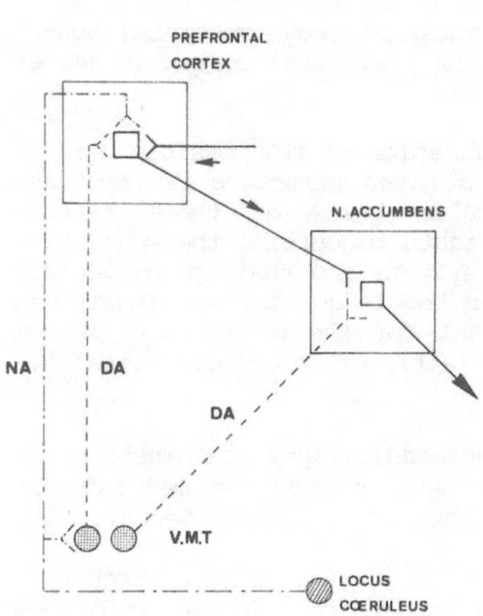

Fig.2: Schematic diagram of the modulatory actions of NA and DA fibers on the expression of locomotor activity. Black arrows represent the hypothetical pathway which conveys the information triggering locomotor behavior. Signals which transit through this polysynaptic neuronal pathway are under the regulation of aminergic neurons both in the cerebral cortex and the nucleus accumbens. Indeed, an increased locomotor activity is observed following the destruction of the DA cell bodies in the VMT. However, a superimposed NA lesion, possibly via an action at the cortical level, attenuates the desequilibrium produced by the lesion of DA neurons alone.

5-HT containing neurons which innervate these structures (Costall et al., 1975). In addition, it has also been shown recently that the number of cortical β-adrenergic receptors is enhanced in the rat following the destruction of ascending 5-HT neurons (Stockmeier et al., 1985).

Examples of interactions between excitatory neuronal pathways containing an amino-acid as a neurotransmitter and monoaminergic neurons can be found in the literature. For example, in the striatum, several data indicate that cortico-striatal glutamatergic neurons control presynaptically the release of DA (Giorguieff et al., 1977) and that a reciprocal regulation occurs (Nieoullon et al., 1982). Ultrastructural studies have shown that both types of neurons contact the large population of medium size striatal neurons which innervate the substantia nigra (Freund et al., 1984) and electrophysiological data have indicated that nigro-striatal DA neurons modulate the responses evoked by the cortico-striatal glutamatergic neurons on their target cells (Brown & Arbuthnott, 1983). A similar pattern of organization may occur in the nucleus accumbens, the DA neurons modulating the response evoked by cortico-nucleus accumbens excitatory fibers on target efferent cells. Since the mesocortico-prefrontal DA neurons are involved in the control of the activity of cortical efferent cells innervating subcortical structures including the striatum and the nucleus accumbens, it can be seen easily that these cortical cells contribute to the so-called interdependance between mesocortical and mesolimbic or mesostriatal DA neurons (Carter & Pycock, 1980; Tassin et al., 1982; Simon & Le Moal, 1984; Glowinski et al., 1984). Indeed, we have obtained recent evidence that, in the striatum, the increase of D1 receptors following an unilateral DA denervation, only occurs in areas receiving non-DA projections from the contralateral DA prefrontal cortex (Hervé et al., in preparation).

It is obvious that the identification of the possible hetero-regulations of known receptors in a given structure is important not only to examine the molecular basis of these hetero-regulations but to get further insights concerning the modalities of interaction between identified systems and their physiological significance. In addition, a better knowledge of these functional hetero-regulations could be fruitful for the development of new pharmacological treatments in psychiatric or neurologic diseases.

REFERENCES

. Beckstead, R.M. (1979) An autoradiography examination of cortico-cortical and subcortical projections of the medio-dorsal projection (prefrontal) cortex in the rat. J. Comp. Neurol., 18, 43-62.

. Björklund, A. & Lindvall, O. (1978) The mesotelencephalic dopamine neuron system: a review of its anatomy. In Limbic Systems (eds K.E. Livingston & O. Hornykiewicz) Plenum Press, New-York, pp.297-331.

. Bockaert, J., Prémont, J., Glowinski, J., Thierry, A.M. & Tassin, J.P. (1976) Topographical distribution of DA innervation and of DA receptors in the rat striatum. II. Distribution and characteristics of DA adenylate cyclase. Interaction of D-LSD with DA receptors. Brain Res., 107, 303-315.

. Brown, J.R. & Arbuthnott, G.W. (1983) The electrophysiology of dopamine (D2) receptors: a study of the actions of DA on cortico-striatal transmission. Neuroscience, 10, 349-355.

. Brunello, N., Barbaccia, M.L., Chuang, D.M. & Costa, E. (1982) Down-regulation of 3-adrenergic receptors following repeated injections of desmethylimipramine: permissive role of serotonergic axons. Neuropharmacol., 21, 1145-1149.

. Bunney, B.S. & Aghajanian, G.K. (1976) Dopamine and norepinephrine innervated cells in the rat prefrontal cortex. Pharmacological differentiation using microiontophoretic techniques. Life Sci., 19, 1783-1792.

. Cannon, W.B. & Rosenblueth, A. (1949) The supersensitivity of denervated structures, McMillan, New York.

. Carter, C.J. & Pycock, C.J. (1980) Behavioural and biochemical effects of DA and NA depletion within the medial prefrontal cortex of the rat. Brain Res., 192, 163-176.

. Costall, B., Fortune, D., Naylor, R.J., Marsden, C.D. & Pycock, C.J. (1975) Serotonergic involvement with neuroleptic catalepsy. Neuropharmacology, 14, 859-868.

. Ferron, A., Thierry, A.M., Le Douarin, C. & Glowinski, J. (1984) Inhibitory influence of the mesocortical dopaminergic system on spontaneous activity or excitatory response induced from the thalamic medio-dorsal nucleus in the rat medial prefrontal cortex. Brain Res., 302, 257-265.

. Fonnum, F., Storm-Mathisen, J. & Divac, I. (1981) Biochemical evidence for glutamate as neurotransmitter in cortico-striatal and cortico-thalamic fibres in rat brain, Neuroscience, 6, 863-873.

. Freund, T.F., Powell, J.F. & Smith, A.D. (1984) Tyrosine hydroxylase-immunoreactive boutons in synaptic contact with identified striato-nigral neurons with particular reference to dendritic spines. Neuroscience, 13, 1189-1215.

. Galey, D., Simon, H. & LeMoal, M. (1977) Behavioural effects of lesion in the A10 dopaminergic area of the rat. Brain Res., 124, 83-97.

. Galey, D., Jaffard, R. & LeMoal, M. (1979) Alternation behaviour spatial discrimination and reversal after electrocoagulation of the ventral mesencephalic tegmentum in the rat. Behav. Neural. Biol., 26, 81-88.

. Giorguieff, M.F., Kemel, M.L. & Glowinski, J. (1977) Presynaptic effect of L-glutamic acid on the release of dopamine in rat striatal slices. Neuroscience Letters, 6, 73-77.

. Glowinski, J., Tassin, J.P. & Thierry, A.M. (1984) The meso-cortico-prefrontal DA neurons, TINS, 7, 11, 415-418.

. Hervé, D., Tassin, J.P., Studler, J.M., Dana, C., Kitabgi, P., Vincent, J.P., Glowinski, J. & Rostène, W. (1986) Dopaminergic control of ^{125}I-labeled neurotensin binding site density in

corticolimbic structures of the rat brain. Proc. Natl. Acad. Sci., 83, 6203-6207.
. Kebabian, J.W., Petzold, G.L. & Greengard, P. (1982) Dopamine sensitive adenylate cyclase in caudate nucleus of rat brain and its similarity to the dopamine receptor. Proc. Nat. Acad. Sci. USA, 69, 2145-2149.
. Kelley, A..E, Domesick, V.B. & Nauta, W.J.H. (1982) The amygdalostriatal projection in the rat: an anatomical study by anterograde and retrograde tracing methods. Neuroscience, 7, 615-630.
. Krettek, J.E. & Price, J.L. (1978) Amygdaloïd projections to subcortical structures within the basal forebrain and brainstem in the rat and cat. J. Comp. Neurol., 178, 225-280.
. Krueger, B.K., Forn, J., Walters, J.R., Roth, R.H. & Greengard, P. (1976) Stimulation by dopamine of adenosine cyclic 3'-5'-monophosphate formation in rat caudate nucleus: effect of lesions of the nigro-neostriatal pathway. Mol. Pharmacol., 12, 639-648.
. LeMoal, M., Cardo, B. & Stinus, L. (1969) Influence of ventral mesencephalic lesions on various spontaneous and conditioned behaviours in the rat. Physiol. Behav., 60, 567-573.
. LeMoal, M., Stinus, L. & Galey, D. (1976) Radiofrequency lesion of the ventral mesencephalic tegmentum: neurological and behavioural considerations. Exp. Neurol., 50, 521-535.
. Nieoullon, A., Kerkerian, L. & Dusticier, N. (1982) Inhibitory effects of dopamine on high affinity glutamate uptake from rat striatum. Life Sciences, 30, 1165-1172.
. Reibaud, M., Blanc, G., Studler, J.M., Glowinski, J. & Tassin, J.P. (1984) Non-DA prefronto-cortical efferents modulate D1 receptors in the nucleus accumbens. Brain Res., 305, 43-50.
. Sakanaka, M., Shiosaka, S., Takatsuki, K. & Tokyama, M. (1983) Evidence for the existence of a substance P-containing pathway from the nucleus laterodorsalis tegmenti (castaldi) to the medial frontal cortex of the rat. Brain Res., 259, 123-126.
. Simon, H. & LeMoal, M. (1984) In Catecholamines Neuropharmacology and central nervous system. Theoretical aspects (eds. E. Usdin, A. Carlsson, A. Dahlström & J. Engel) Alan R. Liss, Inc. New York, pp. 293-307.
. Stockmeier, C.A., Martino, A.M. & Keller K.J. (1985) A strong influence of serotonin axons on 3-adrenergic receptors in rat brain. Science 230, 323-325.
. Studler, J.M., Simon, H., Cesselin, F., Legrand, J.C., Glowinski, J. & Tassin, J.P. (1981) Biochemical investigation on the localization of the cholecystokinin octapeptide in DA neurons originating from the ventral tegmental area of the rat. Neuropeptides, 2, 131-139.
. Taghzouti, K., Simon, H., Hervé, D., Blanc, G., Studler J.M., Glowinski, J., LeMoal, M. & Tassin, J.P. (1987) Disturbances of motor and cognitive functions following lesions of the mesencephalic dopaminergic cell bodies are corrected by lesions of the dorsal noradrenergic ascending system (submitted).
. Taghzouti, K., Simon, H. & LeMoal, M. (1986) Disturbances in exploratory behaviour and functional recovery in the Y and radial

mazes following DA depletion of the lateral septum, Behav. Neural Biol., 45, 48-56.

. Tassin, J.P., Chéramy, A., Blanc, G., Thierry, A.M. & Glowinski, J. (1976) Topographical distribution of DA innervation and of DA receptors in the rat striatum. I. Microestimation of (^{3}H)DA uptake and DA content in microdiscs. Brain Res., 107, 291-301.

. Tassin, J.P., Bockaert, J., Blanc, G., Stinus, L., Thierry, A.M., Lavielle, S., Prémont, J. & Glowinski, J. (1978) Topographical distribution of DA innervation and DA receptors of the anterior cerebral cortex of the rat. Brain Res., 154, 241-251.

. Tassin, J.P., Stinus, L., Simon, H., Blanc, G., Thierry, A.M., LeMoal, M., Cardo, B. & Glowinski, J. (1978b) Relationships between the locomotor hyperactivity induced by A10 lesions and the destruction of the fronto-cortical dopaminergic innervation in the rat. Brain Res. 141, 267-281.

. Tassin, J.P., Simon, H., Hervé, D., Blanc, G., LeMoal, M., Glowinski, J. & Bockaert, J. (1982) Non-dopaminergic fibers may regulate dopamine-sensitive adenylate cyclase in the prefrontal cortex and the nucleus accumbens. Nature, 295, 696-698.

. Tassin, J.P., Simon, H., Glowinski, J. & Bockaert, J. (1982) Modulations of the sensitivity of dopaminergic receptors in the prefrontal cortex and nucleus accumbens: relationships with locomotor activity. In Brain Peptides and Hormones (eds. R. Collu et al.) Raven Press, New York, pp.17-30.

. Tassin, J.P., Reibaud, M., Blanc, G., Studler, J.M. & Glowinski, J. (1984) Regulation of the sensitivity of D1 receptors in the prefrontal cortex and the nucleus accumbens by non dopaminergic pathways. In Catecholamines: Neuropharmacology and Central Nervous System. Theoretical Aspects (eds. Usdin, A., Carlsson, A., Dahlström & J. Engel) Alan R. Liss, Inc, New York, pp.103-111.

. Tassin, J.P., Studler, J.M., Hervé, D., Blanc, G. & Glowinski, J. (1986) Contribution of noradrenergic neurons to the regulation of dopaminergic (D1) receptor denervation supersensitivity in rat prefrontal cortex. J. Neurochem. 46, 243-248.

. Von Voigtlander, P.F., Boukma, S.J. & Johnson, S. (1973) Dopaminergic denervation supersensitivity and dopamine stimulated adenylcyclase activity. Neuropharmacol., 12, 1081-1086.

. Wang, R.Y., de Montigny, C., Gold, B.I., Roth, R.H. & Aghajanian, G.K. (1979) Denervation supersensitivity to serotonin in rat forebrain: single cell studies. Brain Res., 178, 479-497.

. Waszczak, B.L. & Walters, J.R. (1983) DA modulation of the effects of GABA on substantia nigra pars reticulata neurons. Science, 220, 218-221.

. Waterhouse, B.D., Moises, H.C., Yoh, H.H. & Woodword, D.J. (1982) NE enhancement of inhibitory synaptic mechanisms in cerebellum and cerebral cortex: mediation by 3-adrenergic receptors, J.P.E.T., 221, 495-506.

. Woodward, D.J., Moises, H.C., Waterhouse, B.D., Hoffer, B.J. & Freedman, R. (1979) Modulatory actions of norepinephrine in the central nervous system. Fed. Proc. 38, 2109-2116.

22

Interactions between Vasoactive Intestinal Peptide and Norepinephrine, Ergot Alkaloids and Prostanoids in Mouse Cerebral Cortex

Pierre J. Magistretti, Michel Schorderet, Patrick R. Hof and Nicolas Schaad

There is growing evidence that two, or possibly more, neurotransmitters can coexist within the same neuron (Hökfelt *et al.*, 1980). In particular, the presence of a peptide and a biogenic amine has been demonstrated in the same terminals of central and peripheral neurons (Hökfelt *et al.*, 1980). These findings have led to the hypothesis that neurotransmitters, coexisting within the same neurons, can interact at pre- or postsynaptic sites in a functionally coordinated manner. Interactions between neurotransmitters contained in distinct neuronal systems terminating within the same region of the central nervous system (CNS) can also be envisaged. We have examined this last possibility in the cerebral cortex, an area of the CNS where the two neurotransmitters vasoactive intestinal polypeptide (VIP) and norepinephrine (NE) are contained in separate neuronal systems and where they both stimulate the formation of cAMP (Morrison and Magistretti, 1983; Magistretti, 1986a,b).

MORPHOLOGY

VIP-containing neurons in rodent cerebral cortex represent a rather homogeneous population of cells. Thus, over 95 % of VIP-containing neurons are intracortical, bipolar and radially oriented (i.e. oriented in a plane perpendicular to the pial surface) neurons with minimal arborization in the horizontal plane, except in layers I and IV-V where the arborization span of dendrites extends horizontally for 60-100 µm (Loren *et al.*, 1979; Morrison *et al.*, 1984; Magistretti and Morrison, 1985). Each VIP neuron can therefore be identified with a unique radial cortical volume with a diameter of 60-100 µm. From these morphological observations it is apparent that the actions of VIP molecules released from activated VIP neurons, will be restricted locally, within radially oriented cortical domains (Morrison *et al.*, 1984; Magistretti and

Research is supported by Fonds National Suisse de la Recherche Scientifique Grants N° 3.357-0.86 and 3.969-0.84. PRH is supported by "Fondation Centre de Recherches Médicales Carlos et Elsie de Reuter". NS is supported by Fellowships from "Fondation Suisse de Recherche sur l'Alcool" and "Fondation Pierre Mercier".

Morrison, 1985). A quantitative immunohistochemical analysis of VIP neurons in rat visual cortex has revealed that, on the average, there is one VIP-immunoreactive cell per radial cortical volume of 30 µm diameter, and that the largest cortical volume without a VIP bipolar cell has a diameter of approximately 60 µm (Morrison *et al.*, 1984). This last figure, when compared to those of the mean density of VIP neurons (i.e. one per 30 µm) and of their dendritic span (i.e. 60-100 µm) indicates that the radial cortical domains defined by the arborization pattern of each VIP cell partially overlap (Morrison *et al.*, 1984).

The morphological characteristics of the NE-containing cortical neuronal system are strikingly different. NE is contained in a highly divergent neuronal circuit which originates in the nucleus locus coeruleus in the brainstem, from where it projects to several regions of the CNS (Dahlström and Fuxe, 1964; Morrison *et al.*, 1981). In the cerebral cortex noradrenergic fibers are organized predominantly in a plane parallel to the pial surface (thus orthogonally to the radially oriented VIP-containing neurons), and span across several functionally distinct cortical regions (Morrison *et al.*, 1981). Therefore, in contrast to the locally-acting VIP neurons, NE-containing fibers can influence their target cells globally, throughout a vast rostro-caudal expanse of neocortex simultaneously.

SYNERGISTIC INTERACTION IN STIMULATING cAMP FORMATION

In the cerebral cortex, VIP stimulates in a concentration-dependent manner the formation of cAMP, with an EC_{50} of 800 nM (Quik *et al.*, 1978; Magistretti and Schorderet, 1984). We have recently observed that an interaction exists, under the form of a synergism, between VIP and NE in eliciting cAMP increases in mouse cerebral cortical slices (Magistretti and Schorderet, 1984; 1985). Thus, in the absence of phosphodiesterase inhibitors, NE at 10 µM, when tested alone, moderately increases cAMP levels (approximately 50-60 pmoles/mg of protein above basal levels). This action in mouse cerebral cortex is mediated by adrenergic receptors of the ß type. VIP alone, at 1 µM, under identical experimental conditions, is a far more efficient stimulator of cAMP formation with levels reaching 200-250 pmoles/mg of protein (Magistretti and Schorderet, 1984). When both neurotransmitters are tested in combination (NE at 10 µM and VIP at 1 µM), cAMP levels increase to 600-700 pmoles/mg of protein, thus far in excess of a simple additivity (Magistretti and Schorderet, 1984) (Figure 1). This synergism is antagonized in a concentration-dependent manner by prazosin, a selective α_1-adrenergic antagonist (Magistretti and Schorderet, 1985) (Figure 2) and the rank-order of potencies of several adrenergic agonists in potentiating the effect of VIP on cAMP levels is characteristic of the activation of α_1-adrenergic receptors (Magistretti and Schorderet, 1985) (Figure 3). These observations indicate that NE, via the activation of α_1-adrenergic receptors can potentiate the stimulatory action of VIP on cortical cAMP levels.

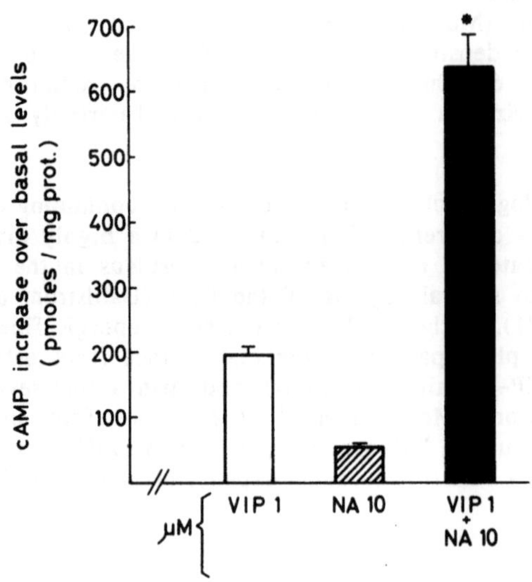

Figure 1. Synergism between VIP and NA in the stimulation of cAMP formation by mouse cerebral cortical slices. In all conditions where drugs were tested, cAMP levels were significantly increased over basal levels ($P < 0.001$). Absolute cAMP levels were (pmol per mg protein ± s.e.m., n = 8-17) : basal, 13.8 ± 0.91; VIP 1 μM, 211.9 ± 16.7; NA 10 μM, 67.9 ± 4; VIP 1 μM + NA 10 μM, 648.7 ± 54.5. *, Significantly different from the sum of VIP and NA tested separately ($P < 0.001$) by Student's t-test (From Magistretti and Schorderet, 1984).

This synergism between VIP and NE in stimulating cAMP formation indicates that VIP and NE can act physiologically in concert within discrete cerebral cortical domains; the spatial coordinates of such domains would be delineated by the intersection of the tangentially-organized noradrenergic fibers and a group of activated, radially oriented VIP-containing neurons (Figure 4). The synergism would lead to the generation of a "hot spot" within the cortex where cAMP levels would be drastically increased over those in surrounding regions to activate those intracellular mechanisms that are regulated by cAMP dependent phosphorylations.

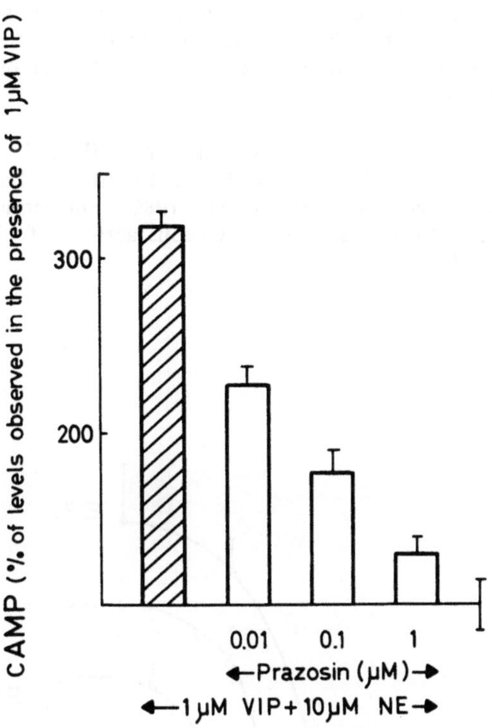

Figure 2. Effect of prazosin on the increases in cAMP elicited by 1 µM VIP in the presence of 10 µM NE. Basal cAMP levels ± SEM were 5.59 ± 0.2 pmol/mg of protein. cAMP values (picomoles per milligram of protein) ± SEM in the presence of 1 µM VIP were 146.5 ± 239. The *error bar* at the *far right* represents SEM of VIP tested alone. Results represent the mean ± SEM of quadruplicate determinations. The antagonism by prazosin of the synergism between 1 µM VIP and 10 µM was significant at all three concentrations testes ($p < 0.001$). Statistical significance was assessed by the Student's t-test (From Magistretti and Schorderet, 1985).

The question arises of when VIP- and NE-containing neuronal systems would be simultaneously active in cerebral cortex. Electrophysiological recordings in behaving rats have demonstrated that NE-containing neurons exhibit pronounced responses to non-noxious, unexpected auditory, visual, and somato-sensory stimuli (Aston-Jones and Bloom, 1981a,b). The firing pattern and activatory inputs to VIP intracortical neurons are not known. However, at the light microscopic level, the overall morphological characteristics and pattern of dendritic arborization of VIP-immunoreactive neurons resemble those of the "bipolar cell" described in Golgi-impregnated material by Peters and Kimerer (1981). These authors, by using a combined Golgi-electron microscopy technique, have demonstrated that in rat visual cortex "bipolar cells" receive direct thalamocortical inputs. These observations would therefore suggest that

VIP-bipolar neurons could also be activated by sensory inputs relayed by specific thalamocortical afferents and, hence, that sensory stimulation could constitute a behavioral event whereby the synergistic interaction between VIP and NE would become operative within the terminal field of thalamocortical afferents in primary sensory cortex.

Similar synergistic interactions between VIP and NE have been confirmed in guinea pig cerebral cortex (Hollingsworth and Daly, 1985) and in discrete hypothalamic nuclei (Redgate *et al.*, 1986). Furthermore VIP and NE interact synergistically to depress the spontaneous activity of identified neurons in the rat neocortex (Ferron *et al.*, 1985).

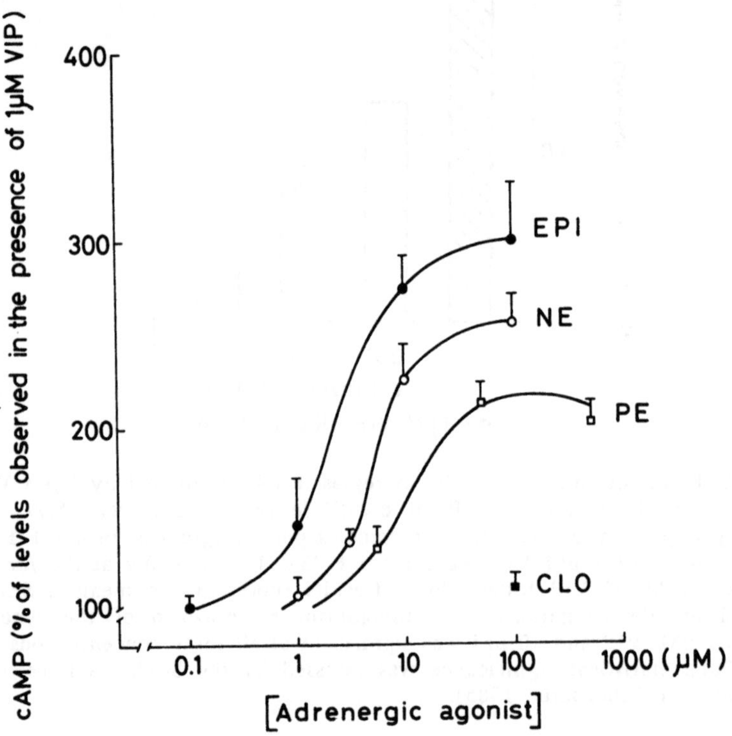

Figure 3. Concentration-response curves of the stimulatory effects of various adrenergic agents on cAMP accumulation elicited by 1 μM VIP. Results are from one experiment repeated twice, with similar results. Each *point* represents the mean ± SEM of four determinations. Basal cAMP levels ± SEM were 10.9 ± 1.63 pmol/mg of protein. cAMP levels ± SEM in the presence of 1 μM VIP were 193 ± 10.3 pmol/mg of protein. The stimulatory effect of EPI and NE was assessed in the presence of 100 μM dl-propranolol to block the ß-adrenergic-mediated increases in cAMP. cAMP (picomoles per milligram of protein) ± SEM in the presence of 100 μM EPI was 79.8 ± 4; for 100 μM EPI plus 100 μM dl-propranolol 20.3 ± 1.1; for 100 μM NE it was 50.1 ± 4.6; and for 100 μM NE plus 100 μM dl-propranolol 18.3 ± 1.5 (From Magistretti and Schorderet, 1985).

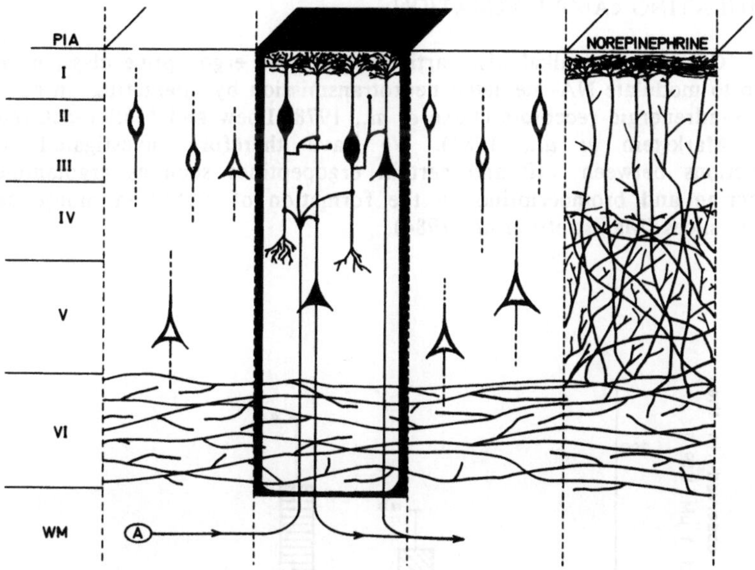

Figure 4. Diagrammatic representation of a "hot spot" resulting from the concomitant activation of noradrenergic cortical afferents and a group of VIP-containing neurons. *Bottom* (layer VI) : Tangentially organized noradrenergic fibers. *Right* : Detail of the arborization pattern of noradrenergic fibers within the cortex (Morrison et al., 1981). Ovoid-shaped cells, VIP-intracortical neurons; triangular-shaped cells, pyramidal neurons. Solid symbols refer to activated neurons. *Far left* : Cortical layers. WM, white matter. The concomitant activation of noradrenergic fibers and of a group of VIP-containing intracortical neurons (by specific thalamic or cortico-cortical inputs, A, bottom left) would determine a drastic increase in cAMP levels within a discrete volume of cortex (delineated in this drawing by solid, thick lines). For graphic clarity, only the VIP-containing and the pyramidal cells have been represented here. In particular, the principal target cells of the thalamo-cortical afferents, i.e., the small stellate cells in layer IV, have been omitted. However, any cell with the capacity of dendritic reception in layer IV may receive thalamic inputs (White, 1981) (Taken from Magistretti and Schorderet, 1985).

The effect of histamine (HIS) and serotonin (5-HT), two other monoamines also contained in cortical afferent neuronal systems (Garbarg *et al.*, 1974; Lidov *et al.*, 1980), on VIP-stimulated cAMP formation have been examined. HIS potentiates in a concentration-dependent manner the effect of 1 µM VIP in stimulating cAMP formation (Magistretti and Schorderet, 1985). This effect is antagonized by mepyramine (H_1 antagonist) but not by cimetidine (H_2 antagonist), indicating that the potentiating action of HIS is mediated by receptors of the H_1 type. In contrast to NE and HIS, 5-HT does not interact synergistically with VIP to increase cAMP levels (Magistretti and Schorderet, 1985).

INTERACTION BETWEEN VIP AND ERGOT ALKALOIDS IN STIMULATING cAMP FORMATION

Certain ergot alkaloids, particularly of the ergopeptine class, have been shown to modulate NA-mediated neurotransmission by interacting, in particular, with α-adrenergic receptors (Fuxe *et al.*, 1978; Loew and Müller-Schweinitzer, 1979; Markstein *et al.*, 1983). We have therefore investigated possible interactions between VIP and certain ergopeptines such as ergotamine, co-dergocrine and bromocriptine on the formation of cAMP in mouse cerebral cortical slices (Magistretti *et al.*, 1984).

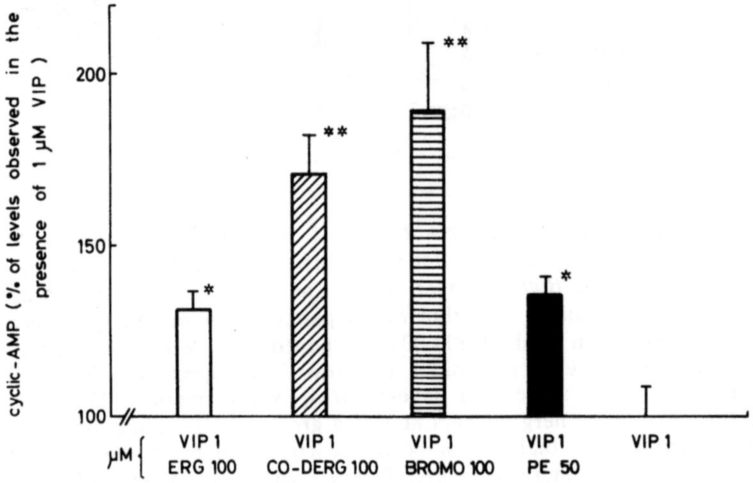

Figure 5. Potentiation by ergopeptines and phenylephrine of the increases in cAMP levels elicited by VIP. ERG : ergotamine; CO-DERG : co-dergocrine; BROMO : bromocriptine; PE : phenylephrine. Results are the mean ± SEM of quadruplicate determinations from one experiment repeated three times with similar results. Error bar on the far right indicates SEM of the effect of 1 μM VIP alone. *P < 0.02; **P < 0.005. Statistical significance assessed by Student's t-test. Basal cAMP levels were (pmol/mg prot ± SEM); n = 4) : 13.4 ± 1.6. cAMP levels in the presence of 1 μM VIP were (pmol/mg prot ± SEM; n = 4) : 124.8 ± 10.2 (From Magistretti *et al.*, 1984).

As shown in Figure 5, bromocriptine, co-dergocrine and ergotamine markedly potentiate the increases in cAMP levels elicited by VIP. It is interesting to note that bromocriptine is the most effective of the three ergopeptines in displaying such a potentiatory effect. For comparison, the effect of the specific α-adrenergic agonist phenylephrine at a 50 µM concentration is also shown in Figure 5. The three ergopeptines and phenylephrine do not increase cAMP levels when tested alone (not shown).

These results indicate that some ergot alkaloids of the ergopeptine class have the capacity to potentiate the effects of VIP in increasing cAMP levels in mouse cerebral cortical slices.

Ergopeptines have been shown to interact with noradrenergic, serotonergic and dopaminergic neuronal systems (Loew and Müller-Schweinitzer, 1979; Markstein *et al.*, 1983; Markstein, 1983). In particular it appears that the effects on noradrenergic neurotransmission are mediated by the interaction with adrenergic receptors of the α subtype (Loew and Müller-Schweinitzer, 1979; Markstein *et al.*, 1983). The fact that bromocriptine, co-dergocrine and ergotamine potentiate the effects of VIP on cAMP levels is consistent with the previously demonstrated α_1-adrenergic mediation of norepinephrine's effect on VIP-stimulated cAMP formation (Magistretti and Schorderet, 1984, 1985).

ROLE OF PROSTANOIDS IN THE EXPRESSION OF THE SYNERGISM BETWEEN VIP AND NE

In a series of experiments directed at an attempt to elucidate the molecular mechanisms mediating the synergistic interaction between VIP and NE, we have observed that the α_1-adrenergic potentiation by NE of the increases in cAMP elicited by VIP was antagonized by indomethacin, a non-steroidal anti-inflammatory drug (Table 1). The degree of inhibition of the synergism by indomethacin was comparable to that observed in the presence of the selective α_1-adrenergic antagonist prazosin (Table 1). Indomethacin, is a rather selective inhibitor of cyclo-oxygenase, the enzyme regulating the synthesis of prostaglandin G_2 (PGG_2) from arachidonic acid. The inhibition of the synergism by indomethacin indicated therefore that a metabolite of arachidonic acid could be formed following α_1-adrenergic stimulation and mediated the potentiating effect of NE on cAMP increases elicited by VIP. We have tested various prostanoids for their potential ability to mimick the effects of NE and initial results indicate that prostaglandins $F_{2\alpha}$ and E_2 selectively potentiate the stimulatory action of VIP on cAMP formation (Schaad *et al.*, in preparation).

Table 1

Effect of indomethacin and prazosin on the synergism between VIP and NE in increasing cAMP levels

Agent added	Concentration (μM)	cAMP (pmol/mg prot)
None		18.5 ± 3.7
VIP	1	144.2 ± 9.4
VIP plus	1	
NE	10	406 ± 19
VIP plus	1	
NE plus	10	
Indomethacin	100	222 ± 9[a]
VIP plus	1	
NE plus	10	
Prazosin	10	210 ± 3.8[a,b]

Results represent the mean ± SEM of 4-7 separate determinations.
[a] Significantly different from cAMP levels in the presence of VIP and NE ($p < 0.005$)
[b] Not significantly different from cAMP levels in the presence of VIP and NE plus indomethacin.

CONCLUSIONS

The observations on the interaction between VIP and NE summarized in this article suggest that VIP- and NE-containing neuronal systems in the neocortex may converge, at least in part, on the same target cells, where activation by NE of α_1-adrenergic receptors will potentiate the effect of VIP in stimulating cAMP formation. This facilitatory action may constitute a feature of α_1-adrenergic receptors present in various CNS areas. For example in guinea pig cerebral cortical slices, HIS and NE display greater than additive effects in stimulating cAMP formation, and this synergism appears to be mediated by the concomitant activation of α-adrenergic and H_1-histaminergic receptors (Daly, 1975). Sattin *et al.* (1975) have reported that the synergistic interaction between NE and adenosine in stimulating the formation of cAMP required the activation of α-adrenergic receptors by NE. Single-cell recordings from the lateral geniculate nucleus have shown that activation of α_1-adrenergic receptors increases the efficacy of excitatory synaptic inputs rather than eliciting direct excitatory effects (Rogawski and Aghajanian, 1980a,b).

α-Adrenergic activation facilitates the responses of somatosensory cortical neurons to excitatory synaptic inputs and to iontophoretically applied acetylcholine (Waterhouse *et al.*, 1981). Furthermore α-adrenergic agonists potentiate the increases in cAMP mediated ß-adrenergic receptors in rat cerebral cortex (Duman *et al.*, 1986), striatum (Leblanc and Ciaranello, 1984) and pineal gland (Klein *et al.*, 1983). Thus, the results summarized in this article taken together with previous physiological and pharmacological studies suggest that activation of α-adrenergic receptors, particularly of the α_1-subtype, facilitates certain cellular effects elicited by other neurotransmitters.

REFERENCES

Aston-Jones, G. and Bloom, F.E. (1981a). Activity of norepinephrine-containing locus coeruleus neurons in behaving rats anticipates fluctuations in the sleep-walking cycle. **J. Neurosci.**, 1, 876-886.

Aston-Jones, G. and Bloom, F.E. (1981b). Norepinephrine-containing locus coeruleus neurons in behaving rats exhibit pronounced responses to non-noxious environmental stimuli. **J. Neurosci.**, 1, 887-900.

Dahlström, A. and Fuxe, K. (1964). Evidence for the existence of monoamine-containing neurons in the central nervous system. I. Demonstration of monoamines in the cell bodies of brain stem neurons. **Acta Physiol. Scand.** 62, 1-55.

Daly, J. (1975). Role of cyclic nucleotides in the nervous system. **In** Handbook of Psychopharmacology. (eds. L.L. Iversen, D.S. Iversen and S.H. Snyder). pp. 47-130. Plenum Press, New York.

Duman, R.S., Karbon, E.W., Harrington,C. and Enna, S.J. (1986). An examination of the involvement of phospholipases A_2 and C in the α-adrenergic and γ-aminobutyric acid receptor modulation of cyclic AMP accumulation in rat brain slices. **J. Neurochem.**, 47, 800-810.

Ferron, A., Siggins, G.R. and Bloom, F.E. (1985). Vasoactive intestinal polypeptide acts synergistically with norepinephrine to depress spontaneous discharge rate in cerebral cortical neurons. **Proc. Natl. Acad. Sci. USA**, 82, 8810-8812.

Fuxe, K., Fredholm, B.B., Ögren, S.V., Agnati, L.F., Hökfelt, T. and Gustafsson, J.A. (1978). Ergots drugs and central monoaminergic mechanisms : a histochemical, biochemical and behavioral analysis. **Fed. Proc.**, 37, 2181-2191.

Garbarg, M., Barbin G., Feger, J. and Schwartz, J.C. (1974). Histaminergic pathway in rat brain evidenced by lesions of the medial forebrain bundle. **Science**, 186, 833-835.

Hökfelt, T., Johansson, O., Ljungdahl, A., Lundberg, J.M. and Schultzberg, M. (1980). Peptidergic neurones. **Nature**, 284, 515-521.

Hollingsworth, E.B. and Daly, J.W. (1985). Accumulation of inositol phosphates and cyclic AMP in guinea-pig cerebral cortical preparations. Effects of norepinephrine, histamine, carbamylcholine and 2-chloroadenosine. **Biochim. Biophys. Acta**, 847, 207-216.

Klein, D.C., Sugden, D. and Weller, J.L. (1983). Postsynaptic α-adrenergic receptors potentiate the ß-adrenergic stimulation of pineal serotonin *N*-acetyltransferase. **Proc. Natl. Acad. Sci. USA**, 80, 599-603.

Leblanc, G. and Ciaranello, R.D. (1984). α-Noradrenergic potentiation of neurotransmitter-stimulated cAMP production in rat striatal slices. **Brain Res.** 293, 57-65.

Lidov, H.G.M., Grzanna, R. and Molliver, M.E. (1980). The serotonin innervation of the cerebral cortex in the rat : An immunohistochemical analysis. **Neuroscience,** 5, 207-227.

Loew, D.M. and Müller-Schweinitzer, E. (1979). Alcaloïdes de l'ergot de seigle, récepteurs adrénergiques, sérotoninergiques et dopaminergiques. **J. Pharmac., Paris,** 10, 383-399.

Loren, I., Emson, P.C., Fahrenkrug, J., Björklund, A., Alumets, J., Hakanson, R. and Sundler, F. (1979). Distribution of vasoactive intestinal polypeptide in the rat and mouse brain. **Neuroscience,** 4, 1953-1976.

Magistretti, P.J. and Schorderet, M. (1984). VIP and noradrenaline act synergistically to increase cyclic AMP in cerebral cortex. **Nature,** 308, 280-282.

Magistretti, P.J., Hof, P. and Schorderet, M. (1984) The increase in cyclic-AMP levels elicited by vasoactive intestinal peptide (VIP) in mouse cerebral cortical slices is potentiated by ergot alkaloids. **Neurochem. Int.,** 6, 751-753.

Magistretti, P.J. and Morrison, J.H. (1985). VIP neurons in the neocortex. TINS, 8, 7-8.

Magistretti, P.J. and Schorderet, M. (1985). Norepinephrine and histamine potentiate the increases in cAMP elicited by Vasoactive Intestinal Polypeptide in mouse cerebral cortical slices : mediation by α_1-adrenergic and H_1-histaminergic receptors. **J. Neuroscience,** 5, 362-368.

Magistretti, P.J. (1986a). VIP-containing neurons in the cerebral cortex : Cellular actions and interactions with the noradrenergic system. **In** Channels in Neural Membranes. (eds. J.M. Ritchie, C.L. Bolis and R.D. Keynes). pp. 323-331. Alan R. Liss, Inc.

Magistretti, P.J. (1986b). Intercellular communication mediated by VIP in the cerebral cortex. **Peptides,** 7, 169-173.

Markstein, R. (1983). Dopamine receptor profile of co-dergocrine Hygergine[R] and its components. **Eur. J. Pharmac.,** 86, 145-155.

Markstein, R., Closse, A. and Frick, W. (1983). Interaction of ergot alkaloids and their combination (co-dergocrine) with α-adrenoceptors in the CNS. **Eur. J. Pharmac.,** 93, 159-168.

Morrison, J.H., Molliver, M.E., Grzanna, R. and Coyle, J.T. (1981). The intra-cortical trajectory of the coeruleo-cortical projection in the rat : A tangentially organized cortical afferent. **Neuroscience,** 6, 139-158.

Morrison, J.H. and Magistretti, P.J. (1983). Monoamines and peptides in cerebral cortex. **TINS,** 6, 146-151.

Morrison, J.H., Magistretti, P.J., Benoit, R. and Bloom, F.E. (1984). The distribution and morphological characteristics of the intracortical VIP-positive cell : an immunohistochemical analysis. **Brain Res.,** 292, 269-282.

Peters, A. and Kimerer, L.M. (1981). Bipolar neurons in rat visual cortex : A combined Golgi-electron microscpe study. **J. Neurocytol.,** 10, 921-946.

Quik, M., Iversen, L.L. and Bloom, S.R. (1978). Effect of vasoactive intestinal peptide (VIP) and other peptides on cAMP accumulation in rat brain. **Biochem. Pharmacol.,** 27, 2209-2213.

Redgate, E.S., Deupree, J.D. and Axelrod, J. (1986). Interaction of neuropeptides and biogenic amines on cyclic adenosine monophosphate accumulation in hypothalamic nuclei. **Brain Res.,** 365, 61-69.

Rogawski, M.A. and Aghajanian, G.K. (1980a). Modulation of lateral geniculate neurone excitability by noradrenaline microiontophoresis or locus coeruleus stimulation. **Nature,** 287, 731-734.

Rogawski, M.A. and Aghajanian, G.K. (1980b). Activation of lateral geniculate neurone by norepinephrine : Mediation by an alpha-adrenergic receptor. **Brain Res.**, 182, 345-359.

Sattin, A., Rall, T.W. and Zanella, J. (1975). Regulation of cyclic 3',5'-monophosphate levels in guinea-pig cerebral cortex by interaction of alpha-adrenergic and adenosine receptor activity. **J. Pharmacol. Exp. Ther.**, 192, 22-32.

Waterhouse, B.D., Moises, H.C. and Woodward, D.J. (1981). Alpha-receptor-mediated facilitation of somatosensory cortical neuronal responses to excitatory synaptic inputs and iontophoretically applied acetylcholine. **Neuropharmacology,** 20, 907-920.

White, E.L. (1981). Thalamocortical synaptic relations. **In** The Organization of the Cerebral Cortex. (eds. F.O. Schmitt, F.G. Worden, G. Adelman and S.G. Dennis). pp 153-161. MIT Press, Cambridge.

Part C
Interactions between Membrane Receptors in Endocrine and Peripheral Organs

Chairman: E. Kandel

23
Multiple Transduction Mechanisms Activated by the Neuropeptide Somatostatin

Agnes Schonbrunn

INTRODUCTION

Because somatostatin inhibits secretion in a wide variety of target cells, studies probing its mechanism of action have focused on the role played by two of the intracellular messengers known to regulate secretory processes: cyclic AMP and calcium (see reviews 1-5). Thus, somatostatin has been proposed to both regulate the concentrations of these intracellular messengers and also to modify their effectiveness. Although direct evidence showing that somatostatin is able to regulate the potency of either cyclic AMP or calcium is still lacking, recent studies have begun to clarify how somatostatin alters the concentrations of these two intracellular mediators and the extent to which such changes are involved in eliciting somatostatin's effects on secretion. This review summarizes studies on the mechanisms by which somatostatin inhibits growth hormone (GH) and prolactin (PRL) secretion from the GH_4C_1 pituitary cell line. These cells have two major advantages for elucidating the biochemical mediators involved in somatostatin action. First, the effects of somatostatin to inhibit GH and PRL secretion in GH_4C_1 cells parallel its actions in estrogen primed pituitary cells both in primary culture and <u>in vivo</u> (4,6-9). Second, GH_4C_1 cells are clonal in origin and therefore both hormonal and biochemical responses are produced by the same population of target cells and can be quantitatively correlated.

RESULTS

Biological Actions of Somatostatin in GH_4C_1 Cells

Somatostatin inhibits hormone secretion by GH_4C_1 cells at physiological concentrations (IC_{50} = 0.7 nM)(10,11). Following addition of 100 nM somatostatin, maximal inhibition of hormone secretion is observed within 5 minutes, the earliest time examined, and inhibition remains unabated for hours in the continued presence of peptide (10, B.D. Koch and A. Schonbrunn, unpublished observations). Somatostatin decreases hormone release under a variety of conditions: it not only inhibits basal secretion, but it also reduces secretion stimulated by secretagogues which utilize different intracellular messengers (10-12). Thus, somatostatin noncompetitively inhibits the stimulatory effect of vasoactive intestinal peptide (VIP), which acts by increasing the activity of adenylate cyclase and elevating intracellular cyclic AMP levels (11,13). In addition, somatostatin inhibits stimulation by thyrotropin releasing hormone (TRH) and bombesin which act by increasing phospholipase C activity and thereby elevate intracellular levels of diacylglycerol, inositol polyphosphates and cytosolic free calcium ($[Ca^{2+}]_i$) (12,14-16).

Binding studies with $[^{125}I\text{-}Tyr^1]$somatostatin have shown that intact GH_4C_1 cells contain a single class of specific, high-affinity receptors for somatostatin which appear to mediate the biological actions of this peptide (10,17). Unlike most other peptide hormones, receptor-bound somatostatin is not rapidly internalized but remains at the cell surface during the time that secretion is inhibited (18). Therefore, somatostatin must elicit its intracellular effects from the plasma membrane.

Somatostatin Inhibits Hormone Secretion by Multiple Mechanisms

As with most secretory cells, increasing intracellular cyclic AMP concentrations with any of several pharmacological agents markedly stimulates hormone release by GH_4C_1 cells (11,19,20). Therefore, one potential mechanism for inhibiting secretion is to reduce cyclic AMP levels. In fact, several lines of evidence indicate that somatostatin does inhibit hormone secretion by regulating the intracellular concentration of cyclic AMP. Although somatostatin does not significantly reduce basal intracellular cyclic AMP levels, it consistently inhibits stimulation of cyclic AMP accumulation by VIP (13). Inhibition of VIP-stimulated cyclic AMP accumulation by somatostatin is maximal within 15 seconds, the earliest time tested (13,21). Therefore, this effect is sufficiently rapid to initiate somatostatin's inhibition of VIP-stimulated hormone release. The concentrations of somatostatin which cause half-maximal inhibition of VIP-stimulated cyclic AMP accumulation (IC_{50} = 1.2 nM) and prolactin secretion (IC_{50} = 0.7 nM) are similar (11,13). In addition, somatostatin reduces the effect of maximal concentrations of VIP on both cyclic AMP accumulation and hormone secretion, but it does not significantly alter the dose-response characteristics of VIP for either response (11,13). The close parallel between the effects of somatostatin on VIP-stimulated cyclic AMP accumulation and VIP-stimulated hormone release indicates that changes in cyclic AMP levels mediate the inhibitory effect of somatostatin on VIP-stimulated hormone secretion. In further support of this conclusion, forskolin, which activates adenylate cyclase and can therefore increase intracellular cyclic AMP concentrations to supermaximal levels, attenuates somatostatin inhibition of prolactin release in the presence of VIP (11). These results indicate that inhibition of VIP-stimulated cyclic AMP accumulation is at least partially responsible for somatostatin's reduction of VIP-stimulated hormone secretion.

Despite the parallel between the effects of somatostatin on VIP-stimulated cyclic AMP accumulation and VIP-stimulated hormone release, there is a clear discrepancy between the ability of somatostatin to inhibit basal hormone secretion and its lack of effect on basal cyclic AMP levels (11,13). Two hypotheses can explain this apparent discrepancy. Somatostatin may cause undetectably small changes in basal cyclic AMP levels which are nonetheless biologically significant and therefore it may inhibit both basal and VIP-stimulated secretion by the same mechanism. Alternatively, somatostatin may activate two different mechanisms: inhibition of VIP-stimulated hormone release could be mediated by alterations in cyclic AMP levels, while inhibition of basal hormone release would occur independently of changes in cyclic AMP concentrations. To distinguish between these two hypotheses, we determined whether elevating intracellular cyclic AMP levels pharmacologically could completely overcome somatostatin inhibition of hormone secretion.

We used the cyclic AMP analog 8-(4-chlorophenylthio)-cyclic AMP [(Cl-ø-S)cAMP] to directly stimulate cyclic AMP-dependent protein kinases and thereby elicit cyclic AMP inducible effects at a site distal to the activation of adenylate cyclase. As expected, (Cl-ø-S)cAMP caused a dose-dependent stimulation of PRL release in GH_4C_1 cells (22, 23). Somatostatin did not significantly affect the potency of (Cl-ø-S)cAMP to increase secretion, but reduced the stimulation produced by maximum concentrations of this secretagogue (22,23). Thus, somatostatin can inhibit hormone secretion in the presence of maximally effective intracellular cyclic AMP levels. In this situation, the high (Cl-ø-S)cAMP concentrations would overwhelm any undetectably small reduction in cyclic AMP levels which somatostatin might produce in the absence of a stimulator of adenylate cyclase. Therefore, when maximal concentrations of a cyclic AMP analog are present, inhibition of hormone secretion by somatostatin must be independent of changes in cyclic AMP levels: i.e., somatostatin must have a cyclic AMP-independent mechanism of action. This conclusion is supported by similar experiments in which cyclic AMP levels were pharmacologically increased by another cyclic AMP analog, 8-Br-cAMP, and by forskolin (4,11).

In summary, our results show that somatostatin acts by at least two different mechanisms: inhibition of VIP-stimulated hormone secretion is largely due to inhibition of VIP-stimulated cyclic AMP accumulation, whereas inhibition of (Cl-ø-S)cAMP-stimulated hormone secretion, and probably also basal hormone secretion, are independent of changes in intracellular cyclic AMP levels.

Somatostatin Reduces Intracellular Cyclic AMP by Inhibiting Adenylate Cyclase

Somatostatin could decrease cyclic AMP accumulation either by stimulating cyclic AMP degradation or by inhibiting cyclic AMP synthesis. To identify the primary site of somatostatin action in intact GH_4C_1 cells, we determined the degree to which somatostatin inhibited VIP-stimulated cyclic AMP accumulation in the presence of varying concentrations of the phosphodiesterase inhibitor 3-isobutyl-1-methylxanthine (IBMX) (13). Although increasing IBMX concentrations markedly elevated cyclic AMP levels, both in the absence and presence of VIP, the degree to which somatostatin inhibited VIP stimulation was not altered (13). If somatostatin acts to stimulate phosphodiesterases, pharmacological inhibition of these enzymes should block reduction of cyclic AMP levels. Since somatostatin's effect was not reduced by IBMX, the peptide must decrease cyclic AMP concentrations primarily by inhibiting adenylate cyclase.

To investigate the mechanism by which somatostatin regulates adenylate cyclase, we examined the interaction between the somatostatin receptor and components of the adenylate cyclase system (21,24). In parallel with the lack of effect of somatostatin on basal cyclic AMP levels in cells, somatostatin did not significantly alter unstimulated adenylate cyclase activity in GH_4C_1 cell membranes (24). However, in both membranes and cells somatostatin decreased the maximal stimulatory effect of VIP, without changing its potency. Half-maximal inhibition of adenylate cyclase was produced by 2.3 nM somatostatin, which is close to the IC_{50} for somatostatin inhibition of cyclic AMP accumulation in intact cells (1.2 nM) (13,24). Furthermore, several somatostatin analogs were shown to have similar relative potencies to inhibit cyclic AMP production in cells and adenylate cyclase activity in membrane preparations (13,24). Thus, the effects of somatostatin on cyclic AMP accumulation in intact cells parallel its actions on adenylate cyclase activity in membranes.

Like other receptors which regulate adenylate cyclase activity (25), the somatostatin receptor is coupled to this enzyme by a guanine nucleotide binding protein. Thus, both GTP and the nonhydrolyzable GTP analog Gpp(NH)p, decreased the saturable binding of [^{125}I-Tyr^1]somatostatin to GH_4C_1 cell membranes, whereas the nonhydrolyzable ATP analog App(NH)p had no effect (24). GTP did not alter the number of somatostatin receptors in GH_4C_1 cell membranes; rather, it reduced the affinity of these sites for somatostatin (24). Furthermore, the decrease in binding affinity induced by guanine nucleotides could be explained by an increase in the rate constant for somatostatin dissociation (24). In fact, results from kinetic experiments indicated that somatostatin receptors exist in two interconvertible forms with different affinities for agonists. Guanine nucleotides regulate conversion of the receptor between a slowly dissociating, high affinity form and a rapidly dissociating, low affinity form (24). This behavior mimics that of beta-adrenergic receptors, which stimulate cyclase (26), and indicates that a guanine nucleotide binding protein mediates inhibition of adenylate cyclase by the somatostatin receptor in a manner analogous to the action of stimulatory receptors (24).

The guanine nucleotide binding protein which couples the somatostatin receptor to adenylate cyclase is inactivated by the Bordetella pertussis toxin, islet-activating protein (21). This toxin catalyzes the transfer of ADP-ribose from NAD to several membrane-associated guanine nucleotide binding proteins with molecular weights between 39,000 and 41,000 daltons (27-29). One of these proteins (M.W. = 41,000 daltons) is the alpha subunit of N_i, which has been proposed to mediate inhibition of adenylate cyclase. Covalent modification of N_i by pertussis toxin prevents inhibition of adenylate cyclase by agonists acting at several different receptors (27-29). In accord with results from other systems, pretreatment of GH_4C_1 cells with pertussis toxin completely blocked somatostatin inhibition of VIP-stimulated cyclic AMP accumulation (21). Furthermore, somatostatin inhibition of VIP-stimulated adenylate cyclase activity was reduced in membranes prepared from pertussis toxin pretreated cells (21). The only substrate for pertussis toxin catalyzed ADP-ribosylation in GH cell membranes appears to be a 41,000-Da protein (30,31, J.B. Fischer, E. Neer and A. Schonbrunn, unpublished observations). Thus, it is likely that inhibition of VIP-stimulated adenylate cyclase by the occupied somatostatin receptor is mediated by N_i.

Pertussis Toxin Blocks All Mechanisms of Somatostatin Action

Since pertussis toxin treatment prevented somatostatin inhibition of VIP-stimulated adenylate cyclase and cyclic AMP accumulation, we expected that it should also block those biological actions of somatostatin which result from this inhibition. As predicted, pertussis toxin pretreatment did prevent somatostatin inhibition of VIP-stimulated PRL release (21). However, pertussis toxin also blocked somatostatin inhibition of basal and cyclic AMP analog stimulated hormone secretion (21). In fact, somatostatin inhibition of TRH, bombesin and K^+ stimulated secretion was also prevented (21,31). The ability of pertussis toxin to block somatostatin inhibition of hormone secretion stimulated by a variety of pharmacological and physiological secretagogues which do not affect adenylate cyclase, suggests that somatostatin's cyclic AMP independent mechanism requires N_i or another pertussis toxin sensitive guanine nucleotide binding protein.

It is not known how many pertussis toxin sensitive GTP-binding proteins interact with the somatostatin receptor. Although pertussis toxin ADP-ribosylates only a single 41,000 dalton protein in GH_4C_1 cell membranes, it is possible that the toxin has multiple targets in intact cells. In fact, even though the alpha subunit of N_i is either the sole or the major protein modified by pertussis toxin in most membrane preparations, other GTP-binding proteins can be ADP-ribosylated by pertussis toxin (32-34). Furthermore, two different pertussis toxin substrates, N_i and N_o, have been shown to couple to the same receptor in reconstitution experiments (35). Unfortunately, there is no practical method for detecting those pertussis toxin substrates which are ADP-ribosylated only in whole cells. The assay used to determine whether a particular protein is modified during incubation of cells with toxin involves preparing membranes from control and toxin-treated cultures and then comparing the extent to which pertussis toxin can transfer [^{32}P]ADP-ribose from [^{32}P]NAD to different molecular weight proteins in the two membrane preparations (27-29). Therefore, the only _in vivo_ effects of the toxin which can be detected are those which prevent pertussis toxin catalyzed ADP-ribosylation in membranes.

To explore the possibility that the somatostatin receptor was coupled to more than one type of GTP-binding protein, we took advantage of the observation that different guanine nucleotide binding proteins are ADP-ribosylated with different efficiencies by pertussis toxin (32). Thus, we determined the concentration dependence for pertussis toxin blockage of somatostatin inhibition of VIP-stimulated as well as K^+-stimulated prolactin secretion (21). High extracellular K^+ stimulates hormone release by depolarizing the plasma membrane, opening voltage sensitive Ca^{2+} channels and increasing intracellular free Ca^{2+} concentrations ($[Ca^{2+}]_i$). Elevated extracellular K^+ does not affect intracellular cyclic AMP levels in GH_4C_1 cells either in the absence or presence of somatostatin (21). Pertussis toxin blocked somatostatin inhibition of both VIP and K^+ stimulated hormone release with the same potency (ED_{50} = 0.3 ng/ml) indicating that the toxin acts at a single, common site to prevent these two actions of somatostatin (21). These results suggest that N_i is involved in both the cyclic AMP dependent and the cyclic AMP independent mechanism by which somatostatin inhibits secretion.

Equilibrium binding studies with [^{125}I-Tyr^1]somatostatin have shown that GH_4C_1 cells contain a single class of high-affinity binding sites for somatostatin (10). Furthermore, the potencies of somatostatin and 19 somatostatin analogs to inhibit VIP-stimulated cyclic AMP production and basal PRL secretion correlated closely with their receptor binding affinities (17). These results indicate that the characterized binding sites represent the biologically relevant receptors that mediate the actions of somatostatin by both the cyclic AMP dependent and cyclic AMP independent mechanisms. Therefore, the simplest hypothesis which explains the currently available data is that somatostatin initiates both its mechanisms by binding to a single class of receptors which interact with a single type of pertussis toxin sensitive GTP-binding protein, probably N_i.

Somatostatin Causes Membrane Hyperpolarization and Decreases Intracellular Free Calcium

The biochemical events involved in somatostatin's inhibition of hormone secretion by the cyclic AMP independent mechanism have recently begun to be investigated. It is unlikely that changes in phosphatidylinositol turnover are involved since somatostatin does not affect basal or TRH stimulated $^{32}PO_4$ incorporation into phospholipids (16) or diacylglycerol production (31). However, regulation of the intracellular free calcium concentration has been implicated in somatostatin's biological actions (1-5). In GH_4C_1 cells, somatostatin decreases the $[Ca^{2+}]_i$ and causes membrane hyperpolarization within seconds (21,36). These ionic effects are observed both under basal conditions, where somatostatin does not alter intracellular cyclic AMP levels, and in the presence of maximal concentrations of cyclic AMP analogs or forskolin (22,37). Thus, somatostatin must produce the hyperpolarization and the reduction in $[Ca^{2+}]_i$ by a cyclic AMP independent mechanism.

Since pretreatment of GH_4C_1 cells with pertussis toxin prevented somatostatin inhibition of hormone secretion under all conditions, the ionic effects of somatostatin should also be inhibited by pertussis toxin if they mediate the peptide's cyclic AMP independent biological actions. In fact, pertussis toxin did block reduction of both basal and K^+-elevated $[Ca^{2+}]_i$ by somatostatin (21,38). In addition, the membrane hyperpolarization produced by somatostatin was also prevented by toxin pretreatment (38). These results indicate that N_i, or another related GTP-binding protein, is required for the somatostatin receptor to elicit both the decrease in $[Ca^{2+}]_i$ and membrane hyperpolarization. Furthermore, because the effects of somatostatin on the $[Ca^{2+}]_i$ and the membrane potential are independent of changes in cyclic AMP concentrations, they may provide the biophysical basis for the cyclic AMP independent mechanism by which somatostatin inhibits hormone secretion.

Somatostatin's Cyclic AMP Independent Actions Require a Transmembrane Potassium Ion Gradient

Since GH_4C_1 cells are electrically active (39) their membrane potential is controlled primarily by the concentration gradients and relative permeabilities of K^+, Na^+, Ca^{2+}, and Cl^-. Therefore, to determine whether the transmembrane movement of these ions is necessary for somatostatin to elicit either the hyperpolarization or the reduction in $[Ca^{2+}]_i$, we examined the integrity of these somatostatin responses when the concentration gradients of these ions were altered (22,23,37). Substitution of choline for all extracellular Na^+ had no effect on either the hyperpolarization or the decrease in $[Ca^{2+}]_i$ induced by somatostatin. Although decreasing the extracellular Cl^- concentration did inhibit the hyperpolarization and the decrease in $[Ca^{2+}]_i$ to some extent, it did not block these responses in a manner consistent with a requirement for a transmembrane Cl^- gradient (22,37). In contrast, decreasing extracellular Ca^{2+} from 1 mM to 250 nM abolished somatostatin's reduction of the $[Ca^{2+}]_i$, but did not block the hyperpolarization response. These results indicate that the somatostatin induced hyperpolarization is not primarily due to changes in the conductances of Na^+, Cl^-, or Ca^{2+}. However, the somatostatin induced decrease in $[Ca^{2+}]_i$ requires Ca^{2+} influx, although it is also independent of the Na^+ and Cl^- gradients.

Somatostatin's actions were most markedly affected by changes in the K^+ gradient. Elevating extracellular K^+ concentrations from 4.6 to 50 mM completely blocked both the somatostatin-induced hyperpolarization and the reduction in $[Ca^{2+}]_i$ (22,37). Furthermore, the magnitude of the hyperpolarization response decreased exponentially with increasing extracellular $[K^+]$, consistent with it being directly dependent upon the K^+ gradient. Together, these results support the conclusion that the primary cyclic AMP independent effect of somatostatin is to increase K^+ conductance, and that the effect on $[Ca^{2+}]_i$ is secondary to this change (22,37). Therefore, our studies indicate that the somatostatin induced hyperpolarization probably causes the decrease in $[Ca^{2+}]_i$ by reducing Ca^{2+} influx through the voltage dependent Ca^{2+} channels which exist in GH cells (39,40). Electrophysiological experiments support this conclusion. Using voltage-clamp techniques, Barker and Dufy found that somatostatin activates a transient outward current in GH cells, consistent with an effect to increase a K^+ conductance (41). Furthermore, somatostatin reduces spontaneous Ca^{2+} action potentials (41). It seems likely that the decrease in the frequency of Ca^{2+} action potentials decreases Ca^{2+} influx sufficiently to account for the fall in $[Ca^{2+}]_i$ produced by somatostatin. Thus, although the electrophysiological studies did not show whether the membrane effects of somatostatin were secondary to changes in cyclic AMP levels, they directly demonstrated that somatostatin regulates both an outward current and Ca^{2+} action potential frequency.

Since a K^+ gradient was required for the cyclic AMP independent effects of somatostatin on both the membrane potential and the $[Ca^{2+}]_i$ (22,37), we explored the importance of this ion in the peptide's cyclic AMP independent inhibition of hormone secretion (23). Somatostatin shifted the dose-response for K^+ stimulation of prolactin secretion to higher concentrations. However, extracellular K^+ concentrations of $\geq$ 50 mM completely reversed somatostatin inhibition (23). In contrast, somatostatin continued to inhibit VIP stimulated hormone release at extracellular K^+

concentrations above 50 mM. Therefore, somatostatin's cyclic AMP mediated mechanism does not require a K^+ gradient whereas its cyclic AMP independent inhibition of hormone release appears to result from a change in K^+ conductance.

SUMMARY

Somatostatin inhibits hormone secretion from GH cells by two mechanisms which appear to be triggered by a single class of receptors (Figure 1). One mechanism involves inhibition of stimulated adenylate cyclase activity. Inhibition of this enzyme by somatostatin reduces the cyclic AMP generated by VIP and other secretagogues which activate adenylate cyclase and consequently decreases stimulation of secretion. Somatostatin also inhibits hormone secretion by a cyclic AMP independent mechanism. In addition, somatostatin causes membrane hyperpolarization and reduces $[Ca^{2+}]_i$ independently of any changes in cyclic AMP levels. Our results indicate that an increase in K^+ conductance is the primary biophysical event responsible for somatostatin's cyclic AMP independent effects on the membrane potential, on the $[Ca^{2+}]_i$, and on hormone secretion. Interestingly, the somatostatin receptor appears to be coupled both to inhibition of adenylate cyclase and to an increase in K^+ conductance by a single pertussis toxin sensitive GTP-binding protein, probably N_i. The ability of the somatostatin receptor to simultaneously activate two separate transduction mechanisms, and thus reduce the concentrations of two intracellular messengers, helps to explain its powerful inhibitory activity.

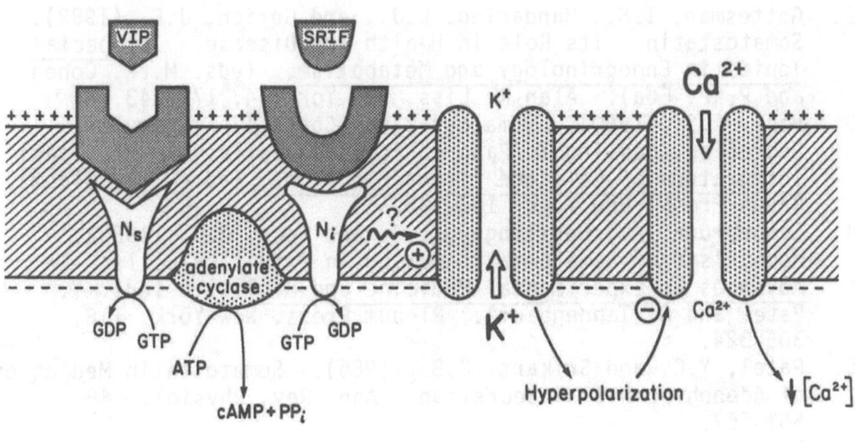

Figure 1. Signal transduction mechanisms activated by somatostatin.
This model shows the fewest components necessary to explain the two independent mechanisms by which somatostatin inhibits

hormone release in GH_4C_1 cells. The receptors for VIP and somatostatin (SRIF) interact with two different GTP-binding proteins: N_s and N_i. These are the guanine nucleotide regulatory subunits which couple adenylate cyclase to stimulatory and inhibitory receptors, respectively. Activation of the somatostatin receptor causes inhibition of VIP-stimulated adenylate cyclase as well as enhancement of a K^+-conductance. Both of these actions are mediated by N_i, or another pertussis toxin sensitive GTP-binding protein. The decrease in intracellular cyclic AMP levels resulting from inhibition of adenylate cyclase produces the cyclic AMP dependent actions of somatostatin. The increase in K^+ conductance results in hyperpolarization of the cell and therefore decreases Ca^{2+} influx through voltage sensitive channels. The resulting decrease in $[Ca^{2+}]_i$ is responsible for the cyclic AMP independent actions of somatostatin.

ACKNOWLEDGEMENTS

The experiments described in this review were supported by a research grant from the NIADDK (AM-32234). I thank Drs. Laneta Dorflinger and Bruce Koch for their many contributions to these studies and Rebecca Siebens for secretarial help.

REFERENCES

1. Reichlin, S. (1983). Somatostatin. New Eng. J. Med., 309, 1495-1501 and 1556-1563.
2. Gottesman, I.S., Mandarino, L.J., and Gerich, J.E. (1982). Somatostatin: Its Role in Health and Disease. In Special Topics in Endocrinology and Metabolism. (eds. M. P. Cohen and P. P. Foa). Alan R. Liss, New York, 4, 177-243.
3. Pace, C.S. (1980). Somatostatin: Control of Stimulus-Secretion Coupling in Pancreatic Islet Cells. In Peptides: Integrators of Cell and Tissue Function. (ed. F. E. Bloom). Raven Press, New York, 163-195.
4. Schonbrunn, A., Dorflinger, L.J. and Koch, B.D. (1985). Mechanisms of Somatostatin Action in Pituitary Cells. In Advances in Experimental Medicine and Biology. (eds. Y. Patel and G. Tannenbaum). Plenum Press, New York, 118, 305-324.
5. Patel, Y.C. and Srikant, C.B. (1986). Somatostatin Mediation of Adenohypophysial Secretion. Ann. Rev. Physiol., 48, 551-567.
6. Vale, W., Rivier, C., Brazeau, P. and Guillemin, R. (1974). Effects of Somatostatin on the Secretion of Thyrotropin and Prolactin. Endocrinology, 95, 968-977.
7. Drouin, J., De Lean, A., Rainville, D., Lachance, R. and Labrie, F. (1976). Characteristics of the Interaction between Thyrotropin-Releasing Hormone and Somatostatin for

Thyrotropin and Prolactin Release. Endocrinology, 98, 514-521.

8. Enjalbert, A., Epelbaum, J., Arancibia, S., Tapia-Arancibia, L., Blute-Pajot, M.-T., and Kordon, C. (1982). Reciprocal Interactions of Somatostatin with Thyrotropin-Releasing Hormone and Vasoactive Intestinal Peptide on Prolactin and Growth Hormone Secretion In Vitro. Endocrinology, 111, 42-47.
9. Cooper, G.R. and Shin, S.H. (1981). Somatostatin Inhibits Prolactin Secretion in the Estradiol Primed Male Rat. Can. J. Physiol., 59, 1082-1088.
10. Schonbrunn, A. and Tashjian Jr., A.H. (1978). Characterization of Functional Receptors for Somatostatin in Rat Pituitary Cells in Culture. J. Biol. Chem., 253, 6473-6483.
11. Dorflinger, L.J. and Schonbrunn, A. (1983). Somatostatin Inhibits Basal and Vasoactive Intestinal Peptide Stimulated Hormone Release by Different Mechanisms in GH Pituitary Cells. Endocrinology, 113, 1551-1558.
12. Westendorf, J.M. and Schonbrunn, A. (1982). Bombesin Stimulates Prolactin and Growth Hormone Release by Pituitary Cells in Culture. Endocrinology, 110, 352-358.
13. Dorflinger, L.J. and Schonbrunn, A. (1983). Somatostatin Inhibits Vasoactive Intestinal Peptide-Stimulated Cyclic Adenosine Monophosphate Accumulation in GH Pituitary Cells. Endocrinology, 113, 1541-1550.
14. Gershengorn, M.C. (1986). Mechanism of Thyrotropin Releasing Hormone Stimulation of Pituitary Hormone Secretion. Ann. Rev. Physiol., 48, 515-526.
15. Williams, J.A. (1984). Regulatory Mechanisms in Pancreas and Salivary Acini. Ann. Rev. Physiol., 46, 361-375.
16. Sutton, C.A. and Martin, T.F.J. (1982). Thyrotropin Releasing Hormone (TRH) Selectively and Rapidly Stimulates Phosphatidylinositol Turnover in GH Pituitary Cells: A Possible Second Step of TRH Action. Endocrinology, 110, 1273-1280.
17. Schonbrunn, A., Rorstad, O.P., Westendorf, J.M., and Martin, J.B. (1983). Somatostatin Analogs: Correlation Between Receptor Binding Affinity and Biological Potency in GH Pituitary Cells. Endocrinology, 113, 1559-1567.
18. Presky, D.H. and Schonbrunn, A. (1986). Receptor-Bound Somatostatin and Epidermal Growth Factor are Processed Differently in GH_4C_1 Rat Pituitary Cells. J. Cell Biol., 102, 878-888.
19. Dannies, P.S., Gautvik, K.M. and Tashjian Jr., A.H. (1976). A Possible Role of Cyclic AMP in Mediating the Effects of Thyrotropin-Releasing Hormone on Prolactin Release and on Prolactin and Growth Hormone Synthesis in Pituitary Cells in Culture. Endocrinology, 98, 1147-1159.
20. Dannies, P.S., and Tashjian Jr., A.H. (1980). Action of Cholera Toxin on Hormone Synthesis and Release in GH Cells: Evidence that Adenosine 3'5'-Monophosphate Does Not Mediate the Decrease in Growth Hormone Synthesis Caused by Thyrotropin-Releasing Hormone. Endocrinology, 106, 1532-1536.
21. Koch, B.D., Dorflinger, L.J. and Schonbrunn, A. (1985).

Pertussis Toxin Blocks Both Cyclic AMP-Mediated and Cyclic AMP-Independent Actions of Somatostatin: Evidence for Coupling of N_i to Decreases in Intracellular Free Calcium. J. Biol. Chem., 260, 13138-13145.

22. Koch, B.D. and Schonbrunn, A. (1986). A Transmembrane K^+ Gradient is Required for Somatostatin to Decrease Intracellular Free [Ca^{2+}] and Inhibit Hormone Release via a cAMP-Independent Mechanism. Proc. of the 16th Meeting of the Society for Neuroscience, p. 734.
23. Koch, B.D. and Schonbrunn, A. Characterization of the Cyclic AMP- Independent Actions of Somatostatin: A Transmembrane Potassium Gradient is Required for Somatostatin Inhibition of Hormone Secretion. Submitted.
24. Koch, B.D. and Schonbrunn, A. (1984). The Somatostatin Receptor is Directly Coupled to Adenylate Cyclase in GH_4C_1 Pituitary Cell Membranes. Endocrinology, 114, 1784-1790.
25. Gilman, A.G. (1984). G Proteins and Dual Control of Adenylate Cyclase. Cell, 36, 577-579.
26. De Lean, A., Stadel, J.M. and Lefkowitz, R.J. (1980). A Ternary Complex Model Explains the Agonist-Specific Binding Properties of the Adenylate Cyclase-Coupled B-adrenergic Receptor. J. Biol. Chem., 255, 7108-7117.
27. Ui, M., Katada, T., Murayama, T., Kurose, H., Yajima, M., Tamura, M., Nakamura, T. and Nogimori, K. (1984). Islet-Activating Protein, Pertussis Toxin: A Specific Uncoupler of Receptor-Mediated Inhibition of Adenylate Cyclase. In Advances in Cyclic Nucleotide and Protein Phosphorylation Research. (ed. P. Greengard). Raven Press, New York, 17, 145-151.
28. Hewlett, E.L., Cronin, M.J., Moss, J., Anderson, H., Myers, G.H. and Pearson, R.D. (1984). Pertussis Toxin: Lessons from Biological and Biochemical Effects in Different Cells. In Advances in Cyclic Nucleotide and Protein Phosphorylation Research. (ed. P. Greengard). Raven Press, New York, 17, 173-182.
29. Sekura, R.D. (1985). Pertussis Toxin: A Tool for Studying the Regulation of Adenylate Cyclase. Methods in Enzymology, 109, 558-566.
30. Wojcikiewicz, R.J.H., Dobson, P.R.M., Irons, L.Q., Robinson, A. and Brown, B.L. (1984). The Relationship between Pertussis-Toxin Induced ADP-Ribosylation of a Plasma-Membrane Protein and Reversal of Muscarinic Inhibition of Prolactin Secretion in GH3 Cells. Biochem. J., 224, 339-342.
31. Yajima, Y., Akita, Y. and Saito, T. (1986). Pertussis Toxin Blocks the Inhibitory Effects of Somatostatin on cAMP-Dependent Vasoactive Intestinal Peptide and cAMP-Independent Thyrotropin Releasing Hormone-Stimulated Prolactin Secretion in GH3 Cells. J. Biol. Chem., 261, 2684-2689.
32. Neer, E.J., Lok, J.M. and Wolf, L.G. (1984). Purification and Properties of the Inhibitory Guanine Nucleotide Regulatory Unit of Brain Adenylate Cyclase. J. Biol. Chem.,

259, 14222-14229.
33. Van Dop, C., Yamanaka, G., Steinberg, F., Sekura, R.D., Manclark, C.R., Stryer, L. and Bourne, H.R. (1984). ADP-Ribosylation of Transducin by Pertussis Toxin Blocks the Light-Stimulated Hydrolysis of GTP and cGMP in Retinal Photoreceptors. J. Biol. Chem., 259, 23-26.
34. Gierschik, P., Falloon, J., Milligan, G., Pines, M., Gallin, J.I. and Spiegel, A. (1986). Immunochemical Evidence for a Novel Pertussis Toxin Substrate in Human Neutrophils. J. Biol. Chem., 261, 8058-8062.
35. Florio, V.A. and Sternweis, P.C. (1985). Reconstitution of Resolved Muscarinic Cholinergic Receptors with Purified GTP-Binding Proteins. J. Biol. Chem., 260, 3477-3483.
36. Schlegel, W., Wuarin, F., Wollheim, C. B. and Zahnd, G.R. (1984). Somatostatin Lowers the Cytosolic Free $Ca2^{+}$ Concentration in Clonal Rat Pituitary Cells (GH3 Cells). Cell Calcium, 5, 223-236.
37. Koch, B.D. and Schonbrunn, A. Characterization of the Cyclic AMP-Independent Actions of Somatostatin. A Transmembrane Potassium Ion Gradient is Required for Somatostatin to Cause Hyperpolarization and Decrease Intracellular Free Calcium. Submitted.
38. Schlegel, W., Wuarin, F., Zbaren, C., Wollheim, C.B. and Zahnd, G.R. (1985). Pertussis Toxin Selectively Abolishes Hormone Induced Lowering of Cytosolic Calcium in GH_3 Cells. FEBS Lett., 189, 27-32.
39. Taraskevich, P.S. and Douglas, W.W. (1980). Electrical Behaviour in a Line of Anterior Pituitary Cells (GH Cells) and the Influence of the Hypothalamic Peptide, Thyrotrophin Releasing Factor. Neuroscience, 5, 421-431.
40. Dubinsky, J.M. and Oxford, G.S. (1984). Ionic Currents in Two Strains of Rat Anterior Pituitary Tumor Cells. J. Gen. Physiol., 83, 309-339.
41. Barker, J.L. and Dufy, B. (1985). Peptide and Amino Acid Electropharmacology of Cultured Mammalian Central Neurons and Clonal Pituitary Cells. Regul. Peptides Suppl., 4, 14-22.

24
Receptor Mediated Activation of Phospholipase A_2: Arachidonic Acid and its Metabolites as Second Messengers

Julius Axelrod, Ronald M. Burch and Carole L. Jelsema

GTP binding proteins, called G or N proteins, are involved in receptor-mediated signal transduction in cells. Certain GTP binding proteins, N_s or N_i, have been shown to cause the activation or inhibition, respectively, of adenylate cyclase (Rodbell, 1980). The transducing N proteins are heterotrimers consisting of α, β, and γ subunits (Gilman, 1984). When a ligand interacts with a membrane receptor, GDP, which is bound to the α subunit of the N protein, is replaced by GTP. This sets off a cascade of biochemical reactions in which the α subunit dissociates from the βγ subunit. The GTP-bound α subunit of N_s then interacts with the adenylate cyclase to enhance conversion of ATP to cyclic AMP. This series of reactions is terminated by the ability of the α subunit to function as a GTPase to hydrolyze the bound GTP to GDP. When this happens the α subunit reassociates with the βγ subunit to terminate the reaction. The actions of the $N_s\alpha$ subunit on adenylate cyclase can be prolonged by the addition of GTPγS, a GTP analogue that is only slowly hydrolyzed by GTPase. The stimulatory $N_s\alpha$ can also be directly activated by cholera toxin which acts by ADP ribosylating the $N_s\alpha$ subunit. This then stimulates adenylate cyclase to generate cyclic AMP. N_i, the GTP binding protein that inhibits adenylate cyclase, acts by the release of βγ subunit which then interacts with the α subunit of N_s to block the stimulatory signal (Gilman, 1984). Pertussis toxin prevents the inhibitory action of N_i by ADP ribosylating the α subunit (Ui, 1984).

In addition to interacting with adenylate cyclase, the GTP binding N proteins can generate other cellular signals. The N protein transducin is highly localized in the rod outer segments of the retina where it couples light activation of rhodopsin to stimulate cyclic GMP phosphodiesterase activity (Fung et al., 1981). N proteins also interact with phospholipases C, some of which are pertussis toxin sensitive (Nakamura and Ui, 1985; Ohta, et al., 1985) and others that are not (Burch, et al., 1986a) as will be described later. GTP binding N proteins also mediate somatostatin (Lewis et al., 1986), GABA and norepinephrine-induced inhibition of calcium channels (Holz et al., 1986) in several tissues. Muscarinic receptors are coupled to potassium channels in cardiac myocytes by a pertussis toxin sensitive N protein (Breitwieser et al., 1985). We have recently found that GTP-binding N proteins are linked to phospholipase A_2 in thyroid and retinal tissues to generate arachidonic acid and its metabolites as second messengers.

α_1-Adrenergic stimulation of arachidonic acid release, metabolism and cell mitogenesis in a rat thyroid cell line.

It has previously been shown that release and metabolism of arachidonic acid by cells are receptor-mediated (Hirata et al., 1979). The chemotatic peptide fMet-Leu-Phe causes the liberation of labelled arachidonic acid previously incorporated into phospholipids of neutrophils. Chemotaxis may then be effected by a leukotriene metabolite of arachidonic acid (Malmsten et al., 1980). α_1-Adrenergic stimulation also causes a release of arachidonic acid in kidney cells (Cooper and Malik, 1985), but no specific biological role has been previously ascribed to receptor-stimulated metabolism of arachidonic acid in that tissue. In our studies on the role of arachidonic acid in receptor mediated events we have used a thyroid cell line, FRTL5, that requires the addition of thyrotropin (TSH) for iodine uptake, release, organification and growth (Kohn et al., 1985). Incubation of these cells with TSH causes a 10-fold increase in α_1 receptors (Corda and Kohn, 1985). Norepinephrine via an α_1 receptor also stimulates iodine release and thyroglobulin organification in FRTL5 cells (Kohn et al., 1985). The effects of α_1 receptor stimulation on iodine release and organification are due to activation of phospholipase C with the formation of inositol-1,4,5-triphosphate (Philp and Grollman, 1986) and elevation of cytosolic calcium (Corda et al., 1985). Indomethacin, a cyclooxygenase inhibitor, blocks TSH-mediated cell proliferation in FRTL5 thyroid cells (Kohn et al., 1985).

In our studies with FRTL-5 cells on arachidonic acid metabolism these cells were first prelabelled with ^{3}H-arachidonic acid and then treated with norepinephrine (Burch et al., 1986b). This caused a rapid release of radioactivity. Phenylephrine, an α_1 agonist, also stimulated the release of radioactivity, while prazosin, an α_1 antagonist, blocked the norepinephrine-induced arachidonic acid release. β adrenergic and α_2 adrenergic agonists and antagonists were without effect.

The norepinephrine stimulated cells were examined for metabolites of arachidonic acid by high pressure liquid chromatography. The major metabolite was identified as prostaglandin E_2 and much smaller amounts of 11-hydroxy-5, 8, 11, 13 eicostatetraenoic acid,

15-hydroxy-5, 8, 11, 13-eicostatetraenoic acid, prostaglandin D_2 and thromboxane B_2 were identified. Formation of all of these metabolites after treating FRTL5 cells with norepinephrine were blocked by the cyclooxygenase inhibitor indomethacin (Burch et al., 1986b).

TSH is required for replication of FRTL-5 cells. Indomethacin partially blocks TSH-induced cell division (Kohn et al., 1985). These observations and those above led to an examination of whether α_1 adrenergic receptor stimulation can induce cell division by examining the uptake of ^{3}H-thymidine into DNA (Burch, et al., 1986b). After starving the cells of TSH, norepinephrine stimulated the uptake of ^{3}H thymidine almost 2-fold. Pharmacological studies showed that the increased uptake of ^{3}H-thymidine is mediated via an α_1-adrenergic receptor but not a β or α_2 receptor. To determine whether arachidonic acid metabolites formed in response to norepinephrine stimulation are involved in FRTL5 cell replication, the cells were first incubated with indomethacin to block the metabolism of the fatty acid. Indomethacin markedly inhibited the norepinephrine-induced ^{3}H-thymidine incorporation. Addition of exogenous PGE_2 at low concentrations reversed the inhibitory effects of indomethacin on ^{3}H-thymidine uptake when the cells were also treated with norepinephrine. None of the other identified metabolites were able to abolish the effect of indomethacin on norepinephrine-induced ^{3}H-thymidine uptake. The endogenous PGE_2 released by norepinephrine corresponded to concentrations of added PGE_2 that stimulated ^{3}H-thymidine incorporation in indomethacin-treated cells after giving norepinephrine. These observations indicated that stimulation of α_1 adrenergic receptors leads to release of arachidonic acid and the formation of the cyclooxygenase metabolite PGE_2. This metabolite by a still unknown mechanism causes DNA replication in FRTL5 thyroid cells.

<u>Phospholipase A_2 linked to GTP binding protein via α_1-adrenergic receptor in thyroid</u>.

The work described above indicated that α_1 adrenergic receptor agonists liberate arachidonic acid and provided a good experimental model to study the transduction mechanism for the activation of lipase involved in the release of this fatty acid. The initial experiment was devised to determine whether a phospholipase could be activated by a GTP binding protein (Burch et al., 1986a). For this, GTPγS, a hydrolysis resistant analogue of GTP which binds to and activates N proteins was used. GTPγS when added to intact FRTL5 cells had no effect on the release of prelabelled ^{3}H-arachidonic acid. After the cells were transiently permeabilized by hypoosmotic treatment in the presence of GTPγS, ^{3}H-arachidonic acid release was increased 2-fold. GTPγS when introduced into FRTL5 cells also stimulated ^{3}H-inositol phosphate, a measure of phospholipase C activity, release several fold. These findings indicated that a GTP binding protein is linked to the α_1

receptor and phospholipase C or phospholipase A_2. Arachidonic acid could be released by a phospholipase C mechanism via a metabolite, diacylglycerol, and the enzyme diacylglycerol lipase. The fatty acid could also be liberated by the action of phospholipase A_2 on phosphatidylcholine or phosphatidylethanolamine. To determine whether the GTPγS stimulation of arachidonic acid release was due to activation to phospholipase A_2, the effect of GTPγS on arachidonic acid liberation was examined after inhibition of phospholipase C (Burch et al., 1986a). Neomycin, which inhibits phospholipase C by binding phosphatidylinositols (Cockcroft and Gomperts, 1985), reduced GTPγS-stimulated ^3H-inositol phosphate formation to near basal levels. However, neomycin did not block the GTPγS-stimulated ^3H-arachidonic acid release. Additional evidence that arachidonic acid release was not mediated by the phospholipase C pathway, diglyceride lipase was inhibited. Inhibition of diglyceride lipase had no effect on the liberation of ^3H-arachidonic acid (Burch et al., 1986a). Membranes of FRTL5 cells were incubated with exogenous phosphatidylcholine labelled with ^3H-arachidonic acid in the 2-position. GTPγS stimulated the release of ^3H-arachidonic acid from these membranes. When membranes were incubated with GTPγS and norepinephrine ^3H arachidonic acid was released in a synergistic manner (Burch, et al., 1986a).

To further establish that phospholipase A_2 is associated with a GTP binding protein, the FRTL5 cells were treated with pertussis toxin. Pertussis toxin catalyzes the ADP ribosylation of certain N proteins reducing their ability to couple receptors to effector proteins (Ui, 1984). Treatment of FRTL5 cells with pertussis toxin blocked the GTPγS stimulated arachidonic acid release. The toxin had no effect on the release of ^3H-inositol phosphate by GTPγS. However, when FRTL5 cells were incubated with pertussis toxin norepinephrine no longer could stimulate the release of ^3H-arachidonic acid. All of these experiments indicated that phospholipase A_2 and phospholipase C are activated by different GTP binding proteins in response to α_1 adrenergic stimulation in FRTL5 thyroid cells.

Phospholipase A_2 has a requirement for calcium and the activation of this enzyme is probably the consequence of the elevation of calcium by receptor agonist. The levels of calcium available in cells (nM) after receptor activation are below that usually found necessary for phospholipase A_2 activation in vitro (mM). The addition of the calcium chelating agent EGTA to FRTL5 cells blocked the norepinephrine stimulated release of ^3H-arachidonic acid. The calcium necessary for the activation of phospholipase A_2 mediated release of ^3H-arachidonic acid could be provided extracellularly or from an intracellular source. Nifedipine, a compound that inhibits the extracellular uptake of calcium into cells through voltage dependent calcium channels, had no effect on norepinephrine stimulated ^3H arachidonic acid release (Burch, et al, 1986a). However, TMB-8, which antagonizes the mobilization of intracellular calcium, prevented the release of ^3H-

arachidonic acid evoked by norepinephrine. Neither the addition of EGTA nor TMB-8 affected the liberation of ^{3}H-inositol induced by norepinephrine. It thus appears that an intracellular source of calcium is necessary for the activation of phospholipase A_2 and the generation of arachidonic acid. It is generally believed that the intracellular source of calcium is initiated by inositol triphosphate, a metabolite of inositol phosphate generated by phospholipase C (Berridge and Irvine, 1984). As has been demonstrated above, phospholipase A_2 can be stimulated in the absence of phospholipase C activity. Where does the calcium necessary for the activation of phospholipase A_2 arise? It has been shown that the activation of an N protein markedly lowers the calcium requirement of phospholipase C to a level similar to the resting free calcium (Smith et al., 1986). It has also been found that arachidonic acid itself releases calcium from intracellular stores independent of inositol phosphate formation (Wolf et al., 1986). It is possible that the activation of the GTP binding protein linked to phospholipase A_2 can lower the requirement for calcium when the α_1 adrenergic receptor is activated by norepinephrine. The lowered requirement for calcium by phospholipase A_2 would cause the release of arachidonic acid which would then generate additional calcium to further activate phospholipase A_2.

Stimulation of an α_1 receptor in other tissues has been shown to activate protein kinase C which catalyzes the phosphorylation of the receptor (Leeb-Lundberg et al., 1985). Phosphorylation then desensitizes the receptor. The effect of activating protein kinase C on the α_1 release of arachidonic acid by norepinephrine was examined by activating the kinase with either phorbol ester or 1-oleoyl-2-acetyl-*rac*-glycerol. Both of these compounds blocked the norepinephrine-stimulated arachidonic acid release (Burch, et al., 1986a).

A model showing that phospholipase A_2 and phospholipase C are activated by two separate GTP binding proteins in response to α_1 adrenergic stimulation in FRTL5 cells is shown in Fig. 1. Upon the addition of norepinephrine the GTP binding proteins N_A for phospholipase A_2 and N_C for phospholipase C are activated. Whether a single receptor is coupled to both N proteins or whether subpopulations of receptor are involved that are linked to one or the other N proteins remains to be determined. N_A via one of its subunits interacts with phospholipase A_2 to release arachidonic acid as a second messenger. The arachidonic is metabolized by cyclooxygenase to prostaglandin E_2 which together with norepinephrine stimulated DNA synthesis and thyroid growth. Intracellular calcium which activates phospholipase A_2 is generated by arachidonic acid by a pathway which is independent of phospholipase C and inositol phosphate. N_C activates phospholipase C to form inositol triphosphate (IP_3) and diacylglycerol (DAG). IP_3 increases intracellular calcium formation to activate phospholipase C and increases iodine flux which in turn iodinates thyroglobulin. Diacylglycerol activates protein kinase C which then inhibits the α_1 receptor or N protein linked to phospholipase A_2 and C.

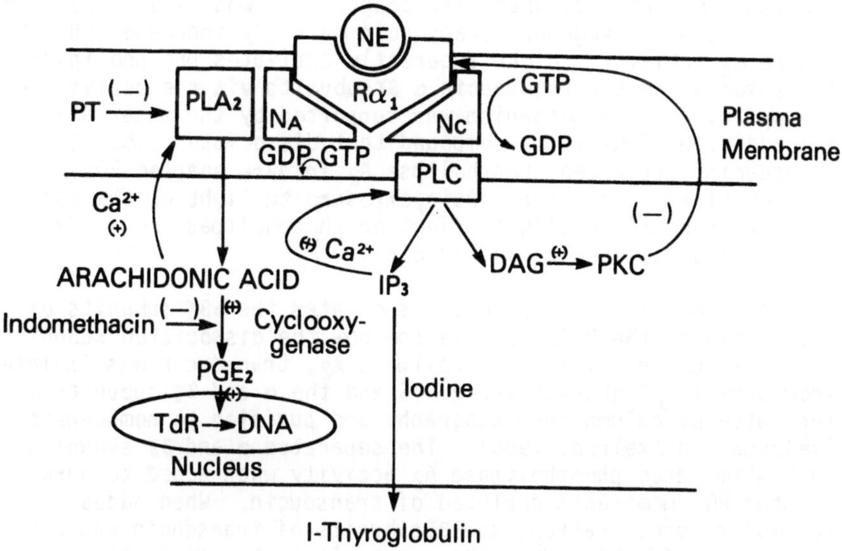

Fig. 1. Biochemical cascade induced by norepinephrine binding to α_1 adrenergic receptors in FRTL5 cells. PT is pertussis toxin, PKC is protein kinase C, PLA_2 is phospholipase A_2, PLC is phospholipase C, DAG is diacylglycerol, NE is norepinephrine, $R\alpha_1$ is α_1 adrenergic receptor.

GTP binding proteins, phospholipase A_2 and arachidonic acid release in the visual system

Another cell system that appears to be linked to phospholipase A_2 and C via GTP binding proteins is the rod outer segment (ROS) of the retina. Transducin, the major N protein of the retina is highly localized in the rod outer segment (Fung, et al., 1981). Like other N proteins transducin is a heterotrimer consisting of α, β and subunits (Fig. 2). When light activates rhodopsin transducin is dissociated into its subunits and cyclic GMP phosphodiesterase is activated (Fung et al., 1981). This enzyme reduces the cyclic GMP level in the ROS thus closing the sodium channels which then sends a signal to the visual areas of the brain. Light has also been shown to activate phospholipases (Jelsema, 1985; Ghalayani, 1984). This prompted a study on the role of transducin in the light activation of phospholipase A_2 (Jelsema, 1986).

Phospholipase A_2 activity was determined by measuring the liberation of ^{14}C-arachidonic acid from 1-palmitoyl-2-[^{14}C arachidonyl] phosphatidylcholine incubated with bovine ROS (Jelsema, in press). There was a marked stimulation of phospholipase A_2 activity in dark adapted ROS when exposed to light. Removal of transducin from the ROS by washing, considerably reduced the phospholipase A_2 activity when dark-adapted ROS was exposed to light. The addition of exogenous transducin markedly increased phospholipase A_2 activity. Light apparently activates phospholipase A_2 by dissociating the transducin α βγ subunits via the activation of rhodopsin. Such a mechanism was supported by the observation that the addition of GTPγS, a compound that dissociates subunits of transducin, activated phospholipse A_2 in dark adapted ROS. In the absence of exogenous transducin, exposure to light or the addition of GTPγS had only a slight effect on phospholipase A_2 in dark adapted ROS depleted of transducin.

As shown in Fig. 2, light dissociates the αβγ subunits of transducin of the ROS. To examine how the dissociated subunits of transducin activate phospholipase A_2, transducin was isolated from purified, light-activated ROS and the α and βγ subunits were separated by column chromatography and purified to homogeneity (Jelsema and Axelrod, 1986). The separated α and βγ subunits each stimulated phospholipase A_2 activity when added to dark adapted ROS membranes depleted of transducin. When added in equimolar concentrations the βγ subunits of transducin was 3 to 5 times more effective than the α subunit in the activation of phospholipase A_2. The addition of α subunit together with the βγ subunit abolished the stimulation of phospholipase A_2. This inhibition is probably due to a removal of the stimulatory βγ subunit by reassociation of the α βγ subunits to a heterotrimer complex (Fig. 2). Additional evidence for such a mechanism is the observation that GTPγS which prevents the reassociation in the dark also prevented the inhibitory action of α subunit on the stimulatory action of the βγ subunit on phospholipase A_2. Pretreatment of transducin by pertussis toxin which stabilizes the N protein to form a heterotrimer complex also inhibited the activation of phospholipase A_2 by exogenously added transducin in the presence of light (Jelsema and Axelrod, 1986).

These findings demonstrate a novel role for the α βγ subunits of transducin. The βγ subunits of transducin are linked to phospholipase A_2 in a stimulatory manner. This is in contrast to the inhibitory role of βγ subunits in the adenylate cyclase system (Gilman, 1984). The α subunit which has previously been found to stimulate cyclic GMP phosphodiesterase can also prevent the stimulatory effects of βγ subunit on phospholipase A_2.

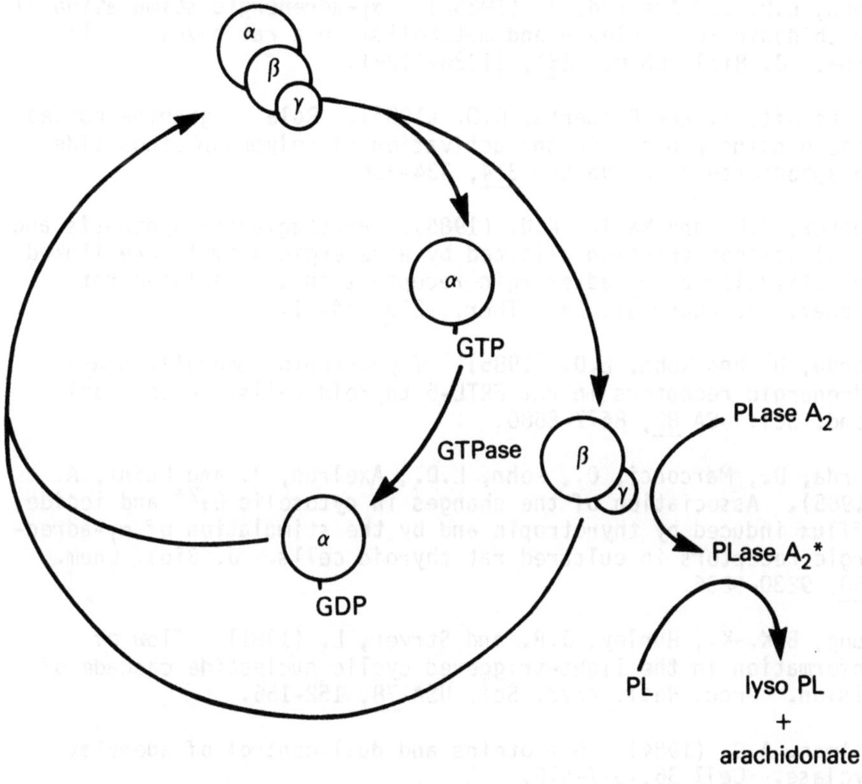

Fig. 2. The activation of phospholipase A_2 by βγ subunits of transducin after exposure of bovine rod outer segments to light.

References

Berridge, M.J. and Irvine, R.F. (1985). Inositol triphosphate, a novel second messenger in cellular signal transduction. Nature 312, 334-336.

Breitwieser, G.F. and Szabo, G. (1985). Uncoupling of cardiac muscarinic and β-adrenergic receptors from ion channels by a guanine nucleotide analogue. Nature 317, 538-540.

Burch, R.M., Luini, A. and Axelrod, J. (1986a). Phospholipase A_2 and phospholipase C are activated by distinct GTP-binding proteins in response to α_1-adrenergic stimulation in FRTL5 thyroid cell. Proc. Natl. Acad. Sci. USA, in press.

Burch, R.M., Luini, A., Mais, D.E., Corda, D., Vanderhoek, J.Y., Kohn, L.D. and Axelrod, J. (1986b). α_1-adrenergic stimulation of arachidonic acid release and metabolism in a rat thyroid cell line. J. Biol. Chem., 261, 11236-11241.

Cockcroft, S. and Gomperts, B.D. (1985). Role of guanine nucleotide binding protein in the activation of polyphosphoinositide phosphodiesterase. Nature 314, 334-336.

Cooper, C.L. and Malik, K.U. (1985). Prostaglandin synthesis and renal vasoconstriction elicited by adrenergic stimuli are linked to activation of α_1-adrenergic receptors in the isolated rat kidney. J. Pharmcol. Exp. Ther., 233, 24-31.

Corda, D. and Kohn, L.D. (1985). Thyrotropin upregulates α_1-adrenergic receptors in rat FRTL-5 thyroid cells. Proc. Natl. Acad. Sci. USA 82, 8677-8680.

Corda, D., Marcocci, C., Kohn, L.D., Axelrod, J. and Luini, A. (1985). Association of the changes in cytosolic Ca^{2+} and iodide efflux induced by thyrotropin and by the stimulation of α_1-adrenergic receptors in cultured rat thyroid cells. J. Biol. Chem., 260, 9230-9236.

Fung, B.K.-K., Hurley, J.B. and Stryer, L. (1981). Flow of information in the light-triggered cyclic nucleotide cascade of vision. Proc. Natl. Acad. Sci. USA 78, 152-156.

Gilman, A.G. (1984). G proteins and dual control of adenylate cyclase. Cell 36, 577-579.

Hirata, F., Corcoran, B.A., Venkatasubramanian, K., Schiffmann, E. and Axelrod, J. (1979). Chemoattractants stimulate degeneration of methylated phospholipids and release of arachidonic acid in rabbit leukocytes. Proc. Natl. Acad. Sci. USA 76, 2640-2643.

Holz, G.G., Rane, S.G. and Dunlap, K. (1986). GTP-binding proteins mediate transmitter inhibition of voltage-dependent calcium channels. Nature 319, 670-672.

Jelsema, C.L. Light activation of phospholipase A_2 in rod outer segments of bovine retina and its modulation by GTP-binding proteins. J. Biol. Chem., in press.

Jelsema, C.L. and Axelrod, J. (1986). The $\beta\gamma$ subunits of GTP binding protein activate and the α subunit inhibits phospholipase A_2. Cell Biol. Abstract.

Kohn, L.D., Aloj, S.M., Tombaccini, D., Rotella, C.M., Toccafondi, R., Marcocci, C., Corda, D. and Grollman, E.F. (1985). The Thyrotropin Receptors. In Biochemical Actions of Hormones. (ed. G. Litwack). Academic Press, NY. 12, 457-512.

Leeb-Lundberg, L.M.F., Cotcchia, J., Lomasney, J.N., DeBernandis, J.F., Lefkowitz, R.J. and Caron, M.B. (1985). Phorbol esters promote α_1-adrenergic receptor phosphorylation and receptor uncoupling from inositol phospholipid metabolism. Proc. Natl. Acad. Sci. USA 82, 5651-5656.

Lewis, D.L., Weight, F.F. and Luini, A. A guanine nucleotide-binding protein mediates the inhibition of voltage-dependent calcium current by somatostatin in a pituitary cell line. Proc. Natl. Acad. Sci. USA, in press.

Malmsten, C.L., Palmblad, J., Uden, M.-M., Radmark, O., Engstedt, L. and Samuelsson, B. (1980). Leukotriene B_4: a highly potent and stereospecific factor stimulating migration of polymorphonuclear leukocytes. Acta Physiol. Scand., 110, 449.

Nakamura, T. and Ui, M. (1985). Simultaneous inhibitions of inositol phospholipid breakdown, arachidonic acid release, and histamine secretion in mast cells by islet-activating protein, pertussis toxin. J. Biol. Chem., 260, 3584-3593.

Ohta, H., Okajima, F. and Ui, M. (1985). Inhibition by islet-activating protein of a chemotactic peptide-induced early breakdown of inositol phospholipids and Ca^{2+} mobilization in guinea pig neutrophils. J. Biol. Chem., 260, 15771-15780.

Philp, N.J. and Grollman, E.F. (1986). Stimulation of inositol phosphate formation in cultured rat thyroid cells. FEBS Lett., 202, 193-196.

Rodbell, M. (1980). The role of hormone receptors and GTP-regulatory proteins in membrane transduction. Nature 6, 17-22.

Smith, C.D., Cox, C.C. and Snyderman, R. (1986). Receptor-coupled activation of phosphoinositide-specific phospholipase C by an N protein. Science 232, 97-100.

Ui, M. (1984). Islet-activating protein, pertussis toxin: a probe for functions of the inhibitory guanine nucleotide regulatory component of adenylate cyclase. Tr. Pharmacol. Sci., 5, 277-279.

Wolf, B.A., Turk, J., Sherman, W.R. and McDaniel, M.L. (1986). Intracellular Ca^{2+} mobilization by arachidonic acid. J. Biol. Chem., 261, 3501-3511.

25
Interactions between the Modulator Adenosine and Classical Transmitters

Bertil B. Fredholm, Christer Nordstedt, Ingeborg van der Ploeg, Eva Lindgren, Janet Ng and Mikael Jondal

INTRODUCTION

There is continuous formation of adenosine in all cells. Methyltransferase reactions lead to the formation of S-adenosylhomocysteine, which is rapidly broken down to form adenosine. The rate of adenosine formation is markedly enhanced by any procedure that leads to a dissociation between the rate of ATP-production and the rate of ATP-utilization. Thus, adenosine levels rapidly increase in tissues whenever there is a threatening depletion of high-energy phosphate compounds from any of the cells in the tissue. Adenosine may be thought of as a signal of the metabolic state of a tissue or a group of cells.

The effects of adenosine are exerted over two types of receptors, designated A_1 and A_2 (see Fredholm 1982). The A_1-receptors are responsible for several effects including inhibition of lipolysis (Londos et al., 1980), inhibition of transmitter release (Fredholm et al., 1983) and inhibition of evoked potentials in e.g. the hippocampus (Dunwiddie and Fredholm, 1984). An inhibition by A_1-receptors of cyclic AMP accumulation can be demonstrated, but it seems certain that not all the effects that are mediated via A_1-receptors do in fact depend on decreases in cyclic AMP-accumulation.

The A_2-receptors are linked via a N_s-protein to stimulation of adenylate cyclase in most cells (Londos et al., 1980). It is possible that there are effects of adenosine, mediated via A_2-receptors that are not effected via cyclic AMP, but the evidence is very weak.

Even though there is some evidence that purines could be transmitters in some circumstances the major role of adenosine is that of a modulator. In the nervous system, the modulation may be pre- and postsynaptic. In other words, adenosine could modulate the release of other transmitters and it could modulate the effect of these other signals. In the present communication, we will focus on the latter possibility.

In slices from the rat hippocampus adenosine and adenosine analogues are able to stimulate the formation of cyclic AMP. This is mediated via adenosine receptors of the A_2-subtype (Fredholm et al., 1982). Provided that the rate of cyclic AMP production is stimulated by forskolin, it is possible to demonstrate also adenosine receptors of the A_1-subtype that are linked to inhibition of adenylate cyclase (Fredholm et al., 1986b). The two types of receptors seem to coexist - perhaps on the same cells - since blockade of A_1-effects enhanced the A_2-receptor mediated cyclic AMP increase (Fredholm et al., 1985) and conversely blockade of A_2-effects revealed an A_1-mediated inhibition by adenosine itself and 2-chloroadenosine, which were not otherwise manifested (Fredholm et al., 1986b).

It has been known for some time that α-receptor stimulating drugs can increase the level of cyclic AMP in brain slices provided that endogenous adenosine and/or prostaglandins are available (cf. Daly 1977). Apparently the α-agonist (or an histamine H_1-agonist) can stimulate cyclic AMP formation if and only if there is a basal rate of receptor-activated cyclic AMP formation. Since α-agonists (and H_1-agonists) are known to activate the breakdown of phosphatidylinostol-4,5-diphosphate to form inositol 1,4,5-trisphosphate and diacylglycerol (Hollingsworth and Daly, 1985), and since these products are known to activate proteinkinase C, we decided to investigate if protein kinase C could be important in the regulation of cyclic AMP accumulation by adenosine analogues.

MATERIALS AND METHODS

The majority of the experiments reported here were performed on slices of the rat hippocampus. Male Sprague-Dawley rats were killed, the hippocampus rapidly dissected out and cut into 400 μm thick slices with a MacIlwain chopper. They were incubated in Krebs Ringer bicarbonate buffer of the following composition (in mM): NaCl (118), KCl (4.85), $MgSO_4$ (1.15), KH_2PO_4 (1.15), $CaCl_2$ (2.5), $NaHCO_3$ (25), glucose (11.1). For the determination of cyclic AMP accumulation the slices were incubated with [^{3}H]-adenine to label the ATP-stores (Fredholm et al., 1982).

NEM-treatment

Some slices were treated with 30 or 100 μmol/l NEM for a total of 15 min after the first 60 min preincubation period. Thereafter the slices were washed and labelled as described above. In a few cases the slices were treated with NEM in the presence of 100-150 μmol/l dithiothreitol.

Pertussis toxin stimulated [^{32}P]-ADP ribosylation

After incubation with phorbolester or NEM, the slices were homogenized in 20 nM Tris/HCl pH 7.5 containing 2 mM EDTA, 0.5 mM EGTA and 2 nM phenylmethylsulfonylfluoride. The pellet after centrifugation was incubated with 1 μM [^{32}P]-NAD and 100 μM GTP and 40 μg/ml pertussis toxin and subjected to gel electrophoresis as described (Laemmli, 1970).

Experiments on Jurkat cells

The Jurkat cells (a human T-cell leukemia line) was maintained in RPMI 1640 medium supplemented with glutamine, calf serum and antibiotics. The cells were resuspended in RPMI 1640 at a concentration of 1 miljon cells /ml. The incubation with drugs was carried out in the presence of a phosphodiesterase inhibitor rolipram (0.03 mmol/l) for a period of 10 min. Hypotheses are tested by analysis of variance procedures and a probability level of 5% is considered significant.

Drugs

The following drugs and reagents were used: 1-[7,8-^{3}H]-noradrenaline (36 Ci/mmol, 1 mCi (37 MBq)/ml) and [2-^{3}H]-adenine (25 Ci/mmol) and [8-^{3}H]-adenosine 3',5'-monophosphate, ammonium salt (28 Ci/mmol) were obtained from the Radiochemical Centre, Amersham, UK. [^{32}P]-NAD (specific activity 30-100 Ci/mmol) was from New England Nuclear, Boston, Mass., USA. Ascorbic acid, EGTA, adenosine, adenosine 3',5'-monophosphate, N-ethylmaleimide and dithiothreitol were obtained from Sigma, St. Louis, Mo., USA. Forskolin and 8-phenyltheophylline were obtained from Calbiochem-Behring, LaJolla, Ca., USA. Pertussis toxin was from List Laboratories, Campbell, Calif., USA. The two diastereoisomers of PIA, NECA and GTP were obtained from Boehringer, Mannheim, BRG, Rolipram (ZK 62,711, 4-(3-cyclopentyloxy-methoxyphenyl)-2-pyrrolidone was a gift from Schering AG, Berlin, BRG. Phytohemagglutinin (PHA) weas from Wellcome laboratories. 12-O-tetradecanoyl-phorbol-13-acetate (TPA) , 12,13-phorboldibutyrate (PDiBu) and 4 α-phorbol-12,13-didecanoate were from BioZac, Stockholm.

RESULTS

The effect of phorbolesters on cyclic AMP accumulation induced by endogenous or exogenous adenosine receptor agonists

The tumorpromoting phorbol ester phorbol 12,13-dibutyrate (PDiBu) had a slight stimulatory effect on basal cyclic AMP accumulation. This effect was blocked -almost completely - by adding the adenosine receptor antagonist 8-p-sulphophenyl theophylline (Nordstedt and Fredholm, 1986). This clearly suggests that the effect of endogenous adenosine on cyclic AMP accumulation is enhanced by the phorbol ester.

The cyclic AMP accumulation induced by the synthetic adenosine receptor agonist NECA was potently enhanced by phorbol esters (Nordstedt and Fredholm, 1986). The effect of the phorbol ester was concentration-dependent and the potency increased with time of incubation with the slices - as one would expect for a very lipophilic molecule slowly equilibrating with the extracellular fluid in a brain slice that is largely composed of lipid material. A threshold effect of PDiBu was observed at 10 nmol/l.

As seen in Fig. 1 the magnitude of the enhancement was substantial. Thus the effect of 1 μmol NECA/ml was increased more than 4-fold by PDiBu (0.1 μmol/l). By contrast, a maximally effective concentration of forskolin (that stimulates the cyclic AMP accumulation about equally as NECA). Forskolin in this concentration apparently acts directly on the cyclase molecule, whereas NECA has to involve three molecules, the adenosine receptor, the N_s-protein complex and the receptor in order to increase the accumulation of cyclic AMP. Interestingly, a lower concentration of forskolin that presumably acts - at least in part - by enhancing the effects of N_s was stimulated by PDiBu to an intermediate extent. This suggests that the effect of the phorbol ester may involve an action on the coupling between the receptor and the effector.

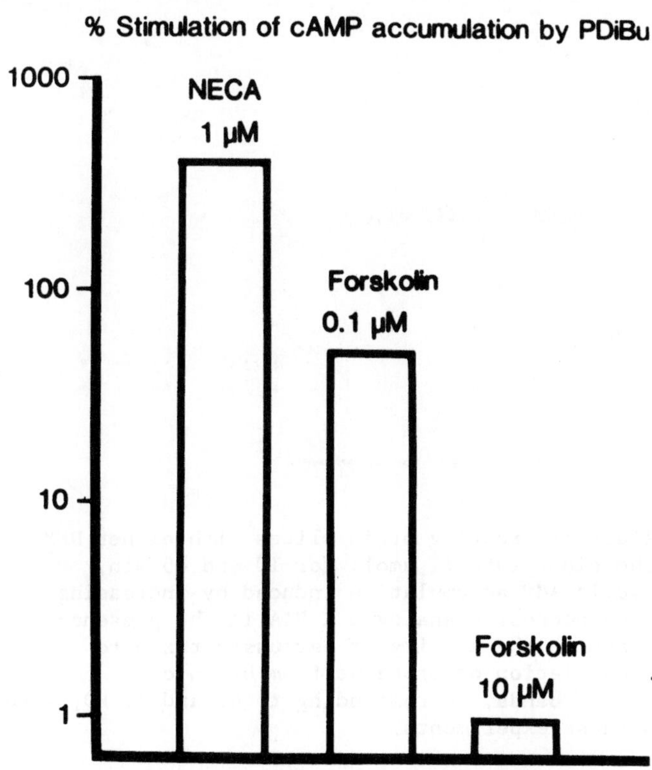

Figure 1. The degree of stimulation by PDiBu of cyclic AMP accumulation induced by NECA and by a low and a high concentration of forskolin. The results are set in relation to the factors involved in producing the increase in cyclic AMP by the different agents. Note that the ordinate is a logarithmic scale!

Mechanism by which phorbol esters stimulate adenosine receptor mediated cyclic AMP accumulation

In a recent paper Katada and co-workers (1985) found that activation of protein kinase C can lead to a phosphorylation of the N_i α-subunit, which is thereby inactivated. This is a possible explanation for the interactions seen in the present experiments. We therefore wanted to examine if the effect of the phorbolester treatment was similar to that of inactivating N_i by other means. Since it is not possible to use pertussis toxin in the slices (unpublished observations), we resorted to the use of low concentrations of N-ethylmaleimide (NEM). NEM has been found to selectively antagonize inhibition of cyclic AMP accumulation in the slices without effecting stimulation of it (Fredholm et al., 1985) and also to inhibit the A_1-receptor mediated effects on transmitter overflow (Fredholm et al., 1986 a, Fredholm and Lindgren 1986). The results of these experiments are summarized in Fig. 2.

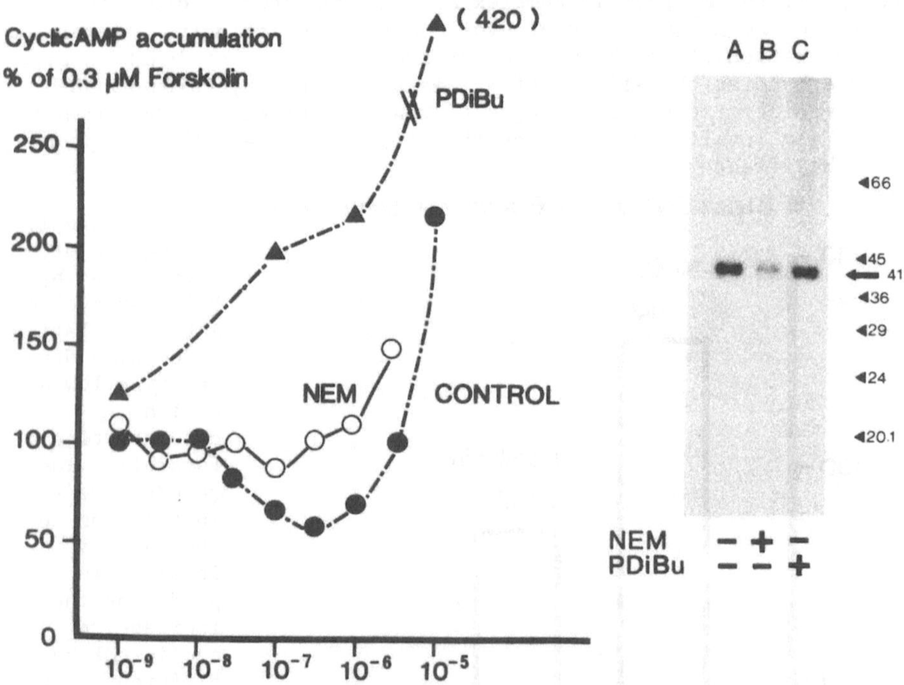

Figure 2. Effect of treating brain slices with either NEM (100 µmol) or phorboldibutyrate (1 µmol) for 10 and 45 min, respectively, on cyclic AMP accumulation induced by increasing concentrations of the adenosine analogue R-PIA in the presence of 0.3 µmol forskolin and on the ability of pertussis toxin to catalyze the ADP-ribosylation of proteins from brain cell membranes. At least two bands, corresponding to 41 and 39 kD, were ADP-ribosylated in these experiments.

It is seen that phorbol dibutyrate treatment shared with NEM the ability to abolish the inhibitory phase of R-PIA action on cyclic AMP formation, and to cause an enhancement of the stimulatory phase. Phorboldibutyrate appeared much more effective than NEM, suggesting that it might have another mechanism of action. Whereas NEM was able to reduce the ATP-ribosylation induced by pertussis toxin, this was not possible with the phorbolester.

We therefore investigated if the phorbolester was able to stimulate cyclic AMP accumulation also in slices treated with NEM. The results presented in Fig. 3 show that both NECA and the β-adrenoceptor agonist isoprenaline produce a larger raise in cyclic AMP in phorboldibuterate treated slices than in slices without phorboldibuterate despite the fact that the N_i-protein was inactivated with NEM. These results suggest that the mechanism behind potentiation is not entirely due to an inactivation of the N_i-subunit. Further studies are required to examine what the exact mechanism is.

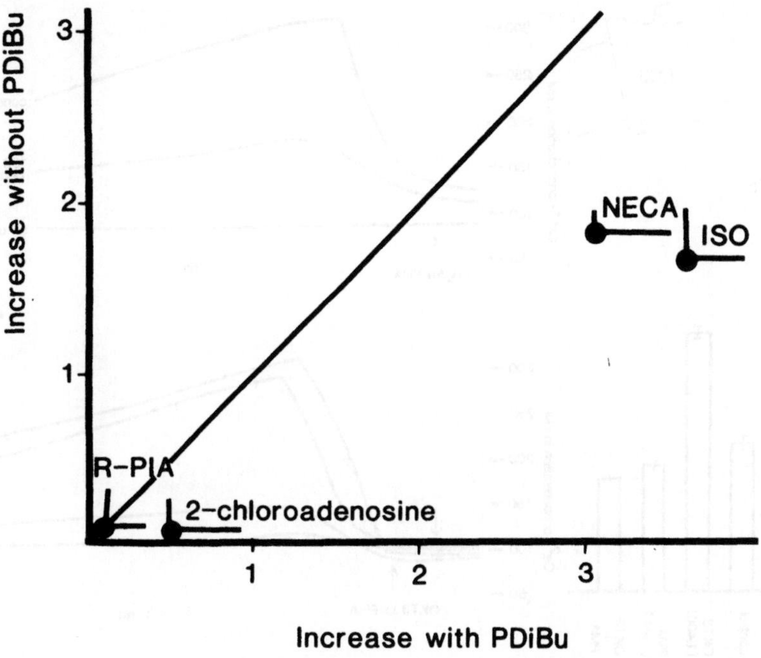

Figure 3. The stimulation of [^{3}H]-cyclic AMP accumulation in rat hippocampal slices treated with NEM by 2-chloroadenosine, NECA and isoprenaline. The results are expressed as the increase in cyclic AMP in the presence (abscissa) or absence (ordinate) of phorboldibutyrate (PDiBu, 1 μM). The line is the line of identity. A point below the line indicates that phorboldibutyrate is able to enhance stimulated cyclic AMP formation even in the presence of NEM.

Interactions between adenosine and agents that raise protein kinase C activity in cultured cells

In order to be able to perform such studies of the exact mechanism by which protein kinase C activation leads to an enhanced accumulation of cyclic AMP was necessary to use isolated cells. Several different cell lines in cultur were therefore screened. In many cells no interaction between the phorbolester and adenosine was seen. However, a marked interaction was observed in a human leukemia cell line Jurkat. In Jurkat cells there are two types of activating receptors, CD2 and CD3 (O'Flynn et al., 1986). The CD2-receptor can be activated by lectins and the CD3-receptor by anti-T3 antibodies (Fig. 4).

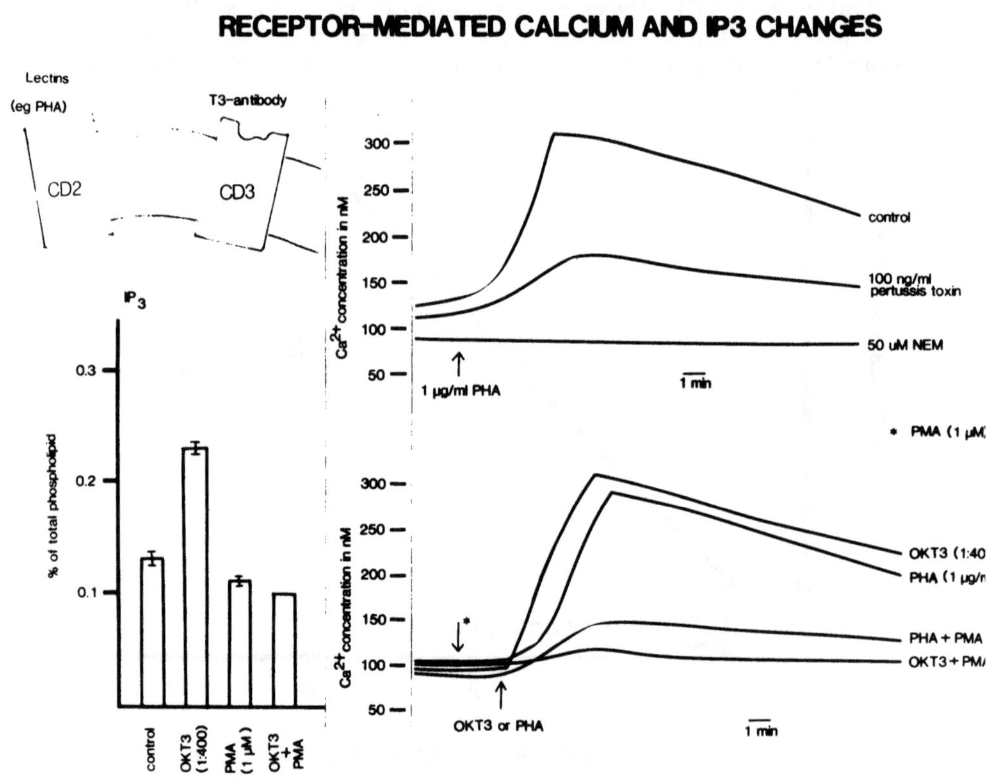

Figure 4. Receptor-mediated calcium and inositoltrisphosphate (IP_3)-changes in Jurkat cells. Jurkat cells were treated with either phytohaemagglutinin (PHA) 1 μg/ml or with antibody to T3 (OKT3). The effect on intracellular calcium levels as measured by Quin2 fluorescence was examined. We have also seen that treatment with the OKT3 antibody leads to an approximately doubling of the IP_3-accumulation.

Activation of either receptor leads to an increase in intracellular calcium, as measured by Quin-2-fluorescence. This rise can be blocked by treatment with either pertussis toxin or ethylmaleimide suggesting that it is mediated via a N-protein that is similar to N_i. Interestingly, the raise in calcium can also be blocked by pretreating the cells with the phorbolester TPA (PMA, 1 μmol). This effect is paralleled by a blockade of the receptor-mediated increase in inositoltrisphosphate accumulation (Fig. 4).

Table 1. Effect of phorbolesters and of phytohemagglutinin (PHA) on the cyclic AMP accumulation induced by the adenosine analogue NECA. Results are mean ± s.e.m. Number of experiments within parenthesis.

	Cyclic AMP pmol/10^6 cells	
Drugs	Control	NECA (10 μM)
No addition	3.7±0.5 (21)	32.8±1.6 (33)
TPA 100 nM	4.5±1.1 (9)	59.8±3.3 (18)
PDiBu 100 nM	4.2±0.8 (6)	75.3±3.0 (12)
PDD 100 nM	2.9±0.9 (6)	38.2±1.6 (6)
PHA 1 μg/ml	1.4±0.3 (6)	36.0±3.3 (9)
10 μg/ml	1.8±0.4 (9)	57.3±2.4 (12)
100 μg/ml	13.4±2.3 (3)	76.4±4.5 (3)

CONCLUSIONS

Adenosine is a physiologically important modulator of several cellular functions. It has one particularly important characteristic, that is not shared by most transmitters or modulators, namely that it is always present in amounts that are sufficient to cause some small stimulation of its own receptors. Furthermore, most cells do have receptors for adenosine. Thus, there is practically always a small background stimulation of adenylate cyclase that is afforded by adenosine.

The present experiments show that in some cells, in the periphery as well as in the central nervous system, drugs that act on receptors that are linked to phosphatidylinositol breakdown and/or increases in intracellular calcium levels can indirectly activate cyclic AMP-formation by potentiating the effect of endogenous adenosine. The mechanism behind this interaction is not known but seems to involve protein kinase C mediated changes in the N-proteins that couple adenosine receptors to the cyclase.

The adenosine levels are increased by hypoxia, ischemia and by cellular activity. Consequently the magnitude of the cyclic AMP response to eg. α-stimulating drugs will be dependent upon the metabolic and functional staste of the cell. It is tempting to speculate that the cyclic AMP response is involved in the fine-tuning of the magnitude of the PI and/or calcium response to the metabolic state of the tissue so that the cells are not overstimulated and their metabolic integrity thereby compromised.

Acknowledgements.

The present studies were supported by the Swedish medical research council (proj. no 2553), by Ostermans fund, by stiftelsen gamla tjänarinnor, by Astra läkemedel and by Karolinska institutet.

References

Daly, J.W. (1977). Cyclic nucleotides in the nervous system. Plenum Press, New York.

Dunwiddie, T.V. and Fredholm, B.B (1984). Adenosine receptors mediating inhibitory electrophysical responses in rat hippocampus are different from receptors mediating cyclic AMP accumulation. Naunyn-Schmiedeberg's Arch. Pharmacol., 326, 294-301.

Fredholm, B.B. (1982). Adenosine receptors. Med. Biol., 60, 289-293.

Fredholm, B.B., Fastbom, J. and Lindgren E. (1986). Effects of N-ethylmaleimide and forskolin on glutamate release from rat hippocampal slices. Evidence that prejunctional receptors are linked to N-proteins but not to adenylate cyclase. Acta Physiol. Scand., 127, 381-386.

Fredholm, B.B, Jonzon, B. and Lindström, K. (1986). Effect of adenosine receptor agonists and other compounds on cyclic AMP accumulation in forskolin-treated hippocampal slices. Naunyn-Schmiedeberg's Arch. Pharmacol., 332, 173-178.

Fredholm, B.B. and Lindgren, E. (1986). Possible involvement of the N_i-protein in the prejunctional inhibitory effect of a stable adenosine analogue (R-PIA) on noradrenaline release in the rat hippocampus. Acta physiol. scand. 126, 307-309.

Fredholm, B.B., Lindgren, E. and Lindström, K. (1985). Treatment with N-ethylmaleimide selectively reduces adenosine receptor mediated decreases in cyclic AMP accumulation in rat hippocampal slices. Br. J. Pharmacol., 86, 509-513.

Fredholm, B.B., Gustafsson, L.E., Hedqvist, P. and Sollevi, A. (1983). Adenosine in the regulation of neurotransmitter release in the peripheral nervous system. In Regulatory Function of Adenosine. Developments in Pharmacology 2. (eds. R.M. Berne, T.W. Rall and R. Rubio). Martinus Nijhoff Publishers, The Hague.

Fredholm, B.B., Jonzon, B., Lindgren, E. and Lindström, K. (1982). Adenosine receptors mediating cyclic AMP production in the rat hippocampus. J. Neurochem., 39, 165-175.

Hollingsworth, E.B. and Daly, J.W. (1985). Accumulation of inositol phosphates and cyclic AMP in guinea-pig cerebral cortical preparations. Effects of norepinephrine, histamine, carbamylcholine and 2-chloroadenosine. Biochem. Biophys. Acta, 847, 207-216.

Katada, T., Gilman, A.G., Watanabe, Y., Bauer, S. and Jakobs, K.H. (1985). Protein kinase C phosphorylates the inhibitory guanine-nucleotide-binding regulatory component and apparently suppresses its function in hormonal inhibition of adenylate cyclase. Eur. J. Biochem., 151, 431-437.

Laemmli, U.K. (1970). Cleavage of structural proteins during the assembly of the head of bacteriophage T4. Nature, 227, 680-685.

Londos, C., Cooper, D.M. and Wolff, J. (1980). Subclasses of external adenosine receptors. Proc. Natl. Acad. Sci. U.S.A., 77, 2551-2554.

Nordstedt, C. and Fredholm, B.B. (1986). Phorbol-12,13-dibutyrate enhances the cyclic AMP accumulation in rat hippocampal slices induced by adenosine analogues. Naunyn-Schmiedeberg's Arch. Pharmacol, in press.

O'Flynn, K., Russul-Saib, M., Ando, I., Wallace, D.L., Beverley, P.C.L., Boylston, A.W. and Linch, D.C. (1986). Different pathways of human T-cell activation revealed by PHA-P and PHA-M. Immunology, 57, 55-60.

26

Interactions between Neuropeptide Y and Noradrenaline Receptor Mechanisms in Sympathetic Vascular Control

J.M. Lundberg, J. Pernow, A. Franco-Cereceda and A. Rudehill

INTRODUCTION

The presence of avian pancreatic polypeptide (APP)-like immunoreactivity (-IR) together with catecholamines in sympathetic cardiovascular nerves and adrenal medulla /1,2/ initiated the search for a novel endogenous peptide. After the isolation of neuropeptide Y (NPY), which has marked structural homology to APP (19 out of 36 amino acids in identical positions) /3/, it was demonstrated that NPY represented the APP-IR initially observed in cardiovascular nerves /4,5/. NPY was found both to exert potent vasoconstrictor activity /4,6/ and to depress the field stimulation-evoked twitch contractions of the vas deferens, suggesting prejunctional inhibition of noradrenaline (NA) release /4,7,8/. Furthermore, electrical stimulation of sympathetic nerves leads to release of not only NA but also of NPY-like immunoreactivity (-LI) /9,10/. Since a considerable part of the vasoconstrictor response to nerve stimulation remained after adrenoceptor blockade /6,9,10/ or NA depletion using reserpine treatment /10/, it has been suggested that NPY could represent a novel transmitter in sympathetic vascular control /see 11/. In this paper we will summarize some recent data which support that NPY fulfils several neurotransmitter criteria and is of importance in peripheral sympathetic vascular control. Furthermore, many drugs used for studies or treatment of hypertension also influence NPY mechanisms in various ways.

METHODS

The tissue content and plasma levels of NPY-LI were analysed using a sensitive radioimmunoassay (RIA), with antiserum N1 raised against synthetic porcine NPY /12/.

To characterize NPY-LI in plasma or tissue extracts, high performance liquid chromatography (HPLC) was performed /12,13/.

The detailed localization of NPY-LI in peripheral organs and adrenal medulla was studied using immunohistochemistry /5/. To establish the presence of NPY in relation to catecholamines, antisera raised against the synthesizing enzymes tyrosine hydroxylase (TH), dopamine-beta-hydroxylase (DBH) or phenylethanolamine N-methyltransferase (PNMT) /14/ were used. Furthermore, animals were pretreated with the neurotoxin 6-hydroxydopamine (6-OHDA), which selectively destroys noradrenergic nerve fibres in the periphery /see 15/. Binding of NPY to possible receptor sites was studied in membrane preparations of the pig spleen in vitro, using ^{125}I-Bolton-Hunter labelled porcine NPY /see 16/.

To study the possible physiological role of NPY, several in vivo models were used both in animals and man. The blood-perfused cat or pig spleen were studied in vivo /9,10. Electrical stimulation of the postganglionic sympathetic nerves was performed or substances were infused locally intra-arterially. Perfusion pressure was monitored as an indication of vascular resistance changes. Veno-arterial plasma gradients of NPY-LI and NA over the spleen were determined to study release. To investigate the influence of reflexogenic activation of the sympatho-adrenal system on NPY release, pigs were subjected to hemorrhagic shock /17/. Local veno-arterial gradients of plasma NPY-LI and NA over the spleen were then monitored.

Human volunteers were subjected to physical exercise either in a graded fashion /13/ or to a prolonged exercise (1 h) at a constant high level (70 % of maximal oxygen uptake) /18/. Systemic plasma levels of NPY-LI and catecholamines were analysed at regular intervals in addition to mean arterial blood pressure and heart rate. In separate experiments, NPY and NA were infused locally into a brachial artery of the human forearm. Local blood flow and venous tone in the forearm were monitored, using non-invasive techniques /19/.

A variety of drugs, which influence autonomic neurotransmission to cardiovascular effector organs and the adrenal, were studied regarding effects on NPY release, tissue content of NPY-LI and functional effects of NPY.

RESULTS and DISCUSSION
NPY and sympathetic vascular control

Localization, tissue levels and subcellular storage

NPY-IR nerve fibres are present in all mammalian species (rats, guinea-pigs, cats, dogs, monkeys and man) and in all vascular beds so far investigated /4,5, 20-23/. These perivascular nerve fibres seem to be sensitive to 6-OHDA treatment and have mostly a similar distribution pattern as TH- and DBH-positive nerves, suggesting a co-existence of NPY with NA. Double labelling experiments on the same sections have shown that NPY-IR nerves around blood vessels also contain TH-IR. A variable portion between different ganglia of principal ganglion cells, however, are TH- and DBH-positive but seem to lack NPY-LI, suggesting that only a population of postganglionic NA nerves contain NPY.

A chemical heterogeneity as the basis for a functional differentiation of the sympathetic nerves seems to be present, since noradrenergic nerves around blood vessels but not close to exocrine elements in the submandibular salivary gland of cat /4/ and pig /11/ and around brown fat cells in the rat /24/ do not contain NPY-LI. Thus, it may be of relevance to induce a sympathetic exocrine response and lipolysis via NA without a concomitant activation of vascular nerves and reduction in metabolic capacity due to diminished blood flow (due to NA and possibly also NPY release). In the pig spleen, there seems to be a complete overlap between NPY- and TH-IR nerve fibres both in the capsule, capsular trabecula and around veins and arteries. In addition, many NPY- and TH-IR nerves are present in the red pulp.

Electron-microscopic and subcellular fractionation studies from the spleen have provided evidence that NA is stored in two types of organelles, small dense-cored vesicles (40-50 nm) and large dense-cored vesicles (80-90 nm) /see 25/. NPY, however, seems to be exclusively localized to the large dense-cored vesicles in the spleen /26/. Being a peptide, most likely NPY is synthesized in the cell-body region and after release resupplied to nerve endings only by axonal transport /see 26/. This feature is likely to limit the amount of NPY being available for release compared to NA, which can be efficiently synthesized locally in the nerve endings by specific enzymes, e.g. TH and DBH. In accordance, the tissue contents of NPY are much lower on a molar basis as compared to NA /27/.

Characteristics for co-release of NA and NPY

In contrast to NA, which is efficiently taken up into sympathetic nerve endings, no difference was observed between arterial and venous levels of plasma NPY-LI in the skeletal muscle of man /18/ or in the isolated spleen from experimental animals /9, 10/. Plasma obtained from experimental animals or man during basal conditions contains usually very low amounts of NPY-LI (20-40 pmol/l). Using the blood-perfused cat spleen as a model, initial experiments showed that electrical stimulation of postganglionic nerves induced release of NPY-LI, as revealed by a positive veno--arterial gradient over the spleen /9/. In the pig spleen, it was clear that the NPY release compared to NA was markedly enhanced at higher stimulation frequencies (Fig. 1) /10/. Furthermore, intermittent stimulation with bursts at a high frequency enhanced the NPY release compared to a continuous stimulation, giving the same total number of impulses (Fig. 1). Possibly these differences may be related to the partially separate storage of NA and NPY (26), whereby the large dense--cored vesicles that contain NPY are releasing their contents mainly upon higher frequencies of stimulation.

A variety of models have been used to induce reflexogenic sympathetic activation in animals and in man. As seen in Fig. 2, hemorrhagic shock is associated with markedly increased plasma levels of NPY-LI in the pig /17/. It has also been possible to detect a positive veno-arterial NPY gradient over the pig spleen upon shock, suggesting a local reflexogenic release. This plasma NPY gradient actually corresponds well under moderate shock conditions with the overflow of NPY-LI induced by the intermittent burst stimulation with an impulse rate of 20 Hz during 2 min (see 10). Under severe shock conditions, plasma NPY levels in the spleen reach levels at which endogenous NPY induces vasoconstriction <u>per se</u> (nM range) (Fig. 2) /10/. In accordance with data obtained with nerve stimulations, the increase of systemic plasma levels of NPY-LI upon hemorrhagic shock is slower compared to NA (Fig. 2)/17/.

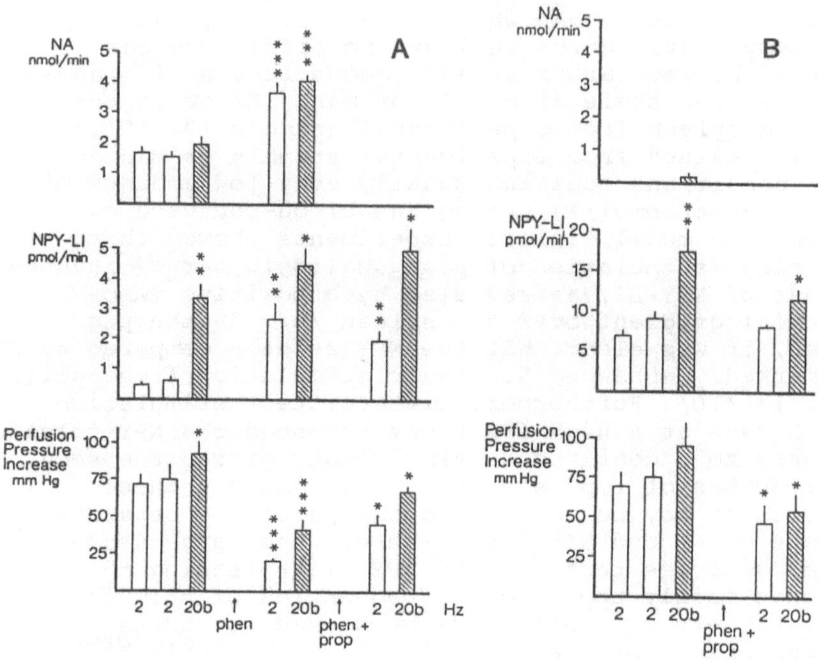

Fig. 1. (A and B) Effects of splenic nerve stimulation for 2 min with a continuous low frequency at 2 Hz (2, open bars) or an intermittent high-frequency burst (b) stimulation at 20 Hz with 10-s intervals (20b, hatched bars), giving the same total number of impulses (240) on perfusion pressure (mmHg) as well as the output of NPY-LI (pmol/min) or NA (nmol/min) from the pig spleen. Stimulations were performed with 15-20-min intervals. (A) untreated control animals; (B) animals which were pretreated with reserpine (1 mg/kg, i.v.) 12-14 h earlier. The responses during basal conditions are compared with the effects 20 min after local intra-arterial infusion of phentolamine (Phen; estimated plasma concentration 5 x 10^{-5} M) or phentolamine plus propranolol (Prop; estimated plasma concentration 10^{-5} M). Data are given as means ± S.E.M. Each value represents observations from 5 animals. Statistical significance between the second 2-Hz and the 20-b stimulations was calculated, using Student's t-test for paired observations, while Kruskal-Wallis analysis of variance with multiple comparisons was used to compare the responses of stimulations before and after adrenoceptor antagonists /from 10/.

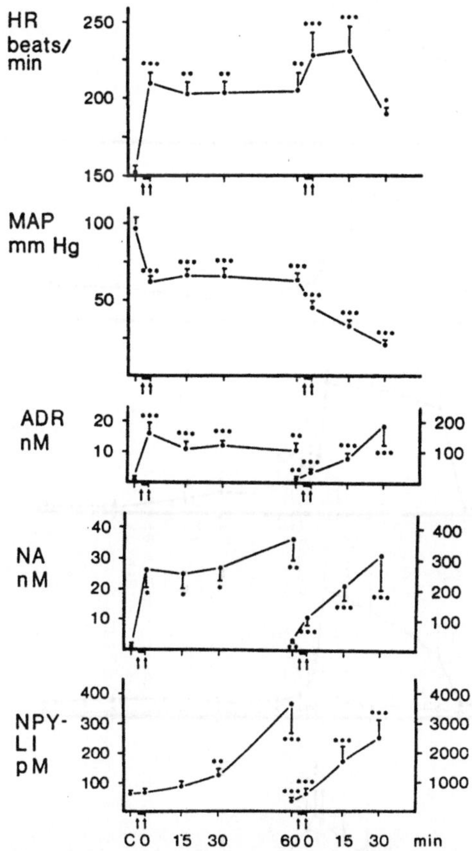

Fig. 2. Effects of two successive bleedings (bleeding 1, 30 ml/kg and bleeding 2, 10 ml/kg) as indicated by arrows on heart rate (HR, beats/min), mean systemic arterial blood pressure (MAP, mmHg), arterial plasma concentrations of adrenaline (ADR, nM), noradrenaline (NA, nM) and neuropeptide Y (NPY-LI, pM). The observation period after bleeding 1 was 60 min and 30 min after bleeding 2. Values are given as means $\pm$ S.E.M. (n = 6). Statistical analysis was performed using Kruskal-Wallis analysis of variance with multiple comparisons. * $p < 0.05$, ** $p < 0.01$, *** $p < 0.001$ /from 17/.

Graded physical exercise in man is associated with progressively increasing plasma levels of both NPY-LI and NA /13/. The NPY levels are highly correlated to plasma NA but not to ADR, suggesting that plasma NPY-LI

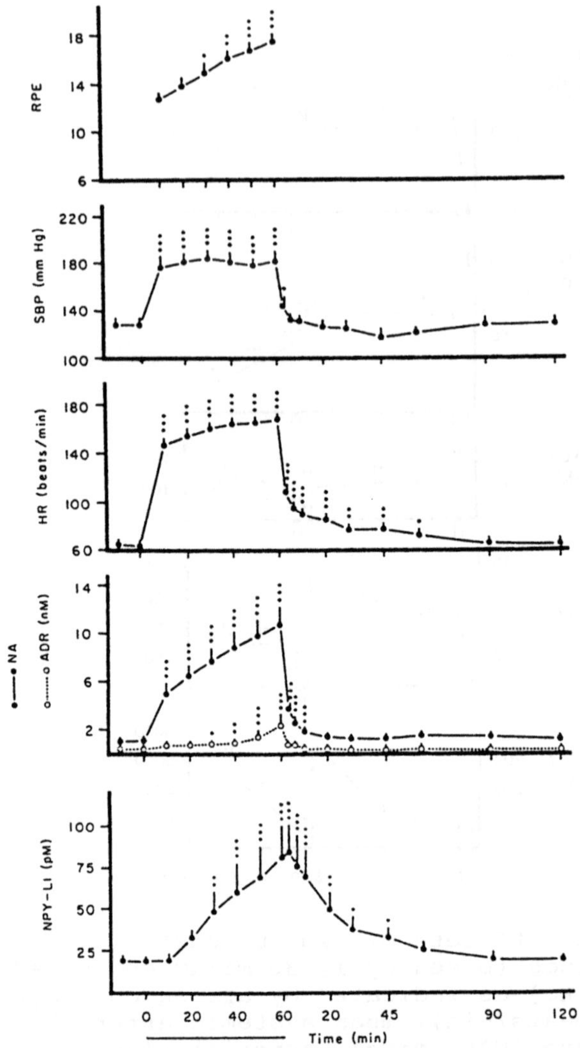

Fig. 3. Effects of prolonged physical exercise during 60 min at a constant work load (70 % of maximum oxygen uptake) on ratings of perceived exertion (RPE), systolic blood pressure (SBP), heart rate (HR), arterial plasma levels of adrenaline (ADR), venous plasma levels of noradrenaline (NA) and neuropeptide Y-like immunoreactivity (NPY-LI). The bar indicates the exercise period, which was followed by 120 min of rest (n = 9 for NA and NPY-LI, n = 5 for ADR). Values are given as means + S.E.M. When compared to basal values, significant differences were calculated using Kruskal-Wallis non-parametric analyses of variance with multiple comparisons. * $p < 0.05$, ** $p < 0.01$, *** $p < 0.001$ /from 18/.

under physical exercise mainly originates from sympathetic nerves and not from the adrenal gland. When using prolonged exercise (at 70 % of maximal oxygene uptake), the rise in plasma NPY-LI is delayed compared to NA, and the decline after completed exercise is much slower than that for NA (Fig. 3) /18/. The highest systemic plasma levels of NPY-LI and catecholamines so far obtained in man occur in newborn infants upon vaginal delivery /28/. Thus, the plasma levels of NPY-LI after an asphyxic period of vaginal delivery was 847 pmol/l. In infants that are delivered by cesarean section, both plasma NPY-LI and catecholamines are considerably lower, again suggesting that the NPY release is dependent on the degree of stress /28/. In patients with NPY-containing phaeochromocytoma tumors, however, plasma NPY-LI upon surgical tumor removal reaches nM levels, which indicates that circulating NPY under these conditions may contribute to the hypertensive response seen in spite of administration of adrenoceptor antagonists /14/.

Receptor activation and second-messenger systems

Initial observations suggested that specific NPY-binding sites were present in membrane preparations from the rat brain /29,30/. Recent studies have shown that a saturable and reversible binding of ^{125}I-NPY occurs to membranes from the pig spleen /16/. Scatchard analysis indicated a single class of binding sites with an equilibrium-dissociation constant of the receptor-^{125}I-NPY complex (Kd) of around 500 pM and a maximal number of specific binding sites (B_{max}) of 25 fmol/mg protein. The peptide with N- and C-terminal tyrosine (PYY), porcine and human NPY all inhibited the specific binding with IC50 values around 100-300 pM (Fig. 4). In contrast, the free acid form of human NPY (i.e. with a non-amidated carboxyl-terminal) was much weaker (IC50 - 100 000 pM) in inhibiting the NPY binding (Fig. 4). The NPY fragments 1-21 or 26-36 did not displace ^{125}I-NPY binding to any significant extent (Fig. 4). This suggests that a large C-terminal portion of the NPY molecule is essential for the receptor binding. Guanyl triphosphate (GTP) and a metabolically stable GTP analogue inhibited ^{125}I-NPY binding in a concentration-dependent fashion (50 % inhibition at 5 x 10^{-4} M GTP and around 10^{-6} M with the stable GTP analogue). No effect on ^{125}I-NPY binding was observed by the presence of adrenoceptor antagonists (prazosin 10^{-6} M, propranolol 10^{-6} M and phentolamine 10^{-5} M, the calcium antagonist nifedipine in concentrations up to 5 x 10^{-7} M or in a Ca^{++}-free medium containing 0.1 mM EGTA.

NPY inhibited cyclic adenosine monophosphate (cAMP) formation in pig-spleen slices induced by forskolin in a

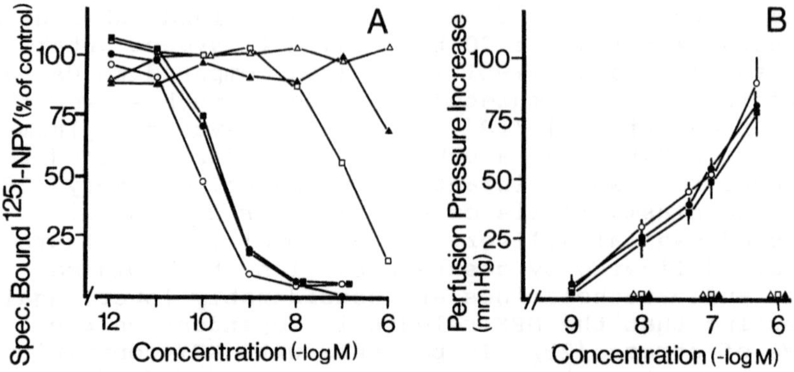

Fig. 4. (A) Displacement of specific ^{125}I-NPY binding to membranes from the pig spleen by increasing concentrations of pNPY (●——●), hNPY (○——○), pPYY (■——■), hNPY-free acid (□——□), pNPY(1-21) (△——△) and pNPY(26-36) (▲——▲). Data are expressed as percentage of ^{125}I-NPY binding in the presence of added peptide. Each point represents mean values from 3-8 separate experiments performed in duplicates. (B) Increase in splenic perfusion pressure (mmHg) in vivo by local intra-arterial infusion of the corresponding peptides as in (A). Each value represents mean $\pm$ S.E.M. of 4-6 experiments /from 16/.

concentration dependent manner with a threshold concentration of 10^{-8} M /16/. The NPY fragments 1-21 and 26-36 or the calcium antagonist felodipine (5×10^{-7} M) did not inhibit the forskolin-stimulated cAMP formation. Thus, one mechanism of action for the vasoconstrictor effect of NPY may be to inhibit cAMP levels in vascular smooth muscle cells. The present data suggest that the NPY receptor in the pig spleen is linked to a regulating, GTP-binding protein which confers GTP sensitivity on the binding and couples the receptor to adenylate cyclase. In comparison, alpha1--adrenoceptor-mediated vasoconstriction in the pig spleen involves the formation of fosfatidyl-inositol /16, see 31/. As mentioned above, although in vitro studies suggest that NPY enhances the contractile effects of NA on vascular smooth muscle, no evidence was obtained that NPY modified binding characteristics for alpha1- or alpha2-ligands in membranes from rat femoral artery /32/.

The calcium antagonists nifedipine, nimodipine, diltiazem and felodipine inhibit the vasoconstrictor effects of NPY in a variety of vascular beds including the splenic /16/, renal, skeletal muscle (Fig. 5)

/33,34/ and cerebral circulations /22/. Nifedipine, however, did not reduce the contractile effects of NPY on human mesenteric veins, indicating some selectivity of action /33/. Felodipine seems to reduce the nerve-evoked as well as the NPY- and NA-induced responses to a rather similar extent in the pig spleen in vivo /16/. When omitting Ca^{2+} in the Krebs medium and adding 0.01 mM EGTA to further reduce the extracellular Ca^{2+} levels, K^+-induced contractile responses of isolated blood vessels (such as the cat middle cerebral artery) in vitro are markedly attenuated (by > 50 %). Recent evidence, however, suggests that the NPY-induced contractions are not reduced in the presence of this "Ca^{2+}-free medium" neither in human skeletal muscle

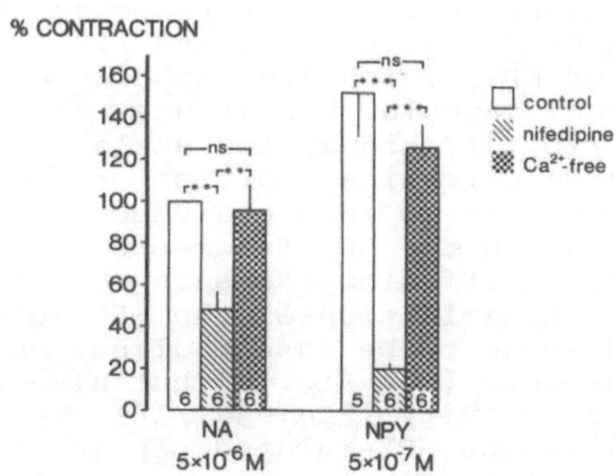

Fig. 5. Contractile effects of noradrenaline (NA) and neuropeptide Y (NPY) on small human skeletal muscle arteries (diameter 0.5 mm) in vitro. The effects were studied, as indicated in standard Krebs solution, after pretreatment with nifedipine (5×10^{-7} M) for 10 min and after 20 min incubation in Ca^{2+}-free Krebs solution (containing 0.01 mM EGTA). Values (mean $\pm$ S.E.M.) are expressed as percentage of contractions evoked by NA during control conditions in each individual preparation. The figures within columns indicate the number of observations. Statistical significances were calculated using Kruskal-Wallis analysis of variance with multiple comparisons; n.s. = not significant, ** $p < 0.01$, *** $p < 0.001$ /from 34/.

arteries nor in the middle cerebral artery of cat (Fig. 5) /33,34/. These seemingly contradictory data suggest that the NPY-induced vasoconstrictor response in vitro is not to any major extent dependent on the influx of extracellular Ca^{2+}. Instead, other actions of nifedipine may underly the inhibition of NPY- and NA-induced contractions in vitro (Fig. 5) /34/. As discussed above, felodipine or nifedipine did not influence the binding of ^{125}I-NPY to splenic membranes. Furthermore, felodipine did not reduce the NPY-induced inhibition of cAMP formation /16/. This suggests that either the fall in cAMP accumulation is unrelated to the functional NPY response in vivo or else there might be a felodipine sensitive step between the inhibitory effect on cAMP and the vasoconstriction /16/.

Cardiovascular effects of NPY in relation to catecholamines

In the initial studies on the cat, systemic bolus administration of NPY caused hypertension and a bradycardia. The latter effect is most likely of reflexogenic nature /6/. Similar responses have been observed in rat /35/, guinea-pig and pig. The systemic plasma levels of NPY, which are associated with an increase in local vascular resistance, are around 1-2 nM in the spleen /10,17/, kidney and skeletal muscle of the pig /17/. A large C-terminal portion of the NPY molecule seems to be necessary to induce a vasoconstrictor response /36/, since the NPY fragments 1-21 and 26-36 did not influence perfusion pressure in the pig spleen (Fig. 4) /16/. The amidation of the NPY molecule at the C-terminal end seems to be essential not only for binding to receptors (see Fig. 4) but also for the vasoconstrictor activity (Fig. 4) /16/. The human form of NPY as well as the NPY-related peptide PYY, which is present in gastrointestinal endocrine cells, causes a rather similar vasoconstrictor response as porcine NPY (Fig. 4). The vasoconstrictor potency of NPY on a molar basis is in the pig spleen about 50-fold higher than for NA /10/.

At plasma levels of NPY similar to those obtained upon, for instance, maximal physical exercise, i.e. around 200 pmol/l, no vasoconstrictor effect was observed in the skeletal muscle of the human forearm /19/. Upon local intra-arterial infusion, however, NPY caused vasoconstriction in the human forearm at plasma concentrations (about 1 nM) as in the pig spleen /19/. The vasoconstrictor effect of NPY in man in vivo compared to NA was very long-lasting (up to 25 min after cessation of the infusion), using the highest NPY and NA doses /19/. When NPY was infused prior to NA or nerve stimulation, an enhanced vasopressor response was observed in the pithed rat in vivo, suggesting interactions at the postjunctional level /37/.

In contrast to NA that contracts most large blood vessels, NPY exerts a variable vasoconstrictor activity in *in-vitro* studies. Thus, certain vessels respond very well to NPY, such as cat /22/ and human mesenteric veins, renal arteries and skeletal muscle arteries (Fig. 5) /33/. On the other hand, only a relatively small increase in tension was induced by NPY on the rat femoral artery and portal vein /32,38/, while no response to NPY could be detected in the mesenteric arteries of rat and man. In low concentrations (nM range), which did not induce any effects *per se*, NPY markedly enhanced the contractile response both of NA and transmural field stimulation in certain vessels, such as femoral artery but not in the femoral vein /21,32/.

NPY also exerts prejunctional effects on autonomic neurotransmission. Thus, transmural nerve stimulation (TNS)-evoked ^{3}H-NA release is inhibited by NPY from a variety of blood vessels *in vitro*, such as rat femoral (Fig. 6), basilar and mesenteric arteries as well as

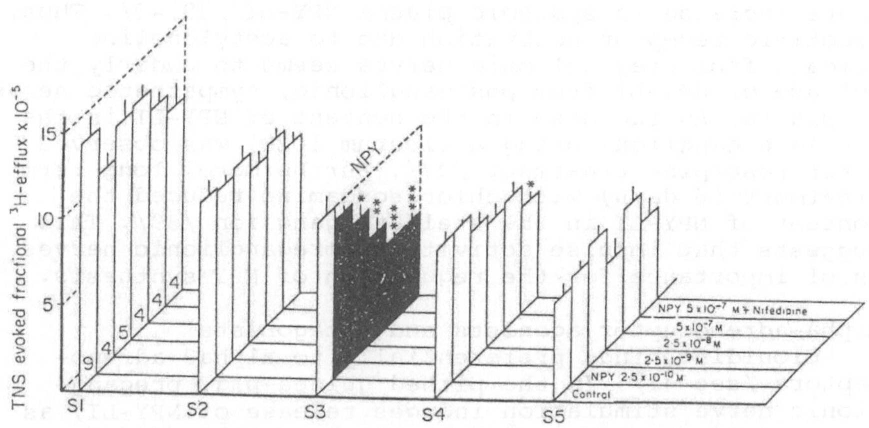

Fig. 6. The TNS (6 Hz for 2 min)-evoked ^{3}H NA efflux (expressed as fractional release per nerve impulse x 10^{-5} M) from the rat femoral artery during five successive stimulations (S1 - S5) with 20 min intervals. NPY was introduced in different concentrations (2.5 x 10^{-10} - 5 x 10^{-7} M) 10 min prior to S3 in separate series of experiments. Nifedipine (5 x 10^{-7} M) was present throughout one series of experiments. Values are given as means ± S.E.M. (n = 4-9 for each series, as indicated in the S1 bars). * $p < 0.05$, ** $P < 0.01$, *** $p < 0.001$, using Kruskal-Wallis analysis of variance compared to the reductions in the parallel control experiments /from 32/.

portal and femoral veins /32,38/. Furthermore, NPY inhibits TNS-evoked ^{3}H-NA release from human mesenteric vein /33/. The NPY-induced depression of NA release is not influenced by nifedipine (Fig. 6).

In general, both the post- (i.e. vasoconstrictor effects) and prejunctional (i.e. inhibition of NA release) actions of NPY are resistant to adrenoceptor--blocking agents and a variety of other receptor antagonists /6,22,36). Furthermore, the postjunctional NPY effects are not reduced after surgical sympathectomy /6/. Thus, as discussed above, it is likely that the NPY effect is a specific, receptor-mediated event, which occurs independently of adrenergic mechanisms, although NPY and NA can interact both at the pre- and post-junctional levels.

Antihypertensive drugs and NPY mechanisms

Ganglionic-blocking agents

Chlorisondamine pretreatment inhibits the reserpine--induced depletion of NPY-LI (see below) from the heart, spleen and skeletal muscle, in addition to the acute increase in systemic plasma NPY-LI /39,40/. Thus, nicotinic receptor activation due to acetylcholine release from preganglionic nerves seems to underly the release of NPY-LI from postganglionic, sympathetic nerve terminals. An increase in the content of NPY-LI in the stellate ganglion, using antiserum 102D, was observed after reserpine treatment /27/. Furthermore, long-term treatment (6 days) with chlorisondamine reduced the content of NPY-LI in the stellate ganglion /39/. This suggests that impulse activity in preganglionic nerves is of importance for the regulation of NPY synthesis.

Alpha-adrenoceptor agonists and antagonists

Clonidine binds preferentially to alpha2-adrenoceptors /see 41/. In the pithed guinea-pig, preganglionic nerve stimulation induces release of NPY-LI, as revealed by elevated systemic plasma levels /42/. Prazosin is considered to be a selective $alpha_1$--antagonist in lower concentrations /see 41/. Prazosin treatment did not change the nerve stimulation-evoked NPY overflow into the systemic circulation of the pithed guinea-pig model /42/. Pretreatment with the preferential alpha2-antagonist yohimbine enhanced markedly the nerve stimulation-evoked NPY overflow into the systemic circulation of the pithed guinea-pig /42/.Clonidine pretreatment inhibited this nerve stimulation-evoked NPY release by about 50 %. This effect was reversed by the preferential $alpha_2$--antagonist yohimbine, suggesting involvement of alpha2-adrenoceptor mechanisms in the regulation of not only NA but also NPY release /42, see also 9,10/.

The tissue content of NPY-LI in the right atrium of the heart had increased significantly 24 h after treatment with clonidine (50 ug x kg^{-1} s.c. every 8 h) /43/, while no change was observed in the vas deferens. The corresponding tissue levels of NA were not changed after clonidine treatment /43/. Whether the clonidine effect on tissue-NPY content is caused by a decreased release due to centrally-mediated reduction in sympathetic nerve-impulse flow or also influenced by peripheral prejunctional alpha2-receptor stimulation remains to be established.

Reserpine

Reserpine disturbs storage mechanisms for monoamines, which leads to a progressive long-lasting depletion of tissue content of dopamine, NA, ADR and serotonin. Reserpine also induces a dose- and time-dependent depletion of NPY-LI from nerves in the heart, spleen and skeletal muscle but not in the vas deferens and uterus of the guinea-pig /27,35,39,40,44/. A higher dose of reserpine is needed to deplete NPY-LI than NA /27/. Furthermore, the depletion of NPY-LI is slower in onset and less complete than that of NA. The reserpine-induced depletion of NPY-LI, but not that of NA, can be blocked by administration of the ganglionic-blocking agent chlorisondamine /39,40/ or surgical preganglionic denervation /45/. This suggests that enhanced release of NPY-LI occurs after reserpine treatment as a result due to increased nerve activity. Furthermore, lack of prejunctional, inhibitory, alpha2-adrenoceptor mediated modulation by NA may account for the finding that the nerve stimulation-evoked release of NPY-LI is markedly enhanced from the spleen in acutely reserpinized animals (Fig. 1) /10,45/. The large remaining splenic functional responses upon nerve stimulation with higher frequencies in reserpinized pigs (Fig. 1) even in the presence of adrenoceptor antagonists clearly indicate that other messengers than NA, such as NPY, are involved in splenic vascular control /10/.

Reserpine treatment combined with interruption of preganglionic nerve-impulse flow creates a situation where virtually no NA is present in the tissue, while the levels of NPY are largely intact. Thus, in the cat spleen where the preganglionic fibres have been cut one week prior to reserpine treatment, NPY release evoked by postganglionic nerve stimulation is markedly enhanced compared to control conditions. A long-lasting functional response is then present in spite of an almost total (99 %) NA depletion and adrenoceptor blockade /45/. These observations provide strong evidence that NPY is a mediator of sympathetic vascular control.

Acknowledgements. The present studies were supported by grants from the Swedish Medical Research Council (14X-6554), the American Council for Tobacco Research, the Swedish Tobacco Company, Petrus och Augusta Hedlunds Stiftelse and funds from the Karolinska Institute. For expert secretarial help we thank Mrs Hilka Lindberg.

REFERENCES

1. Lundberg JM, Hökfelt T, Änggård A, Kimmel J, Goldstein M, Markey K. Co-existence of an avian pancreatic polypeptide (APP) immunoreactive substance and catecholamines in some peripheral and central neurons. Acta Physiol Scand 1980:110: 107-109.
2. Lundberg JM, Hökfelt T, Änggård A, Terenius L, Elde R, Markey K, Goldstein M. Organization principles in the peripheral sympathetic nervous system: subdivision by co-existing peptides (somatostatin-, avian pancreatic polypeptide- and vasoactive intestinal polypeptide-like immunoreactive materials). Proc Nat Acad Sci, USA, 1982:79: 1303-1307.
3. Tatemoto K. Neuropeptide Y: the complete amino-acid sequence of the brain peptide. Proc Nat Acad Sci, USA, 1982:79: 5485-5489.
4. Lundberg JM, Terenius L, Hökfelt T, Martling C-R, Tatemoto K, Mutt V, Polak J, Bloom SR, Goldstein M. Neuropeptide Y (NPY)-like immunoreactivity in peripheral noradrenergic neurons and effects of NPY on sympathetic function. Acta Physiol Scand 1982:116:477-480.
5. Lundberg JM, Terenius L, Hökfelt T, Tatemoto K. Comparative immunohistochemical and biochemical analysis of pancreatic polypeptide-like peptides with special reference to presence of neuropeptide Y in central and peripheral neurons. J Neurosci 1984:4:2376-2386.
6. Lundberg JM, Tatemoto K. Pancreatic polypeptide family (APP, BPP, NPY and PYY) in relation to alpha-adrenoceptor-resistant sympathetic vasoconstriction. Acta Physiol Scand 1982: 116:393-402.
7. Allen JM, Tatemoto K, Polak JM, Hughes J, Bloom SR. Two novel related peptides, neuropeptide Y (NPY) and peptide YY (PYY) inhibit the contraction of the electrically stimulated mouse vas deferens. Neuropeptides 1982:3:71-77.

8. Lundberg JM, Stjärne L. Neuropeptide Y (NPY) depresses the secretion of ^{3}H-noradrenaline and the contractile response evoked by field stimulation in rat vas deferens. Acta Physiol Scand 1984:120:477-479.
9. Lundberg JM, Änggård A, Theodorsson-Norheim E, Pernow J. Guanethidine-sensitive release of NPY-like immunoreactivity by sympathetic nerve stimulation. Neurosci Lett 1984:52: 175-180.
10. Lundberg JM, Rudehill A, Sollevi A, Theodorsson--Norheim E, Hamberger B. Frequency- and reserpine-dependent chemical coding of sympathetic transmission: differential release of noradrenaline and neuropeptide Y from pig spleen. Neurosci Lett 1986:63:96-100.
11. Lundberg JM, Hökfelt T. Multiple co-existence of peptides and classical transmitters in peripheral autonomic and sensory neurons - functional and pharmacological implications. In: Hökfelt T, Fuxe K, Pernow B, eds. Progress in Brain Research, Elsevier, Amsterdam, 1986:68:241-262.
12. Theodorsson-Norheim E, Hemsén A, Lundberg JM. Radioimmunoassay for neuropeptide Y (NPY): chromatographic characterization of immunoreactivity in plasma and tissue extracts. Scand J Lab Invest 1985:45:355-365.
13. Lundberg JM, Martinsson A, Hemsén A, Theodorsson-Norheim E, Svedenhag J, Ekblom E, Hjemdahl P. Co-release of neuropeptide Y and catecholamines during physical exercise in man. Biochem Biophys Res Commun 1985:133:30-36.
14. Lundberg JM, Hökfelt T, Hemsén A, Theodorsson--Norheim E, Pernow J, Hamberger B, Goldstein M. Neuropeptide Y-like immunoreactivity in adrenaline cells of adrenal medulla and in tumors and plasma of phaeochromocytoma patients. Regulatory Peptides 1986:13:169-182.
15. Jonsson G. Chemical lesioning techniques: monoamine neurotoxins. In: Björklund A, Hökfelt T, eds. Methods in chemical neuroanatomy. Elsevier, Amsterdam, 1983:463-507. (Handbook of chemical neuroanatomy; vol. 1).
16. Lundberg JM, Hemsén A, Larsson O, Rudehill A, Saria A, Fredholm BB. Neuropeptide Y: specific binding sites, second messenger formation foundation and vasoconstrictor activity in pig spleen. N S Arch Pharmacol (submitted).
17. Rudehill A, Lundberg JM, Sollevi A, Hamberger B. Release of neuropeptide Y upon hemorrhagic hypovolemia in relation to vasoconstrictor effects in the pig. Anesthesiology (submitted).

18. Pernow J, Lundberg JM, Kaijser L, Hjemdahl P, Theodorsson-Norheim E, Martinsson A, Pernow B. Plasma neuropeptide Y-like immunoreactivity and catecholamines during various degrees of sympathetic activation in man. Clin Physiol 1986:6:561-578.
19. Pernow J, Lundberg JM, Kaijser L. Vasoconstrictor effects in vivo and plasma disappearance rate of neuropeptide Y in man. Life Sci (in press).
20. Lundberg JM, Terenius L, Hökfelt T, Goldstein M. High levels of neuropeptide Y (NPY) in peripheral noradrenergic neurons in various mammals including man. Neurosci Lett 1983:42: 167-172.
21. Ekblad E, Edvinsson L, Wahlestedt C, Uddman R, Håkanson R, Sundler F. Neuropeptide Y co-exists and co-operates with noradrenaline in perivascular nerve fibres. Regul Peptides 1984:8:225-235.
22. Edvinsson L, Emson PC, McCulloch K, Tatemoto K, Uddman R. Neuropeptide Y, cerebrovascular innervation and vasomotor effects in the cat. Neurosci Lett 1983:43:79-84.
23. Edvinsson L, Emson P, McCulloch J, Tatemoto K, Uddman R. Neuropeptide Y: immunocytochemical localization to and effect upon feline pial arteries and veins in vitro and in situ. Acta Physiol Scand 1984:122:155-163.
24. Cannon B, Hökfelt T, Lundberg JM, Terenius L, Goldstein M. Selective distribution of NPY to noradrenergic vascular nerves in brown adipose tissue of the rat. Exp Cell Tissue Res 1986:164:546-550.
25. Fried G. Small noradrenergic storage vesicles isolated from rat vas deferens - biochemical and morphological characterization. Acta Physiol Scand 1980, Suppl. 493:1-28.
26. Fried G, Lundberg JM, Theodorsson-Norheim E. Subcellular storage and axonal transport of neuropeptide Y (NPY) in relation to catecholamines in the cat. Acta Physiol Scand 1985:125:145-154.
27. Lundberg JM, Saria A, Franco-Cereceda A, Hökfelt T, Terenius L, Goldstein M. Differential effects of reserpine and 6-hydroxydopamine on neuropeptide Y and noradrenaline in peripheral neurons. N S Arch Pharmacol 1985:328:331-340.
28. Lundberg JM, Hemsén A, Fried G, Theodorsson-Norheim E, Lagercrantz H. High plasma levels of neuropeptide Y (NPY)-like immunoreactivity and catecholamines in newborn infants. Acta Physiol Scand 1986:126:471-473.
29. Undén A, Bartfai T. Binding of neuropeptide Y to rat brain membranes. FEBS Letters 1984:177:125-128.
30. Saria A, Theodorsson-Norheim E, Lundberg JM. Evidence for specific neuropeptide Y-binding sites in rat brain synaptosomes. Europ J Pharmacol 1985:107:105-107.

31. Taylor C, Merritt JE. Receptor coupling to polyphosphoinositide turnover: a parallel with the adenylate cyclase system. TIPS 1986:7:238-242.
32. Pernow J, Saria A, Lundberg JM. Mechanisms underlying pre- and postjunctional effects of neuropeptide Y in sympathetic vascular control. Acta Physiol Scand 1986:126:239-249.
33. Pernow J, Svendberg T, Lundberg JM. Inhibitory effects of nifedipine on neuropeptide Y and noradrenaline induced contractions of human peripheral blood vessels in vitro. Europ J Pharmacol (submitted).
34. Pernow J, Lundberg JM. Neuropeptide Y contracts human skeletal muscle arteries via a nifedipine--sensitive mechanism independent of extracellular calcium? Acta Physiol Scand (in press).
35. Lundberg JM, Saria A, Änggård A, Hökfelt T, Terenius L. Neuropeptide Y and noradrenaline interaction in peripheral cardiovascular control. Clin Exp Theory Practice 1984:A6: 1961-1972.
36. Rioux F, Bachelard H, Martel JC, StPierre S. The vasoconstrictor effects of neuropeptide Y and related peptides on the guinea-pig isolated heart. Peptides 1986:7:27-31.
37. Dahlöf C, Dahlöf P, Lundberg JM. Neurpeptide Y (NPY): enhancement of blood-pressure increase upon alpha-adrenoceptor activation and direct pressor effects in pithed rats. Europ J Pharmacol 1985:109:289-292.
38. Dahlöf C, Tatemoto K, Lundberg JM. Neuropeptide Y (NPY) reduces field stimulation-evoked release of noradrenaline and enhances contractile force in the rat portal vein. N S Arch Pharmacol 1985:328:327-330.
39. Lundberg JM, Saria A, Franco-Cereceda A, Theodorsson--Norheim, E. Treatment with sympatholytic drugs changes tissue content of neuropeptide Y in cardiovascular nerves and adrenal gland. Acta Physiol Scand 1985:124:603-611.
40. Lundberg JM, Al-Saffar A, Saria A, Theodorsson--Norheim E. Reserpine-induced depletion of neuropeptide Y from cardiovascular nerves and adrenal gland due to neurogenic activation. N S Arch Pharmacol 1986:332:163-168.
41. Langer SZ. Pharmacological characterization of alpha-adrenoceptor subtypes. In: Kobinger W, Ahlquist RP, eds. Alpha- and beta-adrenoceptors in the cardiovascular system. Excerpta Medica, 1984:9-16.

42. Dahlöf C, Dahlöf P, Lundberg JM. Alpha2--adrenoceptor-mediated inhibition of nerve stimulation-evoked release of neuropeptide Y (NPY)-like immunoreactivity in the pithed guinea-pig. Europ J Pharmacol (in press).
43. Nagata A, Franco-Cereceda A, Svensson TH, Lundberg JM. Clonidine treatment elevates the content of neuropeptide Y in cardiac nerves. Acta Physiol Scand 1986:128:321-322.
44. Carlsson A. Drugs which block the storage of 5-hydroxytryptamine and related amines. In: Erspamer, ed. Handbook of experimental pharmacology. Springer Verlag, Berlin-Heidelberg-New York, 1965:19:529-592.
45. Lundberg JM, Fried G, Pernow J, Theodorsson-Norheim E, Änggård A. NPY - a mediator of reserpine-resistant, non-adrenergic vasoconstriction in cat spleen after preganglionic denervation? Acta Physiol Scand 1986:126:151-152.

27
New Paradigm: Autonomic Neurotransmission by Multiple Interaction between Sites Intermittently Secreting Single Mixed Quanta

Lennart Stjärne

INTRODUCTION

Recent findings by novel techniques severely challenge so many key aspects of earlier models of autonomic neurotransmission (Burnstock, 1986a) that mere revision is not enough - an entirely new paradigm is required (Stjärne, 1986,1987). The present paper describes the microanatomical aspects of the new model as applied to the guinea-pig and mouse vas deferens. These tissues are selected for two reasons: First, they are uniquely well suited for combined biochemical and electrophysiological analysis of the secretory activity in single sympathetic nerve varicosities (Blakeley and Cunnane, 1979; Stjärne, 1985). Second, since their smooth muscle cells have a dual autonomic innervation (Burnstock, 1970) they permit study of interactions between the sympathetic and parasympathetic nervous control of smooth muscle. Due to the limited space the discussion is brief and references are often given to review articles rather than to original papers.

CHEMICAL COOPERATIVITY IN AUTONOMIC NEUROTRANSMISSION

Noradrenaline (NA) and acetylcholine (ACh) are not the exclusive mediators of sympathetic (Fig. 1A) and parasympathetic nerves (Burnstock, 1986a; Lundberg and Hökfelt, 1986). In sympathetic nerves some effects are mediated by adenosine 5'-triphosphate (ATP) or neuropeptides such as neuropeptide Y (NPY) or enkephalins (Fig. 1B). Similarly some effects of parasympathetic nerve stimulation are exerted through neuropeptides, e.g. vasoactive intestinal polypeptide (VIP). But the old nomenclature persists: NA- and ACh-containing nerves are still often denoted as 'noradrenergic' (or worse, and for historical reasons, as 'adrenergic') and 'cholinergic', on habit or on the view that NA and ACh are the 'principal', if not the only, transmitters in these autonomic nerves.

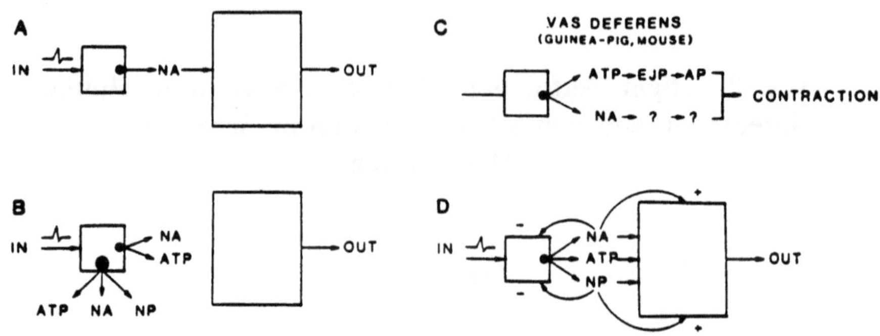

Fig. 1. Illustration of the growing complexity of models of sympathetic neuro-transmission. (A) A simple early linear model: Transmission is mediated exclusively through NA; all effects are exerted intrajunctionally. (B) A more modern model: There are two kinds of transmitter organelles in sympathetic nerve varicosities. The 'small' vesicles contain NA and ATP; the 'large' ones also contain NA and ATP but in addition often NP, i.e a neuropeptide (Fried et al., 1986). Transmitter secretion from a 'small' vesicle may occur from the 'preferred release site' of the varicosity (Cunnane and Stjärne, 1984; Burnstock, 1986b) but 'large' vesicles may secrete transmitter outside these sites ('directed' and 'non-directed' release respectively; Zsu, Thuresson-Klein and Klein, 1986). (C) In some organs during low frequency stimulation sympathetic neuromuscular transmission is mediated exclusively through NA (by a 'metabotropic' action) and ATP (by an 'ionotropic' action). In the guinea-pig and mouse vas deferens ATP triggers a contraction by evoking EJPs and muscle action potentials while NA contributes to the contractile response without depolarizing the smooth muscle cell (Burnstock, 1986a; Stjärne and Åstrand, 1985, Bolton and Large, 1986). (D) Sympathetic neuro-effector throughput is complex. Each transmitter acts in a dual manner on the effector cells; each of them by an indirect (often extrajunctional) action enhances the excitability of the effector to all stimuli (Stjärne, 1986), and by a direct (intra- and extrajunctional) action triggers the (e.g. contractile) response. Further, NA (and possibly the neuropeptide, NP), but not (or only to a lesser extent) ATP, autoinhibit the secretory mechanisms of sympathetic nerve terminals (Stjärne and Åstrand, 1985; Lundberg and Hökfelt, 1986; Stjärne, and Lundberg, 1986; Stjärne, Lundberg and Åstrand, 1986).

The continued use of the old nomenclature should be discouraged. For example, it seems both pointless and misleading to denote the sympathetic innervation of the guinea-pig and mouse vas deferens as 'noradrenergic'. In both tissues the excitatory junction potential (EJP) in smooth muscle cells is caused by ATP (Sneddon and Westfall, 1984; Burnstock, 1986a); hence for this effect ATP is not only the 'principal' but actually the exclusive transmitter

(Fig. 1C). With regard to the twitch contraction to electrical nerve stimulation the situation is more complex: In the guinea-pig vas deferens ATP is the 'principal' and NA the 'auxiliary' transmitter for this response, but in the mouse vas deferens the reverse is true (Stjärne and Åstrand, 1984, 1985).

In the new working hypothesis it is the rule rather than the exception that more than one transmitter is involved in the responses of a tissue to autonomic nerve stimulation. Hence the suffix '-ergic' is not used for the nerve itself but reserved for specific (e.g. 'NA-', 'ATP-' or 'NP-ergic') effects elicited by stimulation of the nerve (Fig. 1D; Stjärne, 1986, Stjärne and Lundberg, 1986). In the autonomic throughput each cotransmitter contributes by complex pre- and postjunctional effects (Fig. 1D). The relative importance of each varies with the tissue. For example pharmacological 'removal' of ATP-mediated EJPs and muscle action potentials hardly alters the contractile response in the mouse vas deferens and many blood vessels but in the guinea-pig vas deferens depresses the contractions by about 75%. Thus in some tissues the 'electrical link' seems to be largely an epiphenomenon while in other tissues it plays a major role as trigger of contractions (Stjärne and Åstrand, 1985; Bolton and Large, 1986). NPs presumably contribute to the mechanical effects of autonomic nerve stimulation in many organs (Lundberg and Hökfelt, 1986), but due to lack of specific blocking agents their relative importance is difficult to assess (Stjärne, Lundberg and Åstrand, 1986).

A CRUCIAL ISSUE: THE MODE OF OPERATION OF INDIVIDUAL VARICOSITIES

One fundamentally important issue about which surprisingly little has been known is the mode and extent to which an impulse in a sympathetic parent axon induces transmitter secretion in the approximately 20000 varicosities in its terminal branches, each with some 500 transmitter vesicles (Hökfelt, 1969; Gabella, 1980). The constraints have been derived from studies of the overflow of NA from a tissue, induced by electrical stimulation of its sympathetic nerves. In a number of tissues the NA output per stimulus corresponds to about 1/50000 of the tissue content, i.e. to an average of 1% of the content of a single vesicle in each varicosity. As clearly realized by the pioneers on this problem (Folkow and Häggendal, 1970) this may imply that each stimulus activates only 1% of the varicosities releasing a single quantum equal to the content of a vesicle. To Folkow and Häggendal neurotransmission by such extremely intermittent secretion of single big quanta seemed wasteful and inefficient, "almost like shooting at a flock of birds with occasional artillery grenades instead of using ... a shotgun". On this common sense argument they proposed the hypothesis (here termed the 'nonintermittent' model) that every nerve impulse activates every varicosity and (contrary to the vesicle hypothesis; see Smith and Winkler, 1972) releases only a small fraction (about 1%) of the transmitter content of a single vesicle. Bevan, Chesher and Su (1969) were the first to propose the opposite hypothesis (here termed the 'intermittent' model). According to them the individual varicosity is only intermittently activated (by 1 out of every 7-8 nerve impulses), each time releasing the entire content of a vesicle. However, phenoxybenzamine was present in their experiments; in the absence of this α-adrenoceptor blocking drug the varicosity would be activated much more rarely, by about 1% of the stimuli (Stjärne, 1985).

By a new technique (here termed the 'discrete event method'; Blakeley and Cunnane, 1979) for electrophysiological analysis of the local output of ATP coreleased with NA from the same vesicle of the sympathetic nerves of the guinea-pig (Sneddon and Westfall, 1984) or mouse vas deferens (Stjärne and Åstrand, 1984) it seems possible to decide between the two models described above. EJPs recorded intracellularly in single smooth muscle cells (Fig. 2A) often reflect the summed effect of many transmitter quanta (Burnstock and Holman, 1966). In the new method the transient peak in the dV/dt of the rising phase of the EJP (the 'discrete event') is shown to resolve the local current caused by impact of individual transmitter quanta on the impaled cell or one of its electrically coupled neighbours (Blakeley and Cunnane, 1979).

The interpretation of results by this method has not been unanimous however. Both groups which have used it agree that the individual varicosity is not activated by each nerve impulse and that the quantum presumably equals the content of a single 'small' vesicle (Blakeley and Cunnane 1979; Cunnane and Stjärne, 1984), in line with the 'intermittent' model described above. The disagreement concerns the degree of secretory intermittency, and the number of quanta released from each varicosity. In one group (Blakeley, Mathie and Petersen, 1984) the view is that the individual varicosity in the guinea-pig and mouse vas deferens is relatively frequently activated (by 2-48%, or 50-100% of the stimuli, respectively) and each time releases 1-10 quanta. This view is incompatible both with the 'nonintermittent' and the 'intermittent' models of NA secretion from a varicosity (Stjärne, 1985). In the other group (Cunnane and Stjärne, 1984; Stjärne, 1985, 1986) it is concluded that the individual varicosity in both species is activated by 0.2-3% of the stimuli and each time releases a single quantum. This view agrees well with the 'intermittent' model and also with the finding in guinea-pig mesenteric arteriole that each stimulus activated only 1-2% of the varicosities (Hirst and Neild, 1980), and with a number of reports from other systems, which suggest that it may be a general biological feature that nerve impulses release only single quanta from a varicosity (Jack, Redman and Wong 1981; Hackett, Jackson and Rubel, 1982; Korn, 1984).

In the new working hypothesis each impulse in the parent axon releases the transmitter content of a vesicle from 1% of the varicosities in the terminal arbour, creating extremely high local concentrations in 1% of the narrow neuro-effector gaps. The principle may apply at least in part also to parasympathetic nerves (see Kadlec, Masek and Seferna, 1984). How can transmission by such intermittent release of single big quanta ("like occasional artillery grenades"; Folkow and Häggendal, 1970) be efficient?

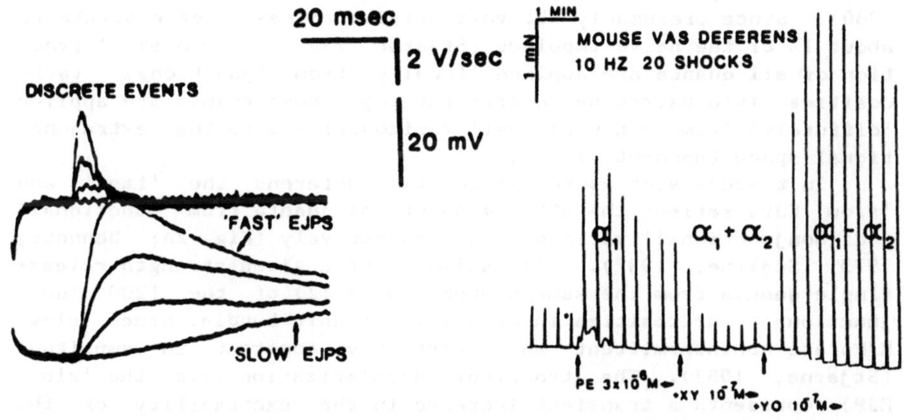

Fig. 2. 'Focal' or 'diffuse' application of transmitter (Bennett, 1973): (A) Some 15 electrically coupled smooth muscle cells make up a functional unit bundle in the mouse vas deferens (Furness, 1970). Hence a microelectrode in a smooth muscle cell may detect transmitter from up to 75 'junctional' and 1200 'nonjunctional' varicosities (Bennett, 1973). The ATP component of single quanta from 1-5 of 75 'junctional' varicosities cause 'fast' EJPs and 'discrete events', which are intermittent and variable in amplitude. Single ATP quanta from up to 12 of 1200 'nonjunctional' varicosities cause 'slow' EJPs, which are nonintermittent and relatively constant in amplitude (Bennett, 1973; Stjärne, 1987). (B) An example of mimicking the effects of the NA components of transmitter quanta from 'nonjunctional' varicosities, on the twitch contrations of the mouse vas deferens to short trains of electrical nerve stimuli (via ring electrodes around the prostatic end). Phenylephrine (PE), which mimics the α_1-face of NA, causes two effects: A transient increase in the resting tension and a long-lasting 3-5fold increase in the twitch response to nerve stimulation. This effect is only to a small extent (less than 10%) due to increased transmitter secretion (Stjärne to be published). Xylazine (XY), which mimics the α_2-face of NA, depresses the twitch responses (due to depression of transmitter secretion). Yohimbine (YO), which blocks the α_2-adrenoceptor mediated autoinhibition dramatically increases the twitch amplitude (due to increase in transmitter secretion).

A TWO-COMPARTMENT MODEL OF SYMPATHETIC NEUROTRANSMISSION

Diagrams of the current model often show only a single varicosity and the effector cell it innervates and hence (unintentionally?) suggest that all varicosities have a clear junctional relationship to effector cells (Langer, 1981). However at most a small proportion of all varicosities (in the mouse vas deferens less than 1%) are clearly 'junctional', and in all tissues the vast majority are 'nonjunctional' (Burnstock, 1970; Gabella, 1980). Since presumably all varicosities release single quanta to about 1% of the nerve impulses (Stjärne, 1987), only a small fraction of all quanta are applied 'focally' from 'junctional' varicosities into narrow neuro-effector gaps; most quanta are applied 'diffusely' from 'nonjunctional' varicosities into the extrajunctional space (Bennett, 1973).

In tissues such as the mouse vas deferens the 'fast' and 'slow' EJPs reflect the ATP components of quanta from 'junctional' and 'nonjunctional' varicosities, respectively (Fig. 2A; Bennett, 1973; Stjärne, 1987). All stimuli of a given strength release single quanta from the same number (up to 12) of the 1200 'nonjunctional' varicosities innervating the unit bundle. Hence 'slow' EJPs are nonintermittent and relatively constant in amplitude (Stjärne, 1987). The transient depolarization (i.e. the 'slow' EJP) represents a transient increase in the excitability of the smooth muscle to all stimuli. Apparently therefore 'diffuse' application of ATP quanta into the extrajunctional space serves as an amplification mechanism. In contrast, many stimuli fail to release transmitter from the much fewer 'junctional' varicosities, explaining why 'fast' EJPs and 'discrete events' are intermittent. Each successful stimulus releases single quanta from 1-5 'junctional' varicosities at different distances from the recording electrode; hence 'fast' EJPs and 'discrete events' are variable in amplitude (Cunnane and Stjärne, 1984; Stjärne, 1987). The enormous depolarization (by up to 30 mV) caused by a single or a few quanta from 'junctional' varicosities often triggers a propagated muscle action potential and a twitch contraction. Apparently therefore 'focal' application of ATP quanta from 'junctional' varicosities serve to trigger rapid effector responses (Stjärne, 1987).

Similar demonstration of the effects of the NA component of quanta from 'junctional' varicosities is not feasible. However, by bath application of exogenous NA (or other α-agonists) the complex and dramatic α_1- and α_2-effects of 'diffusely' applied quanta from 'nonjunctional' varicosities can be mimicked (Fig.2B). The dual effects on the smooth muscle are rapid in onset and excitatory; those on the nerve terminals are slower in onset and inhibitory.

Thus in the new working hypothesis sympathetic neurotransmission is based on interaction between the (excitatory post- and inhibitory prejunctional) <u>modulating</u> effect of transmitters in the extrajunctional space caused by release of quanta from the numerous 'nonjunctional' varicosities, and the <u>triggering</u> effect of quanta from the much fewer 'junctional' varicosities (Fig. 3A; Stjärne, 1986,1987).

AUTONOMIC COOPERATIVITY

In the guinea-pig and mouse vas deferens, which have a dense sympathetic and a sparse parasympathetic innervation (Burnstock, 1970), the contractile response to electrical nerve stimulation is

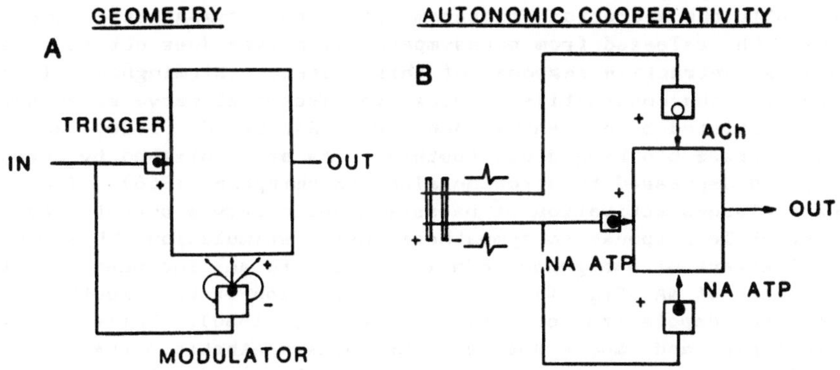

Fig. 3. Amplification mechanisms in sympathetic neurotransmission. (A) The importance of the geometry of varicosities: Most varicosities are 'nonjunctional' and hence most transmitter applied extrajunctionally, potentiating the response of the smooth muscle to transmitter from the much fewer 'junctional' varicosities and autoinhibiting transmitter secretion (Fig. 2B; Stjärne, 1985,1986,1987). (B) Autonomic cooperativity: In the guinea-pig and mouse vas deferens parasympathetic nerves do not antagonize, but potentiate the contractile response to sympathetic nerve stimulation (Birmingham, 1966; Stjärne, 1975; Sjöstrand, 1980; see also Fig. 4).

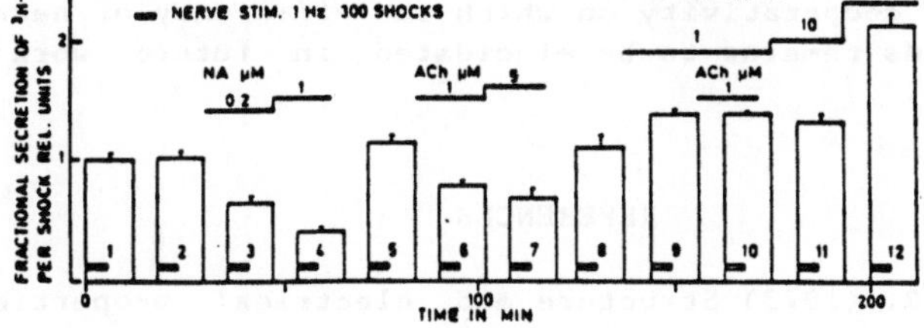

Fig. 4. Upper panel: Representative records of the contractions of the guinea-pig vas deferens to nerve stimulation (the numbers for each stimulation period refer to those in the histogram in the lower panel). Note the different ordinates. Lower panel: Fractional secretion of ^{3}H-NA in response to nerve stimulation. Means + S.E. (3 experiments; Stjärne, 1975).

abolished by guanethidine or bretylium, agents known to block specifically the secretion of transmitter from sympathetic nerves. Thus ACh released from parasympathetic nerves does not in itself cause a contractile response of this tissue (Birmingham, 1966). However, the contractile response to electrical nerve stimulation is potentiated by exogenous ACh (see Sjöstrand, 1980) and by ACh-esterase blocking drugs (both effects are abolished by atropine), and depressed by atropine alone (Birmingham, 1966). Thus in these tissues activation of parasympathetic nerves potentiates the contractile response to sympathetic nerve stimulation. This mechanical effect of endogenous ACh is not due to an increase in the secretion of NA (Fig. 4) but to sensitization of the smooth muscle cells to depolarization (see Sjöstrand, 1980). While in the guinea-pig and mouse vas deferens parasympathetic nerve activity reinforces sympathetic neurotransmission by postjunctional potentiation, in other tissues other forms of pre- and postjunctional interaction occur (see Sjöstrand, 1980).

CONCLUSIONS

A new paradigm is required according to which sympathetic neurotransmission is mediated by highly intermittent secretion of single mixed transmitter quanta from two classes of varicosities, differing in geometry and hence in function. Due to the narrowness of the neuro-effector gap single quanta from a 'junctional' varicosity always saturate the receptor patch on the muscle membrane opposite to the site of release and hence cause a maximal response ('identical' discrete events; Cunnane and Stjärne, 1984; Stjärne, 1986,1987). Still to trigger a contractile response quanta from the small minority of 'junctional' varicosities require the background of enhanced excitability caused by transmitter (at least NA and ATP, possibly NPY also; Stjärne and Lundberg, 1986; Stjärne, Lundberg and Åstrand, 1986) from 'nonjunctional' varicosities (see Vizi, 1985). The nature of the chemical, spatial, temporal and autonomic cooperativity on which the efficiency of neurotransmission depends remains to be elucidated in future work (Stjärne, 1986,1987).

REFERENCES

Bennett M.R. (1973) Structure and electrical properties of the autonomic neuromuscular junction.
Phil. Trans. R. Soc. Lond. B, 265, 25-34.
Bevan J.A., Chesher G.B. and Su C. (1969) Release of adrenergic transmitter from terminal plexus in artery.
Agents and Actions 1, 20-26.
Birmingham, A.T. (1966) The potentiation by anticholinesterase drugs of the responses of the guinea-pig isolated vas deferens to alternate preganglionic and postganglionic stimulation.
Br. J. Pharmac., 27, 145-156.

Blakeley A.G.H. and Cunnane T.C. (1979) The packeted release of transmitter from the sympathetic nerves of the guinea-pig vas deferens: An electrophysiological study.
J. Physiol. Lond., 296, 85-96.

Blakeley A.G.H., Mathie A. and Petersen S.A. (1984) Facilitation at single release sites of a sympathetic neuroeffector junction in the mouse. J. Physiol. Lond., 349, 57-71.

Bolton T.B. and Large W.A. (1986) Are junction potentials essential? Dual mechanism of smooth muscle cell activation by transmitter released from autonomic nerves.
Quart. J. Exp. Physiol., 71, 1-28.

Burnstock, G. (1970) Structure of smooth muscle and its innervation. In Smooth Muscle, (eds. Bülbring, E., Brading, A.F., Jones, A.W. and Tomita, T.), pp. 1-70. Edward Arnold, London.

Burnstock G. (1986a) The changing face of autonomic neurotransmission. Acta physiol. scand., 126, 67-92.

Burnstock G. (1986b) Autonomic neuromuscular junctions: current developments and future directions. J. Anat., 146, 1-30.

Cunnane T.C. and Stjärne L. (1984) Transmitter secretion from individual varicosities of guinea-pig and mouse vas deferens: Highly intermittent and monoquantal. Neuroscience, 13, 1-20.

Folkow B. and Häggendal J. (1970) Some aspects of the quantal release of the adrenergic transmitter. In Bayer-Symposium II, (eds. Schümann, H.J. and Kroneberg, G.), pp.91-97.Springer-Verlag.

Fried, G., Terenius, L., Brodin, E., Efendic, S., Dockray, G., Fahrenkrug, J., Goldstein, M. and Hökfelt, T. (1986) Neuropeptide Y, enkephalin and oradrenaline coexist in sympathetic neurons innervating the bovine spleen. Cell Tissue Res., 243, 495-508.

Furness J.B. (1970) The excitatory input to a single smooth muscle cell. Pflügers Arch., 314, 1-13.

Hackett J.T., Jackson H. and Rubel E.W. (1982) Synaptic excitation of the second and third order auditory neurons in the avian brain stem. Neuroscience, 7, 1455-1469.

Hirst G.D.S. and Neild T.O. (1980) Some properties of spontaneous excitatory junction potentials recorded from arterioles of guinea-pig. J. Physiol. Lond., 303, 43-60.

Hökfelt T. (1969) Distribution of noradrenaline storing particles in peripheral adrenergic neurons as revealed by electron microscopy. Acta physiol. scand. 76, 427-440.

Jack J.J., Redman S.J. and Wong K. (1981) The components of synaptic potentials evoked in cat spinal motoneurones by impulses in single group Ia afferents. J. Physiol. Lond., 321, 65-96.

Kadlec O., Masek K. and Seferna I. (1984) The topography of post-tetanic potentiation in guinea-pig ileum.
Experientia, 40, 404-406.

Korn H. (1984) What central inhibitory pathways tell us about mechanisms of transmitter release.
Exp. Brain Res., Suppl. 9, pp. 201-224.

Langer S.Z. Presynaptic regulation of the release of catecholamines. Pharmac. Rev., 32, 337-362.

Lundberg J.M. and Hökfelt T. (1986) Multiple co-existence of peptides and classical transmitter in peripheral autonomic and sensory neuronsfunctional and pharmacological implications.
Progr. Brain Res., 68, 241-262.
Sjöstrand N.O. (1980) Smooth muscles of vas deferens and other organs in the male reproductive tract. In Smooth Muscle (eds. Bülbring, A., Brading, A.F., Jones, A.-W. and Tomita, T.), pp. 367-376. Edward Arnold, London.
Smith A.D. and Winkler H. (1972) Fundamental mechanisms in the release of catecholamines. In Catecholamines. Handbook of Experimental Pharmacology (eds. Blaschko, H. and Muscholl, E.), Vol. 33, pp. 538-617. Springer Berlin.
Sneddon P. and Westfall D.P. (1984) Pharmacological evidence that adenosine triphosphate and noradrenaline are co-transmitters in the guinea-pig vas deferens. J. Physiol. Lond. 347, 561-580.
Stjärne, L. (1975) Pre- and postjunctional receptor-mediated cholinergic interactions with adrenergic transmission in guinea-pig vas deferens.
Naunyn-Schmiedeberg's Arch. Pharmacol., 288, 305-310.
Stjärne L. (1985) Scope and mechanisms of control of stimulus-secretion coupling in single varicosities of sympathetic nerves.
Clinical Science, 68, (Suppl. 10), 77s-81s.
Stjärne L. (1986) New paradigm: Sympathetic neurotransmission by lateral interaction between secretory units intermittently releasing single mixed quanta?
News in Physiological Sciences, 1, 103-107.
Stjärne L. (1987) New paradigm: A digital model of feedback regulation of sympathetic neurotransmitter secretion. In Proc. 4th Symposium on Vasodilator Mechanisms, Mayo Clinic, Rochester, Minnesota, July 10-12 1986. In press.
Stjärne, L. and Lundberg, J.M. (1986) On the possible roles of noradrenaline, adenosine 5'-triphosphate and neuropeptide Y as sympathetic cotransmitters in the mouse vas deferens.
Progr. Brain Res., 68, 263-278.
Stjärne L. and Åstrand P. (1984) Discrete events measure single quanta of ATP secreted from sympathetic nerves of guinea-pig and mouse vas deferens. Neuroscience, 13, 21-28.
Stjärne L. and Åstrand P. (1985) Relative pre- and postjunctional roles of noradrenaline and adenosine 5'-triphosphate as neurotransmitters of the sympathetic nerves of guinea-pig and mouse vas deferens. Neuroscience, 14, 929-946.
Stjärne L., Lundberg J.M. and Åstrand P. (1986) Neuropeptide Y - a cotransmitter with noradrenaline and adenosine 5'-triphosphate in the sympathetic nerves of the mouse vas deferens? A biochemical, physiological and electropharmacological study.
Neuroscience, 18, 151-166.
Vizi E.S. (1985) Non-synaptic Interactions Between Neurons: Modulation of Neurochemical Transmission. Wiley, Chichester.
Zhu P.C., Thuresson-Klein Å. and Klein R.L. (1986) Exocytosis from large dense cored vesicles outside the active synaptic zones of terminal subnucleus caudalis: a possible mechanism for neuropeptide release. Neuroscience, 19, 43-54.

28
Adenosine Triphosphate: 'Pre-transmitter', Co-transmitter or Modulator at the Skeletal Muscle?

Edith Heilbronn and Johan Häggblad

INTRODUCTION

Burnstock (1982) has summarized evidence suggesting that adenosine triphosphate (ATP) acts as a co-transmitter or neuromodulator in a variety of tissues. This paper concerns ATP and the neuromuscular junction of skeletal muscle and also the embryonic muscle cell, represented by cultured myotubes.

Synaptic vesicles isolated from cholinergic neurons of the electric ray Torpedo marmorata contain ATP together with acetylcholine (ACh) (Dowdall et al, 1974). Very recently, Volknandt and Zimmermann (1986) proved that synaptic vesicles from the motor neurons of rat diaphragm also contain ATP and ACh. The purpose of ATP in cholinergic synaptic vesicles is unknown. Obviously, the compound may serve as an energy source e.g. during uptake of ACh into the vesicle, or ATP may be used when (if?) the vesicles fuse with the presynaptic membrane during evoked transmitter release. Other suggestions include binding of the positively charged ACh to ATP within the vesicles. ATP (Silinsky and Ginsborg, 1983), or rather its degradation product adenosine may act at the presynapse as modulators (reduction) of transmitter release (Ribeiro and Sebastiao, 1986). Scattered observations also suggest a postsynaptic action of ATP at the innervated skeletal muscle or, recently, on cultured myotubes. ATP present in the synaptic cleft originates not only from vesicular release but to an even larger extent from postsynaptic release after membrane depolarization. This paper scrutinizes available evidence for a co-transmitter/modulator role of ATP at the neuromuscular junction, by summarizing the pre- and postsynaptic actions of ATP and metabolites described in the literature and by describing very recent results from our laboratory suggesting that ATP has a postsynaptic function at least at the myotube and perhaps at the non-innervated muscle fiber, perhaps acting as a "pretransmitter" or "conditioner". Our results strongly suggest that ATP at the myotube acts via a P_2-purinoceptor, situated in the myotube membrane, thus causing increased polyphosphoinositide turnover and resulting in raised levels of intracellular inositol phosphates (InsP). The possible implications of found ATP actions are discussed.

REQUIREMENTS FOR ATP TO ACT AS A NEUROTRANSMITTER OR MODULATOR AT THE NEUROMUSCULAR JUNCTION ARE FULLFILLED

Before discussing the evidence suggesting that ATP has a cotransmitter/ modulator action at the skeletal muscle, it seems appropriate to scrutinize if the compound satisfies the requirements for a putative such compound.

The arrival of a nerve impulse at the neuromuscular junction releases ACh and ATP simultaneously into the synaptic cleft (Silinsky and Hubbard, 1973). As mentioned above, the isolation and subsequent analysis of synaptic vesicles from the motor nerve innervating rat diaphragm have demonstrated costorage of ACh and ATP (Volknandt and Zimmermann, 1986; confirmed by Häggblad and Heilbronn). The ratio of ACh/ATP in the vesicles of the mammalian motor neuron is about 10/1. Previous studies have revealed a corresponding costorage in synaptic vesicles from the electromotor nerve endings (Dowdall _et al_, 1974; Tashiro and Stadtler, 1978; Wagner _et al_, 1978) of the electric ray _Torpedo marmorata_, the electric organ of which has long been used as a model of the neuromuscular junction. It should, however, be remembered that differences in the regulation of transmitter release probably exist between the electromotor and the motor neuron. The vesicular ACh and ATP have a correlated but independent metabolism, i.e. depletion and recovery occur at equal rates (Zimmermann and Whittaker, 1974; Zimmermann and Bokor, 1979; Giompres _et al_, 1981). ACh is synthesized in the nerve ending cytoplasm by the enzyme choline acetyltransferase and from there transported into the vesicles; part of it, however, remains in the cytoplasm. ATP probably is synthesized in the mitochondria. Separate uptake systems exist in the neuronal membrane for the ACh precursor choline (Suskiw and Pilar, 1976; Meunier and Morel, 1979) and the ATP precursor adenosine (Zimmermann and Bokor, 1979), respectively as do ecto-phosphatases (ecto-nucleoside di- and triphosphatases, 5'nucleosidases; Grondahl and Zimmermann, 1986) degrading ATP released into the synaptic cleft. These enzymes are considerably slower than acetylcholine esterase (Zimmermann, personal communication). Häggblad (1986) has calculated that, while immediately after release of, say, one quantum of transmitter the concentration of ACh acting at an area of 1 squm in a 50 nm deep synapse would be in the mM region (Kuffler and Nicholls, 1976) and rapidly decrease, that of ATP would decrease about a thousand times slower and thus be ready for a prolonged action.

The literature describes both pre- and postsynaptic responses to externally applied ATP which might be interpreted as those of a co-transmitter or a modulator. Presynaptic effects, as mentioned, seem to be caused mainly by the ATP degradation product adenosine. There are also indications that ATP at the adult neuromuscular junction and in the electric organ acts as a transynaptic messenger, as depolarization of electric organ (Israel <u>et al</u>, 1980), of skeletal muscle (Abood <u>et al</u>, 1962) or of cultured myotubes (Häggblad, 1986) releases considerable quantities of ATP from the postsynaptic cells into the synaptic cleft or to the surrounding medium. This ATP release is blocked by nicotinic antagonists. Further, our recent work (Häggblad <u>et al</u>, 1985a, b; Häggblad and Heilbronn, 1986; Heilbronn and Häggblad, 1986; Häggblad, 1986) shows that externally applied ATP causes myotube contractions, promotes transport of monovalent ions (^{86}Rb) into the cells and also triggers polyphosphoinositide turnover resulting in PIns accumulation in the myotubes (see below). These observations raise the question if ATP also has a role as a "pretransmitter" or "conditioner" at a noninnervated (and denervated?) muscle fiber.

EVIDENCE FOR PURINES AS CAUSATIVE AGENTS AT THE MYOTUBE AND/OR THE NEUROMUSCULAR JUNCTION

Evidence for adenosine.

ATP, released from either pre- or postsynaptic sources and present in the synaptic cleft is evidently metabolized to adenosine. The presence of an adenosine receptor of somewhat unclear nature has been shown (Ribeiro and Sebastiao, 1985; 1986) at the motor neuron ending. Adenosine decreases both basal (Ginsborg and Hirst, 1972; Ribeiro and Domingues, 1978) and stimulated (Ginsborg and Hirst, 1972; Miyamato and Breckenridge, 1974; Ribeiro and Walker, 1975; Silinsky, 1984) ACh release from motor nerve terminals. Decrease in the entry of Ca^{2+} during presynaptic depolarization (Israel <u>et al</u> 1980) or interactions with intracellular Ca-dependent transmitter release processes was suggested (se eg Ribeiro and Sebastiao, 1986).

No clearcut postsynaptic action of adenosine seems to be known at the neuromuscular junction.

Evidence for ATP

ATP as a channel activator.

Several earlier observations suggest synaptic effects of ATP at the striated muscle. Ribeiro & Walker (1975) found that ATP and ADP, on muscle fibers from the rat diaphragm and the frog sartorius,

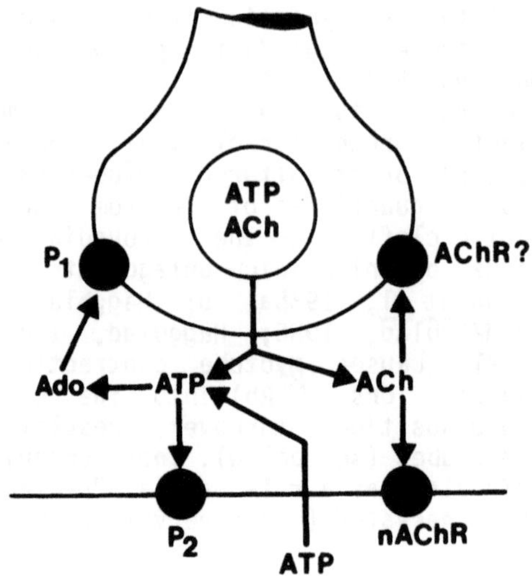

Figure 1: Suggested presynaptic action of ATP and ACh. ATP, released from presynaptic vesicles and postsynaptic cell after depolarization mainly by ACh, is subsequently degraded by ecto-phosphatases to adenosine. Adenosine reduces ACh release via P_1-purinoceptor. Tentatively a nicotinic-like AChR allowing presynaptic facilitation of ACh release is suggested.

decreased the quantum content of the epps and the frequency of the mepps. The potency of ATP equalled that of adenosine, indicating low specificity of a presumtive receptor or rapid hydrolysis of ATP to adenosine. Ewald (1976) suggested that ATP potentiates postsynaptic responses to ACh; later, Akasu et al (1981), using frog sartorius muscle and voltage clamp technique found that ATP increases the sensitivity of the nicotinic acetylcholine receptor (nAChR) to iontophoretically applied ACh, measured as the amplitude of ACh current and suggested that ATP interacts with a nAChR site, without changing the ACh binding affinity. Either the conductance of unit channels or the total number of available channels was suggested to be increased.

ATP has also been suggested to activate cation channels in cultured chick myotubes (Kolb and Wakelam, 1983) These authors discussed an allosteric effect of ATP on nAChR. Two opening events were described electrophysiologically, one with an open time of 3.5 msec and 48pS conductance and another with respectively 1.3 msec and 20 pS. It is not clear if these observations were made on two states of the same channel or on two different channels. The ATP-activated channel(s) did not discriminate between Na- and K-ions.

On the same material ATP was shown to promote influx of $^{86}Rb^+$ in a non-energy-dependent manner. Experimental concentrations below $5x10^{-4}$M were chosen, thus avoiding the mild detergent-like effects of ATP observed both in transformed mouse fibroblast cells (Rozengurt et al, 1977; Weisman et al, 1984) at high pH and absence of external divalent cation (see eg Heppel et al, 1985), and in myotubes by efflux studies with (^{3}H-)deoxyglucose (Häggblad, 1986; Heilbronn & Häggblad, 1986). The slowly hydrolyzable ATP-derivatives, β,γ-imido ATP, β,γ- methylene ATP and α,β-methylene ATP were all non-effective in themselves, but blocked the ATP-induced influx. ADP, AMP and adenosine were all less potent than ATP itself (Häggblad et al, 1985a,b). A weak membrane depolarization was sometimes observed and remains so far unexplained (artefact ?). The ^{86}Rb influx was found to be additive to that induced by cholinergic agonists and could, in contrast to the latter, not be blocked by nAChR antagonists, such as α-bungarotoxin (α-bgt) or d-tubocurarine (Häggblad et al 1985a,b). Thus, it was probably not caused by the opening of the ion channel of the nAChR macromolecule. Tetrodotoxin, amantadine or ouabain also did not affect ATP induced ^{86}Rb influx. The question whether ^{86}Rb passes through a K-channel or not has not yet been answered.

ATP as a trigger of polyphosphoinositide turnover.

ATP is suggested to potentiate the depolarizing effects of ACh in innervated skeletal muscle (Ewald 1976; Akasu et al, 1981) which was proposed to depend on an allosteric interaction of ATP with the nAChR, an idea taken up by Kolb and Wakelam (1983) studying the chick myotube in culture. However, Häggblad et al (1985a,b) found that ATP-induced cation influx into such myotubes is additive to that caused by cholinergic agents and cannot be blocked by α-bgt. Also, ATP was found not to bind to nAChR (Gordon, 1979). Buchthal and Folkow (1948) early described, in a paper that at the time was much disputed because the purity of used ATP was doubted, that ATP promotes muscle contractions. a process known to depend on a transient cytoplasmic increase of the Ca^{2+}-concentration in the muscle fiber.

In several systems, inositol triphosphate ($InsP_3$) has been shown to release Ca^{2+} from intracellular storage organelles (Berridge and Irvine, 1984; Sekar and Hokin, 1986; Volpe et al, 1986). Häggblad and Heilbronn (1986), using (^{3}H)-inositol labelled chick myotubes, observed increase in inositol phosphate accumulation (measured within 10 sec), after application of ATP externally to chick myotubes in culture. A concomittant decrease in membrane-associated polyphosphoinositides was observed. At equal concentrations (0.3mM) none of the ATP-metabolites ADP, AMP and adenosine was as potent as ATP in producing $InsP_3$ accumulation, while the non-hydrolyzable, β,γ-imido ATP was equipotent, suggesting a P_2-purinoceptor-mediated mechanism. The photoaffinity P_2-purinoceptor blocker arylazidoaminopropionyl ATP ($ANAPP_3$, see e g Burnstock and

Kennedy, 1985) inhibited ATP-mediated increase in $InsP_3$ accumulation after illumination of the cultures. No increased $^{45}Ca^{2+}$ influx into the myotubes was observed in the presence of ATP, in contrast to that of carbachol, thus excluding increased polyphosphoinositide-specific phosphodiesterase activity as the cause.

A phosphorothioate derivative of ATP, ATP-γ-S was found to be an even more potent activator of polyphosphoinositide turnover (Häggblad, 1986).

The above described results indicate that, in the chick myotube, the ATP-effect on polyphosphoinositides is mediated by a P_2-purinoceptor. This receptor may tentatively be suggested to be coupled to a polyphosphoinositide-specific phosphodiesterase, via a coupling protein.

Table 1: Pharmacological evidence for P_2-purinoceptor mediated degradation of polyphosphoinositides in cultured chick myotubes.

---

1. Agonist potency order:

$$ATP = \beta,\gamma\text{-imido ATP} > ADP > AMP = \text{adenosine}$$

2. Inhibition of the ATP-induced accumulation of $InsP_3$ by the photo-affinity P_2-purinoceptor blocker $ANAPP_3$.

See Häggblad (1986); Häggblad and Heilbronn (1986).

RELEVANCE OF FINDINGS FOR EMBRYONIC, ADULT INNERVATED AND DENERVATED (MAMMALIAN) SKELETAL MUSCLE

The cultured myotubes used may be regarded as a model of non-innervated embryonic skeletal muscle. They were of non-mammalian type. We have obtained evidence for the occurrence of an ATP-sensitive P_2-purinoceptor there, responsible for the triggering of the turnover of membrane polyphosphoinositides and the accumulation of inositol phosphates, among them $InsP_3$. This compound has been proposed as responsible for release of Ca^{2+} from sarcoplasmic reticulum (SR) into the cytosol (e.g. Volpe et al, 1986). We suggest that, in the absence of an innervation and thus an evoked transmitter release, the ATP- P_2 purinoceptor - $InsP_3$ system might be acting as a "conditioning" mechanism for the myotube. Perhaps ATP acts in the same way at a denervated muscle fiber.

Initiation of muscle activity is a major factor in neural control of intrinsic muscle characteristics (Gilliat and Westgaard, 1978; Lømo, 1976). Stimulation also reduces extrajunctional ACh sensitivity and restores the resting membrane potential of the denervated fiber to normal (Lømo 1976; Desmedt, 1978; Herbison et al, 1983). Electrical stimulation of denervated muscle fiber (Lømo 1976; Lømo and Westgaard, 1975) increases the synthesis of nAChR and improves muscle contraction (Valencic et al, 1986) thus limiting the atrophy of denervated muscle, the loss of oxidative enzymes (Nemeth, 1982) or the change of twitch/tetanus ratio and prevents slowing of contraction and relaxation that follows denervation (Nix, 1982). A purpose of an ATP-induced muscle fiber contraction before innervation or after denervation might be to contribute to the rapid development of the characteristics and properties of the adult muscle fiber, with junctionally localized nAChR and a functioning twitch system, i.e. with a propagated action potential and a triadic system. Such an interpretation of the found ATP effects would mean that in the fully developed, innervated twitch muscle the ATP effect would loose its importance and consequently, the P_2-purinoceptor might disappear. Very preliminary experiments in our laboratory, performed on innervated rat diaphragm speak for such an interpretation of our results (Häggblad, 1986). The mechanism might, however, gain importance again in a denervated muscle fiber or in a muscle fiber where the ACh-release system and/or the nicotinic receptor and the propagated action potential-triadic system are not properly working. This has to be examined. Another system where the P_2-purinoceptor and the polyphosphoinositide-degradation system might work is the slow skeletal muscle fiber.

CONCLUSIONS AND FUTURE WORK

We have suggested that ATP, acting at a myotube, might promote muscle contraction by one of the following two mechanisms: 1) ATP-induced cation influx through a not yet fully identified channel separate from that of nAChR or that of a TTX-sensitive channel; 2) an increase in polyphosphoinositide turnover may, via accumulation of $InsP_3$ contribute to the release of Ca^{2+} from sarcoplasmic reticulum, thus favouring increase in cytosolic Ca^{2+} levels. Both mechanisms may facilitate muscle contraction, especially in non-innervated muscle.

What then is the current evidence that $InsP_3$ triggers Ca^{2+} release from the SR and how would this be achieved in the adult innervated muscle fiber where, so far, we have not proven the existence of a P_2-purinoceptor in the muscle membrane? The very preliminary experiments performed on the rat hemidiaphragm have been performed after a 3 hour loading period with (^{3}H)inositol at 37°C. There may have been considerable proteolysis and thus the result may have due to artefacts. Changes in adult skeletal muscle $InsP_3$ -

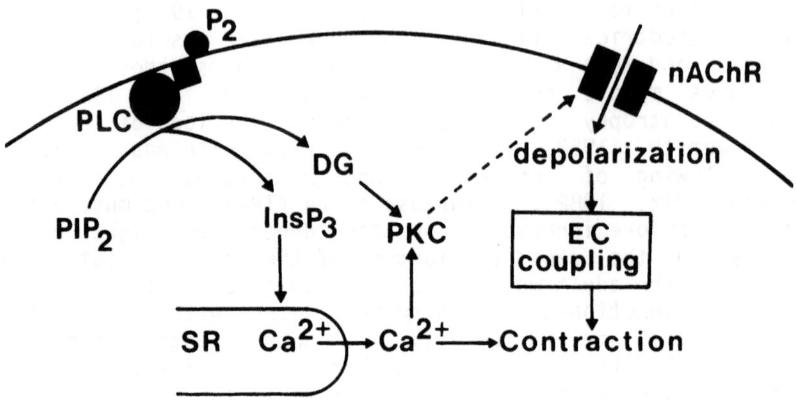

Figure 2: ATP and ACh induced events in the chick myotube. ATP induces membrane polyphosphoinositide (PIP_2) turnover via action on P_2-purinoceptor, coupling protein and phospholipase C (PLC), resulting in an accumulation of $InsP_3$ which may cause a transient Ca^{2+} release from the sarcoplasmatic reticulum (SR). Formed diacylglycerol (and Ca^{2+}) may activate protein kinase C (PKC) to phosphorylate the nAChR from inside of cell.

levels have been observed after vigorous stimulation (50Hz, field stimulation; Vergara et al, 1985) which may have been caused by ATP released from the postsynapse. Adamo et al (1985) found in chick myotubes that ACh stimulates phosphoinositide turnover in cultured myotubes, an effect inhibited by α-bgt and d-tubocurarine. We found, using carbachol, this effect to be very small as compared to that induced by ATP. Adamo et al suggested that their observation might explain why agents known to activate protein kinase C, such as the tumour promotor 12-tetradecanoyl phorbol-13 acetate and diacylglycerol, a product of polyphosphoinositide turnover (see Figure 2) may reduce ACh sensitivity and increase the rate of desensitization in chick embryo cultures myotubes (Eusebi et al 1985). We have, at present, no conclusive evidence that ACh in a twitch muscle causes turnover of polyphosphoinositide and accumulation $InsP_3$ via the nAChR. When studying effects of ACh and carbachol on pieces of rat diaphragm we observe some accumulation of $InsP_3$, but this effect cannot be blocked with 10 uM d-tubocurarine and should therefore not be related to membrane depolarization via the nAChR. We have, as yet, no idea about the situation in slow muscle.

Finally, the observed polyphosphoinositide turnover, the formation of diacylglycerol and the possible increase in cytosolic calcium (Figure 2) might be connected with nAChR phosphorylation and perhaps, nAChR desensitization. Middleton _et al_ (1986) suggested that nAChR phosphorylation can be achieved by three different species of protein kinases and that results obtained with forskolin point to a cAMP-dependent protein kinase (Huganir _et al_, 1986). This is questionable in the light of recent results obtained by Häggblad _et al_ (1986). We suggest that such phosphorylation, _in vivo_, rather is triggered by protein kinase C as a consequence of its activation by diacylglycerol and calcium.

REFERENCES

Adamo, S., Zani, B. M., Nervi, C., Senni, M. I., Molinaro, M. & Eusebi, F. (1985) Acetylcholine stimulation phosphatidylinositol turnover at nicotinic receptors of cultured myotubes. FEBS Lett., 190, 161-164.

Abood, L.G., Koketsu, K. & Miyamoto S. (1962) Outflux of various phosphates during membrane depolarization of excitable membranes. Am. J. Physiol., 202, 469-474.

Akasu, T., Hirai, K. & Koketsu, K. (1981) Increase in acetylcholine receptor sensitivity by adenosine triphosphate: a novel action of ATP on ACh-sensitivity. Br. J. Pharmac., 74, 505-507.

Berridge, M.J. & Irvine, R.F. (1984) Inositol triphosphate, a novel second messenger in cellular signal transduction. Nature, 312, 315-321.

Buchthal, F. & Folkow, B. (1948) Interaction between acetylcholine and adenosine triphosphate in normal, curarized and denervated muscle. Acta Physiol. Scand., 15, 150-160.

Burnstock, G. & Kennedy, C. (1985) Is there a basis for distingguishing two types of P_2-purinoceptor ? Gen. Pharmac., 16, 433-440.

Desmedt, J.E. (1978) Muscular dystrophy contrasted with denervation: different mechanisms underlying spontaneous fibrillations. In _Contemporary clinical neurophysiology_. (eds. WA Cobb & H Van Duijm). Elsevier, Amsterdam.

Dowdall, M.J., Boyne, A.F. & Whittaker, V.P. (1974) Adenosine triphosphate, a constituent of cholinergic synaptic vesicles. Biochem. J., 140, 1-12.

Eusebi, F., Molinaro, M. & Zani, B.M. (1985) Agents that activate protein kinase C reduce acetylcholine sensitivity in cultured myotubes. J. Cell. Biol., 100, 339-1343.

Ewald, D.A.J. (1976) Potentiation of postjunctional sensitivity of rat diaphragm by high-energy-phosphate adenine nucleotides. J. Membr. Biol., 29, 47-65.

Gilliat, R.W. & Westgaard, R.H. (1978) Nerve-muscle interactions. In Contemporary clinical neurophysiology. (eds. WA Cobb & H Van Duijm) Elsevier, Amsterdam.

Ginsborg, B.L. & Hirst, G.D.S. (1972) The effect of adenosine on the release of transmitter from the phrenic nerve of the rat. J. Physiol. (Lond.), 224, 629-645.

Giompres, P.E., Zimmermann, H. & Whittaker, V.P. (1981) Changes in the biochemical and biophysical parameters of cholinergic synaptic vesicles on transmitter release and during subsequent period of rest. Neuroscience, 6, 775-785.

Gordon, A.S., Guillory, R.J., Diamond, I. & Hucho, F. (1979) ATP-binding proteins in acetylcholine receptor enriched membranes. FEBS Lett., 108, 37-39.

Grondal, E.J.M. & Zimmermann, H. (1986) Ectonucleotidase activities associated with cholinergic synaptosomes isolated from Torpedo electric organ. J. Neurochem., 47, 871-881.

Häggblad, J. (1986) Neuromuscular junction revisited: Biochemical studies on mechanisms involved in transmission events. Doctoral dissertation, manuscript. University of Stockholm, Sweden.

Häggblad, J. & Heilbronn, E. (1986) Externally applied ATP causes inositol triphosphate accumulation in cultured chick myotubes. Neurosci Lett., in press.

Häggblad, J., Eriksson, H., Hedlund, B. & Heilbronn, E. (1986) Forskolin blocks agonist mediated permeability of chick myotube nicotinic receptors and inhibits binding of ^{3}H-phencyclidine to Torpedo microsac nicotinic receptors. Naunyn Schm. Arch. Pharmac., submitted.

Häggblad, J., Eriksson, H., & Heilbronn., E. (1985a) ATP induced cation influx in myotubes is additive to cholinergic agonist action. Acta Physiol. Scand., 125, 389-393.

Häggblad, J., Eriksson, H., & Heilbronn, E. (1985b) Effects of extracellular ATP on ^{86}Rb-influx in chick myotubes: Indications of a cotransmitter role in neuromuscular transmission. In Molecular basis of nerve activity. (eds. J-P Changeux, F Hucho, A Maelicke & E Neumann) Walter de Gruyter, Berlin, New York.

Heilbronn, E. & Häggblad, J. (1986) A case for ATP as a neurotransmitter and trigger of postsynaptic Ca-release at the neuromuscular junction? In Cellular and molecular basis of cholinergic funtion. (ed. M J Dowdall), in press.

Heppel, L.A., Weisman, G.A. & Friedberg, I. (1985) Permeabilization of transformed cells in culture by external ATP. J. Membr. Biol., 86, 189-196.

Herbison, G. J., Jaweed, M.M. & Ditunno, J. F. (1983) Acetylcholine sensitivity and fibrillation potentials in electrically stimulated crush denervated rat sekeltal muscle. Arch. Phys. Med. Rehabil. 64, 217-220.

Huganir, R. L., Delcour, A. H., Greengard, P., and Hess, G. P. (1986) Phosphorylation of the nicotinic acetylcholine receptor regulates its rate of desensitization. Nature, 321, 774-775.

Israel, M., Lesbats, B., Manaranche, R., Meunier, F.M. & Frachon, P. (1980) Retrograde inhibition of transmitter release by ATP. J. Neurochem., 34, 923-932.

Kolb, H.A. & Wakelam, M.J.O. (1983) Transmitter-like effects of ATP on patched membranes of cultured myoblasts and myotubes. Nature, 303, 621-623.

Kuffler, S.W. & Nicholls, J.G. (1976) From neurone to brain. Sinauer Ass. Inc., Sunderland, MA, USA.

Lømo, T. (1976) The role of activity in the control of membrane and contractile properties of skeletal muscle. In Motor innervation of muscle. (ed. S Thesleff). Academic Press, London.

Lømo, T. & Westgaard, R. H. (1975) Further studies on the control of ACh sensitivity by muscle activity in the rat. J. Physiol (Lond.), 252, 603-626.

Meunier, F.M. & Morel, N. (1978) Adenosine uptake by cholinergic synaptosomes from Torpedo electric organ. J. Neurochem., 31, 845-851.

Middleton, P., Jaramillo, F. & Schuetze, S.M. (1986) Forskolin increases the rate of acetylcholine receptor desensitization in rat soleus endplates. Proc. Natl. Acad. Sci. USA, 83, 4967-4971.

Miyamoto, M.D. & Breckenridge, B.McL. (1974) A cyclic adenosine monophosphate link in the catecholamine enhancement of transmitter release at the neuromuscular junction. J. Gen. Physiol., 63, 609-624.

Nemeth, P.M. (1982) Electrical stimulation of denervated muscle prevents decreases in oxidative enzymes. Muscle & Nerve, 5, 134-139.

Nix, W. A. (1982) The effect of low-frequency electrical stimulation on the denervated extensor digitorum longus muscle of the rabbit. Acta Neurol. Scand., 66, 521-528.

Ribeiro, J.A. & Dominguez, M.L. (1978) Mechanisms of depression of neuromuscular transmission by ATP and adenosine. J. Physiol. (Paris), 74, 491-496.

Ribeiro, J.A. & Sebastiao, A.M. (1985) On the type of receptor involved in the inhibitory action of adenosine at the neuromuscular junction. Br. J. Pharmac., 84, 911-918.

Ribeiro, J.A. & Sebastiao, A.M. (1986) Adenosine receptors and calcium: Basis for proposing a third (A_3) adenosine receptor. Prog. Neurobiol., 26, 179-209.

Ribeiro, J.A. & Walker, J. (1975) The effects of adenosine triphosphate and adenosine diphosphate on transmission at the rat and frog neuromuscular junctions. Br. J. Pharmac., 54, 213-218.

Rozengurt, E., Heppel, L.A. & Friedberg, I. (1977) Effect of exogenous ATP on the permeability properties of transformed mouse cell lines. J. Biol. Chem., 252, 4584-4590.

Sekar, M.C. & Hokin, L.E. (1986) The role of phosphoinositides in signal transduction. J. Membr. Biol., 89, 193-210.

Silinsky, E.M. (1984) On the mechanism by which adenosine receptor activation inhibits the release of acetylcholine from motor nerve endings. J. Physiol. (Lond.) 346:243-256.

Silinsky, E.M. & Ginsborg, B.L. (1983) Inhibition of acetylcholine release from preganglionic frog nerves by ATP but not by adenosine. Nature, 305, 327-328.

Silinsky, E.M. & Hubbard, J.I. (1973) Release of ATP from rat motor nerve terminals. Nature, 234, 404-405.

Suskiw, J.B. & Pilar, G. (1976) Selective localization of a high affinity choline uptake system and its role in ACh formation in cholinergic nerve terminals. J. Neurochem., 26, 1133-1138.

Tashiro, T. & Stadtler, H. (1978) Chemical compositions of cholinergic synaptic vesicles from Torpedo marmorata based on improved purification. Eur. J. Biochem., 90, 479-487.

Valencic, V., Vodovnok, L., Stefancic, M. & Jelnikas, T. (1986) Improved motor response due to chronic electric stimulation of denervated tibialis anterior muscle in humans. Muscle & Nerve, 9, 612-617.

Vergara, J., Tsien, R.Y. & Delay, M. (1985) Inositol 1,4,5-trisphosphate: A possible chemical link in excitation-contraction coupling in muscle. Proc. Natl. Acad. Sci. USA, 82, 6352-6356.

Volknandt, W. & Zimmermann, H. (1986) Acetylcholine, ATP and proteoglycan are common to synaptic vesicles isolated from electric organs of electric eel and electric catfish as well as from rat diaphragm. J. Neurochem., 47, 1449-1462.

Volpe, P., DiVirgilio, F., Pozzan, T. & Salviati, G. (1986) Role of inositol 1,4,5-trisphosphate in excitation-contraction coupling in skeletal muscle. FEBS Lett., 197, 1-4.

Wagner, J.A., Carlson, S.S. & Kelly, R.B. (1978) Chemical and physical characterization of cholinergic synaptic vesicles. Biochemistry, 17, 1199-1206.

Weisman, G.A., De, B.K., Friedberg, I., Pritchard, R.S. & Heppel, L.A. (1984) Cellular responses to external ATP which precede an increase in nucleotide permeability in transformed cells. J. Cell. Physiol., 119, 211-219.

Zimmermann, H. & Bokor, J.T. (1979) ATP recycles independently of ACh in cholinergic synaptic vesicles. Neurosci. Lett., 13, 319-324.

Zimmermann, H. & Whittaker, V.P. (1974) Effect of electrical stimulation on the yield and composition of synaptic vesicles from cholinergic synapses of the electric organ of Torpedo: a combined biochemical, electrophysiological and morphological study. J. Neurochem., 22, 435-450.

Session III
Receptor–Receptor Interactions in Psychoactive Drug Action

Part A
Anxiolytic Drugs

Chairman: P. Greengard

29
Multiple Signal in GABAergic Transmission: A New Family of Modulatory Peptides

E. Costa, C. Ferrarese, A. Guidotti, M. Miyata and I. Mocchetti

INTRODUCTION

The behavioral and neurological effects of benzodiazepines (BZD) are mediated through high affinity, stereoselective recognition sites which allosterically modulate the probability that γ-aminobutyric acid (GABA) mediates Cl^- channel opening at $GABA_A$ receptors (Costa and Guidotti, 1979).

The $GABA_A$ receptor consists of a supramolecular structure including GABA recognition sites, allosteric modulatory sites recognizing BZD and beta-carbolines (BC) and a polymeric glycoprotein structure composed of α and β subunits that form the dynamic anion-selective pore (Haring et al., 1985; Stephenson et al., 1986) where BZD, BC and $GABA_A$ recognition sites are located. The BZD/BC binding sites do not operate signal transduction by opening Cl^- channels or activating second messengers formation, but negatively (BC) or positively (BZD) modulate GABA mediated Cl^- currents. Thus, their physiological action depends on the presence of GABA, the natural agonist of $GABA_A$ receptors (Haefely et al., 1981). The so-called central BZD and BC recognition sites are allosteric modulatory sites of the $GABA_A$ receptor-gated Cl^- channels located on neuronal membranes.

Two classes of ligands bind to the BZD/BC recognition sites. One class of such ligands (anxiolytic BZD) positively influences the probability that GABA gates the Cl^- channel and the other class of ligands (anxiogenic BC) negatively influences the probability for GABA action. These ligands, cannot be termed "agonists" because they do not produce any effect in absence of the receptor agonist GABA. Provisionally, these ligands could be defined as "allosteric modulators" of GABA receptors with the property of causing positive and negative cooperativity depending on the characteristics of the ligand.

RO-151788 (an imidazobenzodiazepine), CGS-8216 (a pyrazoloquinolin) and ZK-93426 (a beta-carboline-3-carboxylate derivative) virtually have no intrinsic pharmacological activity but block the actions of the positive or negative allosteric modulators of GABA receptors, because they combine with equal affinity to the two conformational states of BZD/BC recognition site but do not modify the action of exogenously applied GABA (Costa et al., 1983). When tested on their ability to modify transynaptically-induced synaptic currents mediated by endogenous GABA, they have a definite action suggesting that an endogenous allosteric modulator of GABA receptors may be coreleased with GABA and have a physiological function (Vicini et al., this book). These experiments were carried out in primary cultures of rat cortical neurons which store GABA and a modulatory peptide of GABA receptors (Alho, personal communication).

In addition to neuronal BZD binding sites associated with GABA recognition sites, radioreceptor binding studies have revealed the presence of another class of BZD binding sites unrelated to GABA receptors. These receptors are located in glial cells and broadly distributed in many non-neuronal tissues including adrenals, testis, kidney, liver and several tumoral cell lines (Braestrup and Squires, 1977; Marangos et al., 1982; Anholt et al., 1984; Anholt et al., 1986). These BZD binding sites have provisionally been defined as "peripheral" or more precisely as mitochondrial (Anholt et al., 1986) BZD recognition sites since structure-activity studies have demonstrated marked differences between them and GABA-associated BZD recognition sites located on the surface of neuronal membranes. For example, RO5-4864 (4-chlorodiazepam) and PK11195 (isoquinoline carboxamide) are the most potent ligands for the BZD binding sites located on intracellular membranes but they are virtually inactive at the central BZD recognition site located on the surface of neuronal membranes. In contrast, clonazepam, one of the most selective ligands for the neuronal BZD binding sites is essentially uncapable of binding to the intracellular binding sites.

The BZD binding sites located on the intracellular membranes are highly concentrated on mitochondria (Anholt et al., 1986) and other intracellular membranes (Ferrarese in preparation) and although their physiological function is unknown, there is sufficient pharmacological evidence that they may be involved in the control of gene expression (Wang et al., 1984; Curran and Morgan, 1985). In fact, stimulation of BZD binding sites on intracellular membranes by appropriate ligands promotes tumor cell differentiation and in PC12 cells facilitates the stimulation of c-fos expression elicited by NGF (Curran and Morgan, 1985).

These studies have suggested that surface and intracellular membrane sites that bind BZD with high affinity might be sites for the action of endogenous physiologically relevant ligands. This possibility has generated great interest in identifying, isolating and purifying endogenous ligands for the BZD recognition sites.

DBI AS ENDOGENOUS ALLOSTERIC MODULATOR OF $GABA_A$ RECEPTORS

A variety of endogenous constituents have been isolated from brain and other tissues and have been reported to displace the neuronal membrane and mitochrondrial membrane binding of BZD. These compounds include purines (Asano and Spector, 1979; Skolnick et al., 1978), nicotinamide (Mohler et al., 1979), BC (Braestrup et al., 1980; Pena et al., 1986), various peptides (Karobat et al., 1978; Guidotti et al., 1978; Davis et al., 1980; Woolf and Nixon, 1981; Colello et al., 1978) and molecules detected by virtue of their affinity to monoclonal antibodies to BZD (Sangameswaren and De Blas, 1985). However, only a few of these substances appear to possess the attributes essential to be ranked as putative endogenous ligands of neuronal membranes BZD recognition site. Purines and nicotinamide displace BZD binding with a very low affinity (K_i in the mmol range). BC occupy BZD recognition sites with high affinity (K_i nmol range) and traces have been detected recently in brain homogenates (Pena et al., 1986), however, it is not yet known if they are present in brain in sufficient quantities to be physiologically relevant (Braestrup et al., 1980). Among the identified endogenous BZD ligands the one that fulfill the criteria for a physiological modulator of BZD recognition site at the GABAergic synapses is a polypeptide called DBI (diazepam binding inhibitor) (Guidotti et al., 1983). This is a 10 KDa polypeptide which may function as a precursor of a family of polypeptides recently sequenced in rat (Ferrero et al., 1986b; Mocchetti et al., 1986), human (Ferrero et al., 1986a; Gray et al., 1986; Marquardt et al., 1986) and cow brain (Marquardt et al., 1986). The human and rat DBI inhibit specifically the binding of BZD and BC to their receptors with a K_i of approximately 5 uM (Guidotti, et al., 1983), but there are probably some other biologically active fragments of DBI that have an even greater biological activity. Pharmacologically, DBI has been shown to act as a putative allosteric modulator of $GABA_A$ receptor regulated Cl^- channels and is endowed with negative cooperativity (Guidotti et al., 1983; Bormann et al., 1985). Immunocytochemical studies show that DBI-like immunoreactivity is unevenly distributed in brain with a region-specific concentration gradient similar to that established for the GABA/BZD receptor complex (Alho, 1985). Furthermore, immunocytochemically, DBI has been shown to colocalize with glutamic acid decarboxylase in primary cultures of rat neurons (Costa et al., 1986) and has been shown to be concentrated in synaptic vesicles by subcellular fractionation studies (Ferrarese et al., 1986). The specific structure of the endogenous ligands derived from DBI is still provisional, pending a clarification of DBI processing.

DBI is present in neurons and glial cells but can be released by depolarization in a TTX- and Ca^{++}-dependent manner only from neurons (Ferrarese, 1986). These studies suggest that the DBI processing products may be coreleased with GABA or released from other neurons in proximity of GABAergic synapses and act as allosteric modulators of GABA receptors in a manner similar to the anxiogenic BC or the anxiolytic BZD.

CHARACTERIZATION OF MULTIPLE GENES CODING FOR DBI-LIKE mRNA.

The amino acid sequence of DBI extracted from rat (Guidotti et al., 1983), human (Ferrero et al., 1986; Marquardt, 1986) and cow (Marquardt, 1986) brain was obtained by sequencing fragments obtained from trypsin digestion and CNBr cleavage. From these structures, oligonucleotide probes were synthesized (Mocchetti et al., 1986; Gray et al., 1986) to hybridize brain cDNA libraries and obtain cDNA clones from which the following amino acid sequences were derived:

```
COW                        SQAEFDKAAEEVKMLKTKPADEEMLFIYSHYKQATVGDINTDRPGML
RAT                        SQADFDKAAEEVKRLKTQPTDEEMLFIYSHFKQATVGDVNTDRPGLL
HUMAN I                    SQAEFEKAAEEVRHLKTKPSDEEMLFIYGHYKQATVGDINTERPGML
HUMAN II WGDLWLLPPASANPGTGTEAEFEKAAEEVRHLKTKPSDEEMLFIYGHYKQATVGDINTERPGML

COW      DFKGKAKWDAWNFLKGTSKEDAMKAYIDKVEELKKKYGI   (Marquardt et al., 1986)
RAT      DLKGKAKWDSWNKLKGTSKENAMKTYVEKVEELKKKYGI   (Mocchetti et al., 1986)
HUMAN I  DFTGKAKWDAWNELKGTSKEDAMKAYINKVEELKKKYGI   (Marquardt et al., 1986)
HUMAN II DFTGKAKWDAWNELKGTSKEDAMKAYINKVEELKKKYGI   (Gray et al., 1986)
```

The DBIs present in various species have a high degree of amino acid sequence homology suggesting that they are produced from the same gene or are encoded in genes very closely related, phylogenetically.

Characterization of human and rat DBI genes by southern blot analyses suggest that DBI is encoded by multiple gene families (Mocchetti et al., 1986; Gray et al., 1986). Both in rat and human brain at least five independent hybridizing bands were observed when the genomic DNA was digested with different restricted endonucleases. Although it cannot be ruled out that some of these genes can be pseudo-genes, good possibilities exist that DBI can be transcribed by more than one gene. Thus, the product of one of these genes or even peptides encoded by more distantly related genes may express a wide spectrum of specific ligands for the different classes.

DYNAMIC EQUILIBRIUM OF DBI FRAGMENTS STORED IN BRAIN AND PERIPHERAL ORGANS

Since DBI-like immunoreactivity is found in neurons, in selective population of glial cells as well as in some peripheral tissues, the discovery of multiple genes raises the possibility that different transcriptional and processing mechanisms exist for DBI produced in different cell types. To unravel this emerging complexity and to obtain more information about the functional role of DBI peptides, we began to examine the DBI dynamic state in neurons and other cells under different experimental conditions.

The availability of a cDNA probe specific for rat brain DBI (Mocchetti et al., 1986), provides the opportunity to measure either the level of DBI mRNA by RNA blot analyses or the location of the DBI mRNA by in situ hybridization techniques. Antibodies raised against DBI (Alho, 1985) can be used to measure changes of DBI content which can then be correlated with changes in mRNA content. Finally, antibodies raised against a biologically active DBI fragment with 18 amino acid residues (ODN) obtained by tryptic digestion of DBI (Ferrero et al., 1986b) help to distinguish whether a change of DBI content can be attributed to an alteration of DBI synthesis, processing or release. In fact, in rats killed with microwaves, RIA with ODN antibodies reveals several molecular forms of ODN-like immunoreactive material which can be separated by HPLC. Similarly, histochemical techniques reveal, in primary cultures of rat brain neurons, the presence of ODN-like immunoreactive material which appears to be the result of a neuron specific processing of DBI.

As shown in Figure 1, there are relatively high levels of DBI mRNA in most brain regions. In all the areas the DBI cDNA hybridizes with a 500 to 600 bases poly A^+ mRNA. The width of the hybridizing band is somewhat broad and could be due to multiple transcripts or less probable to artefacts due to mRNA fragmentation during processing. DBI was present in all the brain areas studied and its concentration correlated with the DBI mRNA content (Figure 1).

The expression of DBI is not limited to the central nervous system; brain DBI cDNA probes hybridize to mRNA of 500 to 600 nucleotides in liver, testis, kidney and other organs. While small amounts of DBI mRNA and DBI are found in gut, lung and muscle, greater amounts of DBI and DBI mRNA are found in liver, adrenal, testis and kidney (Mocchetti et al., 1986; Gray et al., 1986). Also in these organs, in most cases, there is a correlation between the content of DBI-like immunoreactive material and that of DBI mRNA. Moreover, high density of mitochrondrial BZD binding sites have been observed in the peripheral tissues that express DBI and the density of these sites appear to correlate with the concentration of DBI (Gray et al., 1986)). We have recently

FIGURE 1

Correlation between DBI-like immunoreactivity and DBI-mRNA content in different brain areas of rat.

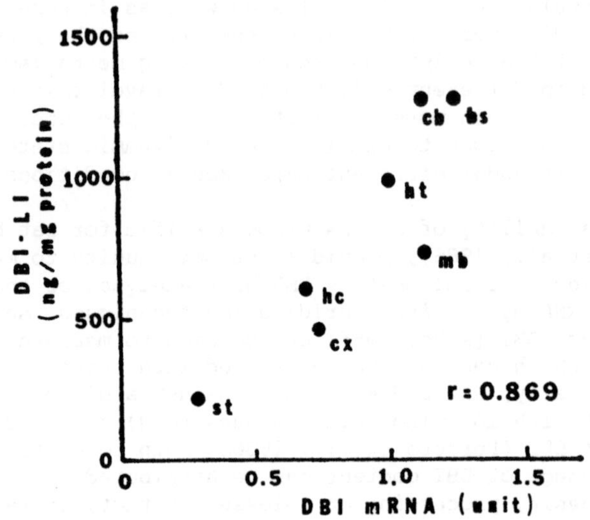

Correlation between DBI-LI and DBI mRNA in Rat Brain

DBI-LI was determined by RIA (Alho et al., 1985). DBI mRNA content was measured in arbitrary units, referred to cerebellum, by hybridization techniques using DBI cDNA (Mocchetti et al., 1986) and a cDNA for an unregulated structural brain protein (Mocchetti et al., 1986). Each value is the mean of 4-5 experiments. st = striatum, cx = cortex, hc = hippocampus, mb = midbrain, ht = hypothalamus, cb = cerebellum, bs - brainstem.

observed a DBI-like peptide purified from rat liver that inhibits the binding of BZD ligands to mitchondrial BZD receptors of adrenal cortex with a K_i of ~ 5 μM. This suggests that a large amount of DBI present in some peripheral organs, such as adrenal, testis and liver, could be related to an unknown regulatory function of DBI on mitochondrial or other intracellular membranes requiring the expression of several DBI genes.

Though preliminary evidence for the existence of high molecular forms of DBI-mRNA exist, these data are not yet consolidated. However, the multiplicity of DBI genes is a strong incentive to persevere in the identification of the multiple molecular forms of DBI mRNA and DBI. While the production and characterization of various molecular forms of cDNA for DBI is now in progress in our laboratory, we are following several lines of independent investigation to foster our present understanding of the function of the DBI polypeptide family. Ferrarese et al., (1986) have recently reported that DBI can be released in vitro upon depolarization from brain slices and neurons in primary culture but not from adrenal and liver slices or from secondary cultures of astrocytes. Subcellular distribution studies also indicate a different intracellular localization of DBI in brain and peripheral organs.

As shown in Table 1, in 0.32 M sucrose tissue homogenates subjected to ultra centrifugation, approximately 70% of brain DBI migrates with the subcellular fraction that contain synaptic vesicles. In contrast, 75% of the DBI present in adrenal, liver and testis remains in the 100,000 x g cytosol fraction. Finally, when brain and adrenal or liver extracts are analyzed by HPLC for the presence of products of DBI processing using ODN antibodies, it is observed that the brain expresses at least three major molecular forms of ODN-like immunoreactive peptides. In contrast, liver and adrenal extracts applied to HPLC fail to show or have only small amounts of ODN-like immunoreactivity.

Table 1. DISTRIBUTION OF DBI-LIKE IMMUNOREACTIVITY IN SUBCELLULAR FRACTIONS OF RAT TISSUES

	% DISTRIBUTION OF DBI-LI			
FRACTION	BRAIN	ADRENAL	TESTIS	LIVER
SUPERNATANT	25	75	70	80
PELLET	75	25	30	20

Homogenates of different tissues were prepared in 0.32 M sucrose. Supernatant and pellet fractions were obtained by centrifugation at 100,000 x 30 min.

CONCLUSION

DBI extracted from rat, human and cow brain belongs to a new family of polypeptides which may have a role in CNS function. Presumably, processing products of DBI may take part in the modulation of GABA/BZD/Cl^- ionophore receptor complex in neuronal membranes or in the regulation of BZD recognition sites located on intracellular membranes.

Genomic DNA blot analysis, cDNA probe hybridization of DBI-mRNA and measurement of DBI and its processing products in neurons, glia and other peripheral tissues suggest that the dynamic state and DBI processing differs in various tissues.

In neurons, DBI is synthesized in a form that can be stored in synaptic vesicles, perhaps this is the molecular form human DBI II that has a "signal peptide" chain. DBI and its fragments can be released upon depolarization from this neuronal storage site.

In peripheral tissues, rich in mitochondrial BZD recognition sites, such as adrenal testis and liver, DBI is expressed in relatively high amount. Lower DBI levels were observed in tissues with lower density of mitochondrial BZD recognition sites. Although a poly A^+ mRNA with an apparent size similar to that of brain can be detected in peripheral tissues the presence of another DBI mRNA in these tissues cannot be excluded.

In all non-neuronal location so far studied, DBI is present in the cytosol fraction and it is not released by depolarization or processed in a manner that can be detected with ODN or DBI antibodies. This suggests that in these non-neuronal locations DBI is not a secretory protein and may act as an intracellular regulatory substance for the control of intracellular functions via the BZD binding sites located on mitochondria or other intracellular membranes.

REFERENCES

Alho, H., Costa, E., Ferrero, P., Fujimoto, M., Cosenza-Murphy, D. and Guidotti, A. (1985). Science, 229, 179-182.
Anholt, R.R.H., Murphy, K.M.M., Mack, G.E. and Snyder, S.H. (1984). J. Neurosci., 4, 593-603.
Anholt, R.R.H., Pedersen, P.L., DeSouza, E.B. and Snyder, S.H. (1986). J. Biol. Chem., 261, 576-583.
Asano, T. and Spector, S. (1979). Proc. Natl. Acad. Sci., 76, 977-981.
Bormann, J., Ferrero, P., Guidotti, A. and Costa, E. (1985). Reg. Peptides, 264, 33-38.
Braestrup, C., Nielsen, M. and Olsen, C. (1980). Proc. Natl. Acad. Sci., 77, 2288-2292.
Braestrup, C. and Squires, R.F. (1977). Proc. Natl. Acad. Sci. 74, 3805-3809.
Coiello, G.D., Hockenbery, D.M., Bosmann, H.B., Fuchs, S. and Folkers, K. (1978). Proc. Natl. Acad. Sci., 75, 6319-6323.
Costa, E., Alho, H., Santi, R., Ferrero, P. and Guidotti, A. (1986). Cotransmission of GABAergic synapses. In Progress in Brain Research, Chapter 24. (eds. T. Hokfolt, X. Fuxe and Pezzow D.). Elsevier Science Publisher. B.V. - Vol. 68, pp. 343-356.
Costa, E., Corda, M.G. and Guidotti, A. (1983). Neuropharmacol., 22, 1481-1492.
Costa, E. and Guidotti, A. (1979). Ann. Rev. Pharmacol. Toxiol., 19, 531-545.
Curran, T. and Morgan, J.I. (1985). Science, 229, 1265-1268.

Davis, L.G. and Cohen, R.K. (1980). Biochem. Biophys. Res. Comm., 92, 141-148.

Ferrarese, C., Vaccarino, F., Alho, H., Mellstrom, B., Costa, E. and Guidotti, A. (1986). J. Neurochem., in press.

Ferrero, P., Conti-Tronoconi, B. and Guidotti, A. (1986a). GABAergic Transmission and Anxiety. (eds. E. Costa and G. Biggio). Advances in Biochem. Psychopharmacol., 41, 177-185, Raven Press.

Ferrero, P., Santi, M.R., Conti-Tronozoni, B. and Costa, E. (1986b). Proc. Natl. Acad. Sci., 83, 827-831.

Gray, P.W., Glaister, D., Seeburg, P.H., Guidotti, A. and Costa, E. (1986). Proc. Natl. Acad. Sci., 83, 7547-7551.

Guidotti, A., Forchetti, C.M., Konkel, D., Bennett, C.D. and Costa, E. (1983). Proc. Natl. Acad. Sci., 80, 3531-3533.

Guidotti, A., Toffano, G. and Costa, E. (1978). Nature, 257, 553-555.

Haefely, W., Pieri, L., Polc, P. and Schaffner, R. (1981). Psychotropic Agents. (eds. F. Hoffmeister and G.M. Stille). Spinger-Verlag, Berlin, pp. 73-262.

Haring, P., Stahli, C., Schoch, P., Takacs, B., Staehelin, T. and Mohler, H. (1985). Proc. Natl. Acad. Sci., 82, 4837-4841.

Karobat, M., Sperk, G. and Schonbeck, G. (1978). Eur. J. Pharmacol., 49, 323-326.

Marangos, P.J., Patel, J., Boulenger, J.P. and Clark-Rosenberg, R. (1982). Mol. Pharmacol., 22, 26-32.

Marquardt, H., Todaro, G.J. and Shoyab, M. (1986). J. Biol. Chem., 261, 9727-9731.

Mocchetti, I., Einstein, R. and Brosius, J. (1986). Proc. Natl. Acad. Sci., 83, 7221-7225.

Mohler, H., Polc, P., Cumin, R., Pieri, L. and Ketter, R. (1979). Nature, 278, 563-565.

Pena, C., Medina, J.H., Novas, M.L., Paladini, A.C. and DeRobertis, E. (1986). Proc. Natl. Acad. Sci., 83, 4952-4956.

Sangameswaran, L. and DeBlas, A.L. (1985). Proc. Natl. Acad. Sci., 82, 5560-5564.

Skolnick, P., Marangos, P.J. and Goodwin, F.K. (1978). Life Science, 23, 1473-1480.

Stephenson, F.A., Casalotti, S.O., Mamalaki, C. and Barnard, E.A. (1986). J. Neurochem., 46, 854-861.

Wang, J.K.T., Morgan, J.I. and Spector, J. (1984). Proc. Natl. Acad. Sci., 84, 3770-3772.

Woolf, J.H. and Nixon, J. (1981). Biochemistry, 20, 4263-4269.

Part B
Antidepressant Drugs

Chairmen: P. Greengard and G. Jonsson

30
Different Presynaptic Receptors Modulate Neuronal Uptake and Transmitter Release in Serotoninergic Nerve Terminals

Salomon Z. Langer and Anne-Marie Galzin

INTRODUCTION

The concept of presynaptic receptors modulating the release of neurotransmitters has arisen from studies on peripheral noradrenergic neurons (Langer, 1974, 1981, Starke, 1981), and has now been extended to the central nervous system for several neurotransmitters, such as noradrenaline, serotonin, acetylcholine and dopamine. It is possible to differentiate between presynaptic inhibitory autoreceptors, acted upon by transmitters which can regulate their own release and in some cases their synthesis, and presynaptic heteroreceptors sensitive to endogenous compounds other than the neuron's own transmitter. Such substances include co-transmitter neuropeptides, transmitters released from adjacent terminals, or locally or blood-borne substances which can facilitate or inhibit the calcium-dependent release of the neurotransmitter.

In addition to the presynaptic receptors modulating transmitter release, presynaptic recognition sites associated with the transporter at the level of the nerve terminal were characterized in serotoninergic, noradrenergic and dopaminergic neurons. The neuronal uptake system for 5HT can be labelled with ^{3}H-imipramine or ^{3}H-paroxetine (Langer et al., 1980b ; Habert et al., 1985), while the noradrenergic transporter can be labelled with ^{3}H-desipramine (Langer et al., 1981a ; Raisman et al., 1982a), and the dopaminergic transporter with ^{3}H-cocaine or ^{3}H-nomifensine (Schoemaker et al., 1985). These presynaptic sites associated with the transporter system differ pharmacologically from the presynaptic autoreceptors which modulate the release of transmitters. The aim of this chapter is to describe the pharmacological characteristics of presynaptic receptors which modulate transmitter release and neuronal uptake in serotoninergic nerve terminals, as well as their functional interaction in the regulation of neurotransmission.

PRESYNAPTIC 5HT AUTORECEPTORS AND THE MODULATION OF THE RELEASE OF SEROTONIN

The experimental evidence available so far supports the view that presynaptic inhibitory 5HT autoreceptors are involved in the modulation of the release of 5HT through a negative feed-back mechanism. The existence of these receptors was first suggested in 1974, when it was reported that in rat brain coronal (Farnebo and Hamberger, 1974) and hippocampal slices (Hamon et al., 1974), the 5HT receptor agonists LSD, ergocornine and 2-brom-α-ergocryptine decrease, while the 5HT receptor antagonist methiothepin increases the ^{3}H-5HT release evoked by high K^+ or electrical stimulation. Among the various 5HT agonists found to inhibit the depolarization evoked release of ^{3}H-5HT, 5-methoxytryptamine is one of the most potent, particularly under conditions of MAO-B inhibition (Galzin et al., 1985 ; Galzin and Langer, 1986 ; see Figure 1). This compound is a putative derivative of tryptophan metabolism in the pineal gland (Pévet, 1983), and it was recently reported that the endogenous levels of 5-methoxytryptamine in the pineal gland of the hamster follow a day-night rhythm in opposition of phase with melatonin (Galzin et al., submitted). It was also found that 5HT itself, in the presence of a 5HT uptake inhibitor, inhibits its own release in rat hypothalamic synaptosomes (Cerrito and Raiteri, 1979) and cortical slices (Göthert and Weinheimer, 1979), and that this inhibitory effect of 5HT is competitively antagonized by methiothepin (Göthert, 1982). These results indicate that both 5HT and methiothepin act at the same site, namely, the presynaptic 5HT inhibitory autoreceptor. In addition, the facilitatory effect of methiothepin on 5HT release is probably due to the disinhibition of the negative feed-back mechanism activated by endogenously released 5HT (Langer and Moret, 1982). The inhibition by LSD and the increase by methiothepin of the release of ^{3}H-5HT is observed for the calcium-dependent, electrically-evoked release, but not for the calcium-independent release of ^{3}H-5HT induced by fenfluramine (Langer and Moret, 1982). Recent _in vivo_ studies also support the view that the presynaptic inhibitory 5HT autoreceptor controls both 5HT release and synthesis in terminal 5HT projection areas (Pettibone and Pflueger, 1984). Chronic treatments with drugs causing activation of 5HT autoreceptors can induce the development of subsensitivity of the presynaptic 5HT autoreceptor while chronic blockade produces supersensitivity of the autoreceptor (Maura and Raiteri, 1984). These results confirm the physiological role of the presynaptic 5HT autoreceptor in the modulation of 5HT neurotransmission, through a negative feed-back mechanism involving the endogenously released transmitter.

Presynaptic inhibitory 5HT autoreceptors were identified in various areas of the rat brain including hippocampus, cerebral cortex, hypothalamus, corpus striatum and cerebellum, and also in other species like the rabbit and the mouse. Recently, inhibitory 5HT autoreceptors were also identified in human cortex slices : the electrically-evoked release of ^{3}H-5HT is inhibited by 5HT, and

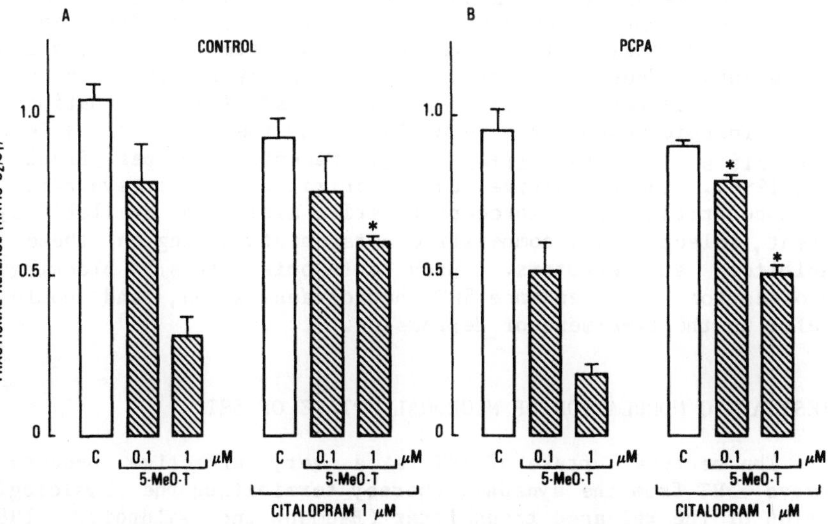

Figure 1 : Antagonism by citalopram of the inhibition by 5-methoxytryptamine of the electrically-evoked release of ^{3}H-5HT from slices of the rat hypothalamus.
Ordinate : Fraction of the total tissue radioactivity released by a 2 min period of electrical stimulation (3 Hz; 20 mA; 2 msec) expressed as the ratio (S_2 / S_1) obtained between the second period of stimulation in the presence of the drug (S_2) and the first period (S_1) carried out within the same experiment.
In A : control rats. In B : PCPA treated rats (300 mg/kg i.p. in 0.4 ml saline, 48 h before the experiment).
C : control ; 5-MeOT : 5-methoxytryptamine. 5-MeOT in the concentrations indicated was added to the medium 20 min before S_2. When used, citalopram (1 μM) was added to the medium 20 min before S_1 and maintained until the end of the experiment. Shown are mean values ± S.E.M of at least three experiments per group.
* $p < 0.01$ when compared to the corresponding concentration of 5-methoxytryptamine in the absence of citalopram.

this effect is antagonized by methiothepin, which increases by itself the release of ^{3}H-5HT (Schlicker et al., 1985).

Pharmacologically, the 5HT autoreceptor clearly differs from the $5HT_2$ or the $5HT_3$ subtype, and possesses the characteristics of a "$5HT_1$-like" receptor subtype (according to the classification of Bradley et al., 1986). It has been suggested that the 5HT autoreceptor localized on 5HT nerve endings in the rat brain is of the $5HT_{1B}$-subtype (Engel et al., 1986), although there is still some controversy on this subject (Vergé et al., 1985). On the other hand the presynaptic inhibitory 5HT autoreceptor localized on cell bodies and/or dendrites of the dorsal raphe nucleus has been identified recently as a $5HT_{1A}$-subtype, with the selective $5HT_{1A}$ ligand ^3H-8-OH-DPAT (Vergé et al., 1985). Such a difference in receptor subtypes between the somatodendritic and the presynaptic autoreceptor does not exist in noradrenergic or dopaminergic neurons (Langer, 1981). This paradoxical situation in serotoninergic neurons may be explained by the lack of selective antagonists for the different $5HT_1$ receptor subtypes (Bradley et al., 1986). The definitive characterization of presynaptic and somatodendritic 5HT autoreceptors still awaits the availability of potent, selective and competitive antagonists acting on these 5HT inhibitory autoreceptors. Such antagonist drugs by blocking the autoreceptors would enhance 5HT neurotransmission, and could be useful in the treatment of depression.

PRESYNAPTIC MODULATION OF NEURONAL UPTAKE OF 5HT

The neuronal uptake of 5HT is a very effective process to remove 5HT from the synapse, thereby terminating the physiological action of the released transmitter (Baumann and Waldmeier, 1981). This process of neuronal uptake involves the sodium-dependent, active transport of the neurotransmitter across the membrane of the nerve terminal. In analogy with the presynaptic receptors which modulate the release of 5HT, noradrenaline or dopamine, it was recently suggested that presynaptic sites may exist, which are involved in the modulation of neuronal uptake of these transmitters (Langer and Raisman, 1983 ; Langer, 1984).

Both ^3H-imipramine and ^3H-paroxetine bind with high affinity to a recognition site which is associated with the neuronal uptake of serotonin, and which is present in the brain as well as in platelets of different species, including humans (Langer et al., 1980a,b, 1981b ; Habert et al., 1985). There is a good correlation between the potencies of drugs at inhibiting ^3H-5HT uptake and at inhibiting ^3H-imipramine binding (Langer et al., 1980b) or ^3H-paroxetine binding (Habert et al., 1985).

Dissociation kinetic experiments indicate that the site labelled by ^3H-imipramine is not identical with the substrate recognition site of the serotonin transporter (Segonzac et al., 1985). It therefore appears that ^3H-imipramine labels a presynaptic modulatory site for the 5HT transporter rather than the substrate recognition site, and that an endocoid different from serotonin may subserve the function of the endogenous ligand acting on the ^3H-imipramine or the ^3H-paroxetine recognition site to modulate 5HT uptake (Langer and Raisman, 1983 ; Barbaccia et al., 1983 ; Langer et al., 1984a,b ; Barbaccia and Costa, 1984).

While investigating the structure activity relationships among known indoleamines which are thought to occur endogenously (Airaksinen and Kari, 1981), it was found that substituted tryptolines inhibited both the binding of ^{3}H-imipramine, as well as the uptake of ^{3}H-5HT (Langer et al., 1984a,b ; Segonzac et al., 1985). These studies led to the proposal that 5-methoxytryptoline (6-methoxy-tetrahydro-betacarboline) or a close analog may be the endogenous modulator acting on the ^{3}H-imipramine or the ^{3}H-paroxetine recognition site (Langer et al., 1984a,b). There is evidence in the literature for the presence of 5-methoxytryptoline in rabbit and pig retinae (Leino et al., 1983) and in cock and post-mortem human pineals (Kari, 1981 ; Kari et al., 1983). However, a recent study revealed only trace amounts of 5-methoxytryptoline in the pineal glands of several species including man (Langer et al., 1985). Therefore, it is still an open question as to whether 5-methoxytryptoline or a close analog plays a physiological role in the modulation of 5HT uptake through an action at the level of the ^{3}H-imipramine or the ^{3}H-paroxetine recognition site (Langer et al., 1986).

The Bmax of ^{3}H-imipramine binding is reduced in platelets from untreated severely depressed patients (Langer et al., 1981b ; Paul et al., 1981 ; Raisman et al., 1982b ; Lewis and McChesney, 1985 ; Poirier et al., 1986). It is therefore possible that the density of ^{3}H-imipramine binding sites in platelets may turn out to be a useful biological marker in depression.

INTERACTION BETWEEN 5HT UPTAKE INHIBITORS AND THE PRESYNAPTIC 5HT AUTORECEPTOR

In the rat hypothalamus, the selective 5HT uptake inhibitor citalopram antagonizes the inhibitory effect of LSD at the level of the presynaptic inhibitory 5HT autoreceptor (Langer and Moret, 1982). This observation was recently extended to other types of antidepressant drugs, such as the tricyclics imipramine and amitriptyline, and the non-tricyclic inhibitor of 5HT uptake paroxetine. These three inhibitors of serotonin uptake antagonized the stimulation of presynaptic inhibitory 5HT autoreceptors by LSD or 5-methoxytryptamine (Galzin et al., 1985 ; Figure 1A). The possibility that this interaction could be due to the fact that the uptake blockers may act as antagonists on the 5HT autoreceptor, was excluded because neither citalopram, imipramine, amitriptyline, paroxetine or 6-nitroquipazine increased, by themselves, the overflow of ^{3}H-5HT elicited by electrical stimulation (Figure 2A). This is in contrast to the findings with the 5HT autoreceptor antagonist, methiothepin which enhances ^{3}H-5HT overflow during electrical stimulation (Langer and Moret, 1982 ; Galzin et al., 1985). The lack of effect of 5HT uptake inhibitors on the electrically-evoked release of ^{3}H-5HT is surprising when compared with the well-known effects of noradrenaline or dopamine uptake inhibitors which increase the electrically-evoked release of these

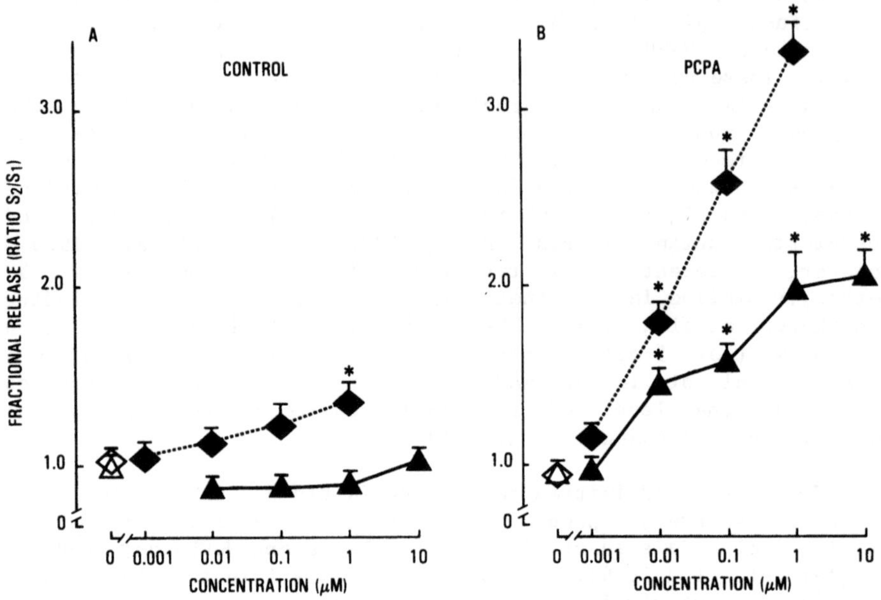

Figure 2 : Effect of 6-nitroquipazine and citalopram on the electrically-evoked release of ^{3}H-5HT from slices of the rat hypothalamus.
Ordinate : Fraction of the total tissue radioactivity released by a 2 min period of electrical stimulation (3 Hz; 20 mA; 2 msec) expressed as the ratio (S_2 / S_1) obtained between the second period of stimulation in the presence of the drug (S_2) and the first period (S_1) carried out within the same experiment.

In A : control rats. In B : PCPA-treated rats (300 mg/kg i.p. in 0.4 ml saline, 48 h before the experiment).

△ : control ; ◆ : 6-nitroquipazine ; ▲ : citalopram.

Abscissae : Concentration of drugs (in micromolar). 6-Nitroquipazine or citalopram in the concentrations indicated were added to the medium 20 min before S_2.
Each point represents the mean values $\pm$ S.E.M. of 3 to 11 experiments per group. * $p < 0.01$ when compared to the corresponding control value in the absence of the drug.
Note that 6-nitroquipazine or citalopram enhanced ^{3}H-5HT release following pretreatment with PCPA but failed to do so in control untreated rats.

catecholamines (Pelayo et al., 1980 ; Kamal et al., 1983). Yet, in rats treated acutely with the tryptophan hydroxylase inhibitor parachlorophenylalanine (PCPA) in a dose producing a 90% decrease of endogenous stores of 5HT (Table 1), exposure to 5HT uptake inhibitors markedly increased the electrically-evoked release of ^{3}H-5HT (Figure 2B). These results suggested that depletion of endogenous 5HT with PCPA may modify the sensitivity of the neuronal 5HT transporter. However, there was no change in the density or affinity of ^{3}H-paroxetine binding to membranes of the rat cerebral cortex after treatment with PCPA (Table 1), indicating that the modulatory site of the 5HT transporter was not directly affected, but that PCPA treatment appears to modify the sensitivity of the 5HT transporter through a mechanism which remains to be elucidated.

Table 1 : ^{3}H-paroxetine binding in control and PCPA-treated rat brain cortical membranes.

	Bmax (fmoles/mg prot)	Kd (pM)	Endogenous levels of 5HT[a)] ng/g (hypothalamus)
Control	453.3 ± 8.5 (8)	72.0 ± 7.0 (8)	1052.0 ± 33.0 (5)
PCPA (300 mg/kg ip 48 h before experiment)	417.7 ± 44.3 (8)	59.0 ± 9.0 (8)	119.0 ± 44.0* (5)

a) endogenous levels of 5HT were measured by HPLC with coulometric detection in hypothalamus of control and PCPA-treated rats.
The Bmax and Kd values were calculated from Scatchard analysis using nine concentrations of ^{3}H-paroxetine from 0.025 to 4 nM.
* $p < 0.01$ when compared to the control value. Number of experiments between brackets.

The interaction between 5HT uptake inhibitors and 5HT autoreceptor agonists does not seem to be due to increased levels of released 5HT in the synaptic cleft under conditions of reuptake inhibition, since a similar interaction is observed following depletion of endogenous stores of 5HT by pretreatment of rats with PCPA (Galzin et al., 1985 ; Figure 1B). The physiological relevance of this interaction between the 5HT transporter and the autoreceptor was recently confirmed by two different studies. Firstly, a comparison of release of ^{3}H-5HT by high K^{+} or by electrical depolarization in the rat hypothalamus indicates that citalopram does not antagonize the inhibitory effect of LSD when the release of ^{3}H-5HT is evoked by K^{+} depolarization, but that the

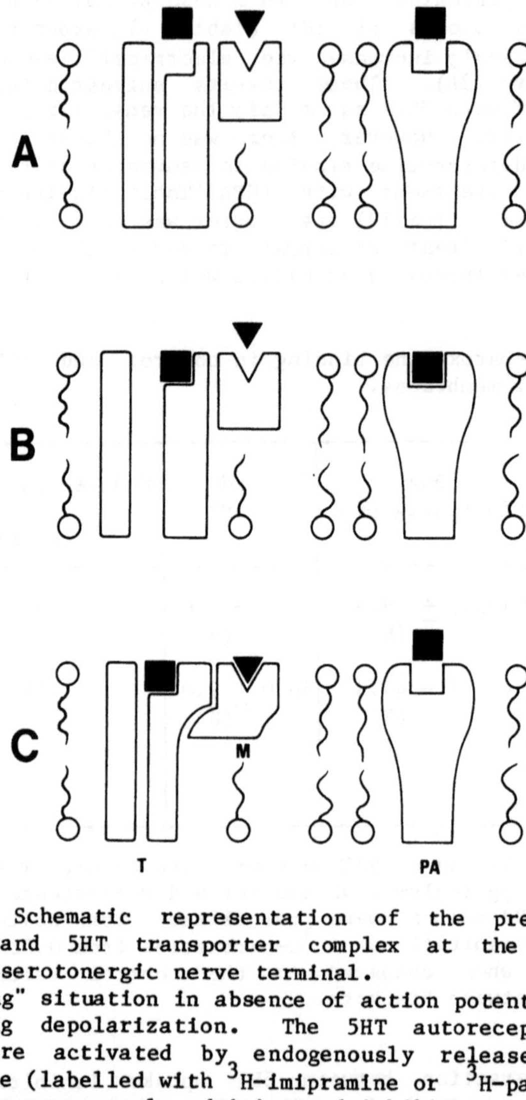

<u>Figure</u> <u>3</u> : Schematic representation of the presynaptic 5HT autoreceptor and 5HT transporter complex at the level of the membrane of a serotonergic nerve terminal.
In A : "resting" situation in absence of action potential.
In B : during depolarization. The 5HT autoreceptor and 5HT transporter are activated by endogenously released 5HT. The modulatory site (labelled with ^{3}H-imipramine or ^{3}H-paroxetine) is not occupied by exogenously administered inhibitors.
In C : same as in B, but the modulatory site is occupied by a 5HT uptake inhibitor or by an endocoid. The 5HT uptake is inhibited through an allosteric mechanism, acting on the transporter while the sensitivity of the 5HT autoreceptor is decreased (through an effect at the level of the membrane or through intraneuronal mechanisms involving second messengers or other phenomena which remain to be elucidated).
■ : 5HT ; ▼ : 5HT uptake inhibitor or endocoid ; T : 5HT transporter ; M : modulatory site ; PA : presynaptic 5HT autoreceptor.

interaction is present only under conditions of transmitter release elicited by electrical stimulation (Passarelli et al., 1987). Secondly, chronic treatment with citalopram induces the development of subsensitivity at the level of the nerve terminal (as well as somatodendritic) 5HT autoreceptor, as measured by electrophysiological techniques in the rat hippocampus (Chaput et al., 1986). It appears that the development of subsensitivity of the 5HT autoreceptor induced by long-term treatment with citalopram is more complex than a simple decrease in the number of receptors due to chronic exposure to an increased level of synaptic serotonin (Chaput et al., 1986).

The interaction between inhibitors of 5HT uptake and the 5HT autoreceptor may be relevant to the antidepressant effects of this class of drugs, which, by producing subsensitivity of the presynaptic 5HT autoreceptors, enhance 5HT neurotransmission. It is tempting to speculate that the time-dependent development of this subsensitivity of 5HT autoreceptors observed under in vivo conditions, could explain the latency period observed for the clinical improvement during antidepressant drug therapy with drugs which inhibit serotonin uptake.

CONCLUSIONS

In addition to the presynaptic autoreceptors involved in the modulation of 5HT release, it is now generally accepted that a separate receptor-mediated process modulates the neuronal uptake of the neurotransmitter through the 5HT transporter. This modulation of the neuronal uptake of 5HT involves a site which possesses most of the properties of a pharmacological receptor, and which is different from the presynaptic inhibitory 5HT autoreceptor modulating the release of the neurotransmitter (Figure 3). Although pharmacologically different, both presynaptic recognition sites may be linked either at the level of the neuronal membrane or through intracellular messengers (Figure 3) and this interaction could be relevant to the mechanism of action of antidepressant drugs which inhibit 5HT uptake.

REFERENCES

Airaksinen, M.M. and Kari, I. (1981). B-Carbolines, psychoactive compounds in the mammalian body. Med. Biol., 59, 190-211.

Barbaccia, M.L. and Costa, E. (1984). Autocoids for drug receptors : a new approach in drug development. New York Acad. Sci., 430, 103-114.

Barbaccia, M.L., Gandolfi, O., Chuang, D.M. and Costa, E. (1983). Modulation of neuronal 5HT uptake by a putative endogenous ligand of imipramine recognition sites. Proc. Natl. Acad. Sci. USA, 80,

5134-5138.

Baumann, P.A. and Waldmeier, P.C. (1981). Further evidence for negative feed-back control of serotonin release in the central nervous system. Naunyn-Schmiedeberg's Arch. Pharmacol., 317, 36-43.

Bradley, P.B., Engel, G., Feniuk, W. Fozard, J.R., Humphrey, P.P.A., Middlemiss, D.N., Mylecharane, E.J., Richardson, B.P. and Saxena, P.R. (1986). Proposals for the classification and nomenclature of functional receptors for 5-hydroxytryptamine. Neuropharmacology, 25, 563-576.

Cerrito, F. and Raiteri, M. (1979). Serotonin release is modulated by presynaptic autoreceptors. Eur. J. Pharmacol., 57, 427-430.

Chaput, Y., de Montigny, C. and Blier, P. (1986). Effects of a selective 5HT reuptake blocker, citalopram, on the sensitivity of 5HT autoreceptors : electrophysiological studies in the rat brain. Naunyn-Schmiedeberg's Arch. Pharmacol., 333, 342-348.

Engel, G., Göthert, M., Hoyer, D., Schlicker, E. and Hillenbrand, K. (1986). Identity of inhibitory presynaptic 5-hydroxytryptamine (5-HT) autoreceptors in the rat brain cortex with $5HT_{1B}$ binding sites. Naunyn-Schmiedeberg's Arch. Pharmacol., 332, 1-7.

Farnebo, L.O. and Hamberger, B. (1974). Regulation of [^{3}H]5-hydroxytryptamine release from rat brain slices. J. Pharm. Pharmacol., 26, 642-644.

Galzin, A.M. and Langer, S.Z. (1986). Potentiation by deprenyl of the autoreceptor-mediated inhibition of [^{3}H]-5-hydroxytryptamine release by 5-methoxytryptamine. Naunyn-Schmiedeberg's Arch. Pharmacol., 333, 330-333.

Galzin, A.M., Eon, M.T., Esnaud, H., Lee, C.R., Pévet, P. and Langer, S.Z. Day-night rhythm of 5-methoxytryptamine levels in the pineal gland of the Golden Hamster (Mesocricetus auratus). Endocrinology, submitted.

Galzin, A.M., Moret, C., Verzier, B. and Langer, S.Z. (1985). Interaction between tricyclic and nontricyclic 5-hydroxytryptamine uptake inhibitors and the presynaptic 5-hydroxytryptamine inhibitory autoreceptors in the rat hypothalamus. J. Pharmacol. Exp. Ther., 235, 200-211.

Göthert, M. (1982). Modulation of serotonin release in the brain via presynaptic receptors. Trends in Pharmacol. Sci., 3, 437-440.

Göthert, M. and Weinheimer, G. (1979). Extracellular 5-hydroxytryptamine inhibits 5-hydroxytryptamine release from rat brain cortex slices. Naunyn-Schmiedeberg's Arch. Pharmacol., 310,

93-96.

Habert, E., Graham, D., Tahraoui, L., Claustre, Y. and Langer, S.Z. (1985). Characterization of [^{3}H]-paroxetine binding to rat cortical membranes. Eur. J. Pharmacol., 118, 107-114.

Hamon, M., Bourgoin, S., Jagger, J. and Glowinski, J. (1974). Effects of LSD on synthesis and release of 5-HT in rat brain slices. Brain Res., 69, 265-280.

Kamal, L., Arbilla, S., Galzin, A.-M. and Langer, S.Z. (1983). Amphetamine inhibits the electrically evoked release of [^{3}H]dopamine from slices of the rabbit caudate. J. Pharmacol. Exp. Ther., 227, 446-458.

Kari, I. (1981). 6-Methoxy-1,2,3,4-tetrahydro-B-carboline in pineal gland of chicken and cock. FEBS Lett., 127, 277-280.

Kari, I., Airaksinen, M.M., Gynther, J. and Huhtikangas, A. (1983). Mass spectrometric identification of 6-methoxy-1,2,3,4-tetrahydro-B-carboline in pineal gland. In Recent developments in biochemistry, medicine and environmental research. (ed. A. Frigerio). 8, Elsevier, Amsterdam.

Langer, S.Z. (1974). Presynaptic regulation of catecholamine release. Biochem. Pharmacol. 23, 1793-1800.

Langer, S.Z. (1981). Presynaptic regulation of the release of catecholamines. Pharmacol. Rev., 32, 337-362.

Langer, S.Z. (1984). [^{3}H]imipramine and [^{3}H]desipramine binding : non-specific displaceable sites or physiologically relevant sites associated with the uptake of serotonin and noradrenaine ? Trends in Pharmacol. Sci., 5, 51-52.

Langer, S.Z. and Moret, C. (1982). Citalopram antagonizes the stimulation by lysergic acid diethylamide of presynaptic inhibitory serotonin autoreceptors in the rat hypothalamus. J. Pharmacol. Exp. Ther., 222, 220-226.

Langer, S.Z. and Raisman, R. (1983). Binding of [^{3}H]imipramine and [^{3}H]desipramine as biochemical tools for studies in depression. Neuropharmacology, 22, 407-413.

Langer, S.Z., Briley, M.S., Raisman, R., Henry, J.F. and Morselli, P.L. (1980a). Specific ^{3}H-imipramine binding in human platelets : influence of age and sex. Naunyn-Schmiedeberg's Arch. Pharmacol., 313, 189-194.

Langer, S.Z., Lee, C.R., Schoemaker, H., Segonzac, A. and Esnaud, H. (1985). 5-Methoxytryptoline and close analogs as candidates for the endogenous ligand of the ^{3}H-imipramine recognition site. In

Endocoids. (eds. H. Lal, F. Labella and J. Lane). Alan R. Liss, New York, pp. 441-455.

Langer, S.Z., Lee, C.R., Segonzac, A., Tateishi, T., Esnaud, H., Schoemaker, H. and Winblad, B. (1984b). Possible endocrine role of the pineal gland for 6-methoxytetrahydro-beta-carboline, a putative endogenous neuromodulator of the [^{3}H]imipramine recognition site. Eur. J. Pharmacol., 102, 379-380.

Langer, S.Z., Moret, C., Raisman, R., Dubocovich, M.L. and Briley, M.S. (1980b). High-affinity [^{3}H]imipramine binding in rat hypothalamus is associated with the uptake of serotonin but not norepinephrine. Science, 210, 1133-1135.

Langer, S.Z., Raisman, R. and Briley, M.S. (1981a). High-affinity [^{3}H]DMI binding is associated with neuronal noradrenaline uptake in the periphery and the central nervous system. Eur. J. Pharmacol., 72, 423-424.

Langer, S.Z., Raisman, R., Tahraoui, L., Scatton, B., Niddam, R., Lee, C.R. and Claustre, Y. (1984a). Substituted tetrahydro-B-carbolines are possible candidates as endogenous ligand of the [^{3}H]imipramine recognition site. Eur. J. Pharmacol., 98, 153-154.

Langer, S.Z., Galzin, A.M., Lee, C.R. and Schoemaker, H. (1986) Antidepressant binding sites in brain and platelets In Antidepressants and receptor function Ciba Foundation Symposium 123, John Wiley and Sons, pp. 3-29.

Langer, S.Z., Zarifian, E., Briley, M.S., Raisman, R. and Sechter, D. (1981b). High-affinity binding of ^{3}H-imipramine in brain and platelets and its relevance to the biochemistry of affective disorders. Life Sci., 29, 211-218.

Leino, M., Kari, I., Airaksinen, M.M. and Gynther, J. (1983). 6-Methoxy-tetrahydro-B-carboline in the retinae of rabbits and pigs. Exp. Eye Res., 36, 135-138.

Lewis, D.A. and McChesney, C. (1985). Tritiated imipramine binding distinguishes among subtypes of depression. Arch. Gen. Psychiatry, 42, 485-488.

Maura, G. and Raiteri, M. (1984). Functional evidence that chronic drugs induce adaptive changes of central autoreceptors regulating serotonin release. Eur. J. Pharmacol., 97, 309-313.

Paul, S.M., Rehavi, M., Skolnick, P., Ballenger, J.C. and Goodwin, F.K. (1981). Depressed patients have decreased binding of tritiated imipramine to the platelets serotonin "transporter". Arch. Gen. Psychiatry, 38, 1315-1317.

Passarelli, F., Galzin, A.M. and Langer, S.Z. Interaction between neuronal uptake inhibitors and presynaptic serotonin autoreceptors in rat hypothalamic slices : comparison of K^+ and electrical depolarization. J. Pharmacol. Exp. Ther., submitted.

Pelayo, F., Dubocovich, M.L. and Langer, S.Z. (1980). Inhibition of neuronal uptake reduces the presynaptic effects of clonidine but not of α-methylnoradrenaline on the stimulation-evoked release of [^{3}H]-noradrenaline from rat occipital cortex slices. Eur. J. Pharmacol., 64, 143-155.

Pettibone, D.J. and Pflueger, A.B. (1984). Effects of methiothepin and lysergic acid diethylamide on serotonin release in vitro and serotonin synthesis in vivo : possible relation to serotonin autoreceptor function. J. Neurochem., 43, 83-90.

Pévet, P. (1983). Is 5-methoxytryptamine a pineal hormone ? Psychoneuroendocrinology, 8, 61-73.

Poirier, M.F., Benkelfat, C., Loo, H., Sechter, D., Zarifian, E., Galzin, A.M. and Langer, S.Z. (1986). Reduced B_{max} of [^{3}H]-imipramine binding to platelets of depressed patients free of previous medication with 5HT uptake inhibitors. Psychopharmacol., 89, 456-461.

Raisman, R., Briley, M.S., Bouchami, F., Sechter, D., Zarifian, E. and Langer, S.Z. (1982b). ^{3}H-Imipramine binding and serotonin uptake in platelets from untreated depressed patients and control volunteers. Psychopharmacol., 77, 332-335.

Raisman, R., Sette, M., Pimoule, C., Briley, M.S. and Langer, S.Z. (1982a). High affinity [^{3}H]desipramine binding in the peripheral and central nervous system : a specific site associated with the neuronal uptake of noradrenaline. Eur. J. Pharmacol., 78, 345-351.

Schlicker, E., Brandt, F., Classen, K. and Göthert, M. (1985). Serotonin release in human cerebral cortex and its modulation via serotonin receptors. Brain Res., 331, 337-341.

Schoemaker, H., Pimoule, C., Arbilla, S., Scatton, B., Javoy-Agid, F. and Langer S.Z. (1985). Sodium-dependent [^{3}H]cocaine binding associated with dopamine uptake sites in the rat striatum and human putamen : decrease after dopaminergic denervation and in Parkinson's disease. Naunyn-Schmiedeberg's Arch. Pharmacol., 329, 227-235.

Segonzac, A., Schoemaker, H., Tateishi, T. and Langer, S.Z. (1985). 5-Methoxytryptoline, a competitive endocoid acting at ^{3}H-imipramine recognition sites in human platelets. J. Neurochem., 45, 249-256.

Starke, K. (1981). Presynaptic receptors. Ann. Rev. Pharmacol. Toxicol., 21, 7-30.

Vergé, D., Daval, G., Patey, A., Gozlan, H., El-Mestikawy, S. and Hamon, M. (1985). Presynaptic 5-HT autoreceptors on serotonergic cell bodies and/or dendrites but not terminals are of the $5HT_{1A}$ subtype. Eur. J. Pharmacol., 113, 463-464.

31

Serotonin/Norepinephrine Receptor Interactions: Sensitivity Changes after Antidepressants and Lesions

Elaine Sanders-Bush and Fridolin Sulser

The morphological organization of monoamine systems in brain and psychopharmacological studies have long suggested a functional linkage between noradrenergic and serotonergic neuronal systems. Recent studies of the molecular mechanism of action of antidepressants have added additional support for an aminergic interaction in the central nervous system (CNS). In the present manuscript, we will review this evidence and evaluate the possible molecular basis for the adrenergic/serotonergic link.

DUAL REQUIREMENT OF NORADRENERGIC AND SEROTONERGIC SYSTEMS IN ANTIDEPRESSANT-INDUCED DOWN-REGULATION OF BETA-ADRENOCEPTORS

Down-regulation of beta-adrenoceptors and desensitization of the noradrenergic beta-receptor linked adenylate cyclase/cyclic AMP system is a well-documented adaptive response to chronic administration of tricyclic antidepressants. This neurochemical effect is common to a number of antidepressant treatments and forms the basis of the "revised" catecholamine hypothesis of affective disorders [see Sulser *et al.* (1983) for a review]. Interestingly, intact noradrenergic *and* serotonergic neurons are both required for the antidepressant-induced down-regulation of beta-adrenoceptors. This co-requirement is illustrated in Figure 1, in which the regional effects of desipramine are compared in control and raphe-lesioned rats. All three brain regions (frontal cortex, limbic forebrain and striatum) receive serotonergic input from the raphe nuclei, but, unlike the other areas, the striatum does not receive noradrenergic projections. Also, unlike those in the limbic forebrain and frontal cortex, beta-adrenoceptors in the striatum are not down-regulated by chronic desipramine treatment, confirming the previous results suggesting that the synaptic availability of the agonist norepinephrine (NE) is a prerequisite for this response. A co-requirement for serotonin (5HT) is also suggested, since selective lesions of serotonergic neurons with 5,7-dihydroxytryptamine (5,7-DHT) prevented the down-regulation of beta-adrenoceptors after chronic administration of desipramine. Studies showing that selective depletion of 5HT stores by inhibition of tryptophan hydroxylase (Manier *et al.*, 1984; Brunello *et al.*, 1985) also prevents the desipramine-induced loss of beta-adrenoceptors are

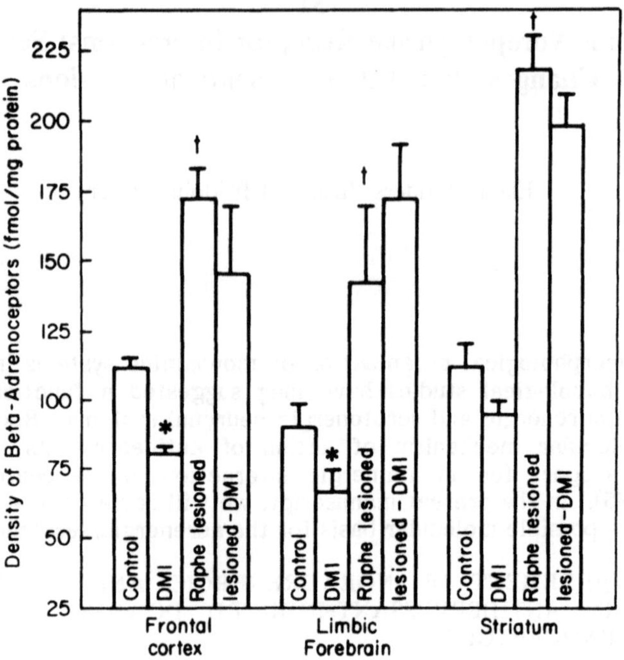

Figure 1. Consequences of 5,7-DHT lesions of serotonergic neurons on regional alteration of beta-adrenoceptors by desipramine (DMI). Ten days following the intraventricular injection of 5,7-DHT, rats were treated daily with desipramine (15 mg/kg, i.p.) or saline for 3 weeks. One day after the last dose, the binding of [^{3}H]-dihydroalprenolol to beta-adrenoceptors was determined. The values are means of 4-9 rats; the vertical bars represent the S.E.M. * $p < 0.05$ versus control; + $p < 0.001$ versus control.

consistent with the interpretation that 5HT is the critical link, rather than some substance that is stored with 5HT in nerve terminals. There is, however, one report (Nimgaonkar et al., 1985) that tryptophan hydroxylase inhibition does not prevent the beta-adrenoceptor down-regulation by antidepressants, although this study confirms that neurotoxic lesions of 5HT terminals do prevent the down-regulation.

Lesions of 5HT neurons with 5,7-DHT results in an increase in the density of beta-adrenoceptors in all three brain areas (Figure 1). Similar increases in the density of beta-adrenoceptors following 5,7-DHT lesions have been previously reported (Janowsky et al., 1982; Stockmeier et al., 1985). However, the cyclic AMP response to a maximal concentration of NE does not altered (Manier et al., 1986), indicating that an increased density does not necessarily predict a parallel change in the functional

Table 1. Non-linear regression analysis of isoproterenol competition binding curves

	Affinity (nM)		Density (fmol/mg protein)	
	K_H	K_L	R_H	R_L
Controls	28 ± 5	36,500 ± 10,200	77 ± 3 (69%)	34 ± 3 (31%)
5,7-DHT	42 ± 10	14,000 ± 1,000	69 ± 6 (36%)	123 ± 14* (64%)

Serotonergic neurons were lesioned by intraventricular injection of 5,7-DHT. Four weeks later, striatal membranes were prepared and (-)isoproterenol competition for [^{3}H]-dihydroalprenolol binding was measured. Competition curves were analyzed using the LIGAND computer program. In both treatments, the data were best fit with a two-site model. R_H and R_L define those receptors which bind isoproterenol with high and low affinity. The dissociation constants for the two affinity states are defined as K_H and K_L (see Manier et al. (1986) for additional details). * $p < .001$ vs. Control.

response. Non-linear regression analyses of agonist competition curves showed that the new sites in 5,7-DHT lesioned rats are largely sites with a low affinity for the agonist, (-)-isoproterenol, while the number of high affinity sites is not changed (Table 1). This finding may explain the discrepancy between increased binding site number and unaltered response, since the ability of a beta-adrenergic agonist to activate adenylate cyclase correlates with the proportion of sites in the high affinity state (Kent et al., 1980).

5HT RECEPTORS AND MEMBRANE TRANSDUCING PATHWAYS

Studies of 5HT receptor function after chronic antidepressants have documented a down-regulation of the 5HT-2 receptor site after chronic administration of tricyclic antidepressants, monoamine oxidase inhibitors and certain atypical antidepressants [see Anderson (1983) for review]. 5HT-1 subtypes are for the most part not altered by these various treatments. The neurochemical changes in 5HT-2 receptor density are not reflected in behavioral or electrophysiological models of 5HT receptors, where chronic antidepressants induce a functional supersensitivity. Although the discrepancy between electrophysiological changes and binding site density is likely explained by different 5HT receptor subpopulations, this is apparently not the case for the behavioral measures. A similar discrepancy between behavioral analyses of 5HT-2 function (supersensitivity) and radioligand binding analyses of the 5HT-2 site (no change) is found after prolonged depletion of brain 5HT [see Conn and Sanders-Bush (1986a) for review]. We hypothesized therefore that regulation of 5HT-2 receptor function may take place at a site(s) distal to the cell surface recognition site (Conn and Sanders-Bush, 1986a). In order to test this hypothesis, it was necessary to characterize the molecular events linked to activation of the 5HT-2 receptor.

The binding of an agonist to a cell surface receptor leads to the production of intracellular second messengers, which then initiate a sequence of biochemical reactions that ultimately lead to specific physiological responses. Two major transmembrane transducing systems which mediate the actions of neurotransmitters and hormones are the adenylate cyclase/cyclic AMP signal cascade and the phospholipase C/phosphoinositide hydrolysis pathway. In the former, specific cell surface receptors are coupled to adenylate cyclase via a guanine nucleotide binding protein; receptor mediated activation of adenylate cyclase causes an increased conversion of ATP to cyclic AMP. Cyclic AMP in turn activates cyclic AMP dependent protein kinases, which phosphorylate specific substrate proteins. Other receptors are linked to phospholipase C, also apparently via a guanine nucleotide binding protein, leading to the hydrolysis of membrane phosphoinositides. The most common substrate of activated phospholipase C is phosphatidylinositol-4,5-bisphosphate, whose two hydrolytic products, inositol-1,4,5-trisphosphate and diacylglycerol, serve as second messengers. Both of these messengers regulate cellular functions via protein kinase activation and protein phosphorylation - IP_3 releases calcium which then activates calcium dependent kinases; diacylglycerol activates another kinase, termed protein kinase C. For more information about these second messenger systems, interested readers are referred to recent reviews by experts (Hirasawa and Nishizuka, 1985; Berridge, 1985).

Central 5HT receptors are positively and negatively linked to adenylate cyclase [see Sanders-Bush and Conn (1986) for review]. Although the specific receptor subtypes that mediate the cyclic AMP responses are unknown, it is clear that 5HT-2 sites are not involved. We therefore initiated a series of studies of the other major transmembrane transducing pathway, phospholipase C/phosphoinositide hydrolysis. In order to study phosphoinositide hydrolysis, rat cerebral cortical slices were labelled with [^{3}H]-inositol and agonist-induced formation of [^{3}H]-inositol-1-phosphate (IP) was measured. The addition of 5HT increases the levels of [^{3}H]-IP, presumably formed by activation of phospholipase C, hydrolysis of phosphatidylinositol-4,5-bisphosphate, and the sequential metabolism of inositol-1,4,5-trisphosphate (IP_3) to inositol-1,4-bisphosphate (IP_2) and then IP (Conn and Sanders-Bush, 1985). Inclusion of lithium markedly and selectively increases the level of [^{3}H]-IP, by inhibiting its hydrolysis to inositol. Efforts to demonstrate 5HT-induced formation of IP_2 and IP_3 in cortical slices have been unsuccessful, possibly because the response is too small. Alternatively, perhaps phosphatidylinositol is the primary substrate for 5HT activated phospholipase C, as has been shown in other systems (Dixon and Hokin, 1985). In the presence of the MAO inhibitor pargyline, 5HT has an EC_{50} value in cerebral cortical slices of 1 μm and the maximum stimulation of IP formation is about two-fold (Conn and Sanders-Bush, 1985). Pharmacological characterization of the cortical 5HT phosphoinositide hydrolysis response suggests that it is mediated by activation of the 5HT-2 receptor. Agonists of the piperazine and the tryptamine series activate phosphoinositide hydrolysis with EC_{50} values

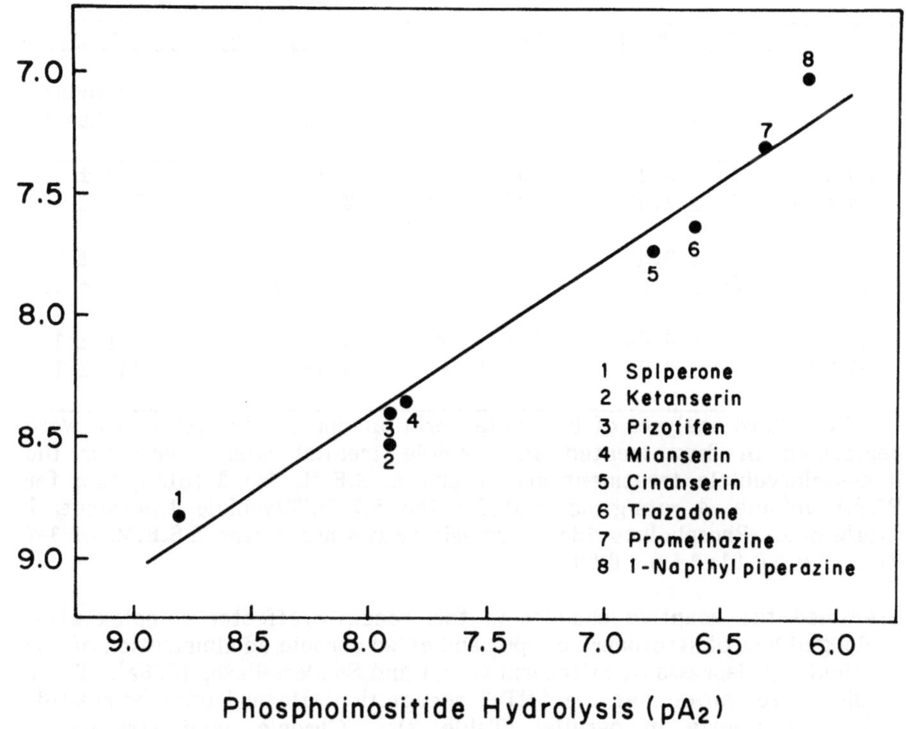

Figure 2. Comparison of Schild determined pA2 values (y-axis; -log K_d) at the cortical phospholipase C-linked 5HT receptor versus K_i values (x-axis, -log K_i) determined from radioligand competition binding with [^{3}H]-ketanserin. The correlation coefficient determined by linear regression analysis is 0.985.

similar to their binding constants for the 5HT-2 site as determined in radioligand binding studies. 5HT-2 antagonists nearly completely inhibit 5HT stimulated phosphoinositide hydrolysis in cortical slices (Conn and Sanders-Bush, 1984; 1985). Furthermore, the affinities of antagonists of 5HT-stimulated phosphoinositide hydrolysis (determined by Schild analyses of dose-response curves) correlated significantly with the K_i values of these antagonists at the [^{3}H]-ketanserin labelled 5HT-2 binding site (Figure 2). On the basis of these results, we concluded that the 5HT-2 recognition site in rat cerebral cortex is linked to phosphoinositide hydrolysis (Conn and Sanders-Bush, 1986a). Other investigators have reached the same conclusion based on studies in peripheral tissues, platelets (de Chaffoy de Courcelles et al., 1985) and vascular smooth muscle (Roth et al., 1986).

The demonstration that cortical 5HT-2 receptor sites are linked to phospholipase C permitted us to investigate the regulation of signal transfer across the membrane at 5HT-2 synapses. To date, we have

Table 2. Effect of chronic treatments on [^{3}H]-ketanserin binding and 5HT stimulated phosphoinositide hydrolysis

	[^{3}H]-Ketanserin binding		Phosphoinositide hydrolysis	
	K_d (nM)	B_{max} (fmol/mg prot)	EC_{50} (μM)	Max response (% of basal)
Vehicle	1.10 ± 0.24	209 ± 25	1.3 ± 0.4	259 ± 8
Mianserin	2.2 ± 0.40	100 ± 1*	2.4 ± 1.0	181 ± 7**
Vehicle	0.93 ± 0.07	172 ± 17	1.9 ± 0.2	199 ± 21
PCPA	0.89 ± 0.07	191 ± 26	2.2 ± 1.0	207 ± 12
Vehicle	0.61 ± 0.08	227 ± 19	0.61 ± 0.1	241 ± 13
5,7-DHT	0.67 ± 0.04	256 ± 18	0.38 ± 0.1	246 ± 15

Scatchard analyses of [^{3}H]-ketanserin binding to the 5HT-2 site were performed in drug treated and vehicle treated rats. Data for the mianserin/vehicle treatment are means ± S.E.M. for 3 rats/group; for PCPA vehicle, 4 rats/group and for the 5,7-DHT/vehicle treatment, 7 rats/group. Phosphoinositide hydrolysis values are means ± S.E.M. of 3-7 rats. * $p < 0.01$; ** $p < 0.001$

examined the adaptive changes in this receptor/effector complex after prolonged loss of serotonergic input and after chronic administration of the atypical antidepressant, mianserin (Conn and Sanders-Bush, 1986a). These studies have shown that the 5HT-2 recognition site and phosphoinositide hydrolysis change in parallel (Table 2). Chronic administration of mianserin results in a 52% decrease in the density of 5HT-2 binding sites, with a corresponding decrease (48%) in the maximum phosphoinositide hydrolysis response to 5-HT. Thus, chronic mianserin-induced down-regulation of the 5-HT-2 site is accompanied by 5-HT-2 receptor subsensitivity. Similarly, chronic administration of the tricyclic antidepressant, imipramine, induces a desensitization of 5-HT-stimulated phosphoinositide hydrolysis in conjunction with a down-regulation of the 5-HT-2 recognition site (Kendall and Nahorski, 1985). Furthermore, the results in Table 2 demonstrate that the 5-HT-2 receptor/effector complex does not develop supersensitivity after the administration of 5,7-DHT or p-chlorophenylalanine, treatments which cause marked and prolonged depletion of brain 5-HT.

Although both the antidepressant-induced beta-adrenoceptor down-regulation and 5HT-2 receptor down-regulation and desensitization of the phospholipase C/phosphoinositide hydrolysis signal are delayed in onset, there is no evidence that these two events are related. Furthermore, the 5HT-2 receptor system is apparently not involved in the regulation of beta-adrenoceptors since chronic treatment with the potent 5HT-2 antagonist ketanserin does not mimic the effects of 5HT depletion (unpublished results).

One potential problem with the studies of the regulation of the 5HT-2 receptor-phospholipase C/phosphoinositide hydrolysis pathway is that the actual formation of second messenger substances is not measured directly; instead the formation of [^{3}H]-IP is monitored as an indirect index of phosphoinositide hydrolysis. There is growing evidence that inositol sugars other than I-1,4,5-P_3 are formed on receptor activation. These include I-1,3,4-P_3 (Irvine et al., 1984), inositoi-1,3,4,5-tetrakisphosphate (Batty et al., 1985) and inositol-1:2-cyclic 4,5-trisphosphate (Wilson et al., 1985), all of which may ultimately be metabolized to IP and hence measured in our assays. Some of these products do not have any known second messenger roles, but if present in significant quantities, they may mask adaptive changes in the specific phosphoinositide hydrolysis pathway responsible for signal transmission. In the cerebral cortex, an investigation of this question is impossible, because the second messenger products cannot be measured directly. We therefore searched for a more robust 5-HT-stimulated phosphoinositide hydrolysis signal in other areas of the CNS. Non-cortical areas of brain have an even smaller 5-HT signal than does the cortex, and the pharmacological profiles of the responses in subcortical regions are different from the 5-HT-2 binding site (Janowsky et al., 1984; Conn and Sanders-Bush, 1985). Efforts to assign a specific receptor classification to the 5-HT responses in non-cortical areas have been unsuccessful because of the weak signal (unpublished results). Fortuitously, during our search for tissues with a strong 5-HT phosphoinositide hydrolysis signal, we looked at the rat choroid plexus (Conn et al., 1986a). In this

Table 3. Potencies at serotonin receptors in cerebral cortex and choroid plexus

	Phosphoinositide response		Radioligand binding	
Compound	Cerebral cortex	Choroid plexus	5-HT_2	5-HT_{1c}
	K_i, nM		K_d, nM	
Mianserin	23	12	6	5
Ketanserin	T2	130	3	196
Spiperone	4	6200	2	4600
	EC_{50}, nM			
Serotonin	1000	50	6200	94
	E_{max} (% of basal)			
Serotonin	180-240	600-1000		

Apparent K_i values at phosphoinositide-linked receptors were estimated from dose-response curve. Apparent K_d values at the two binding sites were estimated from radioligand competition data. E_{max} refers to the maximal response to 5HT in that tissue. See Conn et al. (1986a) for details.

tissue, the response to 5-HT is much more intense (6- and 10-fold stimulation) than in the cerebral cortex and 5-HT is about 10 times more potent as an agonist than in cerebral cortex (Table 3). Interestingly, 5-HT is a unique agonist in this tissue; other neurotransmitters which give strong phosphoinositide hydrolysis signals in other tissues, including brain, do not increase phosphoinositide hydrolysis in the choroid plexus (Conn and Sanders-Bush, 1986b). A comparison of the pharmacological profile of 5-HT-stimulated phosphoinositide hydrolysis in the choroid plexus and in the cerebral cortex is shown in Table 3. The properties of the phosphoinositide hydrolysis-linked receptor in choroid plexus clearly differ from those in cortex. Based on the antagonist profile, we have proposed that the 5-HT-1c binding site, which is highly localized in the choroid plexus (Pazos et al., 1984; Yagaloff and Hartig, 1985), is a functional receptor that is linked to phosphoinositide hydrolysis (Conn et al., 1986a). With the robust response in choroid plexus, it was possible to follow the formation of all three inositol sugars (Conn and Sanders-Bush, 1986b). The studies showed that the rise in IP levels is preceeded temporally by an increase in IP_3 and IP_2, consistent with the interpretation that IP is formed by the breakdown of IP_3 and, perhaps, IP_2. The finding that IP_3 is formed on activation of 5-HT-1c receptors in rat choroid plexus represents the first demonstration of 5HT receptor-mediated phosphatidylinositol-4,5-bisphosphate hydrolysis in mammalian systems. Considering the high density of 5-HT-1c sites in the choroid plexus, the magnitude of the phosphoinositide hydrolysis response mediated by this site, and the apparent absence of other phosphoinositide hydrolysis linked receptors, it was of interest to determine if the 5-HT-1c site receives tonic input from serotonergic neurons. Morphological evidence suggests that the choroid plexus is innervated by projections from the midbrain raphe nuclei, but it is not clear that the projections terminate on the choroidal epithelial cells (Nakamura and Moriyasu, 1978; Moskowitz et al., 1979; Napoleone et al., 1982), where the 5-HT-1c sites are localized. Indirect evidence that the 5-HT-1c/phospholipase C receptor complex is under the tonic influence of 5-HT comes from the finding that lesions of 5-HT neurons by the intraventricular injection of 5,7-DHT causes a supersensitive 5-HT phosphoinositide hydrolysis response in choroid plexus (Conn et al., 1986b). The 5-HT-1c receptor-phospholipase C/phosphoinositide hydrolysis signal cascade thus becomes a prime candidate in the regulation of function in the choroid plexus, including the production of cerebrospinal fluid, which primarily occurs in this tissue.

SUMMARY AND CONCLUSIONS

For years, research on the mechanism of action of antidepressant treatments has centered around the possible roles of central noradrenergic and serotonergic neuronal systems. Most clinically effective antidepressant treatments cause, upon chronic administration, a desensitization of the noradrenergic beta-adrenoceptor coupled adenylate cyclase system that is generally linked to a decrease in the density of beta-adrenoceptors. Delayed supersensitive serotonergic responses are also common to antidepressant treatments. In addition, many antidepressants down-regulate the density of 5HT-2 sites in brain and studies to date show that the functional response (phosphoinositde hydrolysis) changes in parallel. In recent years, research strategies have shifted to the possible interactions between these two aminergic systems in mediating the action of antidepressants. Research reviewed in this paper has demonstrated a functional link between serotonergic and noradrenergic neuronal systems in brain at the level of beta-adrenoceptors. These studies have provided the scientific basis for the clinically relevant "Serotonin/Norepinephrine Link Hypothesis" of affective disorders (Sulser, 1985).

An exciting new area of research is the exploration of the interaction of the multiple second messengers (cyclic AMP, cyclic GMP, IP_3, calcium, diacylglycerol) generated by activation of cell surface receptors. In the noradrenergic system in turkey erythrocytes, for example, activation of protein kinase C (one consequence of diacylglycerol release subsequent to phosphoinositide hydrolysis) causes a desensitization of beta-adrenoceptor-sensitive adenylate cyclase by promoting C kinase-mediated phosphorylation of beta-adrenoceptors (Sibley et al., 1984). Our studies of the privotal role of 5HT in the regulation of beta-adrenoceptors suggest that elucidation of the molecular basis of this linkage may prove valuable in unraveling the complex interactions between the adenylate cyclase/cyclic AMP signal cascade and the phospholipase C/phosphoinositide hydrolysis pathway. One potentially important site of interaction is at the level of the 5HT-1c receptor, which is highly localized on the epithelial cells in the choroid plexus, and is coupled to phospholipase C, causing a breakdown of phosphatidylinositol-4,5-bisphosphate and the generation of two intracellular second messengers, IP_3 and diacylglycerol. These cells also have a high density of beta-adrenoceptors that are coupled to adenylate cyclase (Nathanson, 1979). Choroidal epithelial cells can be readily grown in culture and since the NE and 5HT receptors co-exist on these cells, this system may serve as a useful model for studying the molecular basis of the pivotal role of 5HT in the antidepressant-induced down-regulation of beta-adrenoceptors.

The studies from our laboratories have been supported by USPHS MH29228, MH 26483 and MH34007.

REFERENCES

Anderson, J.L. (1983). Serotonin receptor changes after chronic antidepressant treatments: Ligand binding, electrophysiological and behavioral studies. Life Sci. 32, 1791-1801.

Batty, I.R., Nahorski, S.R. and Irvine, R.F. (1985). Rapid formation of inositol 1,3,4,5-tetrakisphosphate following muscarinic receptor stimulation of rat cerebral cortical slices. Biochem. J. 232, 211-215.

Berridge, M.J. (1985). The molecular basis of communication within cells. Scientific Amer. 253, 142-152.

Brunello, N., Volterra, A., Cagiano, R., Ianieri, G.C., Cuomo, V. and Racagni, G. (1985). Biochemical and behavioral changes in rats after prolonged treatment with desipramine: interaction with p-chlorophenylalanine. Naunyn-Schmiedeberg's Arch. Pharmacol. 331, 20-22.

Conn, P.J. and Sanders-Bush, E. (1984). Selective 5HT-2 antagonists inhibit serotonin stimulated phosphatidylinositol metabolism in cerebral cortex. Neuropharmacology 23, 993-996.

Conn, P.J. and Sanders-Bush, E. (1985). Serotonin-stimulated phosphoinositide turnover: mediation by the S2 binding site in rat cerebral cortex but not in subcortical regions. J. Pharmacol. Exp. Ther. 234, 195-203.

Conn, P.J. and Sanders-Bush, E. (1986a). Regulation of serotonin-stimulated phosphoinositide hydrolysis: Relation to the serotonin 5HT-2 binding site. J. Neurosci. in press.

Conn, P.J. and Sanders-Bush, E. (1986b). Agonist-induced phosphoinositide hydrolysis in choroid plexus. J. Neurochem. in press.

Conn, P.J., Sanders-Bush, E., Hoffman, B.J. and Hartig, P.R. (1986a). A unique serotonin receptor in choroid plexus is linked to phosphoinositide hydrolysis. Proc. Nat. Acad. Sci. USA 83, 4086-4088.

Conn, P.J., Janowsky, A. and Sanders-Bush, E. (1986b). Denervation supersensitivity of 5-HT-1c receptors in rat choroid plexus. Brain Res. in press.

de Chaffoy de Courcelles, D., Leysen, J.E., De Clerck, F., Van Belle, H. and Janssen, P.A.J. (1985). Evidence that phospholipid turnover is the signal transducing system coupled to serotonin-S2 receptor sites. J. Biol. Chem. 260, 7603-7608.

Dixon, J.F. and Hokin, L.E. (1985). The formation of inositol-1,2-cyclic phosphate on agonist stimulation of phosphoinositide breakdown in mouse pancreatic minilobules. J. Biol. Chem. 260, 16068-16071.

Irvine, R.F., Letcher, A.J., Lander, D.J. and Downes, C.P. (1984). Inositol trisphosphates in carbachol-stimulated rat parotid glands. Biochem. J. 223, 237-243.

Janowsky, A., Labarca, R. and Paul, S.M. (1984). Characterization of neurotransmitter receptor-mediated phosphatidylinositol hydrolysis in the rat hippocampus. Life Sci. 35, 1953-1961.

Janowsky, A.J., Steranka, L.R., Gillespie, D.D. and Sulser, F. (1982). Role of neuronal signal input in the down-regulation of central noradrenergic receptor function by antidepressant drugs. J. Neurochem. 39, 290-292.

Kendall, D.A. and Nahorski, S.R. (1985). 5-hydroxytryptamine-stimulated inositol phospholipid hydrolysis in rat cerebral cortex slices: Pharmacological characterization and effects of antidepressants. J. Pharmacol. Exp. Therap. 233, 473-479.

Kent, R.S., DeLean, A. and Lefkowitz, R.J. (1980). A quantitative analysis of beta-adrenergic receptor interactions: resolution of high and low affinity states of the receptor by ligand binding data. Mol. Pharmacol. 17, 14-23.

Manier, D.H., Gillespie, D.D., Steranka, L.R. and Sulser, F. (1984). A pivotal role for serotonin in the down-regulation of beta-adrenoceptors by antidepressants: Reversibility of the action of p-chlorophenylalanine by 5-hydroxytryptophan. Experientia 40, 1223-1226.

Manier, D.H., Gillespie, D.D., Sanders-Bush, E. and Sulser, F. (1986). The serotonin/noradrenaline-link in brain: I. The role of noradrenaline and serotonin in the regulation of density and function of beta adrenoceptors and its alteration by desipramine. Naunyn-Schmiedeberg's Arch. Pharmacol. submitted.

Moskowitz, M.A., Liebman, J.F., Reinhard, Jr., J.F. and Schlosberg, A. (1979). Raphe origin of serotonin-containing neurons within choroid plexus of the rat. Brain Res. 169, 590-594.

Nakamura, S. and Mariyasu, N. (1978). Nerve fibers and nerve endings in the choroid plexus: electron microscopic study. Brain and Nerve 30, 259-266.

Napoleone, P., Sancesario, B. and Amenta, F. (1982). Indoleaminergic innervation of rat choroid plexus: a fluorescence histochemical study. Neurosci. Lett. 34, 143-147.

Nathanson, J.A. (1979). Beta-adrenergic-sensitive adenylate cyclase in secretory cells of choroid plexus. Science 204, 843-844.

Nimgaonkar, V.L., Goodwin, G.M., Davies, C.L. and Green, A.R. (1985). Down-regulation of beta-adrenoceptors in rat cortex by repeated administration of desipramine electronconvulsine shock and clenbuterol requires 5HT neurons but not 5HT. Neuropharmacology 24, 279-283.

Hirasawa, K. and Nishizuka, Y. (1985). Phosphatidylinositol turnover in receptor mechanism and signal transduction. Ann. Rev. Pharmacol. Toxicol. 25, 147-170.

Pazos, A., Hayer, D. and Palacios, J.M. (1984). The binding of serotonergic ligands to the porcine choroid plexus: Characterization of a new type of serotonin recognition site. Europ. J. Pharmacol. 106, 539-546.

Roth, B.L., Nakaki, T., Chuang, D. and Costa, E. (1986). 5-Hydroxytryptamine-2 receptors coupled to phospholipase C in rat aorta: Modulation of phosphoinositide turnover by phorbol ester. J. Pharmacol. Exp. Ther. 238, 480.

Sanders-Bush, E. and Conn, P.J. Neurochemistry of serotonin neuronal systems: Consequences of serotonin receptor activation. In Psychopharmacology, The Third Generation of Progress, in press, 1986.

Sibley, D.R., Nambi, P., Peters, J.R. and Lefkowitz, R.J. (1984). Phorbol diesters promote beta-adrenergic receptor phosphorylation and adenylate cyclase densitization in duck erythrocytes. Biochem. Biophys. Res. Comm. 121, 973-979.

Stockmeier, C.A., Martino, A.. and Kellar, K.J. (1985). A strong influence of serotonin axons on beta-adrenergic receptors in rat brain. Science 230, 323-325.

Sulser, F., Janowsky, A.J., Okada, F., Manier, D.H. and Mobley, P.L. (1983). Regulation of recognition and action function of the norepinephrine receptor-coupled adenylate cyclase system in brain: Implications for the therapy of depression. Neuropharmacology 22, 425-431.

Sulser, F. (1985). The serotonin-noradrenaline link hypothesis of affective disorders. In Psychiatry, Vol. 2, (eds. P. Pichot, P. Berner, R. Wolf, and K. Thau). Plenum Publishing Corp., New York, N.Y., 411-416.

Wilson, D.B., Connolly, T.M., Bross, T.E., Majerus, P.W., Serman, W.R., Tyler, A.N., Rubin, L.J. and Brown, J.E. (1985). Isolation and characterization of the inositol cyclic phosphate products of polyphosphoinositide cleavage by phospholipase C: physiological effects in limulus photoreceptors. J. Biol. Chem. 260, 13496-13501.

Yagaloff, K.A. and Hartig, P.R. (1985). ^{125}I-Lysergic acid diethylamide binds to a novel sertonergic site on rat choroid plexus epithelial cells. J. Neurosci. 5, 3178-3183.

32
Receptor Interactions in Antidepressant Neurochemical Mechanisms: Modulation by Membrane Components

N. Brunello, U. Cicerano, A.C. Rovescalli, M. Riva, R. Galimberti, C. Franzetti and G. Racagni

INTRODUCTION

Recent studies on the mechanism of action of antidepressant drugs have focussed on alterations in monoaminergic neurons following prolonged drug administration, since only those effects elicited after long term antidepressant treatment might be the biochemical substrate accounting for the delayed onset of therapeutic activity (Racagni and Brunello, 1984). Chronic antidepressant administration is associated with several adaptive changes in central monoaminergic systems, which can occur both pre- and postsynaptically (Sugrue, 1983). One of the most significant and constant effect shared by all the treatments that are known to relieve depression, including electroconvulsive therapy, is the down-regulation of beta adrenergic receptors (Banerjee et al., 1977) coupled to the desensitization of noradrenaline (NE) dependent adenylate cyclase (Vetulani and Sulser, 1975). Moreover among the possible mechanisms responsible for the adaptation of central aminergic neurons following antidepressant administration, the interaction between serotonergic and noradrenergic system and changes in the chemico-physical properties of the membranes should be considered.

5HT-NE RECEPTOR INTERACTION

Several observations have indicated that serotonergic and noradrenergic systems in the cerebral cortex are interconnected. It has been widely demonstrated that the ability of tricyclic antidepressants to reduce the sensitivity of the NE-dependent cAMP generating system and the density of beta-receptors does not depend only on an intact noradrenergic neuronal input (Wolfe et al.,

1978), but also upon integrity of serotonergic system. Accordingly it has been shown that chemically-induced lesion of serotonergic fibers prevents the desensitization of the NE-sensitive adenylate cyclase (Brunello et al., 1982) and the down regulation of beta adrenergic receptors in cortex of rats previously treated with desmethylimipramine (DMI) (Janowsky et al., 1982; Racagni et al., 1983; Brunello et al., 1985a). Moreover depletion of serotonin (5HT) content following chronic treatment with para-chlorophenylalanine (pCPA) abolishes the decrease of cortical ß-receptors induced by a prolonged treatment with DMI (Brunello et al., 1985b). Thus the function of noradrenergic system in rat cerebral cortex seems to be dependent on its interaction with serotonin containing neurons.

The regulation of 5HT neurons on $alpha_2$ presynaptic adrenoceptors, throughout antidepressant treatment, has been also investigated using both behavioural and biochemical parameters, such as the clonidine induced reduction of locomotor activity, the concentration of normetanephrine (NMN) the O-methylated metabolite of NE and the density of $alpha_2$-binding sites. Our studies demonstrate that in 5,7 dihydroxytryptamine lesioned animals, the reduced effectiveness of clonidine in suppressing locomotor activity after chronic treatment with DMI is markedly attenuated; moreover, in the same experimental conditions, the changes in NMN level induced by mianserin and clonidine are significantly lower than in sham-lesioned rats (Racagni et al., 1985). On the other hand no changes in ^{3}H-Clonidine binding sites were detected in lesioned animals chronically treated with DMI. Therefore these results suggest that serotonergic system plays a modulatory role also on alpha-2 adrenoceptors even if 5HT denervation did not modify the binding properties of alpha-2-receptors during a prolonged treatment with antidepressants.

MODULATION BY PHOSPHATIDYLSERINE

Chemico-physical properties of plasma neuronal membranes can influence the signal transfer between a neurotransmitter and its receptor and the fluidity characteristics of the membranes could be among the factor that modulate the biochemical action of antidepressant drugs. It has been shown that membrane phospholipid methylation regulates membrane viscosity and that the methylated status of membrane phospholipids modulates the response of hormone receptor (Hirata and Axelrod, 1980). Furthermore we have shown that DMI treatment induces an increase in phospholipid methylation in rat cerebral cortex (Racagni et al., 1983) and we report in the present paper that DMI administration induces a greater accumulation of inositol-1-phosphate in cortical slices either in basal condition or after stimulus with NE (Table 1). Since these processes regulate the fluidity characteristics of the membrane, these DMI-induced effects may influence the functional interaction between a neurotransmitter and its receptor. It was therefore worth of investigating whether the exogenous administration of a membrane component, namely phosphatidylserine, could interfere with signal transduction mechanisms and with those adaptive changes induced by antidepressant administration.

Male Sprague Dawley rats (Charles River) were housed in standard conditions and treated either with DMI (5 or 15 mg/kg i.p. twice daily) or with phosphatidylserine (BC-PS) (15 mg/kg i.p. twice daily) or with the combination of the two drugs for 1 week. 48 hours after the last injection, animals were sacrified by decapitation, brains were excised and cortices dissected out. Cortical slices were immediately prepared for the determination of phosphatidylinositol turnover according to Berridge et al. (1983), whereas tissues for binding studies were kept frozen at -70° C until assay.

Data reported in Table 2 show that the administration of BC-PS induced an increased accumulation of ^{3}H-Inositol-1-phosphate (IP) in cortical slices either in basal condition or after NE stimulation, without affecting the percentage of stimulation.

Table 1. **^{3}H-Inositol-1-phosphate (^{3}H-IP) accumulation in cortical slices from control and desmethylimipramine (DMI) treated rats.**

Treatment	Basal	Stimulated (+ NA 10^{-4} M)
CONTROL	0.0207 ± 0.0021	0.0483 ± 0.0023*
DMI (15 mg/kg i.p. twice daily)	0.0461 ± 0.0030△	0.1122 ± 0.0026△*

*$p < 0.05$ vs basal value
△$p < 0.05$ vs control value

Data are expressed as the ratio between ^{3}H-IP (cpm/mg lipid) and the incorporation in the lipid and aqueous phases (cpm/mg lipid).

Table 2. **^{3}H-Inositol-1-phosphate (^{3}H-IP) accumulation in cortical slices from control and phosphatidylserine (PS) treated rats.**

Treatment	Basal	Stimulated (+ NA 10^{-4} M)
CONTROL	0.021 ± 0.0015	0.045 ± 0.0019 *
PS (15 mg/kg i.p. twice daily for 1 week)	0.043 ± 0.0021 △	0.077 ± 0.0023 * △

* $p < 0.05$ vs basal value

△ $p < 0.05$ vs control value

Data are expressed as the ratio between ^{3}H-IP (cmp/mg lipid) and the incorporation in the lipid and aqueous phase (cpm/mg lipid).

These results indicate that BC-PS is able to modify the basal turnover rate of IP, independently from activation of phospholipase C.

We have then investigated the effect of BC-PS administration on those binding sites known to be altered during a prolonged treatment with antidepressant drugs. Treatment with BC-PS was uneffective on the kinetic constants of ß-adrenergic receptors, measured by Scatchard analysis of ^{3}H-dihydroalprenolol binding (^{3}H-DHA) to cortical membranes according to Bylund and Snyder (1976). Similarly BC-PS, given at the dose of 15 mg/kg i.p. for 1 week, did not modify the kinetic constants of imipramine binding sites labeled with ^{3}H-paroxetine according to Habert et al. (1985) (Table 3). On the contrary, as already reported (Sulser et al., 1978; Raisman et al., 1980; Racagni et al., 1983) DMI, at the highest dose tested, was able to down regulate ß-adrenergic receptors and imipramine binding sites, being though uneffective at the lowest dose of 5 mg/kg. However when the antidepressant was administered toghether with BC-PS, a decreased B_{max} value was detected even at the lowest dose of DMI and a further down regulation was evident at the dose of 15 mg/kg indicating that PS was able to potentiate the DMI-induced down regulation (Table 3).

Table 3. **Effect of 1 week-administration of desmethylimipramine (DMI), phosphatidylserine (PS) and DMI + PS on ^{3}H-DHA and ^{3}H-Paroxetine binding to rat cortical membranes.**

	^{3}H-DHA	^{3}H -PAROXETINE
CONTROL	182.3 ± 15.6	574.7 ± 25.6
DMI 5 mg/kg	165.1 ± 11.5	639.2 ± 31.3
DMI 15 mg/kg	151.2 ± 10.7 *	485.2 ± 20.3 *
PS 15 mg/kg	178.9 ± 13.8	592.3 ± 23.6
PS + DMI 5 mg/kg	140.6 ± 9.3 *	458.7 ± 19.8 *
PS + DMI 15 mg/kg	117.2 ± 8.4 *	345.0 ± 15.6 **

* $p < 0.05$ ** $p < 0.01$ vs control (Dunnett's "t" test)

Data are expressed as fmoles ^{3}H-ligand bound /mg protein and referred to the mean Bmax value ± SEM of at least 4 saturation curves.
No significant changes were detected in the K_d values which ranged between 3.0 and 3.8 nM for ^{3}H-DHA and between 0.4 and 1.0 nM for ^{3}H-Paroxetine binding.

Moreover we have studied the effect of DMI and BC-PS coadministration on the kinetic constants of ^{3}H-ketanserin binding sites, measured according to Leysen et al. (1982) as an index of $5HT_2$ receptors, and the results are shown in Table 4. One week of treatment with DMI did not modify K_d and B_{max} values, whereas the combination of the antidepressant plus BC-PS, which by itself did not induce any change, produced a down regulation of ^{3}H-Ketanserin binding sites.

In all these results indicate that BC-PS without affecting per sè those binding sites altered during a prolonged treatment with antidepressants, can accelerate the onset and potentiate the biochemical changes induced by DMI.

Furthermore we have studied the recovery of ß-adrenergic receptor density after withdrawal from DMI treatment, alone or in combination with BC-PS. Figure 1 shows the time course of ß-adrenergic receptor density after 2,3,4,7,9,11 days of withdrawal from drug administration. ß-receptors recovered faster when BC-PS was given together with either 5 or 15 mg/kg of DMI, even if the

Table 4. **Effect of administration of desmethylimipramine (DMI), phosphatidylserine (PS) and PS + DMI on ^{3}H-Ketanserin binding to rat cortical membranes**

	Kd (nM)	Bmax (fmoles/mg protein)	% reduction
CONTROL	1.91	365.0	---
DMI (15 mg/kg)	2.22	334.8	6%
PS (15 mg/kg)	1.53	358.9	---
PS + DMI	1.42	252.0 *	29.2%

Drugs were daily administered to rats for one week.
* $p < 0.05$ vs control (saline)

down regulation lasted longer in animals receiving the highest dose of DMI.

DISCUSSION

It is well known that one of the major problem associated with the administration of antidepressant drugs is the delayed onset of therapeutic activity. If those biochemical changes induced in rat brain by prolonged treatment with antidepressants are related to clinical efficacy, then any procedure which accelerates the changes in monoaminergic neurons could be of practical importance.

To this purpose we have studied the effect of a membrane phospholipid, phosphatidylserine, on some biochemical parameters, target of antidepressants action since it has been shown that the viscosity characteristics of plasma neuronal membranes can modulate the interaction between a neurotransmitter and its recognition site, influencing thus intracellular events.

Our results demonstrate that BC-PS associated with a typical antidepressant drug can shorten the latency time of DMI-induced changes, such as the down-regulation of $5HT_2$-binding-sites (Table 4).

Moreover BC-PS can potentiate a threshold dose of DMI, since, as shown in Table 3, down regulation of ß-receptors and imipramine binding sites is achieved also at a DMI dose of 5 mg/kg after 1

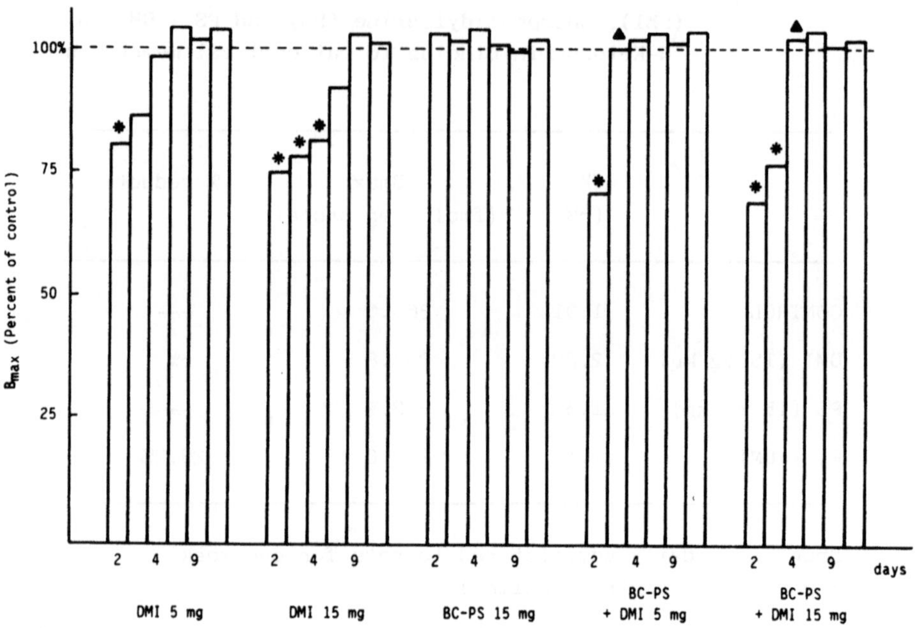

FIG.1 Time course of ß-adrenergic receptors after 2,3,4,7,9,11 days of wash out from desmethylimipramine (DMI) treatment: effect of a concomitant administration of phosphatidylserine (BC-PS).

week of treatment only in combination with BC-PS. It is interesting to note that BC-PS does not effect any binding sites when administered alone, however it enhances PI turnover in basal conditions (Table 2) showing thus a more generalized effect beyond the binding site, at the level of a transducing mechanism.

Therefore in the presence of a stimulus, such as antidepressant treatment, BC-PS, by activating a membrane process, enables a faster response even at lower doses. This is a specific effect on those sites modified by antidepressant administration since no changes have been detected in other receptor system such as dopamine D_1 and D_2 binding sites determined with ^{3}H-SCH 23390 and ^{3}H-spiroperidol respectively (data not shown).

In conclusion we here report that modification of membrane plasticity can modulate antidepressant induced neurochemical changes and these results should be of clinical relevance taking in consideration the latency time of antidepressant therapy and its side effects.

REFERENCES

Banerjee, S.P., Kung, L.S., Riggi, S.J., and Chande, S.K. (1977). Development of beta-adrenergic receptor subsensitivity by antidepressants. Nature (Lond.), 268, 455-456.

Berridge, M.J., Downes, C.P., and Hanley, M.R. (1982). Lithium amplifies agonist dependent phosphatidylinositol responses in brain and salivary glands. Biochem. J., 206, 587-595.

Brunello, N., Barbaccia, M.L., Chuang, D.M., and Costa, E. (1982). Down regulation of beta-adrenergic receptors following repeated desimipramine injections; permissive role of serotonergic axons. Neuropharmacology, 21, 1145-1149.

Brunello, N., Mocchetti, I., Volterra, A., Cuomo, V., and Racagni, G. (1985a). Serotonergic modulation of cortical rat noradrenergic system in the mechanism of action of antidepressant drugs. Psychopharm. Bull., 21, 379-384.

Brunello, N., Volterra, A., Cagiano, R., Ianieri, G.C., Cuomo, V., and Racagni, G. (1985b). Biochemical and behavioural changes after prolonged treatment with desimipramine: interaction with p-chlorophenylalanine. Naunyn-Schmiedeberg's Arch. Pharmacol., 331, 20-22.

Bylund, D.B., and Snyder, S.H. (1976). Beta-adrenergic receptor binding in membrane preparations from mammalian brain. Mol. Pharmacol., 12, 568-580.

Habert, E., Graham, D., Tahraoui, L., Claustre, Y., Langer, S.Z. (1985). Characterization of ^{3}H-paroxetine binding to rat cortical membranes. Eur. J. Pharmacol., 118, 107-114.

Hirata, F., and Axelrod, J., (1980). Phospholipid methylation and the transmission of biological signals through membranes. Sciences, 209, 1082-1090.

Janowsky, A., Okada, F., Manier, D.H., Steranka, L., and Sulser, F. (1982). Role of serotonergic input in the regulation of the beta-adrenoceptor coupled adenylate cyclase system. Science, 218, 900-901.

Leysen, J.E., Niemegeers, C.J.E., Van Nueten, J.M., and Laduron, P.M. (1982). ^{3}H-Ketanserin (R 41468), a selective ^{3}H-ligand for serotonin-2 receptor binding sites: binding properties, brain distribution and functional role. Mol. Pharmacol., 21, 301-314.

Racagni, G., Mocchetti, I., Calderini, G., Battistella A., and Brunello, N., (1983). Temporal sequence of changes in central noradrenergic system of rat after prolonged antidepressant treatment: receptor desensitization and neurotransmitter interactions. Neuropharmacology, 22, 415-424.

Racagni, G., and Brunello, N. (1984). Transynaptic mechanisms in the action of antidepressant drugs. Trends Pharmacol. Sci., 5, 527-531.

Racagni, G., Brunello, N., Mocchetti, I., Cagiano, R., and Cuomo, V. (1985). Presynaptic and transynaptic mechanisms involved in the subsensitivity of rat cortical noradrenergic system after long-term antidepressant treatment. Acta Pharmacol. and Toxicol., 56 (Suppl. 1), 190-197.

Raisman, R., Briley, M., and Langer, S.Z. (1980). Specific tricyclic antidepressant binding sites in rat brain characterized by high affinity ^{3}H-imipramine binding. Eur. J. Pharmacol., 61, 373-380.

Sugrue, M. (1983). Chronic antidepressant therapy and associated changes in central monoaminergic receptor functioning. Pharmac. Ther., 21, 1-33.

Sulser, F., Vetulani, J., and Mobley, P.L. (1978). Mode of action of antidepressant drugs. Biochem. Pharmacol., 27, 257-261.

Vetulani, J., and Sulser, F. (1975). Action of various antidepressant treatments reduces reactivity of noradrenergic cyclic-AMP generating system in limbic forebrain. Nature, 257, 495-496.

Wolfe, B.B., Harden, T.K., Sporn, J.R., and Molinoff, P.B. (1978). Presynaptic modulation of beta-adrenergic receptors in rat cerebral cortex after treatment with antidepressant. J. Pharmacol. Exp. Ther., 207, 446-457.

33
Interactions of Coexisting Neurotransmitters and Effects of Chronic Drug Treatment

Tamas Bartfai, Kerstin Iverfeldt, Peter Serfözö and Sven-Ove Ögren

INTRODUCTION

The phenomenon of coexistence of classical neurotransmitters and neuroactive peptides has been demonstrated in a number of cases in the central and peripheral nervous system (cf. Hökfelt et al., 1980; Cuello, 1982; Lundberg and Hökfelt, 1983). In the descending bulbospinal pathway originating from the raphe nuclei, coexistence of at least three putative neurotransmitters; the undecapeptide substance P (SP), the tripeptide thyrotropin-releasing hormone (TRH) and serotonin (5-hydroxytryptamine, 5-HT) has been shown using immunohistochemical methods (Chan-Palay et al., 1978; Hökfelt et al., 1978; Johansson et al., 1981). Treatment with the serotonergic "neurotoxins" 5,6- or 5,7-dihydroxytryptamine results in an almost complete loss of 5-HT, TRH and SP immunoreactive fibers in the ventral spinal cord (Björklund et al., 1979; Johansson et al., 1981; Gilbert et al., 1982; Marsden et al., 1982) although findings suggesting a limited scope of coexistence only, were also reported (Towle et al., 1986).

The functional consequences of the coexistence have not yet been clarified. SP has been established as an excitatory neurotransmitter in the spinal cord and in some areas of the brain (Konishi and Otsuka, 1974; Otsuka and Konishi, 1976). When applied iontophoretically SP, TRH and 5-HT show similar actions in enhancing glutamate and aspartate-induced excitation of the motoneurons (White, 1985). Trembley et al. (1985) suggested that SP, TRH and 5-HT interact in the control of the motoneuron. Intrathecal administration of SP results in reduced responses of these neurons to both 5-HTP (5-hydroxytryptophan) and TRH. Barbeau and Bédard (1981) showed that TRH can mimic the effects of 5-HTP and 5-HT agonists on spinal reflexes. This effect was blocked with the 5-HT antagonist cyproheptadine, indicating that TRH action involves 5-HT receptors. Also SP can affect 5-HT receptors: The number of available ^{3}H-5-HT binding sites in membrane preparations from the spinal cord increased while the affinity decreased after incubation with SP (10 nM) (Agnati et al., 1983).

Some findings concerning presynaptic regulation of the release of the coexisting transmitters have been reported. The evoked release of SP-like immunoreactivity (SPLI) (induced by high K^+ ion concentration) from slice preparations of the ventral spinal cord increases in the presence of 5-HT (Iverfeldt et al., 1986). This effect is probably mediated by a type-2 5-HT receptor since the enhancement of release could be blocked by ketanserin. 5-HT can inhibit the release of ^{3}H-5-HT from hypothalamic synaptosomes (Cerrito and Raiteri, 1979), cerebral cortex slices (Göthert, 1980) and spinal cord synaptosomes (Monroe and Smith, 1985; Stauderman and Jones, 1986) via an autoreceptor sensitive to methiotepin but not to ketanserin (Monroe and Smith, 1985). The serotonergic autoinhibition could be antagonized by SP in a crude synaptosomal preparation from the ventral spinal cord (Mitchell and Fleetwood-Walker, 1981). According to Tsai et al. (1984) SP can also increase the basal efflux of ^{3}H-5-HT in the spinal cord as well as in the rat cerebral cortex (Solti and Bartfai, 1987).

Treatment with antidepressant drugs with monoamine uptake blocking properties is known to affect the 5-HT system and can also interact with peptides coexisting with 5-HT. Chronic treatment with the specific 5-HT uptake blocker zimelidine results in decreased tissue levels of 5-HT and its metabolite 5-HIAA (5-hydroxyindolacetic acid) in several brain regions and in the ventral spinal cord, while at the same time the tissue levels of SPLI increased in the ventral spinal cord (Brodin et al., 1984). An increase in the evoked release of SPLI (in response to elevated levels of K^+ ions), from slice preparations of the ventral spinal cord, from zimelidine treated animals was also found (Iverfeldt et al., 1986). Chronic treatment with amitriptyline or chlorimipramine, two other antidepressant drugs, led to increases in the tissue levels of TRH-like immunoreactivity (TRH-LI) and in the ex vivo release of TRH-LI from the spinal cord (Lighton et al., 1985).

This paper summarizes the present information of the cross-regulation of the release of the three coexisting neurotransmitters from nervendings in the rat ventral spinal cord.

MATERIALS

Substance P fragment 6-11 (catalogue No. S-1138, batch No. 45F-0262) and bacitracin were purchased from Sigma Chemical Co., St. Louis, MO, USA. Substance P was bought from Peninsula. Antisera directed against the N-terminal of Substance P (SP140) and ^{125}I-labeled Tyr-8 substance P were kindly provided by Ernst Brodin, Department of Pharmacology, Karolinska Institute, S-104 05

Stockholm, Sweden.

All other chemicals were of reagent grade. Solutions were made fresh daily in double distilled H_2O.

RESULTS AND DISCUSSION

The possible regulation of SP release via presynaptic SP receptors was examined utilizing a specific antiserum directed against the N-terminal of SP to measure the release in the presence of SP (6-11) fragment, which was added as a possible ligand to the putative presynaptic SP- or tachykinin-receptor (Figure 1).

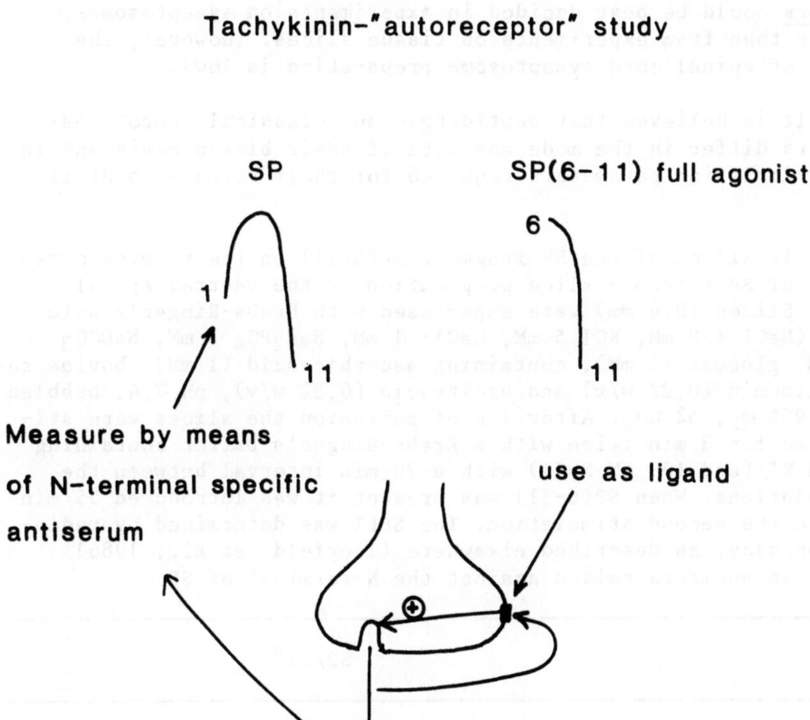

Figure 1

The sketch of the experiment which uses a C-terminal fragment of 98 (SP(6-11)) to study autoregulation of SP-release.

The C-terminal fragment of SP; SP(6-11), which acts as a SP agonist in some biological systems (Bury and Mashford, 1976) enhanced the evoked release of SPLI (induced by high K^+ ion concentration) from slices of the ventral spinal cord (Table 1). In some experiments SP(6-11) caused an increase also in the basal release of SPLI. These results suggest that, there may be a presynaptic tachykinin receptor positively modulating the release of SP. Since the C-terminal fragment of several tachykinins shows strong homology it is not possible to decide on the basis of the experiments with SP(6-11) which tachykinin may act as endogenous ligand in promoting the release of SPLI. The enhancement of the evoked release of SPLI is probably caused by the same SP or tachykinin receptors which are involved in enhancement of 5-HT release (Tsai et al., 1984; Solti and Bartfai, 1987). The scheme which thus emerges, shows that the coexisting neurotransmitters can exert effects on their own release as well as on the release of each other in this system of the rat ventral spinal cord (Figure 2). Whether these interactions involve bonafide autoreceptors could be best decided in experiments on synaptosomes, rather than from experiments on tissue slices. (However, the yield of spinal cord synaptosome preparation is low).

It is believed that peptidergic and classical neurotransmitters differ in the mode and site of their biosynthesis and in the stimulation parameters required for their release: 5-HT is

Table 1. Effect of the SP fragment SP(6-11) on the K^+ evoked release of SPLI from a slice preparation of the ventral spinal cord. Slices (0,4 mm) were superfused with Krebs-Ringer's solution (NaCl 138 mM, KCl 5 mM, $CaCl_2$ 1 mM, NaH_2PO_4 1 mM, $NaHCO_3$ 11 mM, glucose 11 mM), containing ascorbic acid (1 mM), bovine serum albumin (0,2% w/v) and bacitracin (0,3% w/v), pH 7,4, bubbled with 95% O_2, 5% CO_2. After 1 h of perfusion the slices were stimulated for 3 min twice with a Krebs-Ringer's buffer containing 40 mM K^+ (and 103 mM NaCl) with a 70 min interval between the stimulations. When SP(6-11) was present it was introduced 55 min before the second stimulation. The SPLI was determined by radioimmunoassay, as described elsewhere (Iverfeldt et al., 1986) using an antisera raised against the N-terminal of SP.

	S2/S1*	
Control	0.40 ± 0.23	(n = 5)
SP(6-11), (1 μM)	0.66 ± 0.07	(n = 2)
SP(6-11), (0.1 μM)	1.09 ± 0.30	(n = 2)

*S2/S1 stands for the ratio between evoked release of SPLI caused by second (S2) and first (S1) stimulation.

synthesized both in the soma of the bulbospinal neurons in the raphe nuclei and in the terminals in the spinal cord. SP is synthesized on ribosomes in the soma; two mRNA species that code for SP have been isolated from the mammalian brain (Nawa et al., 1983) one of these also codes for another tachykinin; substance K. The small tripeptide TRH was thought to behave as a classical neurotransmitter and a non-ribosomal synthesis of TRH was also suggested (Mitnik and Reichlin, 1972, for review see McKelvy, 1982). In contrast, Richter and collaborators (1982) determined the sequence of a cDNA clone from amphibian skin coding for prepro-TRH. Some evidence for a TRH precursor in the neonatal rat pancreas was also found (Ouafik et al., 1985). In 1986, Lechan and collaborators characterized a preprohormone, containing five copies of TRH, from rat hypothalamus.

Preliminary results from experiments on the release of SPLI, TRH-LI and ^{3}H-5-HT from slices of the ventral spinal cord at different frequencies of stimulation suggest that SP and TRH are released at higher frequencies while ^{3}H-5-HT seems to require a lower frequency of stimulation: TRH-LI and SPLI release is measurable at stimulation frequencies 3-10 Hz while the amounts released at 0.5 Hz are below the level of detection (15 fmol for TRH and 2 fmol for SP). ^{3}H-5-HT on the other hand is released at 0.5 and 1 Hz in readily measurable quantities (data not shown).

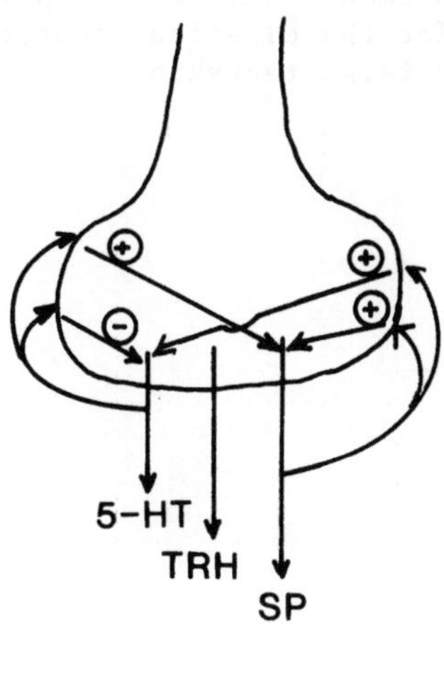

Figure 2

Schematic model indicating cross- and autoregulation in a nerve terminal in the ventral spinal cord containing SP, TRH and 5-HT.

These reports together with the finding that chronic antidepressant treatment cause effects on TRH-LI tissue levels (Lighton et al., 1985) which are similar to the effects on tissue levels of SPLI (Brodin et al., 1984) in the spinal cord suggests that with respect to regulation of release TRH behaves like a peptidergic neurotransmittor (Figure 3).

The question arises whether the changes in the tissue stores of TRH and SP, caused by antidepressant treatment (cf. Figure 3) are reflected in altered release. Data from Lighton et al., 1985 on TRH-LI release and on SPLI-release from this laboratory (Iverfeldt et al., 1986) suggest (cf. Figure 4) that the increased stores are giving rise to increased evoked release of both TRH and SP from tissues from tricyclic antidepressant treated rats.

Chronic treatment with tricyclic antidepressants lowered the tissue levels of the monoamine 5-HT and its metabolite 5-HIAA while TRH-LI and SPLI levels were elevated. These opposite changes in tissue levels of coexisting neurotransmitters may explain several pharmacological paradoxes of chronic treatment and may contribute to both the therapeutic and side effects of the treatment.

In conclusion the data indicate that the concept of autoregulation of the release of neurotransmitters is applicable in case of nervendings with several coexisting neurotransmitters such as those of the descending raphe fibers in the ventral spinal cord. The release of both classical and peptidergic neurotransmitters seems to be regulated via presynaptic (probably autoreceptors) receptors and among these presynaptic receptors are found both receptors for the classical neurotransmitter (5-HT) and for the peptide(s) (e.g. tachykinin).

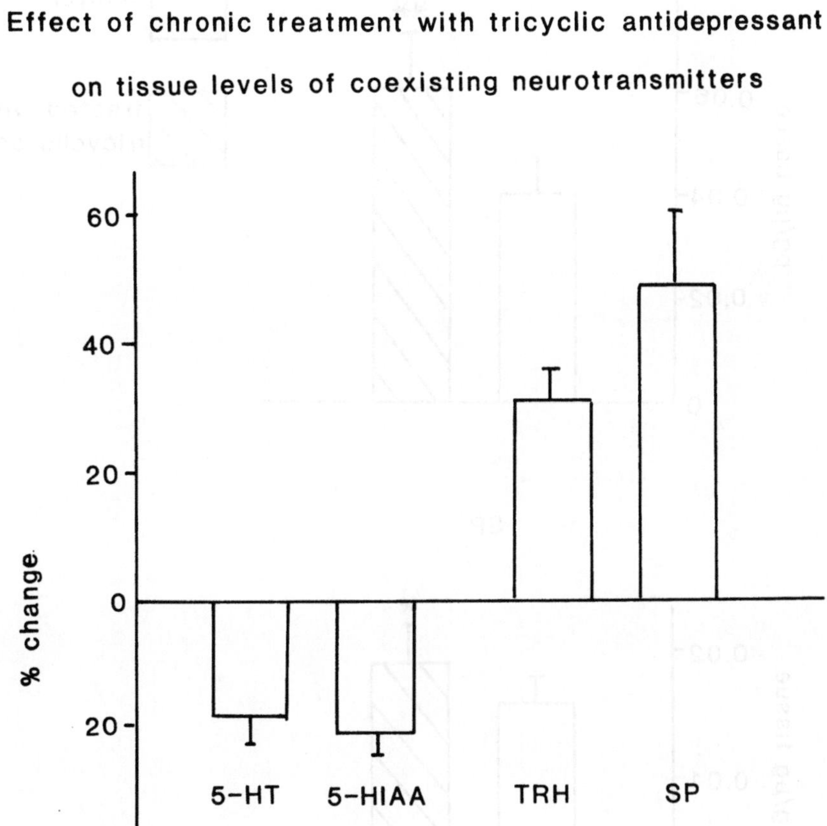

Figure 3

Effects of chronic antidepressant treatment on tissue levels of coexisting neurotransmittors in the spinal cord. Opposite changes in the tissue levels of 5-HT and 5-HIAA on one hand and TRH-LI and SPLI on the other hand. Treatment with imipramine (2 x 10 mole/kg/day p.o. for 14 days) was employed when 5-HT, 5-HIAA and SPLI-levels were measured. Data on TRH-LI levels are from Lighton et al., 1985. The treatment was carried out with another tricyclic antidepressant drug: amitriptyline (2 x 15 mg/kg/day, in the drinking water, for 14 days).

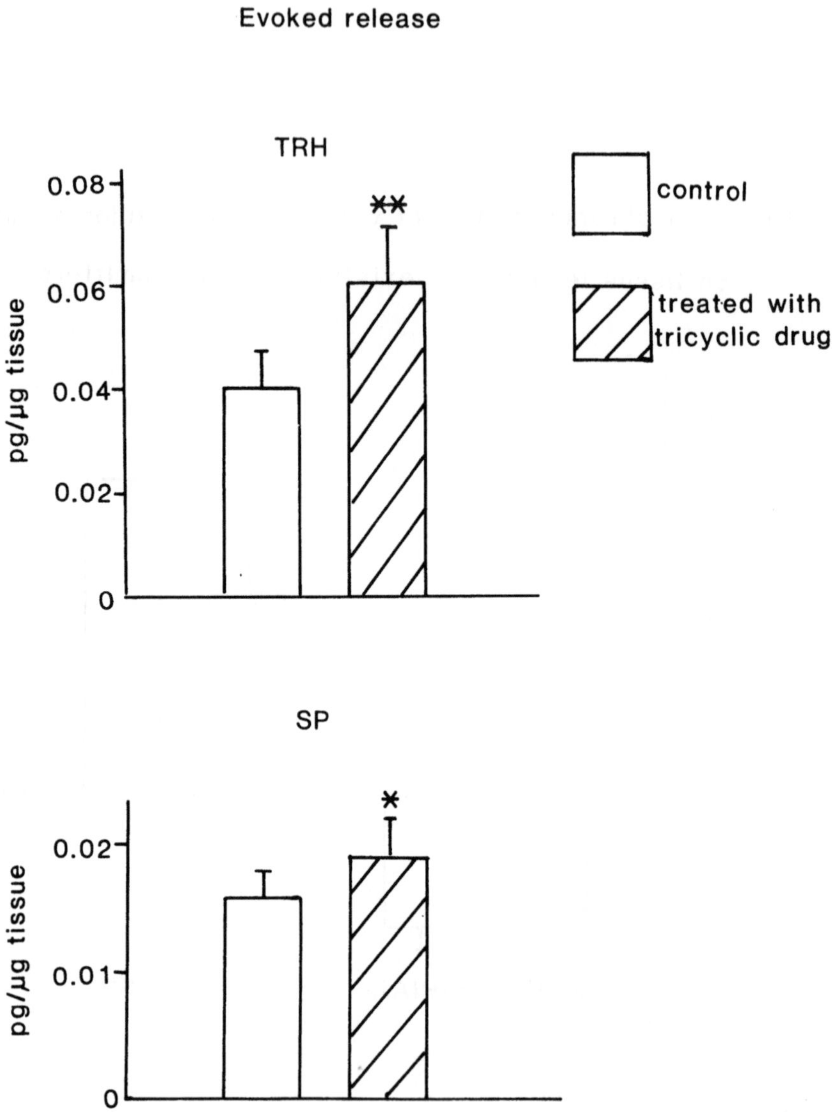

Figure 4

Effect of chronic administration of antidepressant drugs on ex vivo release of SPLI and TRH-LI. The evoked release (40 mM K^+) of SPLI from slices of the ventral spinal cord after chronic imipramine treatment (2 x 10 µmol/kg/day p.o. for 14 days) and the evoked release (56 mM K^+) of TRH-LI (data from Lighton et al., 1985) from lumbar spinal cord slices from rats chronically treated with amitriptyline (2 x 15 mg/kg/day, in the drinking water, for 14 days) were compared to control (vehicle treated rats).

REFERENCES

Agnati, L.F., Fuxe, K., Benfenati, F., Zini, I. and Hökfelt, T. (1983) On the functional role of coexistence of 5-HT and substance P in bulbospinal 5-HT neurons. Substance P reduces affinity and increases density of ^{3}H-5-HT binding sites. Acta Physiol. Scand., 117, 299-301.

Barbeau, H. and Bédard, P. (1981) Similar motor effects of 5-HT and TRH in rats following chronic spinal transection and 5,7-dihydroxytryptamine injection. Neuropharmacol., 20, 477-481.

Björklund, A.J., Emson, P.C., Gilberg, R.F.T. and Shagerberg, G. (1979) Further evidence for the possible coexistence of 5-hydroxytryptamine and substance P in medullary raphe neurons of the rat brain. Br. J. Pharmacol., 66, 112P-113P.

Brodin, E., Peterson, L.-L., Ögren, S.-O. and Bartfai, T. (1984) Chronic treatment with serotonin uptake inhibitor zimelidine elevates substance P levels in rat spinal cord. Acta Physiol. Scand., 122, 209-211.

Bury, R.W. and Mashford, M.L. (1976) Biological activity of C-terminal partial sequences of substance P. J. Med. Chem., 19, 854-858.

Cerrito, F. and Raiteri, M. (1979) Serotonin release is modulated by presynaptic autoreceptors. Eur. J. Pharmacol., 57, 427-430.

Chan-Palay, V., Jonsson, G. and Palay, S.L. (1978) Serotonin and substance P coexist in neurons of the rat's central nervous system. Proc. Natl. Acad. Sci., 75, 1582-1586.

Cuello, A.C. (1982) Cotransmission. MacMillan Press, London.

Gilbert, R.F.T., Emson, P.C., Hunt, S.P., Bennet, G.W., Marsden, C.A., Sandberg, B.E.B., Steinbusch, H.W.M. and Verhofstad, A.A.J. (1982) The effects of monoamine neurotoxins on peptides in the rat spinal cord. Neuroscience, 7, 69-78.

Göthert, M. (1980) Serotonin-receptor-mediated modulation of Ca^{2+}-dependent 5-hydroxytryptamine release from neurons of the rat brain cortex. Naunyn-Schmiedeberg's Arch. Pharmacol., 314, 223-230.

Hökfelt, T., Ljungdahl, A., Steinbuch, H., Verhofstad, A., Nilsson, G., Brodin, E., Pernow, B. and Goldstein, M. (1978) Immunohistochemical evidence of substance P-like immunoreactivity in some 5-hydroxytryptamine-containing neurons in the rat central nervous system. Neuroscience, 3, 517-538.

Hökfelt, T., Johansson, O., Ljungdahl, A., Lundberg, A., Lundberg, J.M. and Schultzberg, M. (1980) Peptidergic neurons. Nature, 284, 515-521.

Iverfeldt, K., Peterson, L.-L., Brodin, E., Ögren, S.-O. and Bartfai, T. (1986) Serotonin type-2 receptor mediated regulation of substance P release in the ventral spinal cord and the effects of chronic antidepressant treatment. Naunyn-Schmiedeberg's Arch. Pharmacol., 333, 1-6.

Johansson, O., Hökfelt, T., Pernow, B., Jeffcoate, S.L., White, N., Steinbusch, H.W.M., Verhofstad, A.A.J., Emson, P.C. and Spindel, E. (1981) Immunohistochemical support for three putative transmitters in one neuron: coexistence of 5-hydroxytryptamine, substance P and thyrotropin releasing hormone-like immunoreactivity in medullary neurons projecting to the spinal cord. Neuroscience, 6, 1857-1881.

Konishi, S. and Otsuka, M. (1974) Excitatory action of hypothalamic substance P on spinal motoneurons of newborn rats. Nature, 252, 734-735.

Lechan, R.M., Wu, P., Jackson, M.D., Wolf, H., Cooperman, S., Mandel, G. and Goodman, R.H. (1986) Thyrotropin-releasing hormone precursor: Characterization in rat brain. Science, 231, 159-161.

Lighton, C., Bennett, G.W. and Marsden, C.A. (1985) Increase in levels and ex vivo release of thyrotropin-releasing hormone (TRH) in specific regions of the CNS of the rat by chronic treatment with antidepressants. Neuropharmacol., 24, 401-406.

Lundberg, J.M. and Hökfelt, T. (1983) Coexistence of peptides and classical neurotransmitters. Trends in Neurosci., 6, 325-333.

Marsden, C.A., Bennett, G.W., Irons, J., Gilbert, R.F.T. and Emson, P.C. (1982) Localization and release of 5-hydroxytryptamine, thyrotropin releasing hormone and substance P in rat ventral spinal cord. Comp. Biochem. Physiol., 72C, 263-270.

McKelvy, J.F. (1982) Biosynthesis of thyrotropin releasing hormone. Front. Horm. Res., 10, 73-84.

Mitchell, R. and Fleetwood-Walker, S. (1981) Substance P, but not TRH modulates the 5-HT autoreceptor in ventral lumbar spinal cord. Eur. J. Pharmacol., 76, 119-120.

Mitnick, M. and Reichlin, S. (1972) Enzymatic synthesis of thyrotropin releasing hormone. Endocrinology, 91, 1145-1153.

Monroe, P.J. and Smith, D.J. (1985) Demonstration of an autoreceptor modulating the release of [^{3}H]5-hydroxytryptamine from a synaptosomal-rich spinal cord tissue preparation. J. Neurochem., 45, 1886-1894.

Nawa, H., Hirose, T., Takashima, H., Inayama, S. and Nakanishi, S. (1983) Nucleotide sequences of cloned cDNAs for two types of bovine brain substance P precursor. Nature (Lond.), 306, 32-36.

Otsuka, M. and Konishi, S. (1976) Substance P and excitatory transmitter of primary sensory neurons. Cold Spring Harb. Symp. Quant. Biol., 40, 135-143.

Ouafik, L.H., Dutour, A., Castanas, E., Boudouresque, F. and Oliver, C. (1985) Evidence for a precursor for TRH in the neonatal rat pancreas. Biochem. Biophys. Res. Commun., 128, 664-669.

Richter, K., Kawashima, E., Egger, R. and Kreil, G. (1984) Biosynthesis of thyrotropin releasing hormone in the skin of Xenopus laevis: partial sequence of the precursor deduced from cloned cDNA. EMBO J., 3, 617-621.

Solti, M. and Bartfai, T. (1987) Tachykinin regulation of serotonin release: enhancement of [^{3}H]serotonin release from rat cerebral cortex by neuromedin K and substance P acting at distinct receptor sites. Brain Res., in press.

Stauderman, K.A. and Jones, D.J. (1986) Presynaptic serotonin receptors regulate [^{3}H]serotonin release from rat spinal cord synaptosomes. Eur. J. Pharmacol., 120, 107-109.

Towle, A.C., Breese, G.R., Mueller, R.A., Hunt, R. and Lauder, J.M. (1986) Early postnatal administration of 5,7-dihydroxytryptamine: Effects on substance P and thyrotropin-releasing hormone neurons and terminals in rat brain. Brain Res., 363, 38-46.

Tremblay, L.E., Maheux, R. and Bédard, P.J. (1986) Substance P in the lumbar spinal cord of the rat affects the motor response to 5-HTP and TRH. Neuropharmacol., 25, 419-424.

Tsai, H.-Y., Maeda, S., Iwatsubo, K. and Inoki, R. (1984) Effect of neuroactive peptides on labeled 5-hydroxytryptamine release from rat spinal cord in vitro. Jpn. J. Pharmacol., 35, 403-406.

White, S.R. (1985) A comparison of the effects of serotonin, substance P and thyrotropin-releasing hormone on excitability of rat spinal motoneurons in vivo. Brain Res., 335, 63-70.

34

Are Opiates Involved in the Adaptation Processes of Dopamine Receptors that Follow Chronic Drug Exposure?

Antonio Groppetti, Luigina Rusconi, Eugenio Parati and Marco Parenti

A number of evidences have appeared over the years in the literature which indicate that opiates alter the activity of dopaminergic neurons and affect the responsiveness to dopaminomimetic or dopamino-lithic drugs. In several animal species opiates, similarly to neuroleptics, increase the rate of dopamine (DA) turnover and the concentrations of homovanillic and dihydroxyphenylacetic acids in the striatum, enhance the firing rate of DA cells in the zona compacta of the substantia nigra and induce a catatonic-like state (Lal, 1975). Cross tolerance with haloperidol has been reported in animals made dependent upon morphine (Puri and Lal, 1974). Laboratory tests have also shown that naloxone potentiated the effects of apomorphine and antagonized the behavioural responses induced by amphetamine (Holtzman, 1974; Dettmar et al., 1978; Agnati et al., 1979; Quock, 1981; Ferrari and Baggio, 1982; Quock et al., 1983).

Several investigations have been undertaken in order to understand the mechanisms by which opiates control DA transmission in the basal ganglia. A population of opiate receptors has been found to be located on DA terminals in the striatum (Pollard et al., 1977); moreover, inhibitory effect of opiates on DA-stimulated adenylate cyclase (AC) activity has also been reported in this brain area (Neff et al., 1981). Therefore it has been suggested that opiates are able to interfere with both pre- and post-synaptic sites of dopaminergic neurons. However, the functional implications of these complex interactions between the opiate and the dopaminergic systems are far from being clarified. The finding that met-enkephalin content in striatum is modified by DA agonists or antagonists does not help to simplify the problem (Hong et al., 1978; Groppetti et al., 1980).

Now we report that opiates are probably involved in the adaptation processes that occur at striatal DA receptors coupled to AC (DA-D1 sites) during prolonged exposure to L-dihydroxyphenylalanine (L-DOPA).

RAT STRIATAL OPIATE AND DOPAMINE RECEPTOR INTERACTIONS UPON ADENYLATE CYCLASE

Inhibition of AC activity by opiates is a well established phenomenon and it has been demonstrated to be present in several tissues, including striatum and cultured cell preparations (Collier and Roy, 1974; Sharma et al., 1975; Blume et al., 1979; Law et al., 1981; Neff et al., 1981; Cooper et al., 1982). In common with other receptor-mediated modulations of AC, opiate-inhibition requires guanine nucleotides and appears to act through transduction mechanisms analogous to those of activators involving an agonist-stimulated hydrolysis of GTP by a low Km GTPase activity (Koski and Klee, 1981; Parenti et al., 1983). As shown in Figure 1 studies carried out mainly with neuroblastoma x glioma NG 108-15 hybrid cells or turkey erythrocytes, have suggested that AC is modulated by at least two systems which operate in opposite directions. One is associated with negatively-coupled receptors, the other with positively-coupled receptors. Both receptors transduce their messages to a common AC moiety through different GTP binding regulatory proteins, termed Gi and Gs. It has been demonstrated that changes in one system result in alterations, usually compensatory in nature, of the other system (Katada and Ui, 1982). There is some evidence suggesting that a similar organization may exist between opiate and prostaglandin E1 receptors positively coupled to AC in cultured NG 108-15 cells (Sharma et al., 1975) and

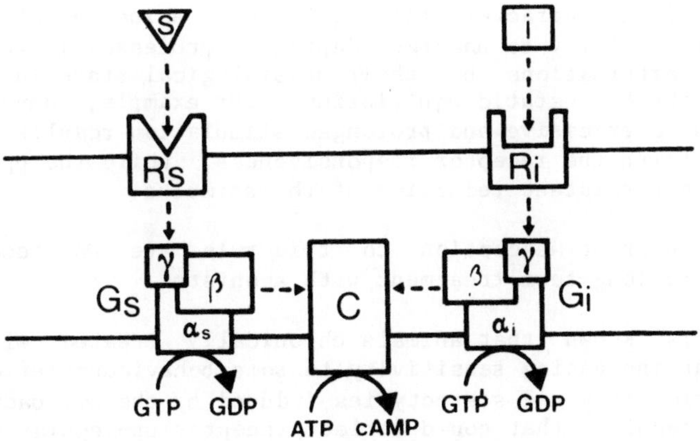

Fig.1 - Schematic organization of a dually regulated AC system. s, i, stimulatory and inhibitory agonists; Rs, Ri, stimulatory and inhibitory receptors; Gs, Gi, stimulatory and inhibitory guanine nucleotide regulatory proteins each composed by three different protein subunits α, β and γ; αs and αi have different molecular weights, respectively 45,000 and 41,000 but they share common properties, i.e. they both possess a high-affinity GTP binding site and a low Km GTPase activity; β and γ are probably common to both Gs and Gi; C, catalytic moiety of adenylate cyclase.

between opiate and DA receptors in the monkey amygdala (Walczak et al., 1979). In 1981 Neff and collaborators reported that also in partially purified striatal membranes the stimulatory effect of DA on AC could be lowered adding morphine to the incubation medium. Also in this preparation changes in one of the components of the AC complex seem to be reflected by alterations of the other. In fact it was found that the increased sensitivity of striatal AC to DA stimulation (Parenti et al., 1981) and the supersensitive behavioural responses to apomorphine treatment (Lal, 1975) present in animals chronically in vivo exposed to morphine were associated and possibly dependent upon desensitization of the opiate inhibitory component of the AC complex (Parenti et al., 1983; Tirone et al., in preparation). Moreover, pertussis toxin, known to inactivate the guanine nucleotide regulatory protein Gi that mediates the inhibition of AC induced by the negatively coupled receptors (Kurose et al., 1983), was able also in striatum to antagonize the inhibitory effect of the opiates on the enzyme, while resulting in an increased sensitivity to DA stimulation (Figure 2).

Thus also in striatum the activity of AC seems to be modulated by at least two opposite systems. Being changes in one of them apparently reflected by the other, it appears that also in this brain area stimulatory (DA) and inhibitory (opiate) receptors coupled to AC are closely functionally related.

ROLE OF THE OPIATES IN THE STRIATAL DOPAMINE-STIMULATED ADENYLATE CYCLASE SUPERSENSITIVITY CAUSED BY PROLONGED L-DOPA TREATMENT

One of the characteristics of biogenic amine receptors in the CNS is the ability to undergo adaptation processes in response to chronic perturbations of their physiological state in order to maintain the homeostatic equilibrium. For example, any procedure that causes excessive and prolonged stimulation results in changes that diminish the receptor responsiveness, while the opposite is true for a persistent reduction of the stimulus.

An apparent exception to this rule are DA receptors in response to long-term treatment with agonists.

It is known that animals chronically treated with L-DOPA develop an increasing sensitivity to some behavioural effects such as hypermotility and stereotypies induced by the aminoacid itself or by DA agonists that could reflect receptor up-regulation instead of desensitization as expected (Klawans et al., 1977; Pycock et al., 1982). Increased supersensitivity to apomorphine stimulation was also found in 6-hydroxydopamine (6-OHDA) lesioned rats chronically treated with L-DOPA (Parenti et al., 1986a).

It has been suggested that these behavioural effects might have their biochemical correlates in the changes of DA-D1 receptors coupled to stimulation of AC (Parenti et al., 1986a). In fact

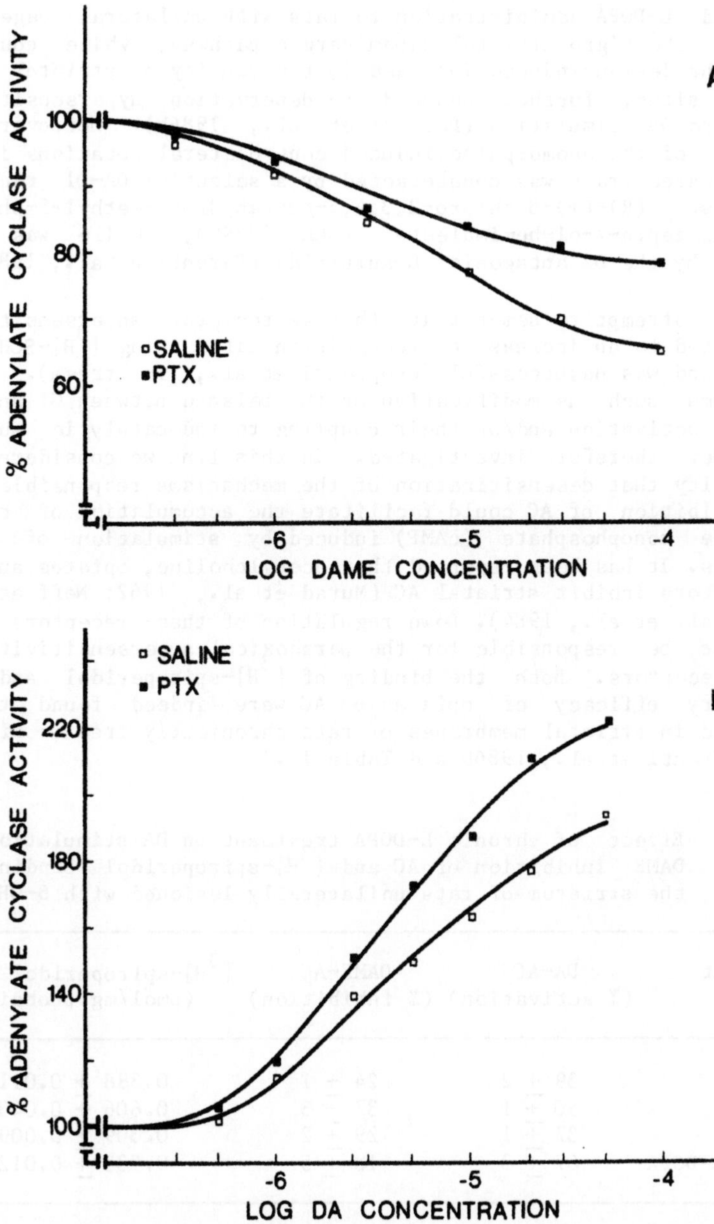

Fig.2 – Effect of pertussis toxin (PTX) on D-Ala^2-Met^5-enkephalinamide (DAME) inhibition (A) and DA stimulation (B) of rat striatal AC. PTX was stereotoxically injected into the striatum at a dose of 5 μg in 2 μl. The animals were killed after two days and washed striatal particulate fractions were prepared for AC assay as described by Parenti et al. (1986b).

prolonged L-DOPA administration to rats with unilateral degeneration of the nigro-striatal dopaminergic pathway, while counteracting the lesion-induced increase in the density of striatal DA-D2 receptor sites, further enhanced the denervation hypersensitivity of AC to DA stimulation (Parenti et al., 1986b). Moreover the increase of the apomorphine-induced controlateral rotations in L-DOPA treated rats was counteracted by a selective DA-D1 receptor antagonist (R)-(+)-8-chloro-2,3,4,5-tetrahydro-3-methyl-5-phenyl-1H-3-benzazepin-7-ol-hemimaleate (SCH 23390), while was not affected by the D2 antagonist L-sulpiride (Parenti et al., 1986a).

The attempt to demonstrate that DA receptor supersensitivity was related to an increase of recognition sites using [^{3}H]-SCH23390 as a ligand was unsuccessful (Groppetti et al., in press). Other mechanisms such as modification of the balance between Gi and Gs receptor activation and/or their coupling to the catalytic subunit of AC were therefore investigated. In this line we considered the possibility that desensitization of the mechanisms responsible for the inhibition of AC could facilitate the accumulation of cyclic adenosine monophosphate (cAMP) induced by stimulation of DA-D1 receptors. It has been reported that acetylcholine, opiates and DA-D2 receptors inhibit striatal AC (Murad et al., 1962; Neff et al., 1981; Onali et al., 1984). Down regulation of these receptors could therefore be responsible for the paradoxical supersensitivity of DA-D1 receptors. Both the binding of [^{3}H]-spiroperidol and the inhibitory efficacy of opiates on AC were indeed found to be decreased in striatal membranes of rats chronically treated with L-DOPA (Parenti et al., 1986b and Table 1).

Table 1. Effect of chronic L-DOPA treatment on DA stimulation and DAME inhibition of AC and [^{3}H]-spiroperidol binding in the striatum of rats unilaterally lesioned with 6-OHDA.

Treatment	DA-AC (% activation)	DAME-AC (% inhibition)	[^{3}H]-spiroperidol bound (pmol/mg protein)
Saline	39 ± 2	24 ± 1	0.388 ± 0.011
6-OHDA	50 ± 1	37 ± 3	0.606 ± 0.043
L-DOPA	37 ± 1	29 ± 2	0.309 ± 0.009
6-OHDA+L-DOPA	67 ± 3	18 ± 5	0.335 ± 0.012

DA-AC was assayed in the presence of 3 μM DA and DAME-AC was measured with 10 μM DAME; [^{3}H]-spiroperidol (0.6 nM) binding was determined in washed striatal particulate fractions as described by Parenti et al. (1986b).
Results are the means ± S.E.M. of three experiments with each value run in duplicate.

It may be speculated that the acute administration of the aminoacid affects DA receptors associated to AC in a dual manner providing more stimulatory transmitter but also activating inhibitory mechanisms coupled to the enzyme system. It is possible that in a system already unbalanced by 6-OHDA lesion the receptor desensitization that develops after prolonged exposure to the agonist is more pronounced for those receptors coupled to the inhibitory component of the AC complex thus resulting in a supersensitivity of the stimulatory counterpart. It may be interesting to note that the data in Figures 3 and 4 show that a remarkably similar situation developed on DA receptors associated to AC when rats unilaterally lesioned with intranigral injection of 6-OHDA are chronically implanted with morphine pellets. In fact, in subjects that received the combined treatment the responses of striatal AC to DA stimulation were greater than with 6-OHDA of morphine alone (Figure 3). Also the rotations induced by apomorphine were considerably greater when treatments with 6-OHDA and morphine were combined than with 6-OHDA alone (Figure 4). Like prolonged treat-

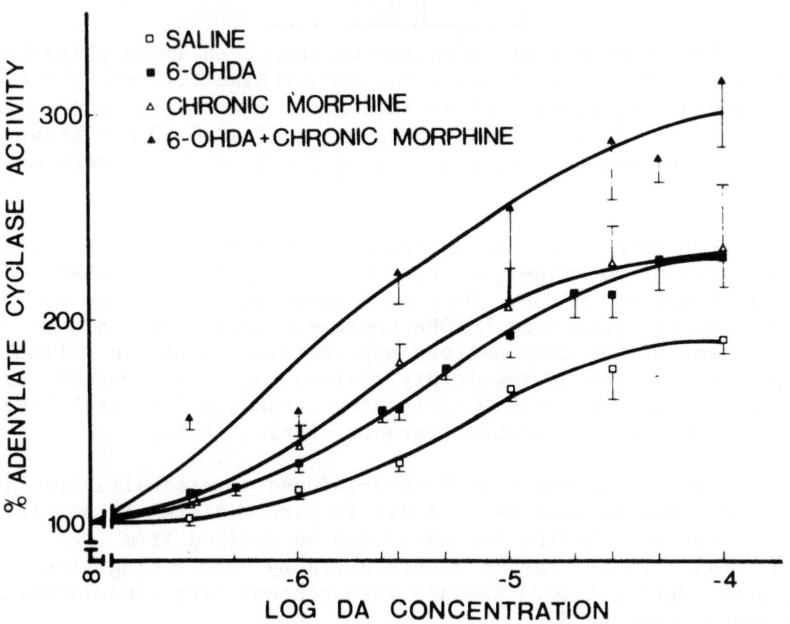

Fig.3 - Effect of chronic morphine treatment on striatal DA-stimulated AC in rats unilaterally lesioned with 6-OHDA. 6-OHDA (8 µg in 2 µl) was stereotoxically injected unilaterally into substantia nigra. Morphine was administered by subcutaneous implantation of two pellets (75 mg morphine base/each) every three days for two weeks starting 15 days after 6-OHDA injection. The animals were killed three days after the last pellet implantation. The results reported are the means $\pm$ S.E.M. of three different experiments with each point run un duplicate.

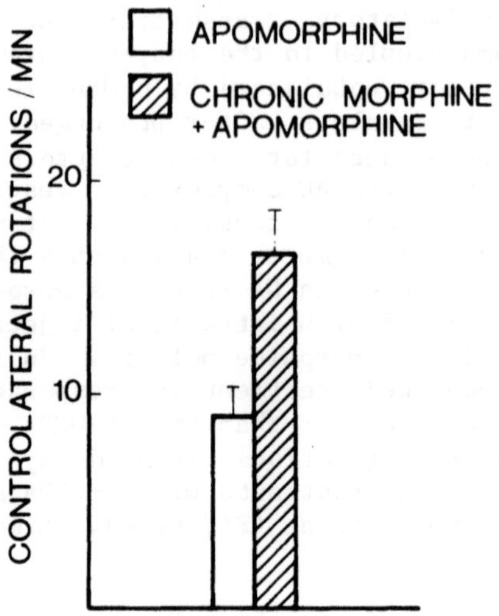

Fig.4 - Effect of chronic morphine treatment on apomorphine-induced rotations in 6-OHDA lesioned rats. Apomorphine (0.5 mg/kg s.c.) was administered to animals three days after the last morphine pellet implantation. Turning behaviour was evaluated 30 min after apomorphine. The results are the means ± S.E.M. of 5 observations.

ment with L-DOPA, chronic morphine in 6-OHDA lesioned striatum led to far greater dopaminergic supersensitivity than was observed with either treatment alone. Chronic activation of endogenous opiate receptors by prolonged L-DOPA treatment could therefore be among the causes of the compensatory hyperactivation of the stimulatory component of the AC regulatory system. In this context it was reported that chronic L-DOPA treatment changes the basal levels of striatal endogenous opiates (Parenti et al., 1986a).

Another evidence that L-DOPA-induced supersensitivity in 6-OHDA lesioned animals is at least in part mediated by an elevation of opiate tone by the aminoacid can be derived from the finding that this effect can be counteracted by protecting the opiate receptors during L-DOPA therapy with contemporary administration of naltrexone (Figure 5).

In conclusion these data indicate that opiate receptors in striatum modulate DA responses on AC and that they may be part of the mechanism responsible for L-DOPA-induced paradoxical supersensitivity.

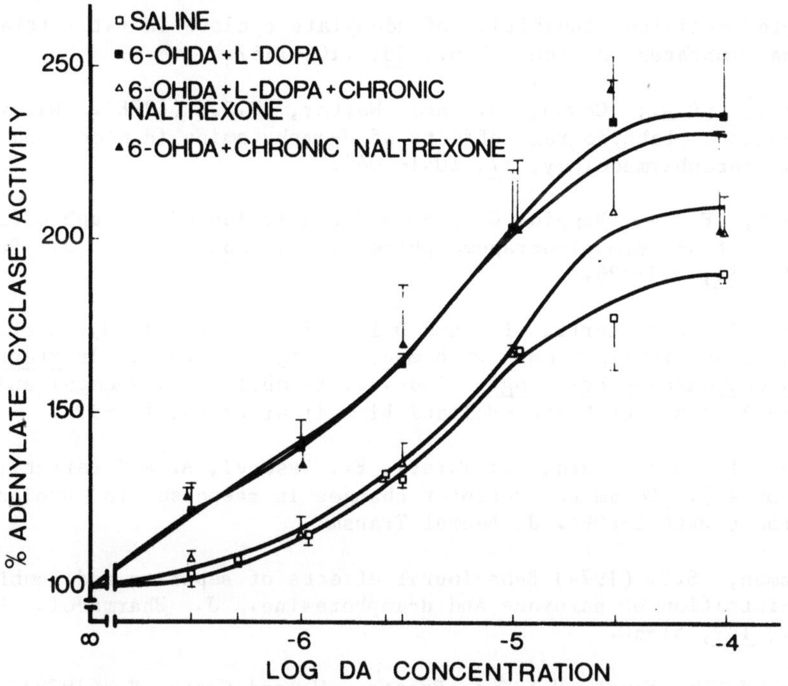

Fig.5 - Effect of chronic naltrexone and L-DOPA treatments on striatal DA-stimulated AC in rats unilaterally lesioned with 6-OHDA. L-DOPA was administered twice a day at the dose of 25 mg/Kg i.p. (plus 2.5 mg/Kg of carbidopa) and naltrexone was injected s.c. at one daily dose of 2 mg/Kg. Both treatments lasted two weeks starting 15 days after 6-OHDA injection. The animals were killed three days after the last naltrexone injection while last L-DOPA + carbidopa treatment was administered 16 hours before killing. The results are the means ± S.E.M. of three different experiments with each point run in duplicate.

REFERENCES

Agnati, L.F.; D'Alessandro, R.; Benfenati, F.; Cortelli, P.; Galassi, E. and Capelli, M. (1979). Naloxone potentiation of 2-Br-alpha-ergocryptine (CB 154) effects on GH secretion in man. Neurosci. Lett., 15, 229-233.

Blume, A.; Lichtschein, D. and Boone, G. (1979). Coupling of opiate receptors to adenylate cyclase: Requirement for Na^+ and GTP. Proc. Natl. Acad. Sci. USA, 76, 5626-5630.

Collier, H.O. and Roy, A.C. (1974). Morphine-like drugs inhibit the stimulation by E prostaglandins of cyclic AMP formation by rat brain homogenate. Nature, 248, 24-27.

Cooper, D.M.; Londos, C.; Gill, D.L. and Rodbell, M. (1982). Opiate receptor-mediated inhibition of adenylate cyclase in rat striatal plasma membranes. J. Neurochem., 38, 1164-1167.

Dettmar, P.W.; Cowan, A. and Walter, D.S. (1978). Naloxone antagonizes behavioural effects of d-amphetamine in mice and in rats. Neuropharmacology, 17, 1041-1044.

Ferrari, F. and Baggio, G. (1982) Potentiation of the aphrodisiac effect of N-n-propyl-norapomorphine by naloxone. Eur. J. Pharmacol., 81, 321-326.

Groppetti, A.; Cerri, L. and Dellavedova, L. (1980). Striatal enkephalins: interactions with monoaminergic neurons. In Progress in Psychoneuroendocrinology. (eds. F. Brambilla; G. Racagni and D. de Wied). Elsevier/North -Holland Biomedical Press, 115-124.

Groppetti, A.; Flauto, C.; Parati, E.; Vescovi, A. and Parenti, M. (in press). Dopamine receptor changes in response to prolonged treatment with L-DOPA. J. Neural Transm.

Holtzman, S.G. (1974) Behavioural effects of separate and combined administration of naloxone and d-amphetamine. J. Pharmacol. Exp. Ther., 189, 51-60.

Hong, J.S.; Yang, H.Y.T.; Fratta, W. and Costa, E. (1978). Rat striatal methionine-enkephalin content after chronic treatment with cataleptogenic and non-cataleptogenic antischizophrenic drugs. J. Pharmacol. Exp. Ther., 205, 141-147.

Katada, T. and Ui, M. (1982). Direct modification of the membrane adenylate cyclase system by islet-activating protein due to ADP-ribosylation of a membrane protein. Proc. Natl. Acad. Sci. USA, 79, 3129-3133.

Klawans, H.L.; Goetz, C.; Nauseida, M.D. and Weiner, W.J. (1977). Levodopa-induced dopamine receptors hypersensitivity. Ann. Neurol., 2, 125-129.

Koski, G. and Klee, W.A. (1981). Opiates inhibit adenylate cyclase by stimulating GTP hydrolysis. Proc. Natl. Acad. Sci. USA, 78, 4185-4189.

Kurose, H.; Katada, T.; Amano, T. and Ui, M. (1983). Specific uncoupling by islet-activating protein, pertussis toxin, of negative signal transduction via alpha-adrenergic, cholinergic and opiate receptors in neuroblastoma x glioma hybrid cells. J. Biol. Chem., 258, 4870-4875.

Lal, H. (1975). Minireview: Narcotic dependence, narcotic action and dopamine receptors. Life Sci., 17, 483-496.

Law, P.Y.; Wu, J.; Koehler, J.E. and Loh, H.H. (1981). Demonstration and characterization of opiate inhibition of the striatal adenylate cyclase. J. Neurochem., 36, 1834-1846.

Murad, F.; Chi, Y.M.; Rall, T.W. and Sutherland, E.W. (1962). Adenylate cyclase. III. Effect of catecholamines and cholinesters on the formation of adenosine 3'-5'-phosphate by preparations from cardiac muscle and liver. J. Biol. Chem., 237, 1233-1238.

Neff, N.H.; Parenti, M.; Gentleman, S. and Olianas, M.C. (1981). Modulation of dopamine receptors by opiates. In Apomorphine and other dopaminomimetics. (eds. G.L. Gessa and G.U. Corsini). Raven Press, New York, 193-200.

Onali, P.F.; Olianas, M.C. and Gessa, G.L. (1984). Selective blockade of dopamine D-1 receptors by SCH-23390 discloses striatal dopamine D-2 receptors mediating the inhibition of adenylate cyclase in rats. Eur. J. Pharmacol., 99, 127-128.

Parenti, M.; Gentleman, S.; Olianas, M.C. and Neff, N.H. (1981). The dopamine receptor complex: biochemical consequences of supersensitivity induced by treatment with 6-hydroxydopamine or morphine. In Apomorphine and other dopaminomimetics. (eds. G.L. Gessa and G.U. Corsini). Raven Press, New York, 201-208.

Parenti, M.; Gazzotti, G., Tirone, F. and Groppetti, A. (1983). Opiate tolerance and dependence is associated with a decreased activity of GTPase in rat striatal membranes. Life Sci., 33, 345-348.

Parenti, M.; Tirone, F.; Flauto, C.; Parati, E.; Vescovi, A. and Groppetti, A. (1986a). Effects of prolonged Levodopa treatment on striatal dopamine receptors. In Modulation of central and peripheral transmitter function. (eds. G. Biggio; P.F. Spano; G. Toffano and G.L. Gessa). Liviana Press, Springer-Verlag, Berlin, 117-124.

Parenti, M.; Flauto, C.; Parati, E.; Vescovi, A. and Groppetti, A. (1986b). Differential effect of repeated treatment with L-DOPA on dopamine-D1 or -D2 receptors. Neuropharmacology, 25, 331-334.

Pollard, H.; Llorens-Cortes, C. and Schwartz, J.C. (1977). Enkephalin receptors on dopaminergic neurons in rat striatum. Nature, 268, 745-747.

Puri, S.K. and Lal, H. (1974). Tolerance to the behavioural and neurochemical effects of haloperidol and morphine in rats chronically treated with morphine or haloperidol. Naunyn-Schmiedeberg's Arch. Pharmacol., 282, 155-170.

Pycock, C.; Dawbarn, D. and O'Shaughnessy, C. (1982). Behavioural and biochemical changes following chronic administration of L-DOPA to rats. Eur. J. Pharmacol., 79, 201-215.

Quock, R.M. (1981). Enhancement of apomorphine-induced climbing in

Sci., 28, 1421-1424.

Quock, R.M.; Lucas, T.S. and Hartl; T.J. (1983) Potentiation of apomorphine-induced stereotypies by naloxone and L-propyl-L-leucyl-glycinamide. Pharmacol. Biochem. Behav., 19, 49-52.

Sharma, S.K.; Nirenberg, M. and Klee, W.A. (1975). Morphine receptors as regulators of adenylate cyclase activity. Proc. Natl. Acad. Sci. USA, 72, 570-594.

Walczak, S.A.; Wilkening, D. and Makman, M.H. (1979). Interaction of morphine, etorphine and enkephalins with dopamine-stimulated adenylate cyclase of monkey amygdala. Brain Res., 160, 105-116.

35

GABA Receptors, Depression and Antidepressant Drug Action

Giuseppe Bartholini, Kenneth G. Lloyd, Bernard Scatton, Branimir Zivkovic and Paolo L. Morselli

INTRODUCTION

The clinical demonstration of the major antidepressant action of GABAmimetic agents (Morselli et al., 1980; 1986; Musch, 1986; Weiss et al., 1986) has led to the formulation of a coherent GABA hypothesis of depression. This hypothesis in addition is supported by a large body of evidence which however, remained inconclusive in the absence of clear cut clinical data with GABAmimetics. The compounds which have stimulated this progress include two benzylidene GABAmimetics, progabide : 4-{[(4-chlorophenyl) (5-fluoro-2-hydroxy-phenyl)-methylene]amino} butanamide, and fengabine : 2-[(Butylimino) (2-chlorophenyl)methyl]-4-chlorophenyl.

Experimental data have shown concomitantly that increase in GABA-mediated transmission by GABAmimetics reverses the behavioural deficit in animal models predictive of antidepressant drug action and causes changes in the turnover of some neurotransmitters and alterations of the corresponding receptors.

This paper focusses on these changes as related to monoamines. In this frame, the role of GABA will be discussed in relation with the pathogenesis of depression and the mode of action of conventional antidepressant drugs.

The pharmacological spectrum of GABAmimetics has been recently reviewed (Bartholini, 1985; Bartholini et al., 1985), as well as their involvement in mood disorders (Bartholini et al., 1986a).

RECEPTOR CHANGES INDUCED BY GABA RECEPTOR STIMULATION

Noradrenergic System

Acute administration of progabide or fengabine as well as other GABAmimetics (e.g. muscimol and THIP) enhances the liberation of norepinephrine (NE) in the rat brain by a picrotoxin-sensitive mechanism (increased rate of NE utilization ; increase in 3,4-dihydroxy-phenylethyleneglycol (DOPEG), 3-methoxy,4- hydroxy-phenylethyleneglycol and normetanephrine levels) (Scatton et al., 1982; 1986; Dennis et al., 1985). This activation occurs in areas innervated by the ventral and the dorsal bundles of the ascending noradrenergic path as well as in the spinal cord (Dennis et al., 1985). The GABA receptor involved in this action seems to be the GABA A subtype (considering the action of muscimol and THIP) ; however, the involvement of the GABA B site is not excluded (Scatton et al., 1986).

The mechanism of the noradrenergic activation by GABA receptor stimulation is probably a primary increase in cell firing. Thus, progabide and muscimol increase the levels of 3,4-dihydroxy-phenylacetic acid (Dennis et al., 1985; Scatton and Serrano, 1986) in the rat locus coeruleus (LC), which reflect activity of noradrenergic neurons (Curet et al., 1985). In addition, the increase by GABAmimetics of DOPEG concentraion in noradrenergic projection areas is blocked by interruption of the impulse flow by injection of tetrodotoxin in the ascending bundle (Dennis et al., 1985; Scatton et al., 1986).

The localization of the GABA receptors involved in the activation of NE neurons is unknown but it is unlikely to be on noradrenergic cell bodies as infusion of muscimol into the rat LC fails to alter DOPEG levels in the projection areas (Dennis et al., 1985). Also, GABA receptors are not localized on NE neuron terminals as muscimol injected into the nucleus accumbens or the hypothalamus does not modify DOPEG levels (Dennis et al., 1985) ; and systemic progabide administration no longer increases hypothalamic DOPEG concentration after acute interruption of noradrenergic impulse flow by tetrodotoxin (see above). It is therefore likely that the GABAergic control of NE neurons occurs indirectly via GABA synapses involved in regional circuits regulating NE cell activity.

In the rat, repeated treatment with progabide (but not fengabine) leads to a reduction in the enhancement of NE turnover seen after a single injection (Scatton et al., 1982). This is not related to induction of enzymes metabolizing progabide as progabide-induced decrease in 5-hydroxytryptamine (5-HT) turnover is even more pronounced after repeated administration of the drug (see below).

It should be expected that the sustained increase in NE synaptic concentration is associated with a down-regulation of β_1 receptors as is the case with conventional antidepressants i.e. monoamine oxidase inhibitors and NE uptake inhibitors (Sellinger-Barnette et al., 1980; see below). In the case of repeated progabide administration, however, no changes in β_1 receptor density occur (Zivkovic et al., 1982). This may be due to the progressive development of tolerance such that the increase in noradrenergic transmission is of insufficient duration and/or intensity to trigger receptor changes. This is however unlikely as fengabine - which causes a more marked activation of NE neurons than progabide and to which tolerance does not develop - also fails to down regulate $ß_1$ receptors (Zivkovic et al., 1986). Alternatively, the lack of adaptive changes in the density of $ß_1$ receptors after progabide could be attributed to the decrease in 5-HT-mediated transmission (see below). Thus, lesion of 5-HT neurons prevents the $ß_1$ receptor down-regulation induced by conventional antidepressants (Brunello et al., 1982; Janowsky et al., 1982). However, fengabine does not affect the 5-HT system (see below) yet is devoid of effects on $ß_1$ adrenoceptor density.

Although progabide and fengabine fail to alter $ß_1$ receptors, measurable changes occur in target cells of NE neurons. Indeed, when progabide - which by itself does not affect the activity of $ß_1$ receptor coupled adenylate cyclase (AC) - is repeatedly administered with a subthreshold dose of desipramine, subsensitivity of the enzyme to isoprenaline develops (Fig. 1; Bartholini et al., 1986b). Moreover, fengabine induces per se desensitization of the $ß_1$ adrenoceptor linked AC (Fig. 1; Zivkovic et al., 1986; Scatton et al., 1987).

These effects, which indicate an uncoupling of the enzyme from the $ß_1$ recognition site, may be the result of direct drug action on NE target cells; thus, sustained fengabine administration activates NE neurons terminating in the hypothalamus and septal areas but not in frontal cortex, whereas the desensitization of AC is observed in all three regions (Zivkovic et al., 1986; Scatton et al., 1987).

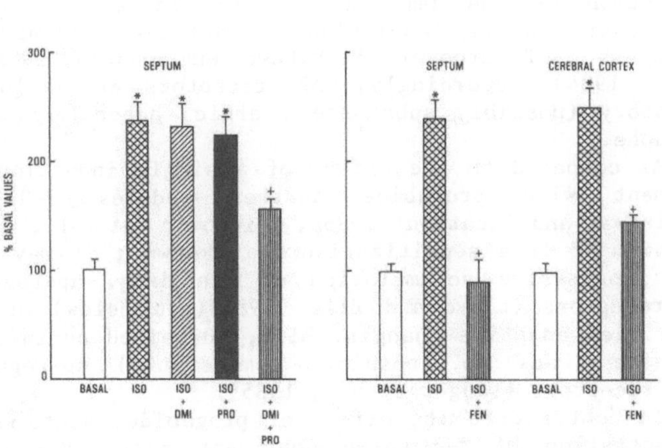

Fig. 1. Effect of repeated treatment with GABAmimetics on isoprenaline (ISO)-stimulated AC activity in rat brain areas. Sacrifice 48 hrs after the last injection of a 14 day treatment with desipramine (DMI, 5mg/kg i.p. once a day) and/or progabide (PRO, 400mg/kg i.p. bid) or with fengabine (FEN, 200mg/kg i.p. bid). (ISO)-stimulated AC activity was measured according to Blumberg et al., 1976. Each value is the mean with S.E.M. of data obtained on 6-16 rats (Zivkovic et al., 1982; 1986; Scatton et al., 1987). *p $<$ 0.01 vs basal; +p $<$ 0.01 vs ISO.

Serotonergic System

In the rat, the turnover of cerebral 5-HT is decreased by GABAmimetics such as progabide or muscimol, as well as by GABA transaminase inhibitors (e.g. γ-acetylenic GABA) via a bicuculline sensitive mechanism [decreased rate of 5-HT utilization and synthesis, reduction of extracellular 5-hydroxyindoleacetic acid (5-HIAA) concentration; Scatton et al., 1986].

This decrease is ubiquitous, occurring in the midbrain raphe nuclei and in the 5-HT projection areas from the raphe pontis, magnus, dorsalis and medianus ; it is probably due to a reduction in 5-HT cell firing as transection of the ascending serotonergic pathway prevents the effect of progabide in striatum and hippocampus (Nishikawa and Scatton, 1985a).

The GABA receptors affecting 5-HT neurons appear to lie in the cell body region (raphe) and not in the terminal areas. In fact, micro-injection of GABA or GABAmimetics in the rat striatum or substantia nigra fails to alter 5-HT related parameters in these same areas; in contrast, infusion of these agents into the corresponding cell body areas (raphe dorsalis and medianus) depresses 5-HT synthesis and release as well as extracellular 5-HIAA concentration in the projection regions (Scatton et al., 1984; Nishikawa and Scatton, 1985a).

The reduction of the 5-HT turnover by GABAmimetics in the anterior raphe of the rat depends on the integrity of the habenulo-raphe path. When this path is destroyed by electro-coagulation or the impulse flow is blocked by injection of tetrodotoxin in the fasciculus retroflexus, GABAmimetics fail to depress 5-HT turnover (Nishikawa and Scatton, 1985b; Scatton et al., 1984). Accordingly, GABA receptors may be localized on excitatory (possibly substance P ergic) habenular afferents to the raphe.

As compared to the effect of a single injection, repeated treatment with progabide further reduces 5-HT turnover (Nishikawa and Scatton 1985a; Zivkovic et al., 1982). The mechanism of this sensitization is unknown ; it may be related to a progressive accumulation of the drug, up-regulation of GABA receptors (Lloyd and Pilc, 1984)(see below) or other yet unclarified adaptive changes. Also, sustained administration of progabide leads to (possibly homeostatic) up-regulation of 5-HT_2 receptors (Langer et al., 1985).

In contrast to the effect of progabide, acute or repeated administration of fengabine does not alter the 5-HT system (5-HT turnover, 5-HT_2 receptors; Zivkovic et al., 1986). This difference is unexplained yet ; it may be connected with the kinetic and metabolic features of fengabine.

GABA System

Repeated administration of progabide affects neither presynaptic parameters related to GABA function e.g. glutamic acid decarboxylase (GAD) activity, GABA levels nor GABA A binding (Lloyd, in preparation). Similarly, fengabine is ineffective on GAD, GABA concentrations and GABA A binding (Scatton et al., 1987).

In contrast, major changes occur in GABA B receptor density by these and other agents. These changes will be discussed below.

GABA, DEPRESSION AND MODE OF ACTION OF ANTIDEPRESSANT AGENTS

In the light of the major therapeutic action of progabide (Morselli et al., 1980; 1986; Weiss et al., 1986) and fengabine (Musch, 1986; Musch, in preparation) in depressive states (see below), the question which arises concerns the mode of action of these drugs as compared to that of conventional anti-depressants. Monoamine uptake inhibitors, by blocking the reuptake of NE and/or 5-HT, enhance monoaminergic transmission; after repeated administration, these drugs cause adaptive changes of the corresponding receptors, i.e. down-regulation of β_1 and/or 5-HT_2 sites (Anderson 1983; Sugrue, 1983) as well as subsensitivity of β_1 receptor linked AC (Sulser, 1984). Mianserine and 5-HT uptake inhibitors however, similarly to GABAmimetics, have been reported to down regulate the β_1 receptor linked AC in the absence of changes in β_1 adreno-ceptors (Mishra et al., 1980). Monoamine oxidase inhibitors by preventing the metabolism of monoamines enhance their intra-synaptic concentrations and, on prolonged treatment, cause down-regulation of monoamine receptors (Sellinger-Barnette et al., 1980).

At variance, GABAmimetics primarily affect the firing rate of monoaminergic neurons and cause adaptive changes in receptors and probably transmission alterations which differ from those of conventional antidepressant drugs (Table 1).

TABLE 1

EFFECTS OF MONOAMINE UPTAKE INHIBITORS AND GABA RECEPTOR AGONISTS ON NE AND 5-HT RELATED PARAMETERS

ANTIDEPRESSANTS	TRANSMISSION		ADAPTIVE CHANGES		
	NE	5-HT	β_1	AD.CYCL.	5-HT_2
MONOAMINE UPTAKE INHIBITORS:					
NE+5-HT	↑	↑	↓	↓	↓
Specific NE	↑	=	↓	↓	=
Specific 5-HT	=	↑	=	=	↓
GABA RECEPTOR AGONISTS:					
Progabide	↑	↓	=	=	↑
Fengabine	↑	=	=	↓	=

↓ : reduction; ↑ : enhancement; = : no alteration;

These heterogeneous patterns challenge the present formulation of the monoamine theory of depression and antidepressant drug action and propose as possible alternative mechanisms a) that the drug-induced monoamine changes are epiphenomena, a GABAergic mechanism being primarily involved or, rather, b) that alterations in monoamine function induced by conventional antidepressants activate the GABAergic system which in turn elicits antidepressant effects.

Evidence for the involvement of GABA.

The clear cut clinical evidence for the antidepressant action of GABAmimetics in major depressive states (as defined by the DSM III diagnostic criteria) (Fig. 2; Morselli et al. 1986; Weiss et al., 1986; Musch, 1986; Musch et al., in preparation) is accompanied by a large body of experimental evidence supporting the involvement of GABA in depression and in the mode of action of antidepressant drugs including monoamine uptake inhibitors.

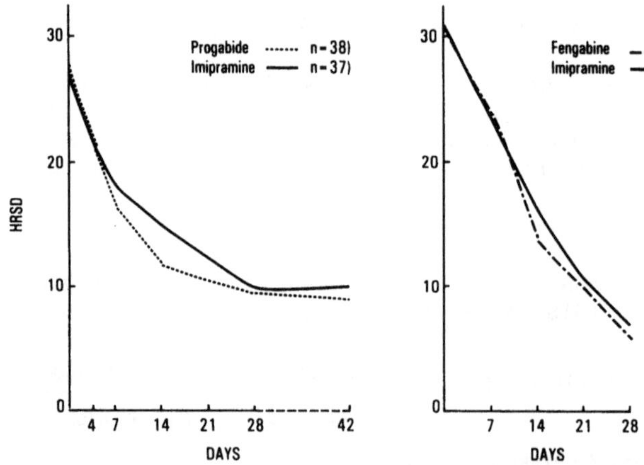

Fig. 2. GABA agonists : two different multicentric double blind trials vs imipramine in depressed patients (DSM III).
Each trial was a multicenter double-blind parallel group study including six centers. Patients received placebo capsules three times daily during a drug-free 48 hour period. Progabide (600-1200mg), fengabine (800-2200mg) and imipramine (50-175mg) were administered daily for the time periods indicated (for details on the trial with progabide cf. Weiss et al., 1986; publication of the trial with fengabine is in preparation by Musch et al.).

A very interesting finding is that GABA B receptor binding in rat frontal cortex membranes is enhanced after repeated (but not acute) administration of GABAmimetics and a wide range of antidepressants (Table 2; Pilc and Lloyd, 1984; Lloyd et al., 1985; 1986). This up-regulation is independant of the chemical structure and the mechanism of action, as selective inhibitors of either dopamine, NE or 5-HT uptake as well as monoamine oxidase inhibitors, GABAmimetics, iprindole, mianserine and electroschock exhibit this effect.

TABLE 2

EFFECT OF REPEATED ADMINISTRATION[1] OF DIFFERENT CLASSES OF PSYCHOTROPIC DRUGS AND ELECTROSHOCK ON GABA B BINDING TO RAT FRONTAL CORTEX MEMBRANES

COMPOUND	mg/kg/day	DURATION days	% CONTROL BINDING
ANTIDEPRESSANTS			
Desipramine	5	18	151*
Maprotiline	10	18	135*
Nomifensine	5	18	146*
Amitryptiline	10	18	154*
Citalopram	10	18	172*
Fengabine	50	6	159*
Progabide	100	18	127*
Iprindole	5	18	129*
Mianserine	10	18	133*
Pargyline	20	18	142*
Electroshock[2]		10	177*
PSYCHOSTIMULANTS			
Amphetamine	2	18	119
SEDATIVES/ANXIOLYTICS			
Diazepam	0.5	18	113
Phenobarbital	40	18	71*
NEUROLEPTICS			
Haloperidol	0.3	18	107
Reserpine	1	18	70*
CHOLINOMIMETICS			
Oxotremorine	4	18	119

1 subcutaneous administration with the exception of fengabine and progabide which were administered i.p.

2 150mA/0.5 sec/48 hrs for 10 days

*binding at 10nM ^{3}H-GABA; $p < 0.01$ vs saline treated control

The only apparent common denominator of these compounds is an antidepressant activity. Representatives of other psychotropic classes do not enhance GABA B binding (Table 2), indicating a specificity of action for this effect. Interestingly, reserpine which is known to trigger depressive states in humans, causes down regulation of GABA B sites.

The observed specificity is also anatomically defined as the up-regulation in the frontal cortex is of greater magnitude then, and occurs prior to that in the hippocampus (Lloyd et al., 1985); at variance, the parietal cortex, occipital cortex, cerebellum and amygdala do not exhibit increased GABA B binding in response to antidepressants (Lloyd and Pichat, unpublished). In the frontal cortex the up-regulation appears to be mainly due to an increase in binding capacity (Bmax) rather than affinity (Kd) (Lloyd et al., 1985).

Similarly to GABAmimetics, conventional antidepressants do not seem to modify GABA A receptors. Thus, the inhibitory response to iontophoretic GABA is unaltered (De Montigny and Aghajanian, 1978; Blier et al., 1984). Also, antidepressants do not alter GABA A binding in vitro or ex vivo (Olsen et al., 1978; Pilc and Lloyd, 1984).

Also, there is no consistent effect of antidepressants on GAD activity in vitro (Olsen et al., 1978) or ex vivo (Lloyd and Pilc, 1984) nor are the GABA levels modified by acute or repeated administration of these agents (Pilc and Lloyd, 1984).

The changes in GABA B receptors induced by antidepressants appear to be relevant to the behavioural activity of these compounds in animal models predictive of antidepressant drug action. Thus, in the olfactory bulbectomized rat, GABA B binding is specifically decreased in the frontal cortex (Lloyd and Pichat, 1986). In this model, antidepressants reverse the deficit in passive avoidance learning (Broekkamp et al., 1980) in some, but not all animals. Antidepressants increase GABA B binding only in those animals in which these compounds reverse the behavioural effect; in the others both the deficit in GABA B binding and passive avoidance learning are retained. Thus, a linkage appears between the behavioural and biochemical events provoked by the antidepressants (Joly et al., 1987).

These data suggest that GABA B receptors are involved in the mode of action of antidepressants ; however, GABA A receptors - although unchanged by antidepressants - appear also to be involved. Thus, in models such as olfactory bulbectomy and/or learned helplessness, bicuculline antagonizes the reversal of the behavioural deficit induced not only by GABAmimetics but also by intracortical, (Sherman and Petty, 1980) or systemically administered imipramine (Fig. 3; Lloyd et al., 1983; Zivkovic et al., 1986). Whether a synergy exists between GABA A and GABA B synapses is an open question ; however such a synergy has been demonstrated between GABA B receptors and β_1-adrenoceptor mediated adenylate cyclase activity (Hill, 1985; Karbon et al., 1984).

Other data further suggest a role of the GABA system in the antidepressant drug action. Thus, in hippocampal slices

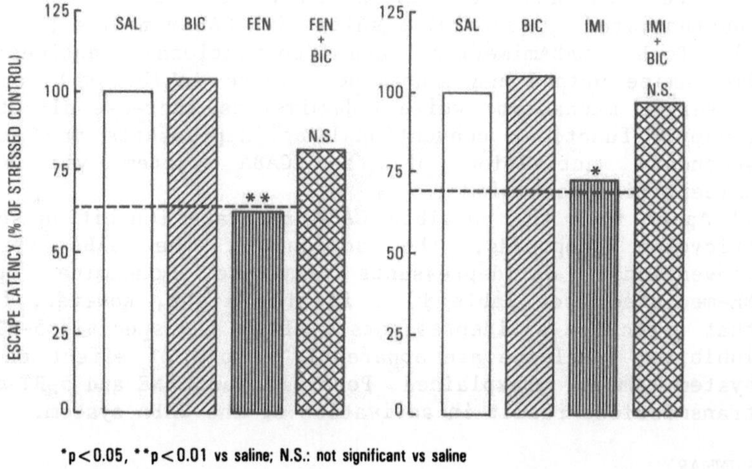

Fig. 3 Reversal by bicuculline of the effects of fengabine and imipramine on escape acquisition in the learned helplessness model. Groups of stressed animals were injected i.p. with saline (SAL; stressed controls), fengabine (FEN; 50 mg/kg daily for two days), imipramine (IMI; 10 mg/kg daily for 5 days) and bicuculline (BIC; 3 mg/kg), respectively. The horizontal dotted line indicates the escape level of rats not submitted to unavoidable shock (for details, cf. Zivkovic et al., 1986 and Bartholini et al., 1986b).

of rat in a learned helplessness situation, depolarization-induced GABA liberation is reduced as compared to controls; imipramine corrects this biochemical alteration and, in parallel, the behavioural deficit (see above) (Petty and Sherman, 1981). Also, injection of GABA into frontal cortex (Sherman and Petty 1980) prevents, whereas hippocampal injection of bicuculline (Petty and Sherman 1981) induces learned helplessness in the rat. Furthermore, in experiments with the push pull cannula, desipramine enhances thalamic GABA liberation (Korf and Venema 1983) ; and the in vitro uptake of GABA is inhibited by antidepressants of tricyclic type (Harris et al., 1973; Snodgrass et al., 1973) at concentrations within the range of those found in human brain (Korf and Venema, 1983).

Finally, GABA levels are reported to be decreased in plasma (Petty and Schlesser 1981) and cerebrospinal fluid (Gold et al., 1980; Gerner and Hare, 1981; Kasa et al., 1982; Gerner et al., 1984) of depressed patients.

From the body of evidence discussed above, the following conclusions can be drawn :

a) A deficit in GABA function is involved in behavioural models predictive of antidepressant drug action ;

b) The relevance of these data to human depression is substantiated by clinical results with GABAmimetics ;

c) Both GABAmimetics and conventional antidepressants (monoamine uptake and monoamine oxidase inhibitors) activate a GABAergic mechanism: while GABAmimetics increase directly GABA synaptic function, conventional antidepressants may act by a secondary activation of the GABA system via a still unidentified mechanism ;

d) Apart from the possible GABA reuptake inhibition shown for tricyclic compounds, the action on the GABA system of conventional antidepressants may be monoamine- possibly NE-mediated (see Table 1) ; in this light, however, the fact that effective antidepressants such as the specific 5-HT uptake inhibitor indalpine are apparently devoid of effect on the NE system remains unexplained. Possibly, both NE and 5-HT-mediated transmissions result in activation of the GABA system.

SUMMARY

In the rat brain, GABAmimetics such as progabide and fengabine enhance norepinephrine turnover. During prolonged treatment these agents do not induce changes in β_1 adrenoceptor density; however, progabide potentiates desipramine-induced desensitization of isoprenaline-stimulated adenylate cyclase and fengabine desensitizes 'per se' this enzyme. Moreover, progabide reduces 5-HT turnover and up-regulates 5-HT_2 receptors whereas fengabine does not affect the 5-HT system. These changes differ substantially from those induced by conventional antidepressants. Furthermore, GABAmimetics as well as all types of conventional antidepressants up-regulate the GABA B receptor subtypes in the absence of changes in GABA A receptors and GABA turnover. GABAmimetics and monoamine uptake inhibitors correct, by a bicuculline sensitive mechanism, the behavioural deficit in animal models predictive of antidepressant drug action. GABAmimetics exhibit a major therapeutic action in depressive states (DSM III criteria). These data indicate a substantial involvement of GABA receptors in depression and in the mode of action of conventional antidepressants.

REFERENCES

Anderson, J.L. (1983). Serotonin receptor changes after chronic antidepressant treatments: ligand binding, electrophysiological and behavioral studies. Life Sci., 32, 1791-1801.

Bartholini, G. (1985). GABA receptor agonists: pharmacological spectrum and therapeutic actions. Med. Res. Rev., 5(1), 55-75.

Bartholini, G., Scatton, B., Zivkovic, B., Lloyd, K.G., Depoortere, H., Langer, S.Z. and Morselli, P.L. (1985). GABA receptor agonists as a new therapeutic class. In Epilepsy and GABA Receptor Agonists: Basic and Therapeutic Research, (L.E.R.S. Monograph Series III) (eds. G. Bartholini, L. Bossi, K.G. Lloyd and P.L. Morselli), Raven Press, New York, 1-30.

Bartholini, G., Lloyd, K.G. and Morselli, P.L. (1986a). GABA and Mood Disorders: Experimental and Clinical Research, (L.E.R.S. Monograph Series, IV). Raven Press, New York.

Bartholini, G., Scatton, B., Zivkovic, B. and Lloyd, K.G. (1986b). On the mode of antidepressant action of GABA receptor agonists and monoamine uptake inhibitors. In GABA and Mood Disorders: Experimental and Clinical Research, (L.E.R.S. Monograph Series, IV)(eds. G. Bartholini, K.G. Lloyd and P.L. Morselli). Raven Press, New York, 105-111.

Blier, P., De Montigny, C., Tardiff, D. (1984). Effects of two antidepressant drugs, mianserine and indalpine, on the serotonergic system: single cell studies in the rat. Psychopharmacol., 84, 242-249.

Blumberg, J.B., Vetulani, J., Stawarz, R. and Sulser, F. (1976). The noradrenergic cyclic AMP generating system in the limbic forebrain: pharmacological characterization in vitro and possible role of limbic noradrenergic mechanisms in the mode of action of antipsychotics. Eur. J. Pharmacol., 37, 357-366.

Broekkamp, C.L., Garrigou, D. and Lloyd, K.G. (1980). Serotonin-mimetic and antidepressant drugs on passive avoidance learning by olfactory bulbectomized rats. Pharmac. Biochem. Behav., 13, 643-646.

Brunello, N., Barbaccia, M.L., Chuang, D.M., and Costa, E. (1982). Down regulation of ß-adrenergic receptors following repeated injections of desmethylimipramine: permissive role of serotonergic axons. Neuropharmacol., 21, 1145-1149.

Curet, O., Dennis, T. and Scatton, B. (1985). The formation of deaminated metabolites of dopamine in the locus coeruleus depends upon noradrenergic neuronal activity. Brain Res., 335, 297-301.

De Montigny, C. and Aghajanian, G.K. (1978). Tricyclic antidepressants: long term treatment increases responsivity of rat forebrain neurons to serotonin. Science, 202, 1303-1306.

Dennis, T., Curet, O., Nishikawa, T. and Scatton, B. (1985). Further evidence for, and nature of, the facilitatory GABAergic influence on central noradrenergic transmission. Naunyn-Schmiedeberg's Arch. Pharmacol., 331, 225-234.

Gerner, R.H., Fairbanks, L., Andersen, G.M., Young, J.G., Scheinini, M., Linnoila, M., Hare, T.A., Shaywitz, B.A., and Chen, D.J. (1984). CSF neurochemistry in depressed, mania and schizophrenia patients compared with that of normal controls. Am. J. Psychiat., 141, 1533-1540.

Gerner, R.H. and Hare, T.A. (1981). CSF GABA in normal subjects and patients with depression, schizophrenia, mania and anorexia nervosa. Am. J. Psychiatry, 138, 1098-1101.

Gold, B.I., Bowers, M.B., Roth, R.H. and Sweeney, D.W. (1980). GABA levels in CSF of patients with psychiatric disorders. 137, 362-364.

Harris,, M., Hopkins, M. and Neal, M.J. (1973). Effect of centrally acting drugs on the uptake of γ-aminobutyric acid (GABA) by slices of rat cerebral cortex. Brit. J. Pharmacol., 47, 229-239.

Hill, D.R. (1985). GABA B receptor modulation of adenylate cyclase activity in rat brain slices. Brit. J. Pharmacol. 84, 249-257.

Janowsky, A., Okada, F., Manier, D., Applegate, C.D. and Sulser, F. (1982). Role of serotonergic input in the regulation of the ß-adrenergic receptor-coupled adenylate cyclase system. Science, 218, 900-902.

Joly, D., Lloyd, K.G., Pichat, P. and Sanger, D.J. (1987). Correlation between the behavioural effect of desipramine and GABA B receptor regulation in olfactory bulbectomised rat. Brit. J. Pharmacol., in press.

Karbon, E.W., Duman, R.S., Enna, S.J. (1984). GABA B receptors and norepinephrine-stimulated cAMP production in rat brain cortex. Brain Res., 306, 327-332.

Kasa, K., Otsuki, S., Yamamoto, M. et al (1982). Cerebrospinal fluid γ-aminobutyric acid and homovanillic acid in depressive disorders. Biol. Psychiatry, 17, 877-883.

Korf, J. and Venema, K. (1983). Desmethylimipramine enhances the release of endogenous GABA and other neurotransmitter amino acids from the rat thalamus. J. Neurochem., 40, 946-950.

Langer, S.Z., Arbilla, S. Scatton, B., Zivkovic, B., Galzin, A.M., Lloyd, K.G. and Bartholini, G. (1985). Progabide and SL 75 102: Interaction with GABA receptors and effects on neurotransmitter and receptor systems. In Epilepsy and GABA Receptor Agonists: Basic and Therapeutic Research, (L.E.R.S. Monograph Series III) (eds. G. Bartholini, L. Bossi, K.G. Lloyd and P.L. Morselli), Raven Press, New York, 81-90.

Lloyd, K.G. and Pichat, Ph. (1986). Decrease in GABA B binding in the frontal cortex of bulbectomized rats. Brit. J. Pharmac., 87, 36P.

Lloyd, K.G. and Pilc, A. (1984). Chronic antidepressants and GABA B binding sites. Neurosci. Abstr. 10, 117.5.

Lloyd, K.G., Thuret, F., and Pilc, A. (1986). GABA and the mechanism of action of antidepressant drugs. In GABA and Mood Disorders: Experimental and Clinical Research, (L.E.R.S. Monograph Series, IV) (eds. G. Bartholini, K.G. Lloyd and P.L. Morselli). Raven Press, New York, 33-42.

Lloyd, K.G., Thuret, F., and Pilc, A. (1985). Up-regulation of γ-aminobutyric acid (GABA) B binding sites in rat frontal cortex: a common action of repeated administration of different classes of antidepressants and electroshock. J. Pharmac. Exp. Ther. 235, 191-199.

Lloyd, K.G., Morselli, P.L., Depoortere, H., Fournier, V., Zivkovic, B., Scatton, B., Broekkamp, C., Worms, P. and Bartholini, G. (1983). The potential use of GABA agonists in psychiatric disorders: evidence from studies with progabide in animal models and clinical trials. Pharmacol. Biochem. Behav., 18, 957-966.

Mishra, R., Janowsky, A. and Sulser, F. (1980). Action of mianserine and zimelidine on the norepinephrine receptor coupled adenylate cyclase system in brain : subsensitivity without reduction in ß-adrenergic receptor binding. Neuropharmacol., 19, 983-987.

Morselli, P.L., Bossi, L., Henry, J.F., Zarifian, E. and Bartholini, G. (1980). On the therapeutic action of SL 76 002, a novel GABAmimetic agent: preliminary observations in neuropsychiatric disorders. Brain Res. Bull. 5 (suppl. 2) 411-414.

Morselli, P.L., Fournier, V., Macher, J.P., Orofiamma, B., Bottin, P. and Huber, P. (1986). Therapeutic action of progabide in depressive illness: a controlled clinical trial. In GABA and Mood Disorders: Experimental and Clinical Research, (eds. G. Bartholini, K.G. Lloyd and P.L. Morselli) (L.E.R.S. Monograph Series, IV) Raven Press, New York, 119-126.

Musch, B. (1986). Antidepressant activity of fengabine (SL 79 229): A critical overview of the present results in open clinical studies. In GABA and Mood Disorders: Experimental and Clinical Research, (L.E.R.S. Monograph Series, IV) (eds. G. Bartholini, K.G. Lloyd and P.L. Morselli), Raven Press, New York, 171-177.

Nishikawa, T. and Scatton, B. (1985a). Inhibitory influence of GABA on central serotonergic transmission. Raphe nuclei as the neuroanatomical site of the GABAergic inhibition of cerebral serotonergic neurons. Brain Res., 331, 91-103.

Nishikawa, T. and Scatton, B. (1985b). Inhibitory influence of GABA on central serotonergic transmission. Involvement of the habenulo-raphe pathways in the GABAergic inhibition of ascending cerebral serotonergic neurons. Brain Res., 331, 81-90.

Olsen, R.W., Ticku, M.K., Van Ness, P.C. and Greenlee, D. (1978). Effects of drugs on γ-aminobutyric acid receptors, uptake, release and synthesis in vitro. Brain Res., 139, 277-294.

Petty, F. and Schlesser, M.A. (1981). Plasma GABA in affective disorders. J. Affective Disord., 3, 339-343.

Petty, F. and Sherman, A.D. (1981). GABAergic modulation of learned helplessness. Pharmacol. Biochem. Behav., 15, 567-570.

Pilc, A. and Lloyd, K.G. (1984). Chronic antidepressants and GABA receptors: a GABA hypothesis of antidepressant drug action. Life Sci., 35, 2149-2154.

Scatton, B., Lloyd, K.G., Zivkovic, B., Dennis, T., Claustre, Y., Dedek, J., Arbilla, S., Langer, S.Z. and Bartholini, G. (1987) Fengabine, a novel antidepressant GABAergic agent.II. Effect on cerebral noradrenergic, serotonergic and GABAergic transmission. J. Pharmacol. Exp. Ther., in press.

Scatton, B., Nishikawa, T., Dennis T., Dedek, J. Curet, O., Zivkovic, B. and Bartholini, G. (1986). GABAergic modulation of central noradrenergic and serotonergic neuronal activity. In GABA and Mood Disorders: Experimental and Clinical Research, (eds. G. Bartholini, K.G. Lloyd and P.L. Morselli), Raven Press, New York, 67-75.

Scatton, B. and Serrano, A. (1986). GABAmimetics increase extracellular DOPAC (as measured by in vivo voltammetry) in the rat locus coeruleus. Naunyn-Schmiedeberg's Arch. Pharmacol. 332, 380-383.

Scatton, B., Serrano, A., Rivot, J.P., and Nishikawa, T. (1984). Inhibitory GABAergic influence on striatal serotonergic transmission exerted in the dorsal raphe as revealed by in vivo voltammetry. Brain Res., 305, 343-352.

Scatton, B., Zivkovic, B., Dedek, J., Lloyd, K.G., Constantinidis, J., Tissot, R. and Bartholini, G. (1982). γ-Aminobutyric acid (GABA) receptor stimulation.III. Effect of progabide (SL 76 002) on norepinephrine, dopamine and 5-hydroxytryptamine turnover in rat brain areas. J. Pharmacol. Exp. Ther., 220, 678-688.

Sellinger-Barnette, M.M., Mendels, J. and Frazer, A. (1980). The effect of psychoactive drugs on ß-adrenergic receptor binding sites in rat brain. Neuropharmacol., 19, 447-454.

Sherman, A.D., and Petty, F. (1980). Neurochemical basis of the action of antidepressants on learned helplessness. Behav. Neurol. Biol., 30, 119-134.

Snodgrass, S.R., Hedley-White, E.T. and Lorenzo, A.V. (1973). GABA transport by nerve ending fractions of cat brain. J. Neurochem., 20, 771-782.

Sugrue, M.F. (1983). Do antidepressants possess a common mechanism of action? Biochem. Pharmacol., 32, 1811-1817.

Sulser, F. (1984). Regulation and function of noradrenaline receptor systems in brain. Neuropharmacol., 23, 255-261.

Weiss, E., Brunner, H., Clerc, G., Guibert, M., Orofiamma, B., Pagot, R., Robert, G., Thilliez, D. and Musch, B. (1986). Multicenter double-blind study of progabide in depressed patients. In GABA and Mood Disorders: Experimental and Clinical Research, (L.E.R.S. Monograph Series, IV) (eds. G. Bartholini, K.G. Lloyd and P.L. Morselli). Raven Press, New York, 127-133.

Zivkovic, B., Scatton, B., Dedek, J. and Bartholini, G. (1982). GABA influence in noradrenergic and serotonergic transmissions: implications in mood regulation. In New vistas in depression (eds. S.Z. Langer, R. Takahashi, T. Segawa and M. Briley) Pergamon Press, Oxford, 195-201.

Zivkovic, B., Scatton, B., Sanger, D., Depoortere, H., Dedek, J., Arbilla, S., Langer, S.Z. and Bartholini, G.. (1986). The pharmacological and neurochemical spectrum of fengabine SL 79 229, a new antidepressant agent. In GABA and Mood Disorders: Experimental and Clinical Research, (L.E.R.S. Monograph Series, IV) (eds. G. Bartholini, K.G. Lloyd and P.L. Morselli). Raven Press, New York, 85-95.

Session IV
Effects of Intracellular Messengers on the Membrane Receptors

Chairman: M.J. Berridge

36
Receptor–Receptor Interactions Mediated by Protein Phosphorylation

Paul Greengard

There is now convincing evidence that protein phosphorylation mediates the physiological effects of a variety of extracellular signals on their target cells, both neuronal and non-neuronal. For many years, the hypothesis (Kuo and Greengard, 1969; Greengard, 1976) that protein phosphorylation mediates certain of the actions of neurotransmitters on target neurons was based largely on correlations between the state of phosphorylation of one or another substrate protein and the state of some physiological response in target neurons. However, during the past five or six years direct evidence has been obtained for a role of protein phosphorylation in the regulation of physiological processes in nerve cells. Thus, there is now evidence, based on intracellular injection experiments, that each of the four major classes of protein kinases listed in Table I mediates the actions of certain neurotransmitters acting on their target nerve cells.

The first protein kinase to be studied in this way was cyclic AMP-dependent protein kinase. This enzyme was studied in two simultaneous series of experiments, one carried out with Kaczmarek and Strumwasser on bag cell neurons of Aplysia (Kaczmarek et al., 1980, 1984), and the other carried out with Castellucci and Kandel on sensory neurons of Aplysia (Castellucci et al., 1980, 1982), both of which involved Angus Nairn of our laboratory. Injection of the catalytic subunit of cyclic AMP-dependent protein kinase into both types of neurons mimicked the effects of physiological activation of these neurons. With both types of nerve cell, the physiological response involved a shutting off of potassium channels, a broadening of the action potential, an increase in calcium influx into the neurons and an enhanced physiological response. Conversely, it was also found with both types of preparation that the injection of a specific protein kinase inhibitor of cyclic AMP-dependent protein kinase, known as the Walsh inhibitor, abolished the response of these

TABLE I. PHYSIOLOGICAL ACTIONS OF PROTEIN KINASES

CLASS OF PROTEIN KINASE	FIRST MESSENGER	PHYSIOLOGICAL RESPONSE	REFERENCES
cAMP-DEPENDENT	UNKNOWN	↓ K^+ CURRENT	KACZMAREK ET AL., 1980, 1984
	SEROTONIN	↓ K^+ CURRENT	CASTELLUCCI ET AL. 1980, 1982
Ca^{2+}/CALMODULIN-DEPENDENT	NERVE IMPULSE	↑ NEUROTRANSMITTER RELEASE	LLINAS ET AL., 1985
Ca^{2+}/DIACYLGLYCEROL-DEPENDENT	UNKNOWN	↑ Ca^{2+} CURRENT	DeRIEMER ET AL, 1985
cGMP-DEPENDENT	SEROTONIN	↑ Ca^{2+} CURRENT	PAUPARDIN-TRITSCH ET AL, 1986B

nerve cells to physiological activation. The ability of the protein kinase to mimic physiological activation of these nerve cells, and the ability of the specific protein kinase inhibitor to abolish the response to physiological activation, demonstrates that cyclic AMP-dependent protein phosphorylation is both necessary and sufficient to produce an appropriate physiological response in these nerve cells.

Calcium/calmodulin-dependent protein kinase II, also known as multifunctional calcium/calmodulin-dependent protein kinase, was studied in a series of experiments carried out in collaboration with Rodolfo Llinás and his colleagues, which involved Teresa McGuiness of our laboratory (Llinás et al., 1985). This kinase was found to be involved in regulation of neurotransmitter release from nerve terminals. These studies were carried out using the squid giant synapse, a preparation which Llinás has exploited effectively for many years to study the regulation of neurotransmitter release. In these studies, it was shown that the injection of the calcium/calmodulin-dependent protein kinase increased the amount of neurotransmitter released from the nerve terminals in response to a constant depolarizing pulse. The results indicated that this protein kinase regulates the effectiveness of nerve impulses in causing neurotransmitter release from nerve endings.

A third protein kinase, calcium/diacylglycerol-dependent protein kinase (or protein kinase C), was studied in a series of experiments carried out in collaboration with Kaczmarek and his colleagues, which involved Katherine Albert of our laboratory (DeRiemer et al., 1985). The injection of this kinase into bag cell neurons of Aplysia caused an increased opening of calcium channels in these neurons, mimicking physiological activation of these nerve cells.

Finally, it was shown in a series of studies carried out in collaboration with Hersch Gerschenfeld and his colleagues, involving Angus Nairn of our laboratory, that cyclic GMP-dependent protein kinase mediates the actions of serotonin on ventral neurons of Helix aspersa (Paupardin-Tritsch et al., 1986b). Specifically, the ability of serotonin to increase the voltage sensitivity of voltage-sensitive calcium channels in these nerve cells was greatly enhanced by the intracellular injection of the non-activated form of cyclic GMP-dependent protein kinase. In addition, the effect of serotonin on calcium channels could be mimicked by the injection of the activated form of cyclic GMP-dependent protein kinase.

Thus, each of the four major classes of protein kinases listed in Table I mimics and presumably mediates the effects of various neurotransmitters acting on their target cells. Moreover, in each of the cases listed, the physiological response studied can be considered to reflect receptor-receptor interaction. I would like in the remainder of this presentation to describe in somewhat greater detail two types of receptor-receptor interaction that involve protein phosphorylation. The first of these concerns the studies with cyclic-GMP dependent protein kinase referred to above, and the second concerns studies of the acetylcholine receptor carried out in our laboratory by Richard Huganir.

Serotonin and Cyclic GMP-Dependent Protein Kinase in Ventral Neurons of Helix Aspersa

Gerschenfeld and his colleagues demonstrated recently (Paupardin-Tritsch et al., 1986a) that serotonin broadens the calcium-dependent plateau of the action potential seen in identified ventral neurons of the abdomino-visceral ganglionic mass of the snail Helix aspersa. They also demonstrated that serotonin raises the levels of cyclic GMP in these nerve cells and that cyclic GMP mimics the effect of serotonin in broadening the action potential. In view of their results, our two research groups carried out a collaborative study to evaluate the possible involvement of cyclic GMP-dependent protein kinase in producing the physiological effects of serotonin on these nerve cells. A low concentration of serotonin (100 nM) by itself produced almost no broadening of the action potential in these nerve cells. However, upon injection of the non-activated form of cyclic GMP-dependent protein kinase, the same concentration of serotonin now became effective in broadening the action potential plateau.

The ionic basis of the effect of cyclic GMP-dependent protein kinase was then investigated. Sodium and potassium currents were blocked pharmacologically, and the effects of serotonin and cyclic GMP-dependent protein kinase were examined on the voltage-sensitive calcium currents in these nerve cells. For this purpose, the cell membrane was voltage-clamped and the calcium current was measured in response to a constant depolarizing pulse. Serotonin (100 nM) caused a slight increase in the calcium current in neurons prior to the injection of cyclic GMP-dependent protein kinase. After the injection of the non-activated form of cyclic GMP-dependent protein kinase into these nerve cells, the response to serotonin was greatly enhanced. The injection of cyclic GMP-dependent protein kinase produced similar results in experiments with zaprinast, a specific inhibitor of cyclic GMP phosphodiesterase. A low concentration (1 μM) of zaprinast by itself had no effect on the voltage-sensitive calcium currents. However, following the injection of the non-activated form of cyclic GMP-dependent protein kinase, this same concentration of zaprinast greatly enhanced the magnitude of the voltage-sensitive calcium current. In control experiments it was demonstrated that the injection of heat-inactivated cyclic GMP-dependent protein kinase had no effect on the responsiveness of these nerve cells to serotonin.

In the series of experiments just described, the injection of the non-activated form of cyclic GMP-dependent protein kinase did not affect the size of the calcium current seen in the absence of serotonin or zaprinast. This can be explained by the low and variable cGMP-content observed in the identified ventral neurons (Paupardin-Tritsch et al., unpublished), which is presumably not high enough to activate the injected cyclic GMP-dependent protein kinase. In contrast, injection of the activated form of cyclic GMP-dependent protein kinase mimicked the ability of serotonin to enhance the voltage-sensitive calcium current in these nerve cells. (The cyclic GMP-dependent protein kinase was activated by preincubation of the enzyme with 1 μM cyclic GMP; it was shown in control experiments that the injection of this amount of cyclic GMP by itself had no effect on the calcium current). In fact, the injection of cyclic GMP-dependent protein kinase increased calcium currents by up to three-fold. This effect is far greater than the 25-30% enhancement of calcium current seen upon application of serotonin. These results suggest that the cyclic GMP-dependent protein kinase present endogenously in nerve cells is rate-limiting in mediating the actions of serotonin. Our interpretation (Paupardin-Tritsch et al., 1986b) of these experiments involving the injection of cyclic GMP-dependent protein kinase into ventral neurons of <u>Helix aspersa</u> is shown in cartoon fashion in Figure 1. One afferent pathway to these

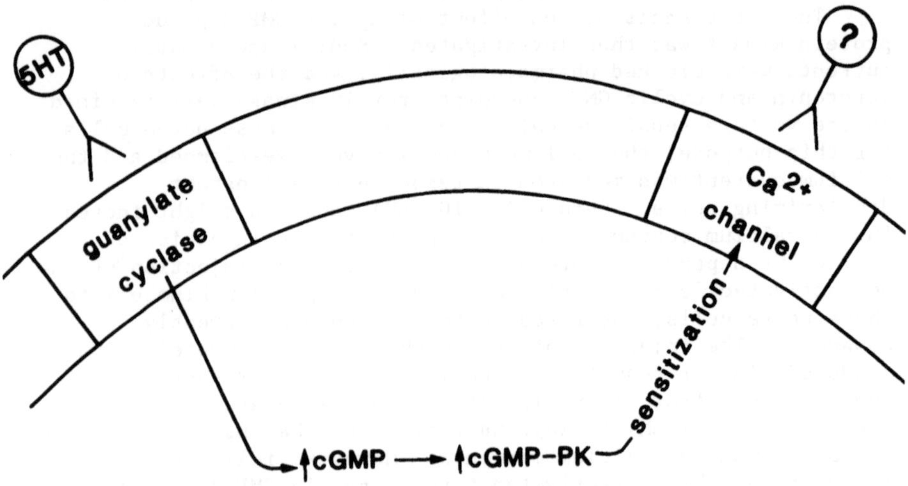

Figure 1. Receptor-receptor interaction mediated by cGMP-dependent protein phosphorylation

neurons releases some excitatory neurotransmitter, the identity of which remains to be determined, resulting in the firing of nerve impulses and the opening of voltage-sensitive calcium channels in the membranes of these neurons. A second afferent pathway releases serotonin, and the serotonin activates a serotonin-sensitive guanylate cyclase, which leads to an increase in the level of cyclic GMP, an increased activity of cyclic GMP-dependent protein kinase and the phosphorylation either of the voltage-sensitive calcium channels or of modulators of these channels. The phosphorylation process results in the voltage-sensitive calcium channels becoming more sensitive to depolarization. Thus, this system represents one example of receptor-receptor interaction mediated by phosphorylation within the membrane of a single neuron.

The Acetylcholine Receptor

Several years ago it was found by Gordon et al. (1977) and by Teichberg et al. (1977) that the acetylcholine receptor is a phosphoprotein. During the past few years, Richard Huganir in our laboratory has carried out an investigation of the biochemical mechanisms that underlie phosphorylation of the acetylcholine receptor and an evaluation of the possible physiological significance of this phosphorylation. Huganir was able to demonstrate that each of three distinct protein kinases is capable of phosphorylating the acetylcholine receptor present in postsynaptic membrane preparations from Torpedo californica. Cyclic AMP-dependent protein kinase causes a rapid and stoichiometric phosphorylation of the γ and δ subunits of the acetylcholine receptor (Huganir and Greengard, 1983). Protein kinase C causes a rapid and stoichiometric phosphorylation of the δ subunit of the receptor (Huganir et al., 1983). It also causes a slow but reproducible phosphorylation of the α subunit of the receptor (Huganir et al., 1984). An endogenous protein tyrosine kinase present in the membranes that contain the acetylcholine receptor causes the rapid and stoichiometric phosphorylation of the β, γ and δ subunits of the acetylcholine receptor. Thus, three distinct protein kinases phosphorylate a total of seven different sites on the four types of subunits of the receptor, with six of these seven phosphorylation reactions occurring rapidly and stoichiometrically.

By comparing the published amino acid sequences of the subunits of the acetylcholine receptor with the published amino acid sequence preferences for phosphorylation by the various protein kinases, we have tentatively assigned amino acid residues for the phosphorylation sites of the receptor (Huganir et al., 1984). It seems very likely that all seven phosphorylation sites occur in a cytoplasmic domain of the acetylcholine receptor molecule. In those cases in which the subunits undergo multisite phosphorylation, the amino acid residues phosphorylated appear to be located within a few amino acids of each other. Moreover, the various subunits appear to be phosphorylated in homologous regions.

What is the physiological significance of phosphorylation of the acetylcholine receptor? Studies carried out by Huganir in collaboration with George Hess and Anne Delcour of Cornell University (Huganir et al., 1986) have indicated that phosphorylation of the δ and γ subunits of the acetylcholine receptor by cyclic AMP-dependent protein kinase accelerates the rate of desensitization of the receptor in the presence of acetylcholine. In these experiments, the acetylcholine receptor was phosphorylated on the δ and γ subunits in the presence of cyclic AMP-dependent protein kinase. The phosphorylated receptor and, for comparison, the non-phosphorylated receptor were purified to homogeneity. The phosphorylated receptor and the non-phosphorylated receptor were then reconstituted into lipid vesicles and the effect of phosphorylation on the rate of desensitization was examined in the presence of the ligand acetylcholine. It was found that, in the presence of acetylcholine, the phosphorylated form of the receptor desensitized with a rate constant approximately ten times greater than that of the dephosphorylated form of the receptor. Interestingly, phosphorylation of the acetylcholine receptor did not desensitize it in the absence of acetylcholine. In some experiments, the rate of desensitization was studied as a function of the stoichiometry of phosphorylation of the receptor. The results of those studies indicated that phosphorylation of either the γ subunit of the receptor or the δ subunit of the receptor or both was sufficient to cause a conversion of the receptor from the slowly desensitizing form to the rapidly desensitizing form.

The results of the studies on regulation of the acetylcholine receptor by phosphorylation are summarized in Table II. Our interpretation of these results is shown in cartoon fashion in Figure 2. One afferent pathway to the target cholinoceptive cells releases acetylcholine, which activates the nicotinic acetylcholine receptor. A second afferent pathway to the same target cells releases a neurotransmitter, the identity of which remains to be determined, which activates adenylate cyclase in the plasma membrane. The activated adenylate cyclase causes an increase in the level of cyclic AMP, leading to an increased activity of cyclic AMP-dependent protein kinase. The active cyclic AMP-dependent protein kinase then phosphorylates the acetylcholine receptor, on either the γ subunit, the δ

TABLE II REGULATION OF ACh RECEPTOR BY PHOSPHORYLATION

(1)	THE ACh RECEPTOR IS PHOSPHORYLATED BY THREE DISTINCT PROTEIN KINASES
(2)	A TOTAL OF SEVEN SITES ARE PHOSPHORYLATED ON THE FOUR SUBUNIT TYPES
(3)	WITHIN EACH SUBUNIT, MULTIPLE PHOSPHORYLATION SITES APPEAR TO BE LOCATED IN CLOSE PROXIMITY TO ONE ANOTHER IN A CYTOPLASMIC DOMAIN
(4)	AMONG DIFFERENT SUBUNITS, THE PHOSPHORYLATION SITES APPEAR TO BE LOCATED IN HOMOLOGOUS REGIONS
(5)	PHOSPHORYLATION OF THE γ AND δ SUBUNITS BY cAMP-DEPENDENT PROTEIN KINASE CAUSES APPROXIMATELY A TEN-FOLD INCREASE IN THE RATE OF DESENSITIZATION OF THE RECEPTOR

subunit, or both, accelerating its rate of desensitization in the presence of acetylcholine. This system provides a second example of receptor-receptor interaction mediated by protein phosphorylation within the membrane of a target cell.

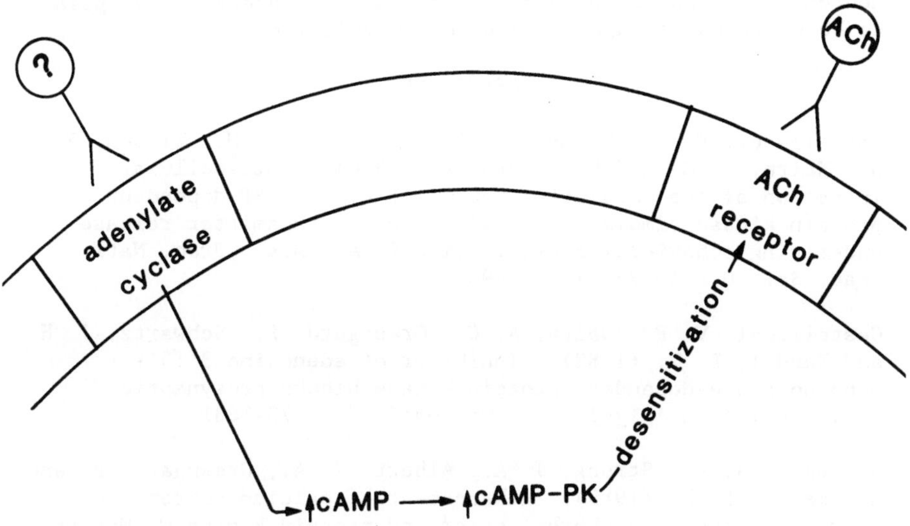

Figure 2. Receptor-receptor interaction mediated by cAMP-dependent protein phosphorylation

As discussed above, two other protein kinases, namely calcium/diacylglycerol dependent protein kinase and an endogenous protein tyrosine kinase, also phosphorylate the acetylcholine receptor in the same regions of the molecule. Experiments are currently being carried out to test whether these other protein kinases also cause the desensitization of the receptor. It is possible that three distinct neurotransmitters acting on three distinct types of protein kinases each regulate the acetylcholine receptor by phosphorylation and desensitization. If so, this system would provide an example of receptor-receptor interactions involving four distinct neurotransmitters within a single membrane.

Concluding Remarks

In conclusion, I have attempted to summarize very briefly recent direct experimental evidence that each of four distinct classes of protein kinases mediates the physiological effects of appropriate neurotransmitters on their target neurons. In addition, I have described in somewhat greater detail two neuronal systems in which receptor-receptor interactions appear to be mediated through protein phosphorylation.

REFERENCES

Castellucci, V. F., Kandel, E. R., Schwartz, J. H., Wilson, F. D., Nairn, A. C. and Greengard, P. (1980). Intracellular injection of the catalytic subunit of cyclic AMP-dependent protein kinase simulates facilitation of transmitter release underlying behavioral sensitization in Aplysia. Proc. Natl. Acad. Sci. U.S.A. 77, 7492-7496.

Castellucci, V. F., Nairn, A. C., Greengard, P., Schwartz, J. H. and Kandel, E. R. (1982). Inhibitor of adenosine 3':5'-monophosphate-dependent protein kinase blocks presynaptic facilitation in Aplysia. J. Neurosci. 2, 1673-1681.

DeRiemer, S. A., Strong, J. A., Albert, K. A., Greengard, P. and Kaczmarek, L. K. (1985). Enhancement of calcium current in Aplysia neurones by phorbol ester and protein kinase C. Nature 313, 313-316.

Gordon, A. S., Davis, C. G., Milfay, D. and Diamond, I. (1977). Phosphorylation of acetylcholine receptor by endogenous membrane protein kinase in receptor-enriched membranes of Torpedo californica. Nature 267, 539-540.

Greengard, P. (1976). Possible role for cyclic nucleotides and phosphorylated membrane proteins in the postsynaptic actions of neurotransmitters. Nature 260, 101-108.

Huganir, R. L. and Greengard, P. (1983). cAMP-dependent protein kinase phosphorylates the nicotinic acetylcholine receptor. Proc. Natl. Acad. Sci. U.S.A. 80, 1130-1134.

Huganir, R. L., Albert, K. A. and Greengard, P. (1983). Phosphorylation of the nicotinic acetylcholine receptor by Ca^{2+}/phospholipid-dependent protein kinase, and comparison with its phosphorylation by cAMP-dependent protein kinase. Abstracts of the 13th Annual Meeting of the Society for Neuroscience, p. 578.

Huganir, R. L., Miles, K. and Greengard, P. (1984). Phosphorylation of the nicotinic acetylcholine receptor by an endogenous tyrosine-specific protein kinase. Proc. Natl. Acad. Sci. U.S.A. 81, 6968-6972.

Huganir, R. L., Delcour, A. H., Greengard, P. and Hess, G. P. (1986). Phosphorylation of the nicotinic acetylcholine receptor regulates its rate of desensitization. Nature 321, 774-776.

Kaczmarek, L. K., Jennings, K. R., Strumwasser, F., Nairn, A. C., Walter, U., Wilson, F. D. and Greengard, P. (1980). Microinjection of catalytic subunit of cyclic AMP-dependent protein kinase enhances calcium action potentials of bag cell neurons in cell culture. Proc. Natl. Acad. Sci. U.S.A. 77, 7487-7491.

Kaczmarek, L. K., Nairn, A. C. and Greengard, P. (1984). Microinjection of protein kinase inhibitor prevents enhancement of action potentials in peptidergic neurons of Aplysia. Abstracts of the 14th Annual Meeting of the Society for Neuroscience, p. 895.

Kuo, J. F. and Greengard, P. (1969). Cyclic nucleotide-dependent protein kinases IV. Widespread occurrence of adenosine 3':5'-monophosphate-dependent protein kinase in various tissues and phyla of the animal kingdom. Proc. Natl. Acad. Sci., U.S.A. 64, 1349-1355.

Llinás, R., McGuinness, T., Leonard, C. S., Sugimori, M. and Greengard, P. (1985). Intraterminal injection of synapsin I or calcium/calmodulin-dependent protein kinase II alters neurotransmitter release at the squid giant synapse. Proc. Natl. Acad. Sci. U.S.A. 82, 3035-3039.

Paupardin-Tritsch, D., Hammond, C., Gerschenfeld, H. M. (1986a). Serotonin and cyclic GMP both induce an increase of the calcium current in the same identified molluscan neurons. J. Neurosci. 6, 2715-2723.

Paupardin-Tritsch, D., Hammond, C., Gerschenfeld, H. M., Nairn, A.C. and Greengard, P. (1986b). cGMP-dependent protein kinase enhances Ca^{2+} current and potentiates the serotonin-induced increase of Ca^{2+} current in identified snail neurones. Nature, in press.

Teichberg, V. I., Sobel, A. and Changeux, J.-P. (1977). In vitro phosphorylation of the acetylcholine receptor. Nature 267, 540-542.

37
The Molecular Logic of Presynaptic Facilitation and Inhibition

Francesco Belardetti, Eric R. Kandel and Steven A. Siegelbaum

INTRODUCTION

A striking feature of the organization of the brain is that some synaptic potentials -- the signals whereby one cell communicates with another -- are fast, lasting only milliseconds, whereas others are slow, lasting many seconds or even many minutes. One of the insights of the last several years is the realization, based on studies of the heart and of molluscan neurons, that the two types of synaptic interactions involve two different molecular mechanisms and serve two different behavioral consequences (fig. 1).

For both types of synaptic actions, the ion channel gated by the transmitter consists of at least two functional components: a receptor site and a channel. The receptor recognizes the transmitter and instructs the channel to open or close. The channel conducts ions only when it is open (fig. 1; for discussion see Hille, 1984). In fast synaptic actions, like those involving the nicotinic ACh channels at the neuromuscular junction and the channels regulated by glutamate, glycine, and GABA in the central nervous system, the receptor is intrinsic, and the channel and the receptor site occupy different domains in a single macromolecule (fig. 1A). By contrast, slow synaptic actions are governed by remote-receptor sites. Remote receptors communicate with their channels by means of a second-messenger system, such as cAMP and the cAMP-dependent protein kinase (fig. 1B).

Second-messenger-mediated actions have generated much interest because they differ markedly from the common fast synaptic actions utilized by the basic neural circuitry that mediate

A

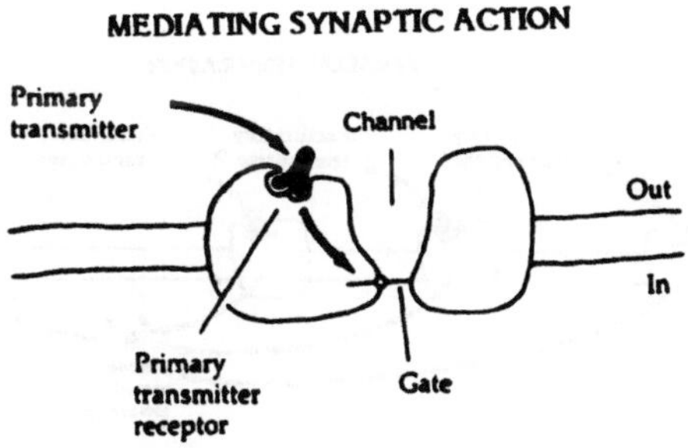

B

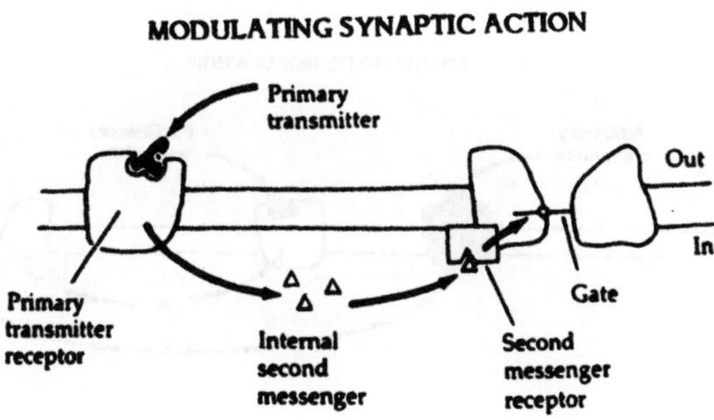

Fig. 1. Schematic diagram illustrating differences between major classes of chemical transmitter actions. **A.** In mediating synaptic actions, the receptor is an intrinsic part of the transmitter-activated ion channel. **B.** In modulating synaptic actions, the receptor is remote from the channel and coupling depends on an intracellular messenger. (Adapted from Hille, 1984.)

behavior. Rather than produce fast synaptic actions, second messengers produce slow actions that modulate the basic neural circuitry mediating behavior. Specifically, modulatory synaptic pathways can serve as reinforcing stimuli in learning.

The finding that the receptor for the action of some transmitters modulates the channel by means of an internal second messenger opens up the possibility for either synergistic or antagonistic interactions of different transmitters on channel function. For example, a small family of transmitters, each acting through its own remote receptors, may instruct the same channel by means of the same or different internal second messengers (fig. 2A and 2B).

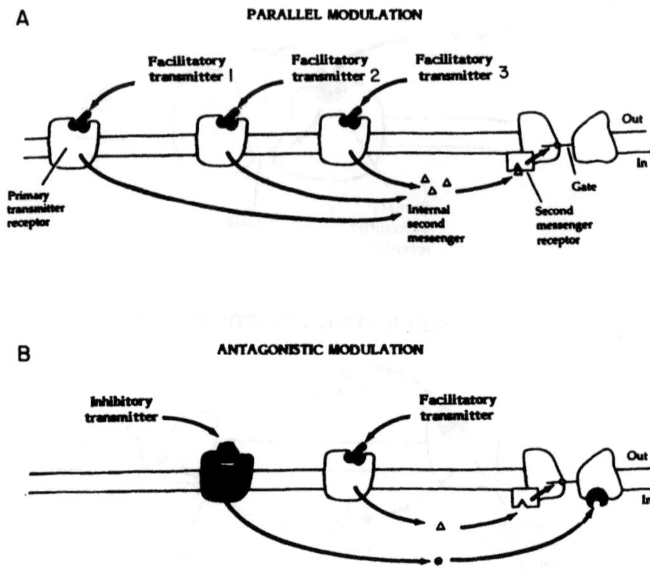

Fig. 2. Possible interactions of modulatory transmitters. **A.** A channel may be linked to several distinct transmitter receptors through a common intracellular messenger. **B.** Different transmitters may produce antagonistic effects on a single ion channel by activating separate second-messenger systems.

We have encountered two such situations in Aplysia. In earlier work, we found that a conventional transmitter, serotonin, and the peptides SCP_A and SCP_B can modulate a specific K^+ channel (the S channel) by means of a common second-messenger system: cyclic AMP (Castellucci et al., 1986). Although the information is still fragmentary, we have indirect evidence that a third class of transmitters, as yet unidentified, works by the same mechanism. More recently, we have encountered an even more interesting case in which two different second-messenger systems activated by different transmitters interact within a single cell to regulate the function of a single class of these ion channels. Here, we describe a first step in examining the interaction of two intracellular cascades by focusing on the reciprocal actions of serotonin (5-HT) and the molluscan neuropeptide, FMRFamide, on individual $S\text{-}K^+$ channels in Aplysia sensory neurons.

The K^+ channels, modulated by serotonin and FMRFamide, are thought to exist not only in the cell body but also on the terminals, where their modulation plays a role in controlling synaptic strength. Specifically, 5-HT produces presynaptic facilitation, whereas FRMFamide produces presynaptic inhibition (fig. 3). Thus, we shall not only be examining the logic whereby these two transmitters interact within the cell to regulate channel function by means of distinct second messengers but, in a larger sense, we shall also be examining how that interaction contributes to the up-and-down regulation of transmitter release.

Previously, 5-HT was shown to cause prolonged all-or-none closure of the S channels through cAMP-dependent protein phosphorylation. The closure of the S channel by 5-HT causes a slow depolarization and broadens the action potential in the sensory neuron cell body, contributing to an increase in transmitter release (presynaptic facilitation) from the sensory neuron terminals (Klein and Kandel, 1980). Recently, the peptide FMRFamide has been found to produce opposite effects on these same neurons, causing a hyperpolarization of membrane potential, a decrease in the action potential duration, and an inhibition of synaptic transmission (Abrams et al., 1984; Belardetti et al., 1986). These opposing effects of 5-HT and FMRFamide at the level of macroscopic neuronal properties led us to investigate whether the transmitters also had opposing action at the microscopic level of S channel function. Using single-channel recording, we have now found that FMRFamide does indeed modulate the same S channels closed by 5-HT and that it increases the opening probability of the S channels via a second-messenger system that is independent of cAMP. In addition, the peptide can reverse the closures of S channels produced by 5-HT or cAMP. Thus, the anatomical pathways of the neuronal modulatory circuits are re-represented within individual cells by the specific molecular circuitry of the second-messenger pathways.

PRESYNAPTIC FACILITATION

Presynaptic facilitation of transmitter release from the sensory neurons of Aplysia is of particular interest becaue it contributes to simple forms of learning. It provides the means whereby a reinforcing stimulus, that increases the activity of one groups of cells, facilitating cells, can modulate the synaptic strength of another group of cells, the sensory neurons. The siphon is innervated by a cluster of about 24 sensory cells. These neurons make excitatory monosynaptic connections with interneurons and motor neurons that produce the withdrawal reflex

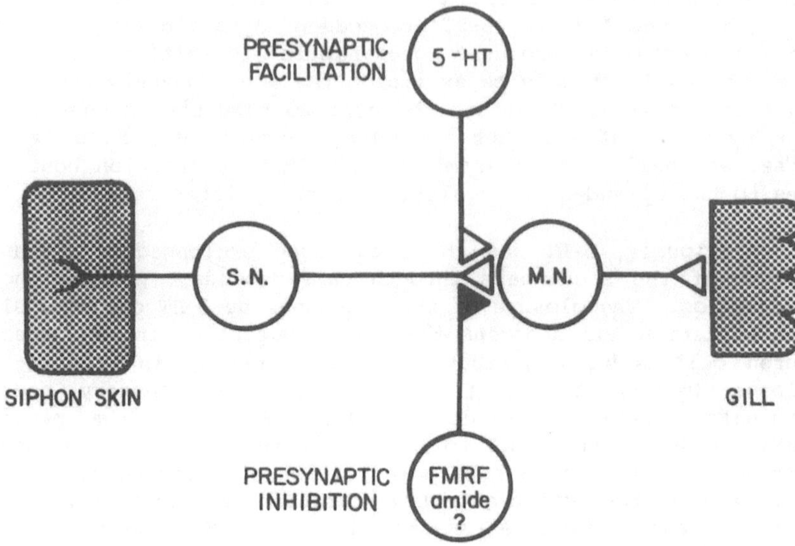

Fig. 3. Circuit diagram for two types of synaptic modulation. In presynaptic facilitation, serotonergic interneurons synapse onto sensory neuron (S.N.) terminals and produce an increase in transmitter release from the sensory neurons onto the follower motor neurons. A second class of modulatory interneurons may release FMRFamide and inhibit transmitter release from the sensory neurons.

(fig. 4). Here, we will focus exclusively on the monosynaptic portion of the reflex circuit consisting of the sensory neuron, the motor neuron, and their connections. Other changes also occur at other loci within the neural circuit, but we will not consider these changes.

A sensitizing stimulus applied to the tail activates facilitating neurons that act on the sensory neurons, including on their terminals, to enhance transmitter release. We have called this process presynaptic facilitation (Kandel and Tauc, 1965a,b; Castellucci et al., 1970; Castellucci and Kandel, 1976). Although we do not know all of the elements of the facilitatory system or the transmitters these elements use, we have identified two groups of cells and have reason to believe there is a third

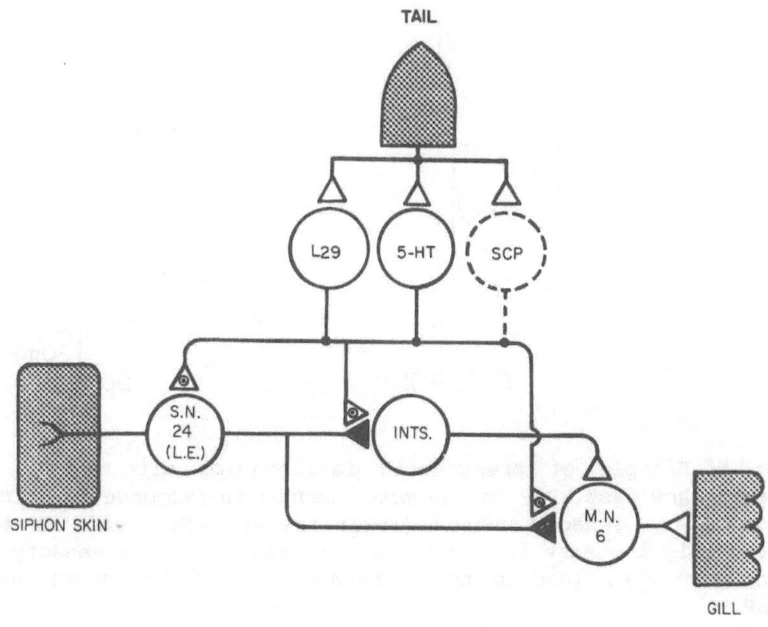

Fig. 4. Simplified diagram of the gill-withdrawal reflex in Aplysia. The mechanoreceptor neurons (**24**) carry the information from the siphon skin to interneurons and gill motor neurons (**6**). Stimulation of the tail (or the neck region) excites subsets of facilitator neurons that increase the synaptic transmission between the sensory neurons and their follower cells (**shaded terminals.** Some facilitators may be serotonergic or peptidergic (SCP). The transmitter of one identified facilitator group of interneurons (L29 cluster) is not yet known.

(figs. 4,5). Each of these three groups uses a different transmitter: (1) serotonin, released by the LC_B and RC_B cells; (2) an as-yet unidentified transmitter that may be related to serotonin, released by the L29 cells, and (3) two small interrelated peptides, SCP_A and SCP_B. Processes containing these peptides come in close apposition to those of the sensory neuron in the neuropile, but their cell bodies have not yet been identified.

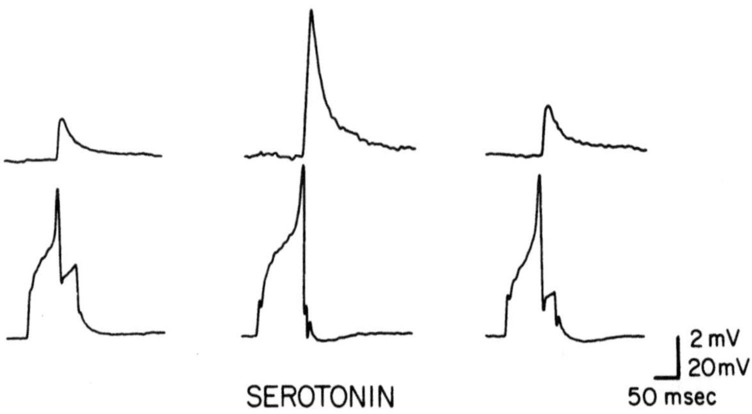

Fig. 5. Example of presynaptic facilitation with 5-HT. Top traces show fast EPSP in the motor neuron in response to stimulation of the sensory neuron (lower trace). Serotonin causes a reversible increase in transmitter release from the sensory neuron terminals, leading to an enhancement of the motor neuron EPSP.

Based on voltage-clamp and biochemical studies of the sensory neuron, both in the intact ganglion and in dissociated cell culture, we have outlined a molecular model for presynaptic facilitation (fig. 6; for an earlier review, see Kandel and Schwartz, 1982). According to this view, serotonin (as well as the other facilitating transmitters), released by the facilitating neurons, activates a transmitter-sensitive adenylate cyclase in the membrane of the sensory neurons, including its presynaptic terminal, that increases the level of intracellular cyclic AMP. Cyclic AMP then activates a protein kinase that phosphorylates a K^+ channel protein or a protein associated with such a protein (Castellucci et al., 1980). This phosphorylation reduces one of the K^+ currents that normally repolarize the action potential. Reduction of this current increases the excitability of the neuron and prolongs the action potential so as to allow more Ca^{++} to flow into the terminals; consequently, more transmitter can be released (fig. 7; Castellucci et al., 1980; Klein and Kandel, 1978; Klein et al., 1982, 1986). This biochemical evidence for the role of cAMP in facilitation is consistent with the physiological effect of serotonin and cAMP on transmitter release (fig. 5). In physiological experiments we found that 5-HT and intracellular injection of cAMP or the cAMP-dependent protein kinase simulate presynaptic facilitation (fig. 7).

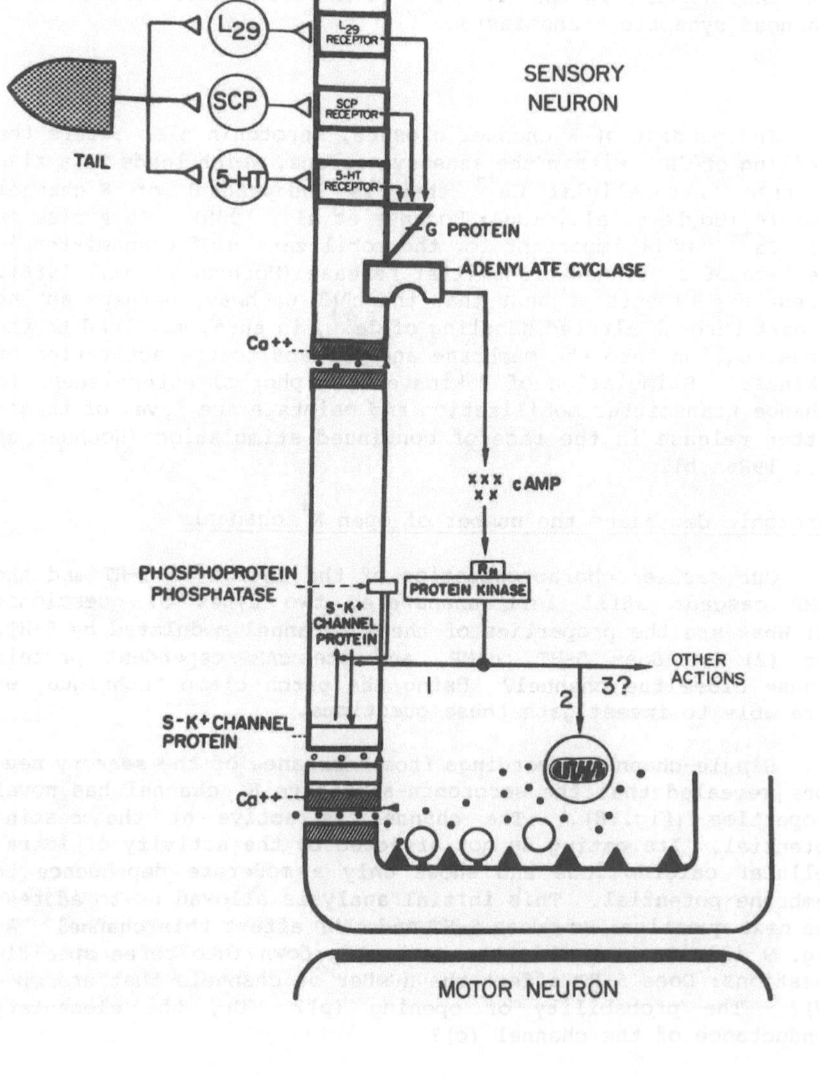

Fig. 6. Molecular model of presynaptic facilitation underlying sensitization. When excited, the various groups of facilitator neurons activate a sequence of events leading to a cAMP-dependent phosphorylation of substrate proteins, which results in the closure of a special K channel and in a change in the Ca^{++} buffering

in the presynaptic terminals of the sensory neuron. The resulting increase in the duration of the sensory neuron action potential leads to an increase of Ca^{++} current. This effect, together with the increase in the free Ca^{++} in the terminals, results in an enhanced synaptic transmission.

Independent of K channel closure, serotonin also alters the handling of Ca^{++} within the sensory neurons, which leads to a rise in free intracellular Ca^{++} that is independent of K channel closure (Boyle et al., 1984; Hochner et al., 1985). This rise in free Ca^{++} may be important for the mobilization of transmitter in the face of continued transmitter release (Hochner et al., 1986). Recent experiments suggest that the cAMP pathway, perhaps acting in part through altered handling of Ca^{++}, in turn, may lead to the translocation into the membrane and perhaps to the activation of C kinase. Stimulation of C kinase with phorbol esters seems to enhance transmitter mobilization and maintain the level of transmitter release in the face of continued stimulation (Hochner et al., 1986a,b).

<u>Serotonin decreases the number of open K^+ channels</u>

Our earlier characterization of the action of 5-HT and the cAMP cascade still left unanswered two types of questions: (1) What are the properties of the K^+ channel modulated by 5-HT, and (2) How does 5-HT, cAMP, and the cAMP-dependent protein kinase close the channel? Using the patch clamp technique, we were able to investigate these questions.

Single-channel recordings from membranes of the sensory neurons revealed that the serotonin-sensitive K^+ channel has novel properties (fig. 8). The channel is active at the resting potential. Its gating is not affected by the activity of intracellular calcium ions and shows only a moderate dependence on membrane potential. This initial analysis allowed us to address the next question: How does 5-HT and cAMP affect this channel? As fig. 9 illustrates, this can be broken down into three specific questions: Does 5-HT affect the number of channels that are open (N)? The probability of opening (p)? Or, the elementary conductance of the channel (c)?

Serotonin appears to reduce the <u>number</u> of functional channels in a patch of membrane and it does so in an all-or-nothing manner (fig. 10). Channels that still open in the presence of 5-HT appear to open and close normally, whereas channels once closed by 5-HT remain closed. The modulatory

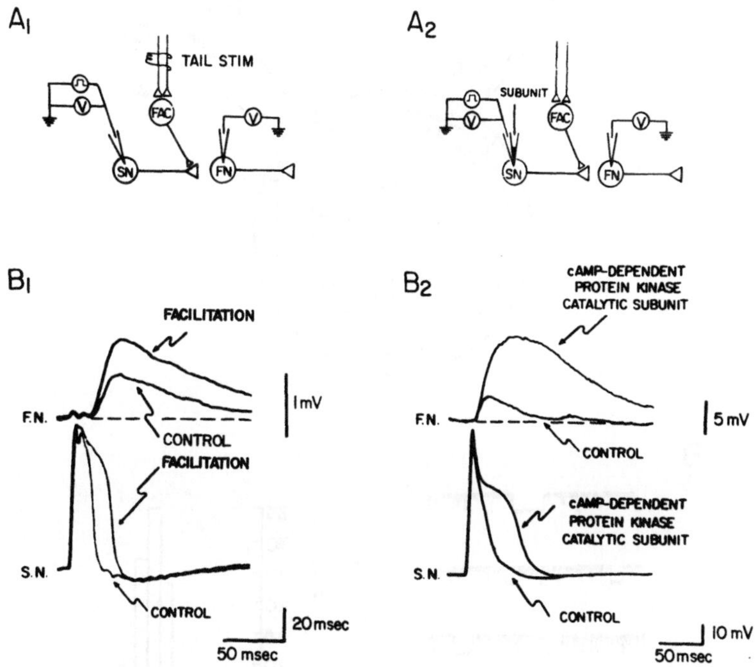

Fig.7. The effects of presynaptic facilitation on duration of the action potential and synaptic strength are replicated by injecting the cAMP-dependent protein kinase into the sensory neuron. **A_1.** Experimental arrangement used to elicit ordinary presynaptic facilitation. An electrode is used to stimulate a sensory neuron intracellularly while a shock is delivered to the animal's tail, exciting a facilitatory interneuron that also synapses on the sensory neuron. The response is recorded in a follower (motor) neuron (**FN**). **A_2.** Experimental arrangement for mimicking the effect of presynaptic facilitation by stimulating the sensory neuron intracellularly, as in **A_1**, but replacing the tail shock with an intracellular injection of the catalytic subunit of the protein kinase directly into the sensory neuron. **B_1.** The presynaptic facilitation produced by activating a single facilitatory interneuron (as in **A_1**) is compared with the response produced in a follower neuron by simple stimulation of a sensory neuron (control). **B_2.** The action potential that results when the catalytic subunit of the protein kinase is injected into the sensory neuron very closely resembles that produced by ordinary presynaptic facilitation. In both cases, the action potential is broadened, leading to an enhancement of transmitter release and a heightened response in the follower neuron. (From Castellucci et al., 1980.)

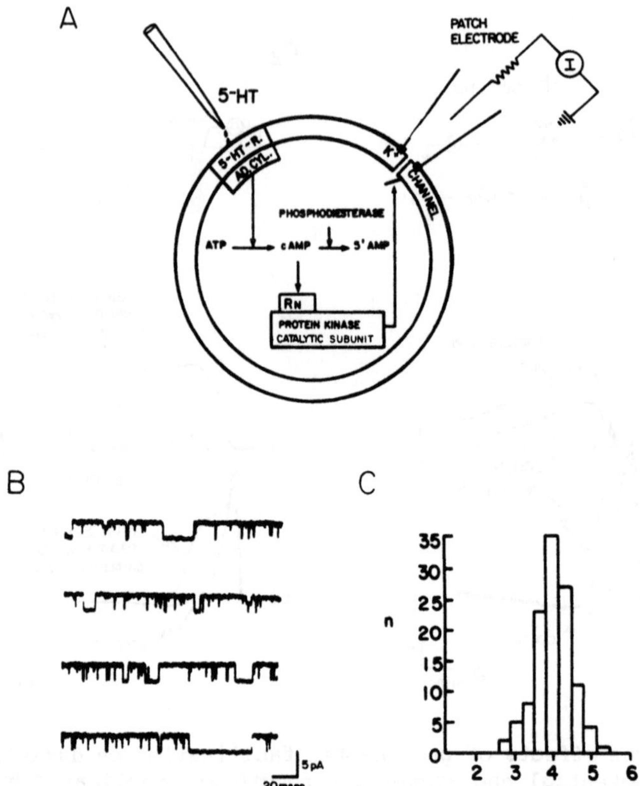

Fig. 8. Patch-clamp recording of single S channel current. **A.** Schematic illustration of the experimental recording protocol and the model for cAMP-dependent action of 5-HT (serotonin). A high-resistance giga-ohm seal is obtained between the extracellular patch electrode and the cell membrane to record single-channel currents in the small patch of membrane under the electrode. This is the cell-attached recording configuration. Serotonin is applied to the cell outside the membrane patch but may still alter channel activity in the patch via the cAMP cascade. The substrate protein for the kinase may or may not be the S channel itself. **B.** Serotonin-sensitive K^+ channel. Single-channel current records from mechanoreceptor sensory neurons (LE cluster) in the abdominal ganglion of Aplysia californica. Both the bath and recording pipette contain artificial sea water. Channel openings appear as step increases in outward current (outward current plotted in upward direction). The current fluc-

tuates between two levels corresponding to the fully closed and fully open channel. The channel shows both brief closures (downward flickers) and longer closures but is in the open configuration for most of the time. Addition of 10 uM serotonin in the bath caused this channel to close (record not shown). C. Histogram of single-channel current amplitudes measured at +11 mV. (From Siegelbaum et al., 1982).

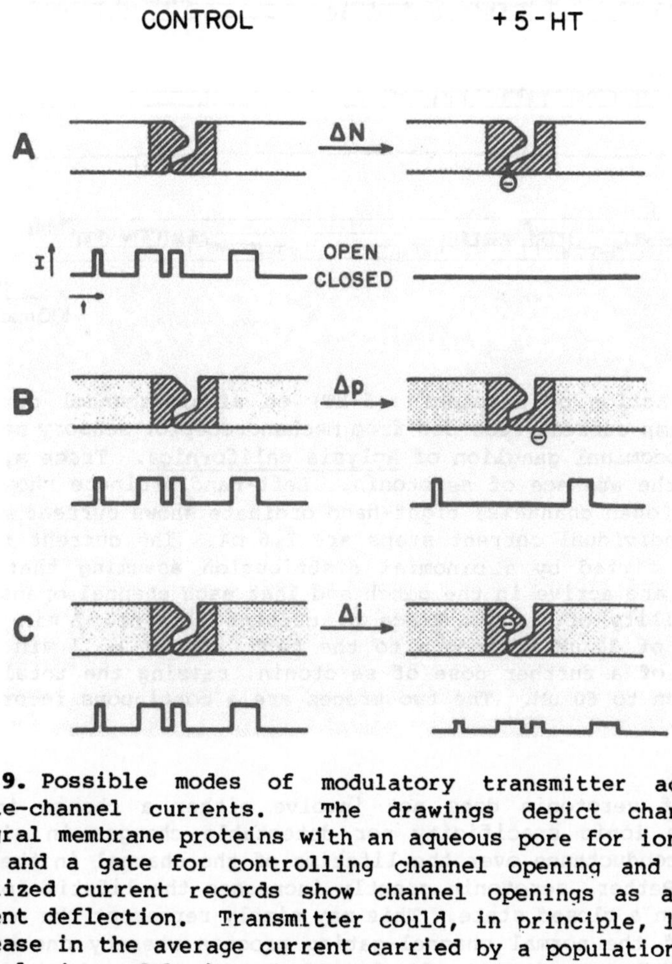

Fig. 9. Possible modes of modulatory transmitter action on single-channel currents. The drawings depict channels as integral membrane proteins with an aqueous pore for ion permeation and a gate for controlling channel opening and closing. Idealized current records show channel openings as an upward current deflection. Transmitter could, in principle, lead to a decrease in the average current carried by a population of such channels by modulating the number of functional channels (A, ΔN), the probability of channel opening (B, Δp) or the amplitude of the single channel current (C, Δi) giving rise to different changes in single-channel function. (Modified from Siegelbaum and Tsien, 1983).

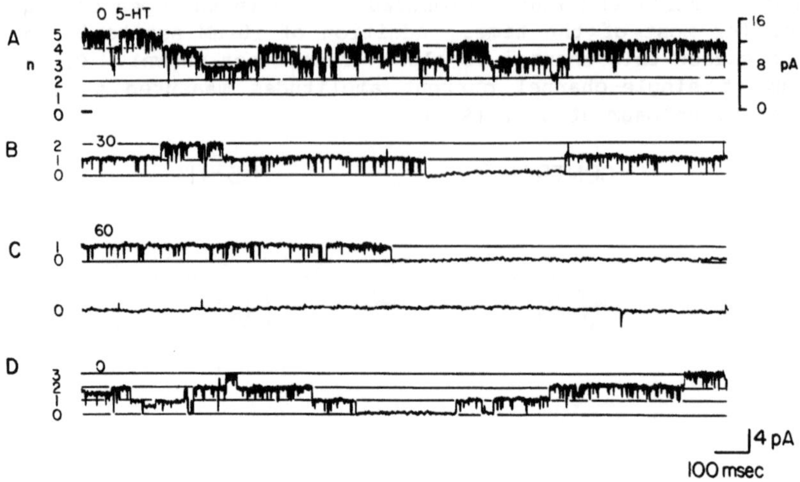

Fig. 10. Action of serotonin (5-HT) on single-channel current. Patch clamp current recorded from mechanoreceptor sensory neurons in the abdominal ganglion of Aplysia californica. Trace **a**, current in the absence of serotonin. Left-hand ordinate shows the number of open channels; right-hand ordinate shows current magnitude. Individual current steps are 2.6 pA. The current record was well fitted by a binomial distribution assuming that five channels are active in the patch and that each channel opens with a probability of 0.84. Trace **b**, current obtained 2 min after addition of 30 uM serotonin to the bath. Trace **c**, 1 min after addition of a further dose of serotonin, raising the total concentration to 60 uM. The two traces are a continuous recording.

action of serotonin does not involve either a change in the channel's ionic specificity nor detectable changes in single-channel conductance over the lifetime of the channel in the open state. Rather, serotonin greatly increases the lifetime of the channel in a closed state. This may simply represent the modification of the normal channel-gating process whereby one of the normal rate constants of channel opening is dramatically decreased. Alternatively, serotonin may induce new closed states of the channel (perhaps corresponding to the channel in a phosphorylated form) in which the channel is not available to the normal gating process.

Cyclic AMP causes channel closure similar to that produced by serotonin

The patch pipette forms, with the membrane, a very stable, high-resistance seal that is thought to prevent diffusion of transmitters or other ligands from outside to inside the pipette. As a result, in the ACh-activated channels of muscle, the transmitter has to be introduced in the patch pipette to affect the channels under the patch pipette. By contrast, serotonin consistently decreased channel opening in the patch when applied to the bathing solution outside the pipette (fig. 10). This finding is consistent with the notion that the modulation of the K^+ channel is mediated by an intracellular messenger that can act on the patch from within the cell. To test this idea directly, we injected the cell with cyclic AMP, the second messenger thought to modulate the serotonin-sensitive K^+ current. Intracellular injection of cAMP, by means of current pulses, causes channels to close in an all-or-none manner similar to that caused by the action of serotonin. In the experiment illustrated by fig. 11A, three channels are active before the start of the current injection (Part A_1). After the onset of the current pulses, these channels drop out, one at a time (Part A_2). Similar effects are seen following application of the catalytic subunit of the cAMP-dependent protein kinase to the inside surface of the isolated membrane patch (fig. 11B). These results suggest that cAMP-dependent protein kinase leads to all-or-none closures of the S channels by acting on a membrane associate substrate protein, either the S channel itself or a regulatory protein that acts on the channel.

PRESYNAPTIC INHIBITION

Presynaptic inhibition is a widely distributed modulatory action, found in both vertebrates and invertebrates (Burke and Rudomin, 1977; Dudel and Kuffler, 1961a,b; Eccles, 1964; Kennedy et al., 1974; Kretz et al., 1984; Nicoll and Alger, 1979; Pearson and Goodman, 1981; Ryall, 1978; Schmidt, 1971; Tauc, 1965; Waziri et al., 1969; Zengel et al., 1983). One mechanism for the reduction in transmitter output during presynaptic inhibition derives from a synaptically mediated increase in the membrane conductance of the presynaptic terminals (frequently to Cl^-), which either reduces the amplitude of the presynaptic spike (Takeuchi and Takeuchi, 1966a,b; Baxter and Bittner, 1981) or causes branch-block at bifurcation points on the presynaptic terminal tree. The increased Cl^- conductance may produce either a depolarization or a hyperpolarization of the terminal, depending on the intracellular concentration of Cl^- (Baxter and Bittner, 1981). This

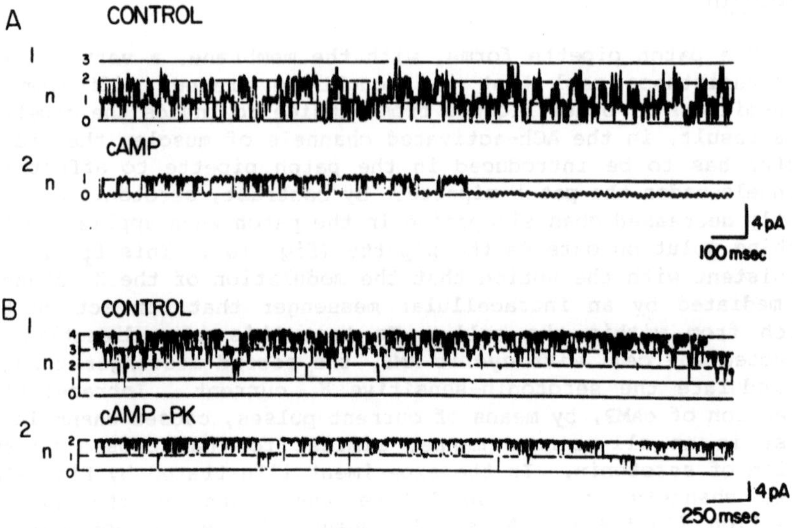

Fig. 11. A. Action of cAMP and the catalytic unit of cAMP-protein kinase on single-channel current recorded from sensory neurons in the abdominal ganglion of Aplysia californica. **A_1.** Current in the absence of cAMP. Left-hand ordinate shows the number of open channels. **A_2.** Effect of intracellular injection of cAMP on single-channel current. Following establishment of the seal between patch electrode and membrane, the cell was impaled with a microelectrode filled with 1 M cAMP (Sigma) and cAMP was injected into cells by hyperpolarizing current pulses. Note that only one active channel is open and after a while this last channel is closed. (From Siegelbaum et al., 1982). **B.** Effect of cAMP-kinase on S channels in inside-out membrane patches from sensory neurons. **B_1.** Before the addition of the kinase, the patch contained four active channels open for a large fraction of the time. **B_2.** Three minutes after addition of cAMP-PK (0.1 uM final concentration) plus 1 mM Mg-ATP, two channels have closed and the current level fluctuates from zero to two channels open. (From Shuster et al., 1985).

mechanism for presynaptic inhibition is generally of short duration, on the order of milliseconds.

A variant of this mechanism is based on the idea that the change in membrane potential (depolarization) of the presynaptic terminals might itself be responsible for reducing transmitter output. Thus, a Cl^--mediated conductance change, or a buildup of extracellular K^+ ions in the restricted spaces around presynaptic release sites due to activity in the inhibitory pathway, would depolarize the presynaptic terminals and reduce transmitter output (as is thought to occur during primary afferent depolarization; for discussion see Schmidt, 1971; Burke and Rudomin, 1977; Ryall, 1978; Nicoll and Alger, 1979; Rudomin et al., 1981). This idea is consistent with the release properties described at the giant synapse in the squid where depolarization reduces transmitter release (Takeuchi and Takeuchi, 1962; Miledi and Slater, 1966; Charlton et al., 1982). However, in Aplysia and certain other neurons, depolarization enhances transmitter release, it does not depress it. At these synapses a different mechanism accounts for presynaptic inhibition (Kretz et al., 1984).

FMRFamide causes a depression of transmitter release and a narrowing of the action potential

Recently, we have begun to study the inhibitory actions of the molluscan peptide FMRFamide on the sensory neurons. FMRFamide depresses the EPSP produced in the motor neuron in response to stimulation of the sensory neuron (fig. 12A). In contrast with the slow, decreased conductance depolarization, and broadening of the action potential broadening produced by 5-HT in these same neurons, application of FMRFamide (2 uM) produces a slow, transient hyperpolarization, an increase in membrane conductance, and a narrowing of the action potential (figs. 12B and 14A).

To produce this hyperpolarization and narrowing of the action potential, FMRFamide alters several macroscopic currents. It decreases a voltage-dependent Ca^{++} current, and it increases a background K^+ current and decreases a Ca^{++}-dependent K^+ current (Belardetti et al., 1986; Brezina et al., 1987). We have here restricted our patch clamp analysis to the actions on the background K current. Specifically we have explored whether FMRFamide modulates the same class of K^+ channels closed by 5-HT? If so, which parameter of S channel function is altered? Finally, how do 5-HT and FMRFamide interact in causing channel modulation?

FMRFamide modulates single S channels

Single-channel current recordings indicate that the slow hyperpolarization produced by FMRFamide is indeed associated with

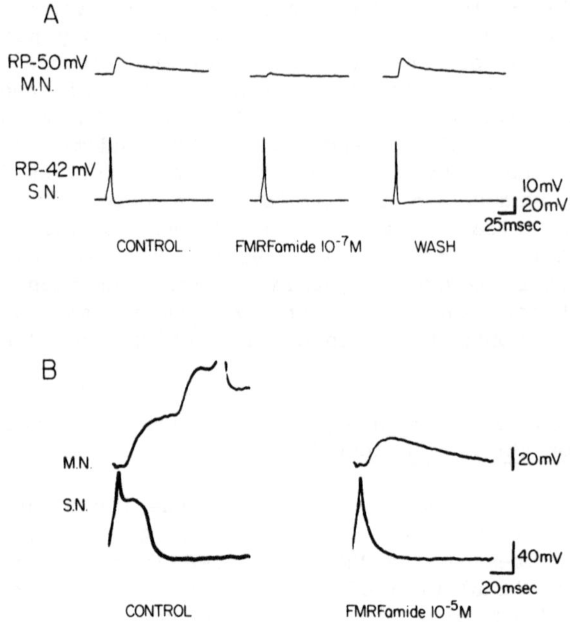

Fig. 12. FMRFamide inhibits synaptic transmission between sensory and motor neurons in culture. **A.** Effect on EPSP. Top trace shows fast EPSP in the motor neuron in response to stimulation of the sensory neuron (bottom trace). FMRFamide produces a reversible decrease in the size of the EPSP in the motor neuron. **B.** Effect of FMRFamide on the duration of the presynaptic (sensory neuron) action potential and on the sensory-motor connection. Top traces show EPSP as recorded from motor neurons; the lower traces show sensory neuron action potential. The left side illustrates control conditioning; the right side, in the presence of FMRFamine (10^{-5} M) -- the bath contains 5 mM TEA.

an increase in the opening of S channels (fig. 13). Fig. 13A shows a channel current trace before addition of FMRFamide from a patch that contains several active channels that display the characteristic properties of the S channels. FMRFamide causes a large increase in the opening of these S channels that is reflected in a two to four-fold increase in the mean outward current ($< I >$) carried by the channels over a wide range of membrane potentials (fig. 13B).

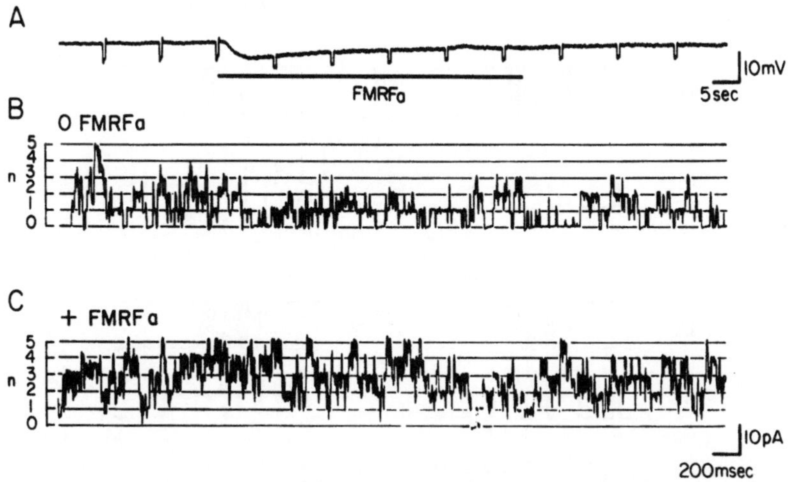

Fig. 13. FMRFamide action on pleural sensory neurons. **A.** FMRFamide produces a slow hyperpolarization of the cell resting potential accompanied by an increase in membrane conductance. The application of 2 uM FMRFamide is indicated by the **solid line.** Constant current hyperpolarizing pulses of 0.1 nA were delivered to the cell to monitor membrane conductance. **B, C.** Effect of FMRFamide (2 uM) on single channel current. Trace B shows channel currents before addition of FMRFamide, and trace C shows the increase in channel current approximately 10 min after addition of 2 uM FMRFamide to the bath. Approximate patch membrane potential was +20 mV, assuming a resting potential of -60 mV.

The mean current, $< I >$, carried by a population of channels, is given by the product $< I > = N_f \, p \, i$, where N_f is the number of functional channels in the membrane, p is the probability that a single channel is open, and i is the amplitude of the elementary current. We have seen above that serotonin decreases the macroscopic S current by causing prolonged all-or-none S channel closures, resulting in an apparent decrease in the number of functional channels in the membrane. We were therefore interested in knowing: Does FMRFamide increase the mean S channel current by increasing N, p or i (fig. 13)? The records of fig. 14 show that FMRFamide produces no marked change in either the single-channel amplitude (i) or in the maximum number of channels that open simultaneously (N_f). A plot of the single-

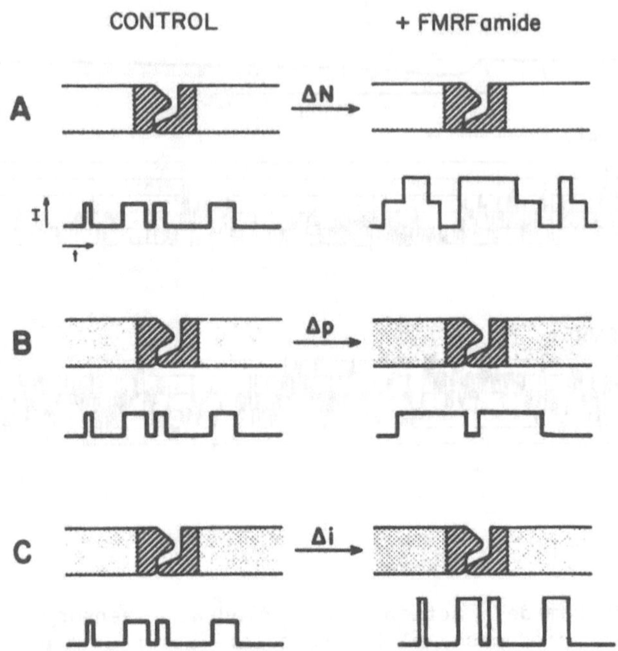

Fig. 14. Possible modes of FMRFamide action at the level of single channel currents. In principle, FMRFamide could lead to an increase in actual K^+ current by: **A**, increasing the functional number of active channels; **B**, increasing the probability of opening a single channel; **C**, increasing single-channel conductance.

channel unit current-voltage relation before and after FMRFamide confirms the lack of change in i, the elementary current (fig. 15A).

A more quantitative test of whether FMRFamide alters N_f or p was obtained from a fit of the binomial distribution to the observed probability distribution that a given number of channels in the patch are open simultaneously. Figs. 15B and 15C compare values for N and p in the absence and presence of FMRFamide, derived from the binomial distribution over a wide range of voltages. These results show that, whereas FMRFamide produces a clear increase in p, N remains essentially the same. The small variation in N among different voltages is within the expected range of uncertainty involved in the binomial fitting procedure.

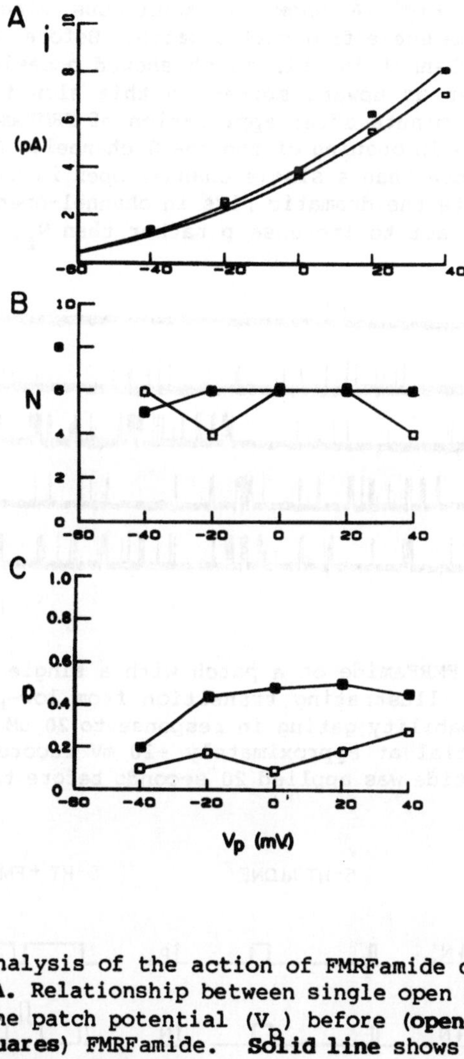

Fig. 15. Analysis of the action of FMRFamide on single S channel current. **A.** Relationship between single open channel current (i) and membrane patch potential (V_p) before **(open squares)** and after **(filled squares)** FMRFamide. **Solid line** shows least-squares fit, of the Goldman-Hodgkin-Katz (Hodgkin and Katz, 1949) constant field current equation, for a pure K current to the data. **B, C.** Maximum likelihood estimates from binomial analysis for N_f and p over a wide range of membrane potentials before **(open squares)** or after **(filled squares)** FMRFamide. The open probability of the S channels shows only a moderate dependence on membrane potential in the absence or presence of FMRFamide, with a 2-3 fold increase in p for an 80 mV depolarization.

The most convincing demonstration that FMRFamide acts to increase p rather than N_f requires a patch that contains only one active S channel. Fig. 16 shows a continuous single-channel record on a slow time scale from such a patch. Before application of FMRFamide, the channel in this patch showed occasional brief openings that appear as upward spikes on this slow time scale. However, within one minute after application of FRMFamide, there is a marked increase in opening of the one S channel. Inasmuch as we never observed more than a single channel open in the presence of FMRFamide, despite the dramatic rise in channel-open probability, FMRFamide must act to increase p rather than N_f.

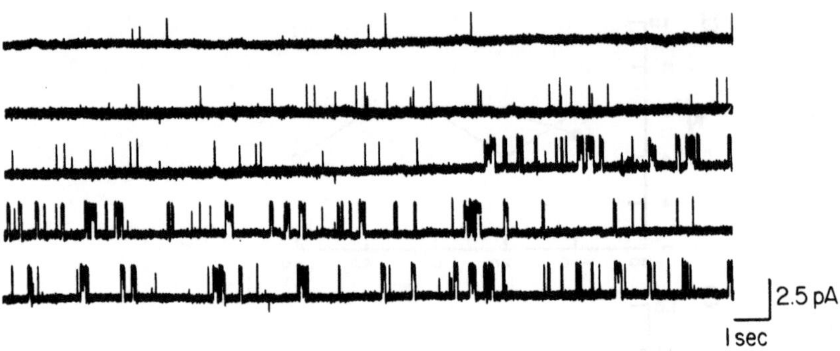

Fig. 16. Effect of FMRFamide on a patch with a single S channel. Consecutive sweeps illustrating transition from low-probability gating to high-probability gating in response to 20 uM FMRFamide. The membrane potential at approximately +20 mV records filtered at 500 Hz. The peptide was applied 20 seconds before the start of the top trace.

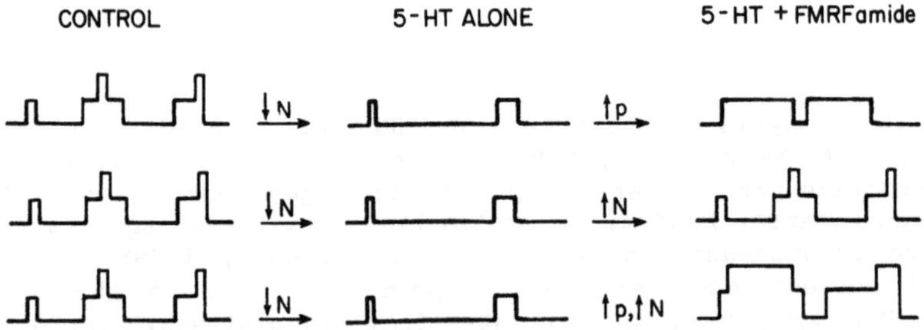

Fig. 17. Diagram of the ways FMRFamide can increase the average currents through the S K^+ channels in the presence of 5-HT. See text for explanation.

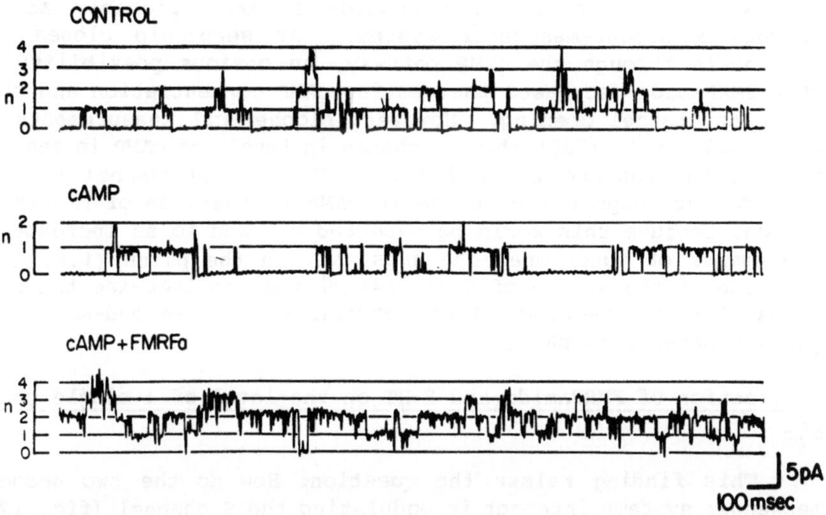

Fig. 18. FMRFamide antagonizes the action of 5-HT and cAMP. On the level of single channels, FMRFamide antagonizes the closure of channels induced by cAMP. Top trace is a current record from patch containing four active S channels before application of cAMP. Middle trace is a current record obtained 10 min after addition of 0.1 mM dibutyryl cAMP (dbcAMP) to bath showing closure of two channels. Closures occurred within 5 min after addition of cAMP and persisted in maintained presence of dbcAMP until addition of FMRFamide. Lower trace is a current record 5 min after application of 2 uM FMRFamide (still in presence of dbcAMP), showing reopening of two channels and increase in open probability. Membrane potential was about 20 mV, current records filtered at 1 kHz.

We observed a similar increase in mean S channel current in 7 out of 8 experiments with FMRFamide. In all of these experiments, both binomial analysis and simple counting of the maximal number of open channels showed a clear increase in p, with no marked effect on N_f. In the eight experiments with FMRFamide, the mean value of p increased five-fold, whereas N remained at 1.05 ± 0.27 (mean ± SEM) of its control value.

Since application of FMRFamide to the bath is capable of opening channels under the patch pipette that have no direct access to the peptide, the peptide is most probably acting through a second-messenger system. As serotonin closes the S channels through the cAMP pathway, an obvious possibility is that FMRFamide decreases the resting cAMP concentration and thus leads to channel opening. However, biochemical measurements of Ocorr and Byrne (1985) show no change in levels of cAMP in sensory neurons in response to FMRFamide. Our single-channel results also do not support a decrease in cAMP as the mode of FMRFamide action, because this would be expected to lead to an increase in the number of functional channels (N_f) in the patch (i.e., the converse of the action of 5-HT) rather than to increase the open probability as observed. Thus, FMRFamide uses a second-messenger pathway other than cAMP.

Interaction of FMRFamide and 5-HT on the level of a single channel

This finding raises the question: How do the two second-messenger systems interact in modulating the S channel (fig. 17)? At the level of the whole cell membrane potential response, FMRFamide antagonizes the slow excitatory action of serotonin (or cAMP). The single-channel basis for this antagonism is shown in figs. 18 and 19. The patch of membrane illustrated in fig. 18 and analyzed in fig. 19 initially displayed four active S channels (control). On application of dibutyryl cAMP to the bath, two of the four active channels close (cAMP), leading to a decrease in N_f as described earlier (figs. 18 and 19; cAMP). The two remaining active channels appear normal, with no change in single-channel amplitude or open probability. We then applied FMRFamide to the cell, still in the presence of cAMP (figs. 18 and 19; cAMP-FMRFamide). Now, in addition to the normal increase in open probability, the peptide also produces an increase in the number of functional channels from 2 up to the original level of 4 (fig. 19). This is in contrast to the single action of FMRFamide to increase p in the absence of cAMP. The ability of FMRFamide to reopen channels closed by cAMP strengthens the argument that the peptide does indeed modulate the same class of S channels closed by cAMP and 5-HT. It also supports the evidence that FMRFamide action does not involve a decrease in adenylate cyclase activity.

THE RECEPTORS FOR PRESYNAPTIC INHIBITION AND PRESYNAPTIC FACILITATION ACTIVATE SEPARATE INTRACELLULAR SECOND MESSENGER SYSTEMS

These results show that the S channel can be modulated in opposite directions during presynaptic facilitation produced by

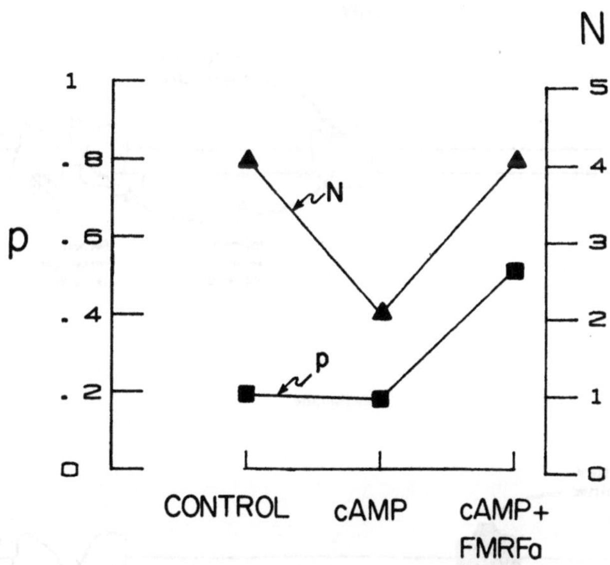

Fig. 19. Results of binomial analysis of several minute-long stretches of data showing maximum likelihood estimates for N_f (**triangles**) and p (**squares**) in control conditions (**leftmost points**), in presence of dbcAMP alone (**middle points**) and in the presence of both dbcAMP and FMRFamide (**rightmost points**).

serotonin and during presynaptic inhibition produced by FMRFamide (fig. 20). However, the modes of action of these modulatory transmitters differ in fundamental ways. Serotonin acts to prolong channel closures, resulting in a decrease in the number of functional channels (fig. 20A); normally, FMRFamide acts to increase channel opening without increasing the number of functional channels (fig. 20B). These observations imply that the S channel has two distinct functional sites of modulation that may correspond to separate structural sites on the channel. Nevertheless, the two transmitter effects are not completely independent. In addition to increasing channel-open probability, FMRFamide is able to reopen channels closed by cAMP (fig. 20C) This suggests that FMRFamide may have several sites of action. One possibility is that in addition to an action at a site on the channel involved in gating, FMRFamide may also stimulate phosphatase activity to reverse cAMP-induced channel closures.

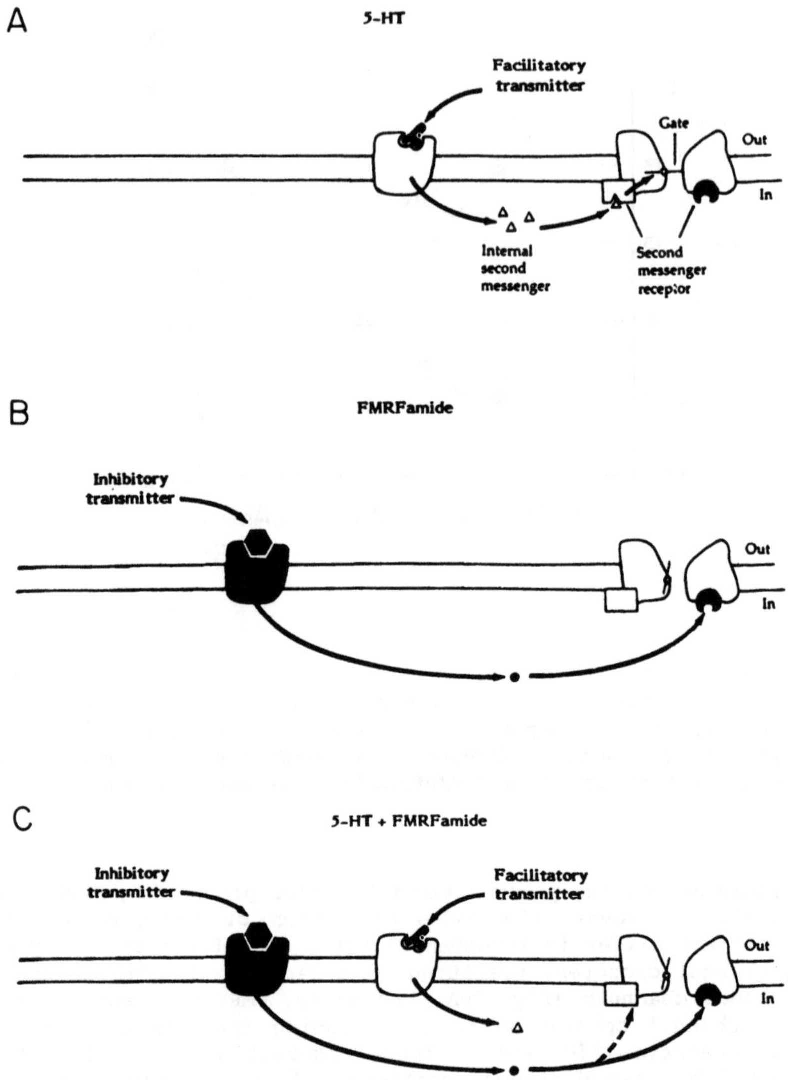

Fig. 20. Summary of modulatory actions on S channel by the presynaptic facilitatory and inhibitory transmitters. **A:** 5-HT, the facilitatory transmitter, leads to S channel closure via cAMP-dependent phosphorylation. **B:** FMRFamide, the inhibitory transmitter, acts via an unidentified second messenger to increase the

probability of an S channel opening. C: FMRFamide is capable of reopening channels closed by 5-HT, suggesting an antagonistic action at either the site on the channel altered with 5-HT or some other step in the phosphorlyation-depolarization reaction.

FMRFamide does not appear to act through cAMP. It does not depress the basal level of cAMP or inhibit the stimulation of cAMP by 5-HT. The transmitters for presynaptic inhibition and presynaptic facilitation therefore seem to act on surface receptors coupled to two independent but interacting second-messenger pathways. This raises the next interesting question: What is the identity of the second-messenger system for presynaptic inhibition produced by FMRFamide?

Independent of the detailed biochemical modulation mechanisms, the experiments on learning indicate that the environment produces substantial effects on behavior by activating different modulatory systems. These modulatory systems act on nerve cells to regulate the cells' excitability and to control synaptic strength. Thus, the action of the environment is represented in the brain by the activity in specific modulatory systems. The experiments we have described indicate that these modulatory systems are, in turn, re-represented in terms of the specific molecular cascades that they activate.

REFERENCES

Abrams, T.W., Castellucci, V.F., Camardo, J.S., Kandel, E.R., and Lloyd, P.E. (1984). Two endogenous neuropeptides modulate the gill and siphon withdrawal reflex in Aplysia by means of presynaptic facilitation involving cyclic AMP-dependent closure of a serotonin-sensitive potassium channel. Proc. Natl. Acad. Sci. USA, 81, 7956-7960.

Baxter, D.A. and Bittner, G.D. (1981). Intracellular recordings from crustacean motor axons during presynaptic inhibition. Brain Res., 223, 422-428.

Belardetti, F., Kandel, E.R., and Siegelbaum, S. Presynaptic inhibition produced by the peptide FMRFamide involves opening of S K^+ channels closed by presynaptic facilitation. Nature, in press.

Brezina, V., Eckert, R., and Erxleben, C. Modulation of potassium conductances by an endogenous neuropeptide in neurones of Aplysia californica. J. Physiol., in press.

Boyle, M.B., Klein, K., Smith, S.J., and Kandel, E.R. (1984). Serotonin increases intracellular Ca^{2+} transients in voltage-clamped sensory neurons of Aplysia californica. Proc. Natl. Acad. Sci. USA, 81, 7642-7646.

Burke, R.E. and Rudomin, P. (1977). Spinal neurons and synapes. In: Handbook of Physiology, Vol. 1: The Nervous System, Part 2. (E. R. Kandel, ed.). American Physiological Society, Bethesda, MD, pp. 877-944.

Castellucci, V. and Kandel, E.R. (1976). Presynaptic facilitation as a mechanism for behavioral sensitization in Aplysia. Science, 194, 1176-1178.

Castellucci, V.F., Kandel, E.R., Schwartz, J.H., Wilson, F.D., Nairn, A.C. and Greengard, P. (1980). Intracellular injection of the catalytic subunit of cyclic AMP-dependent protein kinase simulates facilitation of transmitter release underlying behavioral sensitization in Aplysia. Proc. Natl. Acad. Sci. USA, 77, 7492-7496.

Castellucci, V., Pinsker, H., Kupfermann, I., and Kandel, E.R. (1970). Neuronal mechanisms of habituation and dishabituation of the gill-withdrawal reflex in Aplysia. Science, 167, 1745-1748.

Castellucci, V.F., Schacher, S., Montarolo, P.G., Mackey, S., Glanzman, D.L., Hawkins, R.D., Abrams, T.W., Goelet, P., and Kandel, E.R. (1986). Convergence of small molecule and peptide transmitters on a common molecular cascade. In: Coexistence of Neuronal Messengers: A New Principle in Chemical Transmission, Progress in Brain Research, Vol. 68. (T. Hokfelt, K. Fuxe and P. Pernow, eds.). Elsevier, Amsterdam, pp. 83-102.

Dudel, J. and Kuffler, S.W. (1961a). The quantal nature of transmission and spontaneous miniature potentials at the crayfish neuromuscular junction. J. Physiol. (London), 155, 514-529.

Dudel, J. and Kuffler, S.W. (1961b). Presynaptic inhibition at the crayfish neuromuscular junction. J. Physiol. (London), 155, 543-562.

Eccles, J.C. (1964). The Physiology of Synapses. Springer, Berlin.

Frank, K. and Fuortes, M.G.F. (1957). Presynaptic and postsynaptic inhibition of monosynaptic reflexes. Fed. Proc., 16, 39-40.

Hille, B. (1984). Ionic Channels of Excitable Membranes. Sunderland, Mass.: Sinauer.

Hochner, B., Klein, M., Schacher, S., and Kandel, E.R. (1986a). Action potential duration and the modulation of transmitter release from the sensory neurons of Aplysia in presynaptic facilitation and behavioral sensitization. Proc. Natl. Acad. Sci. USA, 83, 8410-8414.

Hochner, B., Klein, M., Schacher, S., and Kandel, E.R. (1986b). Additional component in the cellular mechanism of presynaptic facilitation contributes to behavioral dishabituation in Aplysia. Proc. Natl. Acad. Sci. USA, 83, 8794-8798.

Hochner, B., Schacher, S., Klein, M., and Kandel, E. R. (1985). Presynaptic facilitation in Aplysia sensory neurons: A process independent of K^+ current modulation becomes important when transmitter release is depressed. Soc. Neurosci. Abstr., 11, 29.

Kandel E.R. and Schwartz, J.H. (1982). Molecular biology of an elementary form of learning: Modulation of transmitter release by cyclic AMP. Science, 218, 433-443.

Kandel, E.R. and Tauc, L. (1965a). Heterosynaptic facilitation in neurones of the abdominal ganglion of Aplysia depilans. J. Physiol. (London), 181, 1-27.

Kandel, E. R. and Tauc, L. (1965b). Mechanisms of heterosynaptic facilitation in the giant cell of the abdominal ganglion of Aplysia depilans. J. Physiol. (London), 181, 28-47.

Kennedy, D., Calabrese, R.L., and Wine, J.J. (1974). Presynaptic inhibition: primary afferent depolarization in crayfish neurons. Science, 186, 451-454,

Klein, M. and Kandel, E.R. (1980). Mechanisms of calcium-current modulation underlying presynaptic facilitation and behavioral sensitization in Aplysia. Proc. Natl. Acad. Sci. USA, 77, 6912-6916.

Klein, M. and Kandel, E.R. (1978). Presynaptic modulation of voltage-dependent Ca^{2+} current: Mechanism for behavioral sensitization in Aplysia californica. Proc. Natl. Acad. Sci. USA, 75, 3512-3516.

Klein, M., Camardo, J.S. and Kandel, E.R. (1982). Serotonin modulates a specific potassium current in the sensory neurons that show presynaptic facilitation in Aplysia. Proc. Natl. Acad. Sci. USA, 79, 5713-5717.

Klein, M., Hochner, B., and Kandel, E.R. (1986). Facilitatory transmitters and cAMP can modulate accommodation as well as transmitter release in Aplysia sensory neurons: Evidence for parallel processing in a single cell. Proc. Natl. Acad. Sci. USA, 83, 7994-7998.

Ocorr, K.A. and Byrne, J.H. (1985). Membrane responses and changes in cAMP levels in Aplysia sensory neurons produced by serotonin, tryptamine, FMRFamide, and small cardioactive peptide B (SCP_B). Neurosci. Letts., 55, 113-118.

Schacher, S. and Proshansky, E. (1983). Neurite regeneration by Aplysia neurons in dissociated cell culture: Modulation by Aplysia hemolymph and the presence of the initial axon segment. J. Neurosci., 3, 2403-2413.

Schmidt, R.F. (1971). Presynaptic inhibition in the vertebrate central nervous system. Ergeb. Physiol. Biol. Chem. Exp. Pharmakol., 63, 20-101.

Shuster, M.J., Camardo, J.S., Siegelbaum, S.A., and Kandel, E.R. (1985). Cyclic AMP-dependent protein kinase closes the serotonin-sensitive K^+ channels of Aplysia sensory neurones in cell-free membrane patches. Nature, 313, 392-395.

Siegelbaum, S.A. and Tsien, R.W. (1983). Modulation of gated ion channels as a mode of transmitter action. Trends Neurosci., 6, 307-313.

Siegelbaum, S.A., Camardo, J.S. and Kandel, E.R. (1982). Serotonin and cyclic AMP close single K^+ channels in Aplysia sensory neurones. Nature 299, 413-417.

Takeuchi, A. and Takeuchi, N. (1966a). A study of the inhibitory action of λ-aminobutyric acid on neuromuscular transmission in the crayfish. J. Physiol. (London), 183, 418-432.

Takeuchi, A. and Takeuchi, N. (1966b). One the permeability of the presynaptic terminal of the crayfish neuromuscular junction during synaptic inhibition and the action of λ-aminobutyric acid. J. Physiol. (London), 183, 433-449.

Takeuchi, A. and Takeuchi, N. (1962). Electrical changes in pre- and postsynaptic axons of the giant synapse of Loligo. J. Gen. Physiol., 45, 1181-1193.

Tauc, L. (1965). Presynaptic inhibition in the abdominal ganglion of Aplysia. J. Physiol. (London), 181, 282-307.

Waziri, R., Kandel, E.R., and Frazier, W.T. (1969). Organization of inhibition in abdominal ganglion of Aplysia. II. Posttetanic potentiation, heterosynaptic depression, and increments in frequency of inhibitory postsynaptic potentials. J. Neurophysiol., 32, 509-519.

Zengel, J.E., Reid, S.A., Sypert, G.W., and Munson, J.B. (1983). Presynaptic inhibition, EPSP amplitude, and motor-unit type in triceps surae motoneurons in the cat. J. Neurophysiol., 49, 922-931.

38
Control of Plasma Membrane Transducing Systems by Second Messengers and by Cellular Metabolism

Howard Rasmussen and Paula Barrett

The focus of this meeting on receptor-receptor interactions is not so much on the direct interaction of one receptor protein molecule with another different protein receptor molecule, but rather on the immediate products of the transducing events mediated by one class of receptors acting as modulators of the functions of a second class of receptors. This is a special case of a much larger phenomenon, that of the interactions between components of one messenger system, e.g., the Ca^{2+} messenger system, with those in another, e.g., the cAMP messenger system (Rasmussen, 1981). It is an intramembrane integrative mechanism. The special feature of the receptor-receptor theme is that the interactions are those which specifically influence other transducing events at the level of the plasma membrane. Even within this category of interactions, receptor-receptor interactions are only one example of this class of phenomena. We will first discuss examples of receptor-receptor interaction in endocrine systems, and then focus attention on two other types of control processes in which a coupling occurs between a transducing event and either a change in ion transport or a change in cellular metabolism. We do so for two reasons: first, to emphasize the fact that signal transducing events other than those directly linked to receptor-transducer interactions are an important feature of plasma membrane function; and second, to emphasize the importance of these membrane-linked events in the regulation of sustained cellular responses.

The three systems which we will use as examples of intramembrane integrative mechanisms are the control of aldosterone secretion from the adrenal glomerulosa cells, the regulation of glycogen metabolism in hepatocytes and of insulin secretion from the beta cell of the islets of Langerhans. The first to illustrate a coupling between the flux of Ca^{2+} through a specific type of channel and the activation of adenylate cyclase, the second to illustrate a coupling of adenylate cyclase activation to alter

Ca^{2+} flux, and the third to illustrate the regulation of a receptor-mediated plasma membrane transducing event by a change in the metabolic state of the cell.

COUPLING OF RECEPTOR-MEDIATED Ca^{2+} FLUX TO ADENYLATE CYCLASE ACTIVATION

Three different agonists will induce a sustained increase in aldosterone production from adrenal glomerulosa cells: the peptide hormones, angiotensin II and ACTH, or an elevation in extracellular K^+.

When ACTH acts, it does so via two separate receptor types: one linked to a stimulation of Ca^{2+} influx, and the other to the activation of adenylate cyclase (Gallo-Payet and Escher, 1985; Kojima et al., 1985a,b; McIlhinney and Schulster, 1975). Activation of the first class leads to a sustained 2-fold increase in Ca^{2+} influx rate and a very transient rise in the intracellular free Ca^{2+} concentration, $[Ca^{2+}]_c$, as measured with the photoprotein, aequorin (Kojima and Ogata, 1986). It is noteworthy that this effect of ACTH is not associated with, nor occurs as a consequence of, an increase in phosphatidylinositol hydrolysis (Kojima et al., 1985a).

During the sustained phase of the ACTH-mediated response, there is a 2-fold increase in rate of Ca^{2+} cycling across the membrane at a time when $[Ca^{2+}]_c$ is back to its basal value, and total cell calcium is not changing (Kojima et al., 1985; Kojima and Ogata, 1986). The ACTH-mediated increase in Ca^{2+} cycling appears to have two effects. First, it plays a role in the activation of adenylate cyclase (Mahaffee and Ontiges, 1980), and second, it somehow acts synergistically with cAMP to mediate cellular response (Kojima et al., 1985a; Sala et al., 1979). Hence, there is direct receptor-receptor interaction in the sense that the immediate consequences of the interaction of the hormone with each of the two receptor types act on a single final transducer, the adenylate cyclase.

COUPLING OF RECEPTOR-MEDIATED Ca^{2+} FLUX TO C-KINASE ACTIVATION

A striking feature of the adrenal glomerulosa cells is that when acted upon by angiotensin II, they also display a sustained 2-fold increase in Ca^{2+} influx rate, but no activation of adenylate cyclase (Albano et al., 1979; Kojima et al., 1984, 1985b,c; Saruta et al., 1972). Rather, the angiotensin II-mediated sustained increase in Ca^{2+} cycling across the plasma membrane regulates the activity of the Ca^{2+}-sensitive, membrane-associated form of protein kinase C which is converted into this form as a consequence of an angiotensin II-induced increase in phosphoinositide hydrolysis and diacylglycerol production (Balla et al., 1984;

Kojima et al., 1984a). In this case, two separate consequences of the activation of a single receptor class act in a cooperative fashion to regulate sustained cellular response. The hormone acting via the hydrolysis of PIP_2 brings about a shift of the C-kinase from its Ca^{2+}-insensitive to its Ca^{2+}-sensitive, membrane-associated form. In this form, the C-kinase is now a transducer in the plasma membrane whose activity is controlled by the rate of Ca^{2+} influx (or Ca^{2+} cycling) across the plasma membrane (Rasmussen and Barrett, 1984; Rasmussen, 1986).

Since a change in the rate of Ca^{2+} cycling across the plasma membrane is a common consequence of the action of each of these peptide hormones, it was of interest to determine whether or not this cycling proceeds via similar or different Ca^{2+} channels, and also to determine whether or not the fates of Ca^{2+} flux through specific types of Ca^{2+} channels may be involved in the regulation, respectively, of the activity of protein kinase C, or adenylate cyclase.

The increased fluxes of Ca^{2+}-mediated by the two hormones differ in their sensitivity to inhibition by nitrendipine: the angiotensin II-mediated flux is blocked by 10^{-6} M nitrendipine, the ACTH-mediated flux is not (Hausdorff et al., 1986; Kojima et al., 1984b, 1985b). Furthermore, when maximal effective concentrations of the two hormones are employed, their effects on Ca^{2+} influx rate are additive (Kojima et al., 1985b). Hence, at this level of analysis, the two hormones appear to regulate Ca^{2+} flux through different types of channels. Nonetheless, the two channels share one common attribute, neither hormone will activate Ca^{2+} influx in cells incubated in a low K^+ (2.0 mM) medium, i.e., the channels seem to be voltage-dependent.

In the case of ACTH action, the consequence of a low external K^+ concentration (2 mM) is an inhibition of ACTH action. This inhibition is due primarily to an inhibition of Ca^{2+} influx which leads secondarily to an inhibition of cyclase activation.

In the case of angiotensin II action, there appears to be a link between the flux of Ca^{2+} through the other type of Ca^{2+} channel and the control of protein kinase C activity. Parenthetically, it is not yet clear how angiotensin II mediates this increase in Ca^{2+} influx. It may occur as a direct consequence of hormone-receptor interaction (as in the case of ACTH) or as a consequence of the generation of a second messenger, possibly inositol 1,3,4,5 tetrakisphosphate (Hansen et al., 1985).

COUPLING OF VOLTAGE-DEPENDENT Ca^{2+} FLUX TO ADENYLATE CYCLASE

Studies of the effect of K^+ on Ca^{2+} flux and adenylate cyclase activation in adrenal glomerulosa cells provide additional evidence that there may be an intimate link between one type of

Ca^{2+} channel and adenylate cyclase activation. The glomerulosa cells are peculiarly sensitive to changes in extracellular K^+ concentration. An increase from 3.5 to 7.0 mM K^+ leads to a 3-4-fold increase in the rate of aldosterone production, and a 3-4-fold increase in Ca^{2+} influx rate (Hausdorff _et al_., 1986; Hyatt _et al_., 1986; Kojima _et al_., 1985d; Saruta _et al_., 1972). However, simply increasing Ca^{2+} influx rate 4-fold with a Ca^{2+} ionophore (A23187) is not a sufficient stimulus to induce a sustained increase in aldosterone production rate (Kojima _et al_., 1984a). These data imply that, in the case of the natural agonist, K^+, an additional messenger or pathway of information flow must supplement that provided by the sustained change in Ca^{2+} influx rate. The additional messenger is cAMP. A rise in $[K^+]$ induces a 3-fold increase in adenylate cyclase activity (Hyatt _et al_., 1986). Furthermore, if the Ca^{2+} ionophore is paired with forskolin (an activator of adenylate cyclase), then a sustained increase in aldosterone production rate is observed (Kojima _et al_., 1985d).

Even though an increase in extracellular $[K^+]$ leads to a 3-fold, Ca^{2+}-dependent increase in adenylate cyclase activity, application of the calcium ionophore (A23187) leads to no increase in cyclase activity although it produces a similar increase in Ca^{2+} influx rate. The paradox to be explained is that K^+ brings about a voltage-dependent, Ca^{2+}-dependent activation of adenylate cyclase in the adrenal glomerulosa cell, but neither angiotensin II, nor A23187 activate cyclase even though they too increase Ca^{2+} influx rate. One might argue that the failure of the angiotensin II-induced increase in Ca^{2+} flux to activate cyclase is due to the fact that there is no angiotensin II receptor linked directly to the cyclase, and that in the case of ACTH one needs inputs from both receptor types in order to activate the cyclase. A possible reason that a change in $[K^+]$, but not a change in angiotensin II concentration, can lead to an activation of adenylate cyclase may simply be the quantitative differences in their effects on Ca^{2+} influx rate: K^+ induces a 4-fold but angiotensin II only a 2-fold increase. This argument, however, is untenable because when A23187 is employed, a 4-fold increase in Ca^{2+} influx rate is also seen, but no increase in adenylate cyclase activity is observed.

The data argue that the flux of Ca^{2+} through a specific class of voltage-dependent and receptor-operated channels is necessary to bring about the activation of adenylate cyclase. Either high K^+ or ACTH can activate this class of channels, increase Ca^{2+} influx, and thereby activate the cyclase, but neither angiotensin II acting on a second class of channels, nor A23187, which induces non-protein ion channels, can activate cyclase even though each can increase Ca^{2+} influx rate across the membrane.

If this model is a correct one, it argues that specific ion channels may be linked directly to particular transducing elements within the membrane, i.e., there is a direct association between these channels and components of the transducing system.

The analysis of the changes which occur in both the Ca^{2+} and cAMP messenger systems when K^+ activates the adrenal glomerulosa cell to secrete aldosterone provides an additional insight into the organization of messenger systems. This insight relates to the nature of plasma membrane transducing events during sustained cellular activation. In the classical second messenger model of cAMP action, there is a single transducing event, activation of adenylate cyclase, which leads via a single pathway of information flow from cell surface to cell interior, the diffusion of cAMP from its site of synthesis into the cell interior (Sutherland _et al_., 1968). Implicit in this model is the notion that following activation of the system, the intracellular [cAMP] rises and remains elevated as long as extracellular messenger is present. However, when K^+ acts on glomerulosa cell, there is no sustained increase in the concentration of either messenger, Ca^{2+} or cAMP (Kojima _et al_., 1985a). Even though there is a sustained 4-fold increase in Ca^{2+} influx rate and of Ca^{2+} cycling across the plasma membrane, the $[Ca^{2+}]_c$ rises only transiently, and then falls so that during the sustained phase of the response, $[Ca^{2+}]_c$ is only slightly higher than its basal value (Apfeldorf and Rasmussen, unpublished). Likewise, even though a sustained, 3-fold increase in cAMP production rate is seen (Hyatt _et al_., 1986), the $[cAMP]_c$ during the sustained phase of cellular response is not measurably different from its basal value (Kojima _et al_., 1985d). These results imply that the information necessary to sustain the response is not the concentrations of specific messengers within the cytosol, but the rates of cycling or turnover of these messengers. This, in turn, implies that the Ca^{2+} cycling and the turnover of cAMP are taking place in a restricted cellular domain, a submembrane domain at or just beneath the plasma membrane. If this is the case, then each of these processes must serve to transduce the respective message into an appropriate form so as to convey this information from cell surface to cell interior.

COUPLING OF RECEPTOR-MEDIATED ACTIVATION OF ADENYLATE CYCLASE WITH RECEPTOR-MEDIATED Ca^{2+} FLUX

Although, as originally developed by Sutherland and coworkers (1968), cAMP was conceived as the second messenger that linked hormone-receptor interaction to the activation of hepatic glycogenolysis, later work has shown that another group of hormones, including vasopressin, angiotensin II, and norepinephrine, activate glycogen breakdown in isolated hepatocytes via the calcium messenger system (Charest _et al_., 1985; Thomas _et al_., 1983; Williamson _et al_., 1981), and do not alter the cAMP content of hepatocytes. Recent work has shown that although activation of hepatocytes either by glucagon alone (an agent which activates adenylate cyclase) or vasopressin alone (an agent which activates the hydrolysis of polyphosphatidylinositides) will lead to a stimulation of glycogen breakdown, the two types of agents act synergistically to stimulate this process (Assimacopoulos _et al_., 1982).

The basis for their synergistic effects seems to lie in a common action: the stimulation of Ca^{2+} influx across the plasma membrane. Each of the three hormones which induces a cellular response by stimulating the hydrolysis of polyphosphatidylinositides also increases the Ca^{2+} influx rate via a common pool of calcium channels (Mauger et al., 1984). The mechanism by which this effect is produced is not completely clear. It may be a direct consequence of hormone-receptor interaction, or may involve inositol 1,3,4-5 tetrakisphosphate as the messenger (Hansen et al., 1986). In contrast, glucagon acts by stimulating adenylate cyclase leading to an increase in cAMP concentration. The rise in cAMP concentration acts at the level of the plasma membrane to enhance Ca^{2+} influx rate (Mauger et al., 1985; Poggioli et al., 1986). The combination of glucagon with either vasopressin or angiotensin II, leads to a more than additive increase in Ca^{2+} influx rate, i.e., the paired hormones have a synergistic effect on plasma membrane Ca^{2+} influx rate. This effect of glucagon, mediated by cAMP, is not due to a cAMP-dependent increase in the turnover of phosphatidylinositides, but is exerted at a step beyond messenger generation. In all likelihood, cAMP brings about a phosphorylation either of the Ca^{2+} channel itself, or a closely-associated regulatory protein (Presti et al., 1985). A similar cAMP-dependent regulation of Ca^{2+} channel activity has been reported in cardiac cells (Brum et al., 1984; Reuter, 1983). The only difference between the two tissues lies in the fact that cAMP influences voltage-dependent Ca^{2+} channels in heart cells, and receptor-operated ones in hepatocytes. In the hepatocytes, inputs from two different classes of hormone receptors, linked to different transducing systems appear to regulate the functions of a common pool of Ca^{2+} channels, another plasma membrane transducing element.

METABOLIC REGULATION OF INTRAMEMBRANE EVENTS IN THE CALCIUM MESSENGER SYSTEM

A variety of agents are known to control insulin secretion from the β cells of the islets of Langerhans (Hedeskov, 1980; Henquin, 1985; Hermanssen, 1984; Malaisse, 1973; Sharp, 1979). Our discussion will focus principally on two of these, glucose and cholecystokinin-8-sulfate (CCK8S). A study of their interrelated effects provides evidence for another type of an intramembrane integrative mechanism (Zawalich et al., 1987).

When islets are perifused with 2.75 mM glucose and then treated with CCK8S, there is no change in the rate of insulin secretion, but if perifused with 7 mM glucose and then treated with CCK8S, a typical biphasic pattern of increased insulin secretion is seen. Likewise, if the glucose concentration is raised from 7 to 10 mM glucose, a similar biphasic insulin secretory response is seen. Yet, CCK8S and glucose do not act by the same mechanisms.

The hormone, CCK8S, induces approximately the same increases in the hydrolysis of phosphatidylinositol 4,5-bisphosphate, the production of inositol 1,4,5-trisphosphate, and the efflux of radiocalcium from prelabeled (with [^{45}Ca]) cells whether the glucose concentration is 2.75 or 7 mM. Hence, the failure of CCK8S to elicit an insulin secretory response from islets incubated in low glucose is a post-receptor and post-transduction event: the same messengers are produced under both circumstances (low and high glucose) but the messages are only read at the higher glucose concentration.

When the glucose concentration is raised from 7 to 10 mM glucose, there are only small changes in PIP_2 hydrolysis, $InsP_3$ production, and Ca^{2+} efflux. These findings imply that the effect of glucose on insulin secretion is not mediated by a glucose-dependent activation of PIP_2 hydrolysis (although this effect may contribute in a minor way to the overall action of glucose). There are two other well established effects of glucose on islet cell function: an increase in plasma membrane Ca^{2+} influx rate, and an increase in the cAMP content of the cells (Hedeskov, 1980; Henquin, 1985; Sharp, 1979).

Drugs which specifically mimic these two actions of glucose are tolbutamide (increase Ca^{2+} influx rate) and forskolin (increase adenylate cyclase activity). Using these two agents, it is possible to examine the question of whether or not either of these actions of glucose plays a role in coupling the CCK8S receptor-mediated events to insulin secretion. When islets perifused with 2.75 mM glucose are exposed to 1 μM forskolin, and then treated with CCK8S, the response of the tissue to CCK8S is altered. This peptide hormone now induces a very large first phase of insulin secretion, but no second phase. If similar islets (perifused with 2.75 mM glucose and 1 μM forskolin) are exposed to both CCK8S and tolbutamide, then an even larger first phase of insulin secretion is seen, but still no second phase is observed. Thus, if one mimics the two well established actions of glucose, one can restore responsiveness of islets to CCK8S. However, this restoration is only partial because no sustained or second phase of insulin secretion occurs. Nonetheless, under any of these circumstances, raising the glucose concentration to 7 mM leads to the reestablishment of a second, sustained phase of insulin secretion.

These results indicate that glucose has an additional effect on islet cell metabolism. We believe that this effect is that of regulating the metabolism of diacylglycerol (DG) by a mechanism other than stimulating PIP_2 hydrolysis. For technical reasons, it is extremely difficult to measure the DG content of islets so final proof of this postulate is not available. On the other hand, Dunlop and Larkins have reported that in cultured neonatal rat islets, glucose has two effects on DG metabolism. First, it increases the de novo synthesis of DG and phosphatidic acid (PA) (Dunlop and Larkins, 1985), and second, it increases the activity of phosphatidate hydrolase, the enzyme which converts PA to DG (PA → DG) (Dunlop et al., 1985). In view of this evidence, and the results of our studies, we postulate that when the medium glucose concentration is low, CCK8S activates PIP_2 hydrolysis leading to the generation of both $InsP_3$ and DG, but no increase in insulin secretion in response to either of these messengers. Raising the cAMP content of the cells allows the Ca^{2+} message (generated by the CCK8S-induced release of intracellular calcium) to induce the first phase of insulin secretion, but does not allow DG to activate C-kinase. Hence, in low glucose, although DG production rate increases in response to CCK8S, the DG content of the membrane does not increase sufficiently to activate C-kinase, and hence, there is no second phase of insulin secretion. When the glucose concentration is raised to 7 mM, then the same CCK8S-induced increase in DG production rate leads to an increase in the DG content of the plasma membrane and thereby to an activation of C-kinase. This activated C-kinase, whose activity is controlled by the glucose-dependent Ca^{2+} cycling across the plasma membrane, mediates the sustained phase of insulin secretion. Glucose by regulating DG metabolism determines whether the receptor-coupled generation of DG is an effective messenger or not.

References

Albano, J.D., Brown, B.L., Ekins, R.P., Tait, S.A.S. and Tait, J.F. (1974). The effect of potassium, 5-hydroxytryptamine, adrenocorticotrophin and angiotensin II on the concentration of adenosine 3':5'-cyclic monophosphate in suspensions of dispersed rat adrenal zona glomerulosa and zona fasiculata cells. Biochem. J., 142, 391-400.

Assimacopoulos-Jeannet, F.D., Blackmore, P.F. and Exton, J.H. (1982). Studies of the interaction between agonists in the control of hepatic glucose output. J. Biol. Chem., 257, 3759-3765.

Balla, T., Enyedi, P., Hunyady, L. and Spät, A. (1984). Effects of lithium on angiotensin II-stimulated phosphatidylinositol turnover and aldosterone production in adrenal glomerulosa cells: a possible causal relationship. FEBS Lett., 171, 179-182.

Brum, G., Osterrieder, W. and Tautwein, W. (1984). β-adrenergic increase in the calcium conductance of cardiac myocytes studied with the patch clamp. Pflügers Arch., 401, 111-118.

Charest, R., Prpic, V., Exton, J.H. and Blackmore, P.F. (1985). Stimulation of inositol trisphosphate formation in hepatocytes by vasopressin, adrenaline and angiotensin II and its relationship to changes in cytosolic free Ca^{2+}. Biochem. J., 227, 79-90, 1985.

Dunlop, M.E. and Larkins R.G. (1985). Pancreatic islets synthesize phospholipids de novo from glucose via acyl-dihydroxyacetone phosphate. Biochem. Biophys. Res. Commun., 132, 467-473.

Dunlop, M.E., Larkins, R.G. and Court, J.M. (1982). Endogenous ionophoretic activity in the neonatal rat pancreatic islets. FEBS Lett., 144, 259-267.

Gallo-Payet, N. and Escher, E. (1985). Adrenocorticotropin receptors in rat adrenal glomerulosa cells. Endocrinology, 117, 38-46.

Hansen, C.A., Mah, S. and Williamson, J.P. (1986). Formation and metabolism of inositol 1,3,4,5 tetrakisphosphate in liver. J. Biol. Chem., 261., 8100-8103.

Hausdorff, W.P., Aguilera, G., and Catt, K. (1986) Selective enhancement of angiotensin II and potassium-stimulated aldosterone secretion by the calcium channel agonist BAY K 8644. Endocrinology 118, 869-974.

Hedeskov, C.J. (1980). Mechanism of glucose-induced insulin secretion. Physiol. Rev., 60, 442-509.

Henquin, J.C. (1985). The interplay between cAMP and ions in the stimulus-secretion coupling in pancreatic β-cells. Arch. Intern. Phys. Biochem., 93, 37-48.

Hermansen, K. (1984). Effects of cholecystokinin (CCK)-4,non-sulfated CCK-9 and sulfated CCK-8 on pancreatic somatostatin, insulin and glucagon secretion in the dog: studies in vitro. Endocrinology, 114, 1770-1775.

Hyatt, P.J., Tait, J.F. and Tait, S.A.S. (1986). The mechanism of the effect of K^+ on the steroidogenesis of rat zona glomerulosa cells of the adrenal cortex: role of cyclic AMP. Proc. Roy. Soc. Lond., B227, 21-42.

Kojima, I., Kojima, K., Kreutter, D. and Rasmussen, H. (1984a). The temporal integration of the aldosterone secretory response to angiotensin occurs via two intracellular pathways. J. Biol Chem., 259, 14448-14457.

Kojima, K., Kojima, I. and Rasmussen, H. (1984b). Dihydropyridine calcium agonist and antagonist effects on aldosterone secretion. Am. J. Physiol., 247 (Endocrinol. Metab. 10), E645-650

Kojima, I., Kojima, K. and Rasmussen, H. (1985a). Role of calcium and cAMP in the action of adrenocorticotropin on aldosterone secretion. J. Biol. Chem., 260., 4248-4256.

Kojima, I., Kojima, K. and Rasmussen, H. (1985b). Characteristic of angiotensin II-, K^+-, and ACTH-induced calcium influx in adrenal glomerulosa cells. J. Biol. Chem., 260, 9171-9176.

Kojima, I., Kojima, K. and Rasmussen, H. (1985c). Role of calcium fluxes in the sustained phase of angiotensin II-mediated aldosterone secretion from adrenal glomerulosa cells. J. Biol. Chem., 260, 9177-9184.

Kojima, I., Kojima, K. and Rasmussen, H. (1985d). Intracellular calcium and adenosine 3',5'-cyclic monophosphate as mediators of potassium-induced aldosterone secretion. Biochem. J., 228, 69-76.

Kojima, I. and Ogata, E. (1986). Direct demonstration of adrenocorticotropin-induced changes in cytoplasmic free calcium with aequorin in adrenal glomerulosa cell. J. Biol. Chem., 261, 9832-9838.

Mahaffee, D.D. and Ontiges, D.A. (1980). The role of calcium in the control of adrenal adenylate cyclase. J. Biol. Chem., 255, 1565-1571.

Malaisse, W.J. (1973). Insulin secretion: multifactorial regulation for a single process of release. Diabetologia, 9, 167-173.
Mauger, J.-P., Poggioli, J. and Claret, M. (1985). Synergistic stimulation of the Ca^{2+} influx in rat hepatocytes by glucagon and the Ca^{2+}-linked hormones vasopressin and angiotensin II. J. Biol. Chem. 260, 11635-11642.
Mauger, J.-P., Poggioli, J., Guesdon, F. and Claret, M. (1984). Noradrenaline, vasopressin and angiotensin increase Ca^{2+} influx by opening a common pool of Ca^{2+} channels in isolated rat liver cells. Biochem. J. 221, 121-127.
McIlhinney, J. and Schulster, D. (1975). Studies on the binding of 1,25-I-labeled corticotrophin to isolated rat adrenocortical cells. J. Endocrinol., 64, 175-184.
Poggioli, J., J.-P. Mauger, and M. Claret. (1986). Effect of cyclic AMP-dependent hormones and Ca^{2+}-mobilizing hormones on the Ca^{2+} influx and polyphosphoinositide metabolism in rat hepatocytes. Biochem. J., 235, 663-669.
Presti, C.F., Jones, L.R. and Lindenmann, J.P. (1985). Isoproterenol-induced phosphorylation of a 15-kilodalton sarcolemmal protein in intact myocardium. J. Biol. Chem., 260, 3860-3867.
Rasmussen, H. (1981) Calcium and cAMP As Synarchic Messengers. John Wiley and Sons, New York.
Rasmussen, H. (1986). The calcium messenger system. N. Engl. J. Med., 314, 1094-1101, 1164-1170.
Rasmussen, H. and Barrett, P.Q. (1984). Calcium messenger system: an integrated view. Physiol. Rev., 64, 938-984.
Reuter, H. (1983). Calcium channel modulation by neurotransmitters, enzymes and drugs. Nature, 301, 569-574.
Sala, G.B., Hayashi, K., Catt, K.J. and Dufau, M.L. (1979). Ad-Adrenocorticotropin action in isolated adrenal cells. J. Biol. Chem. 254, 3861-3865.
Saruta, T., Cook, R. and Kaplan, N.M. (1972). Adrenalcortical steroidogenesis: studies on the mechanism of action of angiotensin and electrolytes. J. Clin. Invest., 51, 2239-2245.
Sharp, G.S. (1979). The adenylate cyclase-cyclic AMP system in islets of Langerhans and its role in the control of insulin secretion. Diabetologia, 16, 287-296.
Sutherland, E.W., Robison, G.S. and Butcher, R.S. (1968). Some aspects of the biological role of adenosine 3':5'-monophosphate (cyclic AMP). Circulation, 37, 279-306.
Thomas, A.P., Marks, J.S., Coll, K.E. and Williamson, J.R. (1983). Quantitation and early kinetics of inositol lipid changes induced by vasopressin in isolated and cultured hepatocytes. J. Biol. Chem., 258, 5716-5725.
Williamson, J.R., Cooper, R.H. and Hoek, J.R. (1981). Role of calcium in the hormonal regulation of liver metabolism. Biochim. Biophys. Acta 639, 243-295.
Zawalich, W.S., Takuwa, N., Takuwa, Y., Diaz, V.A. and Rasmussen, H. (1987) Interaction of cholecystokinin and glucose in rat pancreatic islets. Diabetes. In press.

Supported by grants (AM19813, AM34381, AM33001) from the National Institutes of Health and the Diabetes Research and Education Foundation.

Session V

Physiological and Pharmacological Studies on Interactions between Receptors in Synaptic Membranes

Chairman: G.R. Siggins

Part A
Neurophysiological Studies

39

The Relationships of Receptors for Phencyclidine and Sigma Opiates in Rat Cerebellum: An Electrophysiological Analysis

Barry Hoffer, Michael Palmer, Elizabeth Moore, Martha Kim and Yun Wang

INTRODUCTION

Two groups of opioid compounds seem to share many pharmacological properties with PCP, even though their structures are markedly different: benzomorphans such as cyclazocine and N-allylnormetazocine (SKF-10,047; Zukin et al., 1984; Teal and Holtzman, 1979; Mendelsohn, et al., 1985) and dioxolanes such as dexoxadrol (Cone et al., 1985; Shannon, 1983). This finding has led some to the postulate that the "sigma receptor", one of the multiple opiate receptors reported in the CNS, is identical to the PCP receptor. Supporting this conjective, both cyclazocine and SKF-10,047, which have been shown to have high affinity for sigma opioid receptors (Neil 1985), displace [^{3}H]-PCP binding in rat brain homogenates (Zukin et al., 1984; Mendelsohn et al., 1985). Although dexoxadrol has no analgesic actions in mice, it produces PCP-like catalepsy in the pigeon (Zimmerman et al. 1983) and also displaces [^{3}H]-PCP binding in rat brain homogenates (Zukin, 1982; Hampton et al., 1982). However, other types of receptor-mediated interactions could also account for the behavioral parallelisms of these agents. Thus, PCP could augment release of sigma opiates or visa versa. Alternatively, PCP and sigma opiates could interact with separate receptors which possess synergistic effector mechanisms postsynaptically.

Previous experiments from our laboratory have provided evidence about the electrophysiological effects of PCP and sigma opiates on central neurons. Initial studies were carried out in rat cerebellum because (1) a significant population of PCP receptors are localized in this region (Itzhek et al., 1985; Largent et al., 1984; Zukin and Zukin, 1983; Sircar and Zukin, 1985), and (2) the various subpopulations of cerebellar neurons can be readily identified during acute recording (Eccles et al., 1967). Local applications of PCP by pressure ejection have been reported to slow the discharge of cerebellar Purkinje neurons

(Marwaha et al, 1980a,b; Palmer et al., 1983). We have subsequently shown that metaphit, a PCP antagonist which has been reported to irreversibly acylate PCP receptors (Rafferty et al., 1985; Wang et al., 1986a), completely and irreversibly antagonizes the electrophysiological actions of PCP in cerebellum without affecting the depressions caused by locally applied NE or GABA in that brain area (Wang et al., 1986a). Interestingly, (+)-cyclazocine-induced depressions are also blocked by metaphit, suggesting that the depressant actions of this benzomorphan on cerebellar Purkinje neurons are primarily mediated through PCP receptors (Wang et al., 1986b).

In the present study, we investigated the stereoselectivity and the pharmacological specificity of metaphit interactions to shed light on the possible identity of PCP and sigma opiate receptors. The metaphit sensitivity of dexoxadrol and levoxadrol was compared to the previously reported stereoselectivity of these two latter compounds for PCP-like effects in biochemical and behavioral assays (Cone, et al., 1985; Shannon, 1983). Metaphit actions were also compared with kappa-blocking doses of naloxone (Kosterlitz et al., 1981), a classical opiate antagonist. All drugs were administered locally from multibarrel micropipettes by pressure microejection while recording from identified cerebellar Purkinje neurons in urethane-anesthetized rats using previously described methods (Wang et al., 1986a).

RESULTS:

Dexoxadrol and levoxadrol were equipotent for causing reversible depressions of cerebellar Purkinje neuron discharge when locally applied by pressure ejection. Purkinje neuron firing rates were inhibited 50% by a mean pressure dose of 35.6 $\pm$ 7.0 (S.E.M) PSI X SEC of dexoxadrol while 21.3 $\pm$ 4.4 PSI X SEC levoxadrol was required to cause similar effects on the same neurons (n=27). Metaphit, which acylates PCP receptors and blocks PCP induced electrophysiological responses in cerebellum (Wang et al., 1985, 1986a), also blocked dexoxadrol-induced depressions in 9 of 10 neurons studied (Figures 1,2; Table 1). In the same 10 neurons, the responses of levoxadrol, when applied alternately with dexoxadrol, were not affected by the metaphit administration (Table 1; Figures 1,2). Similar to that previously reported for PCP (Wang et al., 1986a), blockade of dexoxadrol actions was only achieved after sufficient metaphit was given to transiently abolish Purkinje neuron discharge.

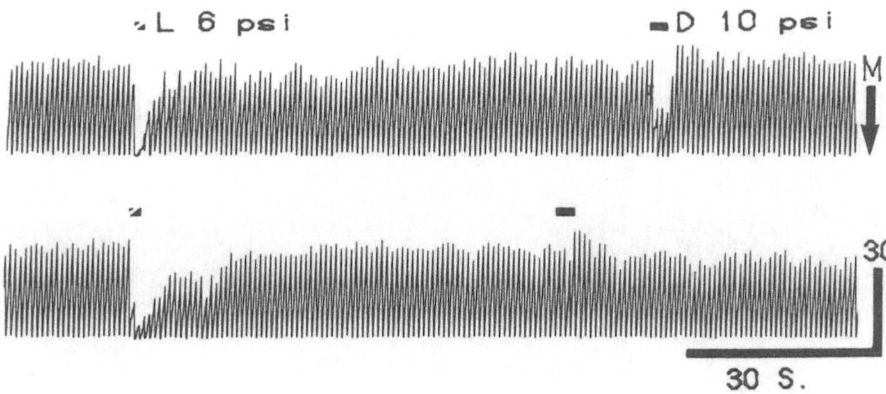

Figure 1. Lack of antagonism of levoxadrol (L) by locally applied metaphit (M) is shown at the doses that block dexoxadrol (D)-induced depression in a single Purkinje neuron. Upper trace represents a ratemeter record of control responses to levoxadrol and dexoxadrol. Lower trace shows responses to levoxadrol and dexoxadrol after metaphit administration. The arrow indicates local application of 0.2% metaphit at 5-10 psi for 90 seconds. After metaphit, levoxadrol-induced depression was unaffected whereas the response to dexoxadral was blocked. In this and all succeeding ratemeter records, the duration of drug ejection is indicated by the bar overlying the trace and number after the drugs shows ejection pressure in pounds per square inch (psi).

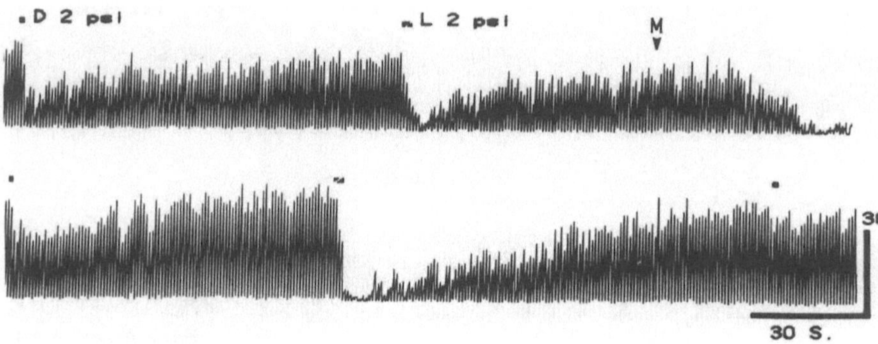

Figure 2. Progressive diminution of dexoxadrol responses occurs after local metaphit application. Upper trace illustrates the control responses to dexoxadrol and levoxadrol. The onset of local application of metaphit is indicated by the M(4-30 psi X 90 sec). Note that metaphit abolished spontaneous Purkinje neuron activity. Ten minutes after metaphit application, the dexoxadrol induced depression was gradually diminished although the levoxadrol response was still intact (lower trace).

Once spontaneous firing had recovered, 5 to 20 minutes after cessation of metaphit application, dexoxadrol-elicited depressions were gradually and irreversibly antagonized (Figure 2).

TABLE 1.

Antagonism of Levoxadrol- or Dexoxadrol-Induced Depressions by Locally Applied Metaphit in Cerebellar Purkinje Neurons.

Antagonism	Levoxadrol (0.1 mM)	Dexoxadrol (0.1 mM)
Strong (>90%)	*1	9
Weak	0	0
None	9	1

* Number of Neurons

Although it is known from biochemical studies that dexoxadrol has 100-300 times higher potency than levoxadrol for displacement of PCP (Zukin, 1982; Hampton et al., 1982), we found that these two dioxolanes were equipotent for producing half-maximal inhibitions of Purkinje neuron activity. The electrophysiological equipotency observed in this study, together with the observation that levoxadrol-induced responses were insensitive to metaphit, might indicate that levoxadrol causes depressions of cerebellar Purkinje neurons through a mechanism different from that of PCP. To further test this possiblilty, local applications of high doses of naloxone were used. Levoxadrol-induced depressions were partially or totally antagonized, in a reversible fashion, by 5-10 minutes of local application of naloxone in 9 of 10 neurons studied (Figure 3; Table 2). In contrast, dexoxadrol effects were not altered by the same naloxone treatments (Figure 3; Table 2).

TABLE 2.

Antagonism of Levoxadrol- or Dexoxadrol-Induced Depressions by Locally Applied Naloxone in Cerebellar Purkinje Neurons.

Antagonism	Levoxadrol (0.1 mM)	Dexoxadrol (0.1 mM)
Strong (>90%)	*5	0
Weak	4	2
None	1	8

* Number of Neurons

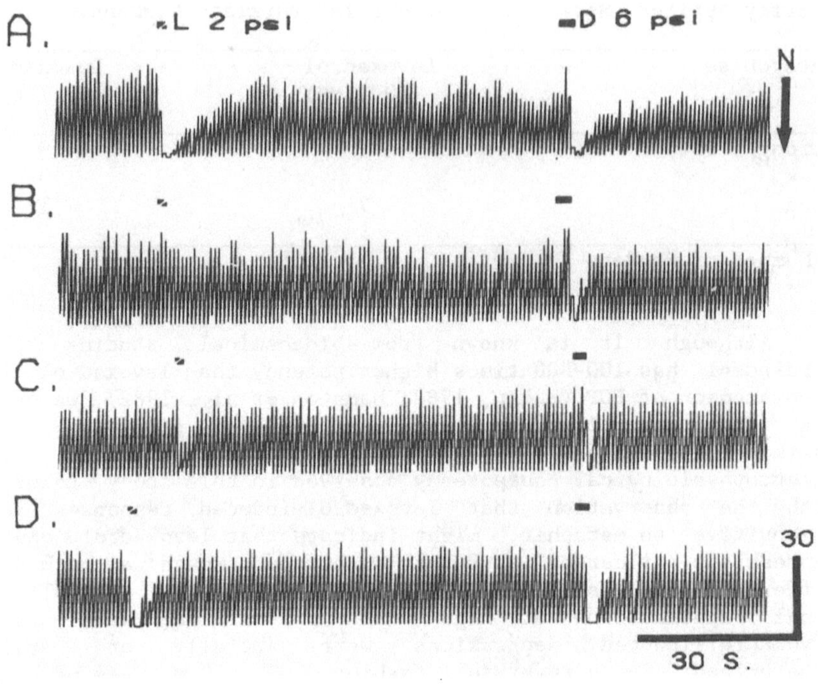

Figure 3. Naloxone reversibly antagonizes levoxadrol but not dexoxadrol responses on a single Purkinje neuron. (A) shows the depressant actions of dexoxadrol and levoxadrol under control condition. After the local naloxone application (N-vertical arrow, 3 psi X 5 min), levoxadrol responses were blocked (B). This blockade was reversible as shown in traces C and D. The levoxadrol response began to recover by 10-20 minutes after the cessation of the naloxone application (C).

DISCUSSION:

It has been previously shown, in general, that mu and kappa opioid receptor mechanisms are stereoselective for (-)-isomer of the various ligands (Zukin and Zukin, 1981; Zukin et al., 1984; Martin, 1984; Hampton, 1982). PCP and sigma actions, on the other hand, are more selective for (+)-isomers. In rats, dexoxadrol produces PCP-like discriminative effects and miosis, while its (-)-isomer, levoxadrol, does not manifest such actions up to toxic doses (Shannon, 1983). Biochemical studies have also shown that dexoxadrol is around two orders of magnitude more potent for displacing 3H-PCP binding than is levoxadrol in rat brain homogenates (Zukin, 1982; Hampton, et al., 1982).

In this study, we found that local applications of both dexoxadrol and levoxadrol caused potent inhibitions of the spontaneous firing of cerebellar Purkinje neurons. Only the response to dexoxadrol, however, was blocked by local application of metaphit. Previously studies from our laboratory showed that metaphit also irreversibly blocked responses to PCP and (+)-cyclazocine, while (-)cyclazocine-mediated depressions of cerebellar Purkinje neurons were unaffected by this treatment alone (Wang et al., 1985, 1986a,b). Furthermore, metaphit acylated 100% of the PCP receptors in cerebellum (Wang et al., 1985), but mu, muscarinic, and benzodiazopine receptors were not similarly affected (Rafferty, et al., 1985). Taken together, these data suggest that metaphit yields irreversible antagonism of only PCP receptor-mediated mechanisms in the cerebellum.

Although local applications of levoxadrol were equipotent with dexoxadrol for inducing depressions of Purkinje neurons, the levoxadrol effects were not sensitive to metaphit. Furthermore, we found that responses to levoxadrol were partially or totally antagonized by high doses of naloxone. However, there are very few mu and delta opioid receptors in cerebellum (Kuhar et al., 1973). Moreover, electrophysiological (Wang et al., 1986b; Kim et al., 1985), biochemical (Neil, 1985), and behavioral (Kosterlitz et al., 1981) studies have shown that high doses of naloxone are required to block kappa opioid responses. Thus, the naloxone antagonism of the levoxadrol-induced depressions of Purkinje neurons reported here may be mediated through kappa opioid mechanisms. Interestingly, the depressant responses of Purkinje neurons to (-)-cyclazocine were blocked only by the combination of metaphit and high doses of naloxone, suggesting that this benzomorphan has mixed PCP-kappa agonist properties (Wang et al., 1986b). Similar conclusions have been reached in biochemical (Zukin and Zukin, 1981) and behavioral (Teal and Holtzman, 1979) studies.

Our suggestion that levoxadrol has no obvious "PCP-receptor component" in its response while (-)-cyclazocine was found to possess both kappa and PCP-receptor properties (Wang et al., 1985c) might be the result of a stereochemical difference between the PCP receptor affinities for these compounds. Thus, while there are strict steric requirements for the binding of dioxolane isomers to the PCP receptor, there can be up to a 20 degree angular difference between stereoisomers of monoaromatic agonists, such as cyclazocine, which are acceptable ligands for the binding site (Cone, 1985). This may explain our electrophysiological findings; both (+)- and (-)- cyclazocine have PCP-like actions at concentrations that can be reached by local application. However, the more rigid steric requirements for dioxolanes allow only the (+)-isomer to have effects.

Biochemical studies also suggest a congruence of PCP and sigma opiate receptors. By using autoradiography, the anatomical localization of PCP receptors in rat brain was found to be markedly similar to that of (+)-SKF-10,047 (Sircar and Zukin 1985). Furthermore, in vitro radioligand studies show that PCP binding is displaced by sigma agonists (SKF-10,047, (+)cyclazocine, and dexoxadrol). Sigma agonist binding is also antagonized by PCP (Zukin, 1982; Mendelsohn et al.,1985; Sethy and McCall, 1984; Johnson and Snell, 1985).

In conclusion, our data show that the irreversible antagonism of the depressant effects of PCP, dexoxadrol, and benzomorphans in cerebellum may be specifically mediated through a metaphit-induced blockade of PCP receptors in that brain area. Thus, our data lend physiological evidence to the concept that the actions of sigma agonists and PCP agonists are mediated through a common mechanism, rather than via an interaction of separate receptors.

REFERENCES

Cone, E.J., McQuinn, R.L., and Shannon, H.E. (1985). Structure-Activity relationship studies of phencyclidine derivatives in rats. J. Pharmacol. Exp. Ther., 228, 147-153.

Eccles, J.C., Ito, M., and Szentagothai, J. (1967). The cerebellum as neuronal machine, New York, Springer Verlag.

Hampton, R.Y., Medzihradsky, F., Woods, J.H., and Dahlstrom, P.J. (1982). Stereospecific binding of ^{3}H-phencyclidine in brain membranes. Life Sci., 30, 2147-2154.

Itzhak, Y., Hiller, J.M., and Simon, E. (1985). Characterization of specific binding sites for [^{3}H](d)-N-Allylnormetazocine in rat brain membranes. Molecular Pharmacol., 27, 46-52.

Johnson, K.M., and Snell, L.D. (1985). Effects of phencyclidine (PCP)-like drugs on turning behavior, ^{3}H-dopamine uptake, and ^{3}H-PCP binding. Pharmacol. Bichem. and Behavior., 22, 731-735.

Kim, M., Pang, K., Freedman, R., and Palmer, M.R. (1985). Electrophysiological effects of cyclazocine on rat cerebellar Purkinje neurons: Comparison with phencyclidine. Alcohol and Drug Res., 6: 23-36.

Kosterlitz, H.W., Paterson, S.J., and Robson, L.E. (1981). Characterization of the k-subtype of the opiate receptors in guinea-pig brain. Br. J. pharmacol., 73, 939-949.

Kuhar, M.J., Pert, C.B., and Snyder, S.H. (1973). Regional distribution of opiate binding in monkey and human brain. Nature 245, 447-451.

Largent, B.L., Gundlach, A.L., and Snyder, S.H. (1984). Psychotomimetic opiate receptors labeled and visualized with (+)-[H3]-3-(3-hydroxyphenyl)-N-(1-propyl) piperidine. Proc. Natl. Acad. Sci., 81, 4983-4987.

Martin, B.R., Katzen, J.S., Woods, J.A., Tripathi, H.L., Harris, L.S., and May, E.L. (1984). Stereoisomers of [H3]-N-Allylnormetazocine bind to different sites in mouse brain. J. Pharmacol. Exp. Ther., 231, 539-544.

Marwaha, J., Palmer, M.R., Hoffer, B., and Freedman, R. (1980a). Phencyclidine-induced depressions of cerebellar Purkinje neurons. Life Sci., 26, 1509-1515.

Marwaha, J., Palmer, M.R., Woodward, D.J., Hoffer, B.J., and Freedman, R. (1980b). Electrophysiological evidence for presynaptic actions of phencyclidine on noradrenergic terminals in rat cerebellum. J. Pharmacol. Exp. Ther., 215, 606-613.

Mendelsohn, L.G., Kalra, V., Johnson, B.G., and Kerchner, G.A. (1985). Sigma opioid receptor:characterization and co-identity with the phencyclidine receptor. J. Pharmacol. Exp. Ther., 233, 597-602.

Neil, A. (1985). Detection and characterization of four binding sites for opioids in mouse brain. Acta Pharmacol. et Toxicol., 56, 108-116.

Palmer, M.R., Bickford, P.C., Hoffer, B.J., and Freedman, R. (1983). Electrophysiological evidence for presynaptic actions of phencyclidine on noradrenergic transmission in rat cerebellum and hippocampus. In Phencyclidine and Related Arylcyclohexylamines: Present and Future Applications, ed. by J.M. Damenka, E.F. Domino, and P. Geneste, pp 443-467, NPP press, Ann Arbor.

Rafferty, M.F., Mattson, M., Jacobson, A.E., and Rice, K.C. (1985). A specific acylating agent for the [^{3}H]phencyclidine receptors in rat brain. FEBS Letters, 181, 318-322.

Sethy, V.H., and McCall, J.M. (1984). High-affinity (^{3}H)-dexoxadrol binding to rat brain membranes. Drug Develop. Res., 4, 635-645.

Shannon, H.E. (1983). Discriminative stimulus effects of phencyclidine: Structure-activity relationships In: Phencyclidine and Related Arycyclohexylamines: Present and Future Application, J.M. Kamenka, E.F. Domino and P. Geneste, eds., Ann. Arbor. NPP press pp 311-335.

Sircar, R., Zukin, S.R. (1985b). Quantitative localization of [^{3}H]TCP binding in rat brain by light microscopy autoradiography. Brain Res. 344, 142-145.

Teal, J.J., and Holtzman, S.G. (1980). Discriminative stimulus effects of cyclazocine in rat. J. Pharmacol. Exp. Ther., 212, 368-379.

Wang, Y., Palmer, M.R., Freedman, R., Hoffer, B.J., Mattson, M., Lessor, R.A., Rafferty, M.F., Rice, K.C., and Jacobson, A.E. (1986a). Antagonism of phencyclidine action by metaphit in rat cerebellar Purkinje neurons: An electrophysiological study. Proc. Nat. Acad. Sci. U.S.A., 83, 2724-2727.

Wang, Y., Palmer, M.R., Freedman, R., Mattson, M., Lessor, R., Rafferty, M., Rice, K., Jacobson, A., Hoffer, B.J. (1985). Electrophysiological and biochemical study of the antagonism of PCP action by metaphit in rat cerebellar Purkinje neurons. In: Abstracts Part II, Society for Neuroscience, 15th annual meeting, Dallas, Texas, p. 910.

Wang, Y., Palmer, M.R., Freedman, R., Rice, K.C., Lessor, R.A., Jacobson, A.E., Hoffer, B.J. (1986b). Electrophysiological interactions of isomers of cyclazocine with the phencyclidine antagonist metaphit in rat cerebellar Purkinje neurons. J. Neurosci., (In press).

Zimmerman, D.M., Woods, J.H., Hynes, M.D., Cantrell, B.E., Reamer, M., and Leander, J.D. (1983). Discovery and characterization of the phencyclidine-like actions of a new series of bens(f)isoquinoline derivatives. In Phencyclidine and Related Arylcyclohexylamines: Present and Future Applications, ed. by J.M. Damenka, E.F. Domino, and P. Geneste, pp 59-69, NPP press, Ann Arbor.

Zukin, R.S., and Zukin, S.R. (1981). Demonstration of [H3] cyclazocine binding to multiple opiate receptor sites. Mol. Pharmacol., 20, 246-254.

Zukin R.S., and Zukin, S.R. (1983). A common receptor for phencyclidine and the sigma receptor. In: Phencyclidine and Related Arylcyclohexylamines: Present and Future Application, J.M. Kamenka, E.F. Domino and P. Geneste, eds., Ann. Arbor. NPP Press, pp 107-124.

Zukin, S.R. (1982). Differing stereospecificities distinguish opiate receptor subtypes. Life Sci., 31, 1307-1310.

Zukin, S.R., Brady, K.T., Slifer, B.L., and Balster, R.L. (1984). Behavioral and biochemical stereoselectivity of sigma opiate/PCP receptors. Brain Res., 294, 174-177.

40
Modulation of Synaptic Potentials *in situ* by Specific Agonists of Excitatory Amino Acids

Paul L. Herrling

INTRODUCTION

The discovery and synthesis of specific agonists and antagonists of excitatory amino acid receptors (Watkins and Evans, 1981) has sparked a renewed interest in the neurobiology of glutamate, aspartate and other endogenous excitatory amino acids.

Work in this laboratory has focussed on the interaction of such agents with the membrane potential of cat caudate neurons in situ in order to assess their physiology and pharmacology. The caudate nucleus was chosen because it receives a powerful excitatory projection from the frontal cortex which is thought to use glutamate or aspartate as transmitter (for a recent rewiew, see Fagg and Foster, 1983).

In the present article, evidence will be presented indicating that at least two distinct excitatory amino acid systems coexist on cat caudate neurons. Furthermore, hypotheses relating to a possible functional significance for the two systems will be discussed.

METHODS

Intracellular recordings from caudate neurons in halothane anaesthetized cats were performed during simultaneous microiontophoretic application of drugs and stimulation of the cortico-caudate pathway at varying frequencies. A detailed description of materials and methods can be found in Herrling et al. (1983) and Do et al. (1986).

RESULTS

Excitation patterns- and pharmacology. Specific agonists of excitatory amino acids elicited two distinct types of excitation in cat caudate neurons (Herrling, 1983; Herrling et al. 1983;): i) agonists such as quisqualate (QUIS) and kainate (KA) elicited depolarizations that could have fast onsets and which were accompanied by regularly spaced action potentials. This pattern is termed type 1 (Fig. 1a); ii) N-methyl-D-aspartate (NMDA) and quinolinate (QUIN), however, induced large depolarizations with abrupt onset and end which were accompanied by bursts of action potentials. This effect is termed type 2 pattern (Fig. 1b). Other excitatory amino acids such as L-aspartate (L-ASP) and L-cysteine sulfinic acid (L-CSA) caused in some cells a mixture of these two patterns, i.e. a bursty type 2 pattern at the beginning of iontophoretic applications that converted to the more regular type 1 pattern when the application was prolonged (Fig. 1c).

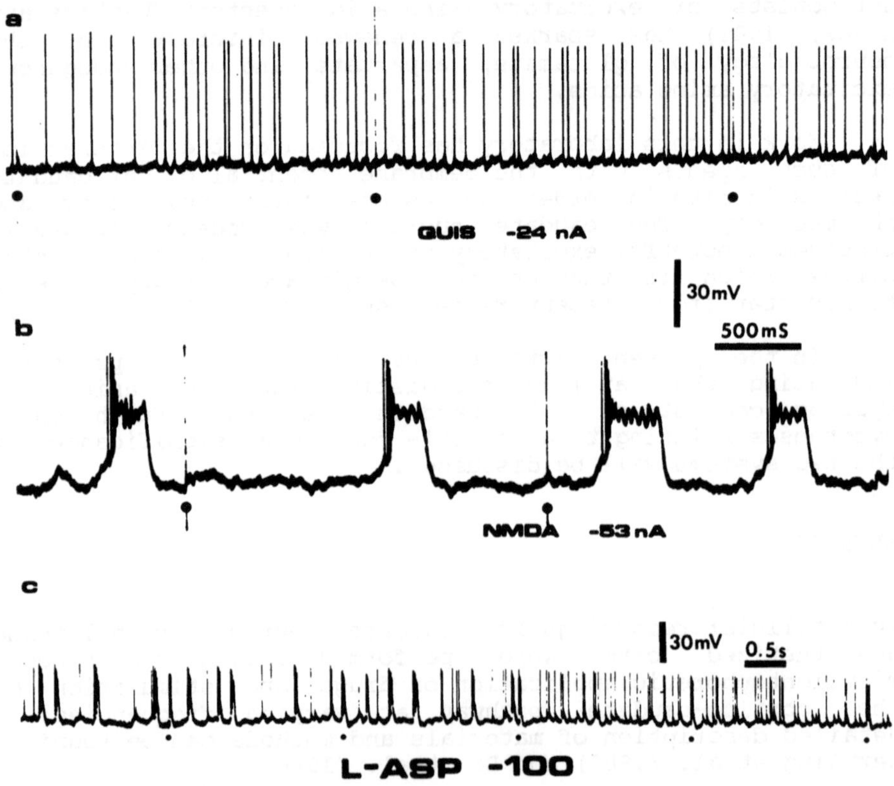

Figure 1. Excitation patterns induced by excitatory amino acids on the membrane potential of caudate neurons. a) type 1 excitation pattern elicited by QUIS; b) type 2 excitation pattern elicited by NMDA; c) mixed pattern elicited by ASP.

In this and the following pictures of intracellular recordings dots indicate time of cortical stimulation; membrane potential is on the ordinate, positive upwards; iontophoretic currents in nA are indicated by the number following the compound symbol.

Type 2 effects could be inhibited selectively by specific NMDA-antagonists such as 2-amino-7-phosphonoheptanoate (AP-7; Perkins et al., 1982) as shown in Fig. 2 a-c. Both type 1 and 2 effects could be inhibited by broad-spectrum excitatory amino acid antagonists such as kynurenate (KYAC; Perkins and Stone, 1982; Herrling, 1985; Fig. 2 d-f).

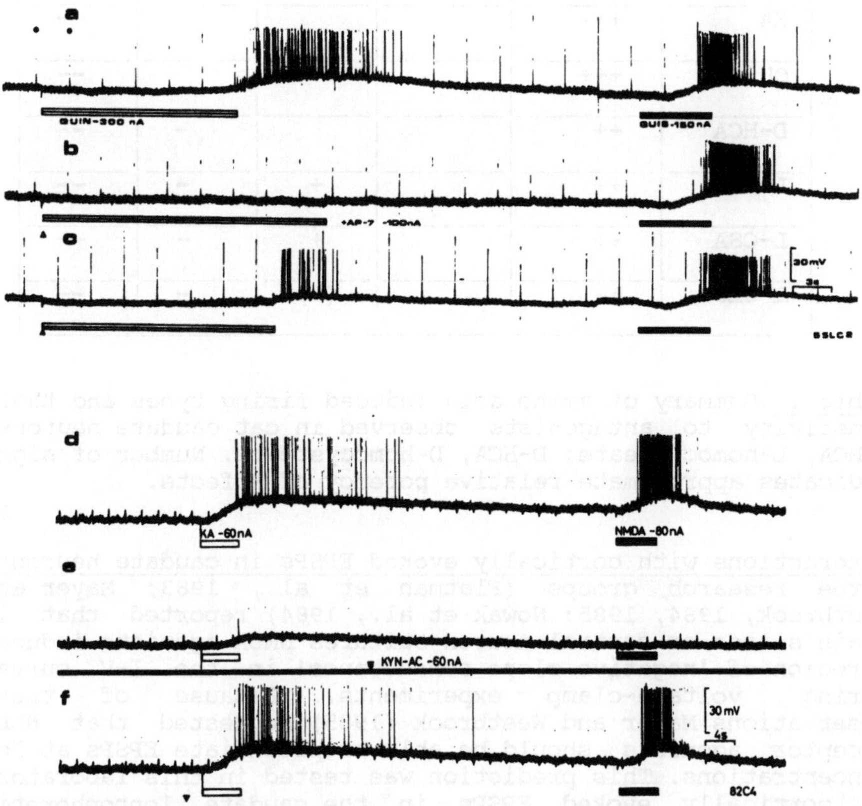

Figure 2. Inhibition of type 1 and 2 effects by different antagonists of excitatory amino acids. a) Control applications of QUIN and QUIS. b) Iontophoretically applied AP-7 abolishes QUIN-induced excitations but not the effects of QUIS. c) Partial recovery of QUIN effects after the end of AP-7 application. d) Control application of KA and NMDA. e) Kynurenic acid (KYN-AC) inhibits excitations induced by both agents. f) recovery.

These pharmacological results are summarized in Table 1.

AGONIST	EXCITATION PATTERN			INHIBITED BY	
	Type 1	Type 2	Mixed	AP-7	KYAC
NMDA		+++		---	--
QUIN		+++		---	--
L-HCA		++		---	--
KA	+++				--
QUIS	+++				--
D-HCA	++			-	--
L-ASP	++		+	-	--
L-CSA	++		+	-	--
L-GLU	++		(+)	-	--

Table 1. Summary of amino acid induced firing types and their senstivity to antagonists observed in cat caudate neurons. L-HCA, L-homocysteate; D-HCA, D-homocysteate. Number of signs indicates approximate relative potency of effects.

Interactions with cortically evoked EPSPs in caudate neurons. Three research groups (Flatman et al., 1983; Mayer and Westbrook, 1984, 1985; Nowak et al., 1984) reported that in brain slices or central neuron cultures NMDA agonists induced a region of 'negative slope conductance' in the I-V curves during voltage-clamp experiments. Because of these observations Mayer and Westbrook (1985) suggested that NMDA receptor agonists should be able to potentiate EPSPs at low concentrations. This prediction was tested in this laboratory on cortically evoked EPSPs in the caudate. Iontophoretic application of NMDA and L-HCA at currents that caused a membrane depolarization of 10 to 15mV resulted in a clear potentiation of cortically evoked EPSPs (Do et al., 1986). Such potentiations are shown in Fig. 3.

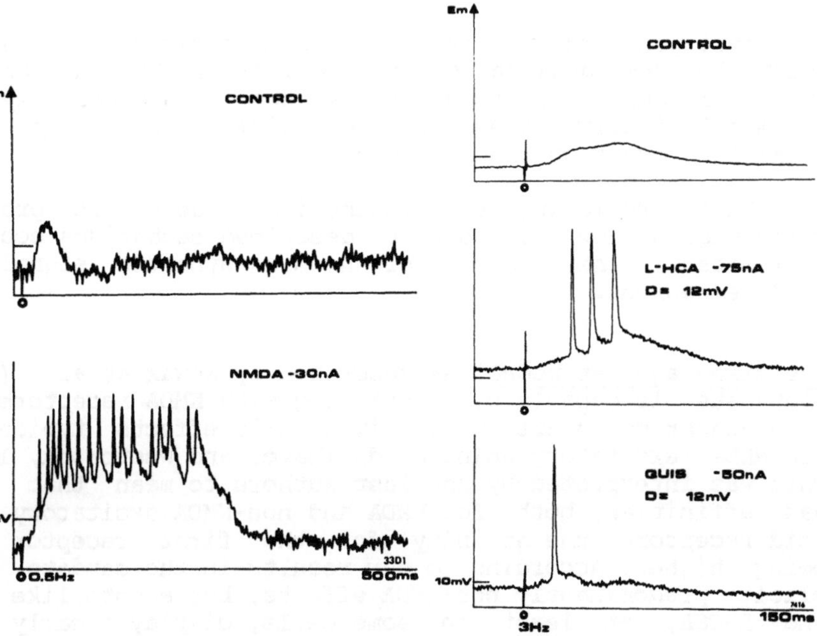

Figure 3. Potentiation of cortically evoked EPSPs elicited by NMDA and L-HCA in cat caudate neurons. Left panel, upper trace: as is observed in most cases in caudate neurons (Buchwald et al., 1973; Herrling et al., 1983; Herrling, 1985) stimulation of the precruciate gyrus in cats results in distinct EPSPs that however rarely reach threshold for generation of action potentials in the absence of drugs. Left panel, lower trace: in the presence of iontophoretically applied NMDA, stimulation of the cortico-caudate pathway at exactly the same parameters as in the upper trace results in the same neuron in a stimulus coupled abrupt depolarization with a burst of action potentials. Such stimulus coupled bursts are never seen in presence of non-NMDA excitatory amino acid receptor agonists. Right panel, upper trace: in another neuron cortical stimulation again results in a subthreshold EPSP. Right panel, middle trace: the same stimulus in presence of L-HCA results in an EPSP with three action potentials; the membrane was depolarized by 12mV respective to control. Right panel, lower trace: if the same neuron is depolarized to the same degree by a non-NMDA excitatory amino acid receptor agonist like QUIS, the relative EPSP-amplitude is much smaller than during L-HCA and only one AP is elicited.

DISCUSSION

The above presented results indicate that cat caudate neurons react to the application of agonists of NMDA and non-NMDA excitatory amino acid receptors by two physiologically and pharmacologically distinct mechanisms and that both mechanisms occur on the same neurons.

In the following, two hypothetical models are proposed which suggest how the two above described mechanisms could be implicated in the treatment of incoming synaptic signals in caudate neurons.

The mixed agonist model. As described by Nowak et al. (1984) glutamate is capable of interacting with NMDA receptors, but when higher doses are applied it elicits effects typical of non-NMDA excitatory amino acids (Mayer and Westbrook, 1984). This was interpreted by the last authors to mean that L-GLU has affinities both for NMDA and non-NMDA excitatory amino acid receptors, the affinity for the first receptor type being higher. According to our results in the caudate, L-GLU elicits predominantly non-NMDA effects, but agents like L-ASP and L-CSA, at least in some cells, display clearly mixed effects. The NMDA-like effects occur at lower concentrations, i.e. at the beginning of the iontophoretic application, and non-NMDA effects occur at higher concentrations, i.e. following prolonged iontophoresis.

In the model depicted in Fig. 4 a synapse that is weakly active, i.e. releasing low amounts of a mixed agonist like L-ASP, would activate mainly the few NMDA receptors present in the postsynaptic membrane because of the higher affinity of the agonist for NMDA receptors. This activation of NMDA receptors would result in a current voltage relationship (I-V curve) with a negative slope (Fig. 4, lower panel, curve labelled 'GLU/ASP low dose') as described by Flatman et al. (1983), Mayer and Westbrook (1984,1985) and Nowak et al. (1984). In this state the postsynaptic neuron would be sensitive to very small variations of currents flowing through its membrane. In particular, it would mean that small inward currents would elicit relatively large depolarizations. When the synapse becomes more active, more agonist is released into the synaptic cleft and consequently the more numerous QUIS receptors are stimulated. Because they are more numerous, their effect swamps the effect of the NMDA receptors and the I-V curve becomes linear. In this state the membrane is less sensitive to fluctuations of currents than it was in the 'negative slope' state. This means that the same inward currents will elicit smaller depolarizations than during the state where only NMDA receptors were activated.

In this model the physiological role of the two excitatory amino acid systems would be to make the synapse very sensitive to inward currents when the presynaptic neuron is weakly active and less sensitive when the presynaptic activity augments. The agonist concentration at which the sensitivity decreases would be determined by the proportion of NMDA to QUIS receptors present in the postsynaptic membrane.

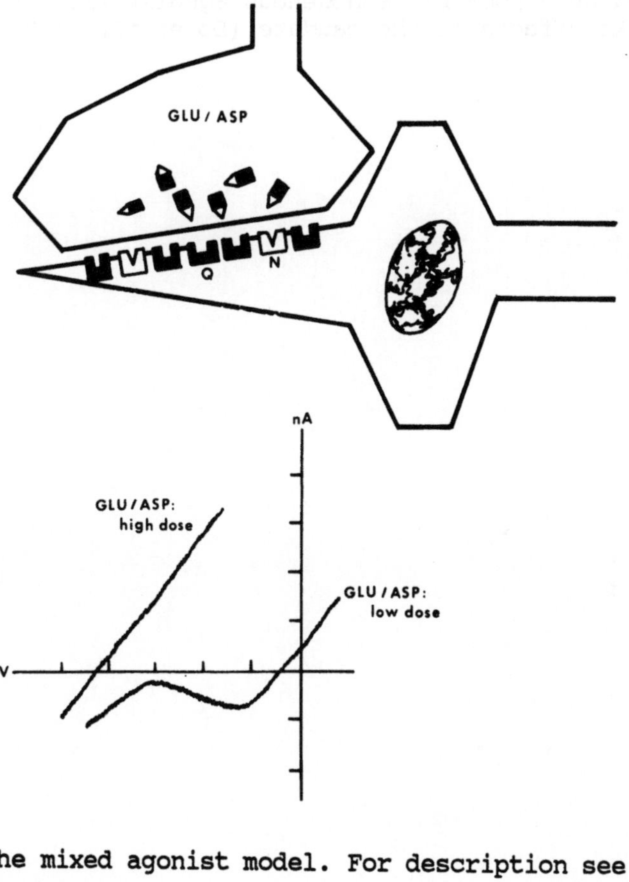

Figure 4. The mixed agonist model. For description see text.

The two agonists model. In this model, two different agonists are released onto the same neuron (Fig. 5). One agonist elicits predominently QUIS-like effects, the other predominantly NMDA-like effects. Activation of synapse A in Fig. 5 elicits release of a transmitter with predominantly QUIS-like effects, as is observed with GLU and ASP on caudate neurons. This sets up a inward synaptic current through the membrane which is in the linear state as indicated by the I-V curve depicted on the left side of the middle panel in Fig. 5. This results in a subthreshold EPSP (Fig. 5, lower left trace). If, however, synapse A displays the same activity as before, but at the same time neuron B or a iontophoretic pipette releases an agonist with predominanty NMDA-like effects, the same inward synaptic current will result in a greatly enhanced depolarization which reaches threshold (Fig. 5, lower right trace), because the postsynaptic membrane is in the 'negative slope' state as indicated by the I-V curve at the right of the middle panel in Fig. 5. L-HCA can be proposed as a possible endogenous agonist with predominantly NMDA-like effects in the caudate (Do et al., 1986).

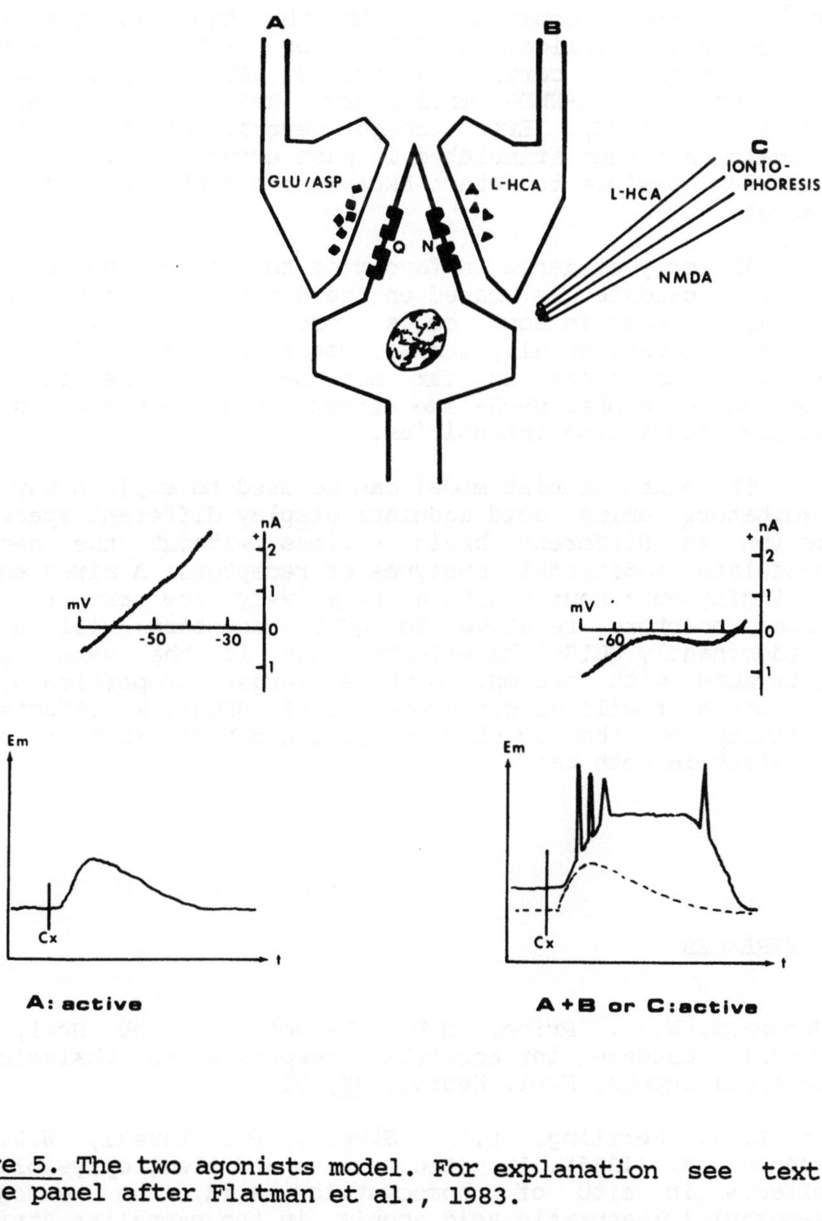

Figure 5. The two agonists model. For explanation see text. Middle panel after Flatman et al., 1983.

In this model, the physiological function of the NMDA system would be to increase the gain of the QUIS system which is responsible for fast EPSPs, such that a subthreshold EPSP is tranformed into a suprathreshold EPSP. A circuit of this type could be imagined to be responsible for some forms of associative learning.

Conclusion. The results obtained to date in the cat caudate nucleus are compatible with the two agonists model: iontophoretic application of L-HCA or NMDA does result in potentiation of cortically evoked EPSPs which seem to be mediated by non-NMDA excitatory amino acid receptors (Herrling, 1985). Even though several afferents to the caudate have been stimulated in past experiments, it has not yet been possible to detect NMDA-like synaptic effects in the caudate.

The only evidence in favour of the mixed agonist concept in the caudate is based on the observation that L-ASP and L-CSA, at least in some cells, are able to elicit mixed effects (Turski et al., subm.). Possibly because of technical difficulties it has so far not been possible to detect cortically evoked NMDA-like effects in caudate neurons, even at low stimulation intensities.

The mixed agonist model can be used to explain why some excitatory amino acid agonists display different spectra of action in different brain regions without the need to postulate additional subtypes of receptors: A mixed agonist impinging on a neuron with a relatively low proportion of NMDA receptors relative to QUIS receptors will display predominantly QUIS-like effects, but if the same agonist interacts with neurons with a larger proportion of NMDA receptors it will elicit predominantly NMDA-like effects. The affinity of the agonist for QUIS and NMDA receptors can be constant in both cases.

REFERENCES

Buchwald, N.A., Price, D.D., Vernon, L. and Hull, C.D. (1973). Caudate intracellular responses to thalamic and cortical inputs. Expl. Neurol. 38, 311-322.

Do, K.Q., Herrling, P.L., Streit, P., Turski, W.A. and Cuenod, M. (1986). In vitro release and electrophysiological effects in situ of homocysteic acid, an endogenous N-methyl-(D)-aspartic acid agonis, in the mammalian striatum. J. Neuroscience, 6, 2226-2234.

Fagg, G.E. and Foster, A.C. (1983). Amino acid neurotransmitters and their pathways in the mammalian central nervous system. Neuroscience, 9, 701-719.

Flatman, J.A., Schwindt, P.C., Crill, W.E. and Stafstrom, C.E. (1983). Multiple actions of N-methyl-D-aspartate on cat neocortical neurons in vitro. Brain Res. 266, 169-173.

Herrling, P.L. (1983) Two types of excitatory amino acid responses in the cat caudate nucleus. In Excitotoxins. (eds. K. Fuxe, P. Roberts and R. Schwarcz) Macmillan Press, London.

Herrling, P.L., Morris, R. and Salt, T.E. (1983).Effects of excitatory amino acids and their antagonists on membrane and action potentials of cat caudate neurones. J. Physiol. (Lond.) 339, 207-222.

Herrling, P.L. (1985). Pharmacology of the cortico-caudate excitatory postsynaptic potential in the cat: Evidence for its mediation by quisqualate- or kainate receptors. Neuroscience 14, 417-426.

Mayer, M.L. and Westbrook, G.L. (1984). Mixed-agonist action of excitatory amino acids on mouse spinal cord neurones under voltage clamp. J. Physiol. (Lond.) 354, 29-53.

Mayer, M.L. and Westbrook, G.L. (1985). The action of N-methyl-D-aspartic acid on mouse spinal neurones in culture. J. Physiol (Lond.) 361, 65-90.

Nowak, L., Bregestovski,P. and Ascher, P. (1984). Magnesium gates glutamate-activated channels in mouse central neurones. Nature 307, 462-465.

Perkins, M.N. and Stone, T.W. (1982). An iontophoretic investigation of the actions of convulsant kynurenines and their interaction with the endogenous excitant quinolinic acid. Brain Res. 247, 184-187.

Perkins, M.N., Collins, J.F. and Stone, T.W. (1982). Isomeres of 2-amino-7-phodphonoheptanoic acid as antagonists of neural excitants. Neurosc. Lett. 32, 65-68.

Watkins, J.C. and Evans, R.H. (1981). Excitatory amino acid transmitter. Annu. Rev. Pharmacol., 21, 165-204.

Part B
Behavioural and Cardiovascular Studies

41

On the Role of Receptor–Receptor Interactions in Central Cardiovascular Regulation: Functional Studies on the Interactions between α_2-adrenergic and Neuropeptide Y Receptors in the Rat Medulla Oblongata

Anders Härfstrand, Kjell Fuxe and Luigi F. Agnati

INTRODUCTION

α_2-adrenergic receptors and neuropeptide Y (NPY) receptors have been shown to overlap within the cardiovascular subnuclei of the nucleus tractus solitarius (nTS) of the male rat (see Fuxe et al., 1986, Härfstrand et al., 1986a and Fuxe et al., this symposium). Large numbers of phenylethanolamine-N-methyltransferase (PMNT) (Hökfelt et al., 1973, 1974, 1984) and NPY immunoreactive (IR) nerve terminals have also been demonstrated within the nTS (Hökfelt et al., 1983a,b; Everitt et al., 1984; Härfstrand et al., 1986b; Chronwall et al., 1985) and many PNMT/NPY costoring nerve terminals exist in this region (Fuxe et al., 1985). Hökfelt and colleagues (Hökfelt et al., 1983a,b) have shown coexistence of NPY in CA neurons in both rat and man. Furthermore, in the dorsal part of the nTS complex (dorsal strip and dorsal subnucleus) PNMT/NPY costoring neurons (Fuxe et al. 1986; Härfstrand et al. 1987a) are intermingled with the baroreceptor afferents (Kalia, 1981). In physiological experiments it has been shown that centrally administered adrenaline (A) (Zandberg et al., 1979; Burkowski and Finch, 1980; Brazenor and Bentley, 1981; Kubo and Misu, 1981), clonidine (CL) (Kobinger and Walland, 1967; Bolme and Fuxe, 1971; Haeusler, 1973) and NPY (Fuxe et al., 1983; Härfstrand et al., 1984) produce a lowering arterial blood pressure, heart and respiratory rate . Furthermore, these effects of NPY and A are also present in the awake unrestrained rat (Härfstrand et al., 1986c,d). Both NPY and CL may inhibit the adenylate cyclase in the nTS area (Härfstrand et al., 1987b). Based on these observations the hypothesis has been introduced that the cardiovascular depression induced by NPY, A and CL is at least in part mediated via activation of NPY and α_2-adrenergic receptor mechanisms, respectively within the nTS leading to enhancement of the efficacy of the central component of the baro-receptor reflex arc (Härfstrand et al., 1986e, Fuxe et al., 1986). In the present paper we will present evidence for the existence of an antagonistic interaction between α_2-adrenergic and NPY receptors in the regulation of cardiovascular function in awake animals. Thus, the present paper represents the physiological counterpart of the biochemical and the receptor autoradiographical experiments of Fuxe et al. (see this volume) giving support for the existence of an intramembrane interaction between α_2-adrenergic and NPY receptors within the nTS of the male rat.

MATERIAL AND METHODS

Male specific pathogen free Sprague-Dawley rats (200-250 g body weight, ALAB, Stockholm, Sweden) have been used.

Experiments on Anaesthetized Rats

The anaesthesia was initiated by Halothan (3% in air 5') and thereafter α-chloralose (100 mg/kg) was given i.v. NPY, CL or A were dissolved in mock CSF (Härfstrand et al. 1986c) and stereotaxically injected into the cisterna magna (i.c.) in a volume of 10 µl (see Härfstrand et al. (1987a). All rats were tracheotomized to ensure free airways. The body temperature was continuously measured via a rectal probe and kept at 37.0 $\pm$ 0.5°C by means of an automatic heating device.

Experiments on Awake Unrestrained Rats

NPY, A or CL was administered into the lateral ventricle on freely moving unrestrained rats. The stainless steel cannulas were stereotaxically implanted in to the lateral ventricle (ivt) two days before the experiment (see Härfstrand et al., 1986c). NPY, A or CL were dissolved in 30 µl mock CSF and administered during a 3 min period by an automatic micro-injection pump unit.

Mean arterial blood pressure (MAP), and heart rate (HR) were measured by means of a heparinized catheter positioned in the common carotide artery, connected to a Statham PC23 DC transducer which was linked to a Grass polygraph (model 7, Grass Instruments, Mass., USA). Registration was performed as previously described (Härfstrand et al., 1986c,d). Registration time was two hours in the awake animals and one hour in the α-chlorolase anaesthetized rats. Peak effects and vasodepressor (DPA), bradycardic (BCA) areas under the curves were calculated for the mean arterial blood pressure and heart rate for each rat and presented as area values expressed in arbitrary units based on the percent change from the respective mean basal value. The Wilcoxon non-parametrical multiple comparison test and Mann-Whitney U-test were used in the statistical analysis (Snedecor and Cochran 1980). The dose effect relationship was tested with the Joncheere-Terpstra test (Hollander and Wolfe, 1973).

RESULTS

α-Chloralose Anaesthetized Rats

The cardiovascular effects of ic administration of NPY, CL and A are shown in fig. 1. All three substances produce a dose-

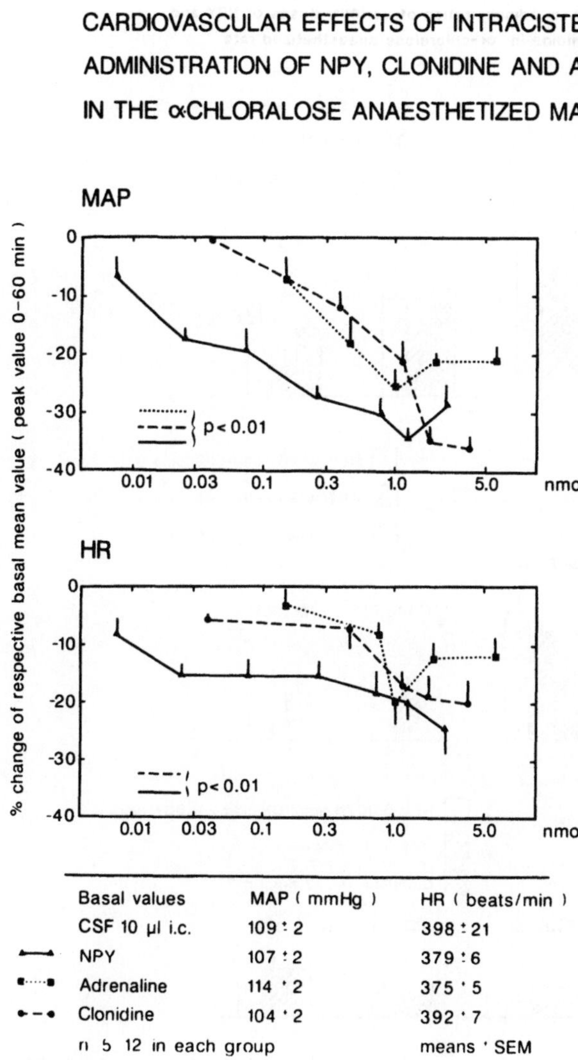

Fig. 1. Maximal responses after intracisternal administration of NPY, clonidine and adrenaline are shown in the α-chloralose anaesthetized rat. Means ± s.e.m. are shown. p values indicate a dose response relationship.

dependent lowering of arterial blood pressure. NPY has the highest potency, while the peak actions of NPY and CL are similar, while somewhat less pronounced for A. Furthermore, CL and NPY produce a dose-dependent lowering of heart rate, again with NPY having the highest potency. In fig. 2 the results obtained by combined treatment of maximal, submaximal or subthreshold doses

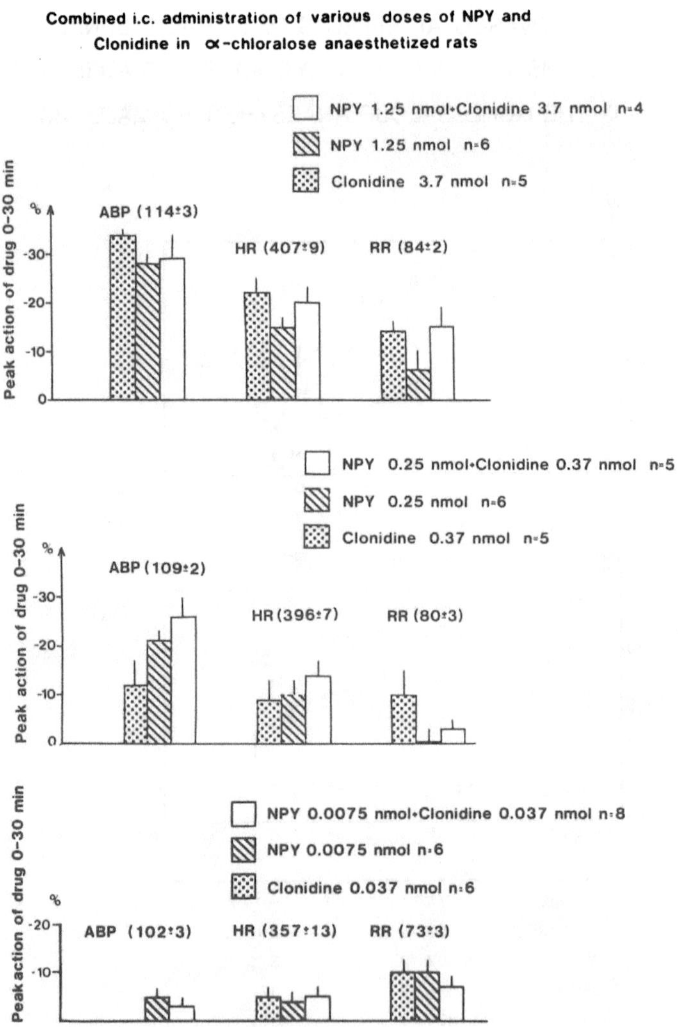

Fig. 2. Maximal responses within the time interval 0-30 min is shown after combined administration of NPY and clonidine in the α-chloralose anaesthetized rats. Maximal doses, submaximal doses (around ED 50 value) and subthreshold doses were combined. No significant additive effects were observed. Basal values are given in parenthesis.

of NPY and CL are shown. In all experiments it is found that the simultaneous administration of NPY and CL does not result in any additional action from that seen after each compound alone.

To further analyze the possible interaction between NPY and α_2-adrenergic receptors in central cardiovascular regulation the α_2-adrenergic receptor antagonist idazoxan (208 nmol; Doxey et al. 1983) was administered ic. 20 min before the administration of various doses of NPY. As seen in fig. 3 the α_2-adrenergic receptor blockade produces an enhancement of the vasodepressor actions when a low dose of NPY was used (0.25 nmol) as evaluated in the analysis of the vasodepressor areas. However, no enhancement of the bradycardic action of NPY was noted.

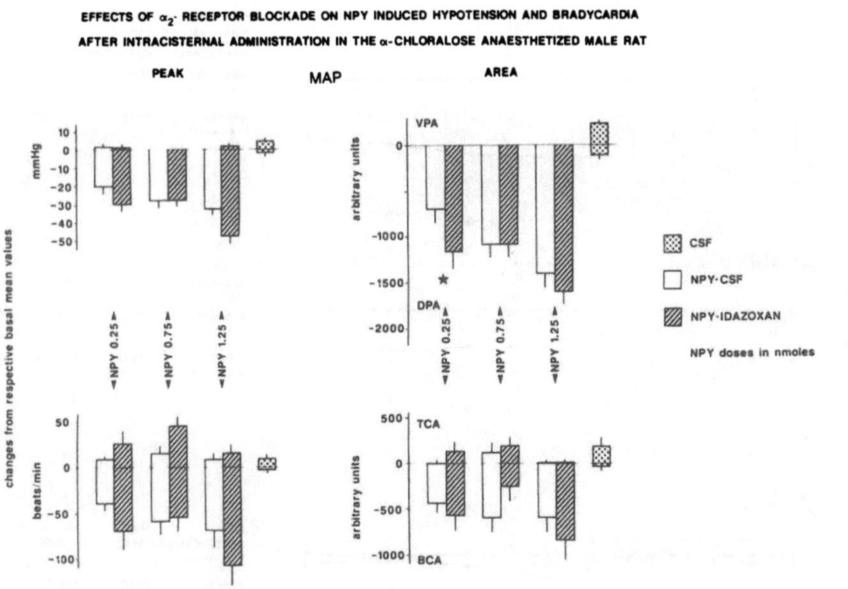

Basal values:	n=6-7 in each group	
	MAP mmHg	HR beats/min
CSF 10 µl i.c.	108 ± 5	393 ± 25
NPY+CSF 10 µl i.c.	106 ± 6	387 ± 12
NPY+IDAZOXAN 208 nmol i.c.	119 ± 6	422 ± 15 means ± SEM

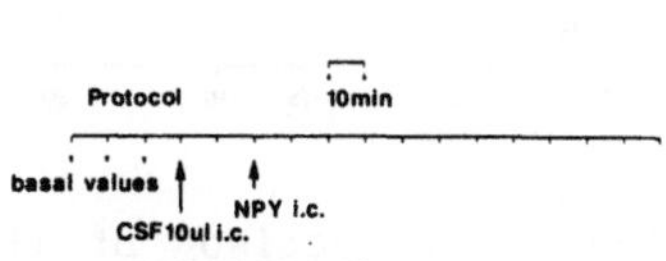

Fig. 3. The effects of i.c. administration of NPY in the presence of α_2-adrenergic receptor blockade induced by idazoxan are shown. A further enhancement of the hypotensive effect is observed after idazoxan treatment (208 nmol ic + NPY 0.25 nmol ic). Means $\pm$ s.e.m. are shown. * = $p < 0.05$. Abbreviations used: MAP= mean arterial blood pressure, HR= heart rate, VPA= vasopressor area, DPA= vasodepressor area, TCA tachycardic area, BCA= bradycardic area .

Studies in Awake Unrestrained Animals

In fig. 4 the time course of the actions of A and NPY has been compared in the awake and in the α-chlorolase anaesthetized male rat. It is found that in the awake animal there is a prolongation of the peak action of NPY compared with the action of A. Instead, the peak actions are similar when comparing the effects of the NPY and A. In fig. 5 the cardiovascular effects of the combined administration of submaximal doses of NPY and A were

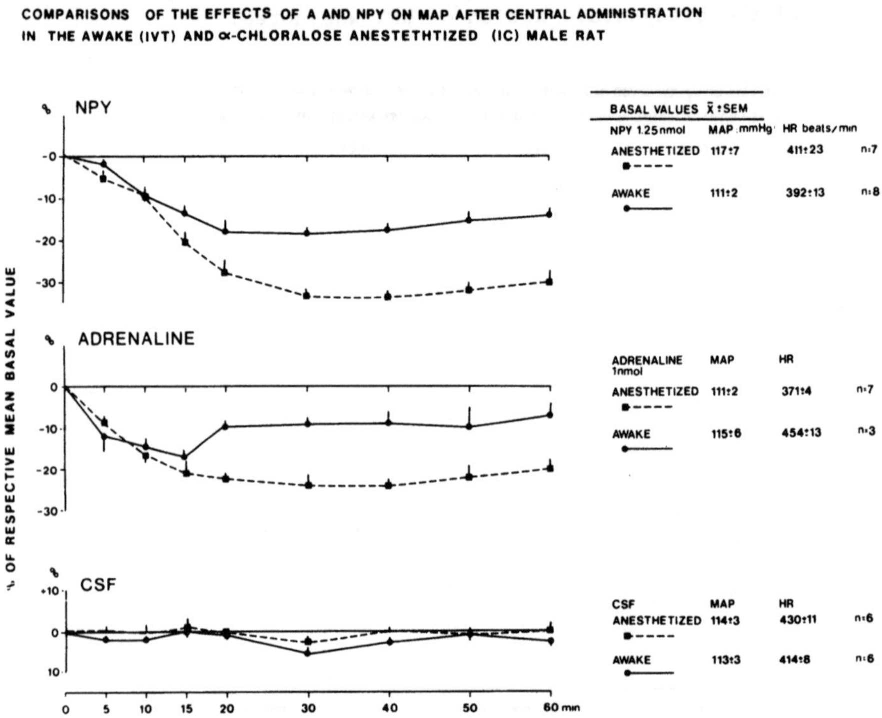

Fig. 4. Comparison of the effects of central administration of NPY and adrenaline in α-chloralose anaestheticed and awake freely mowing rats. In both NPY (1.25 nmol) and adrenaline (1.0nmol) treated rats α-chloralose anaesthesia increased the peak actions. The maximal responses are of the same magnitude for NPY and A. In the awake rats A induces a more rapid response but the marked hypotensive action is partly reversed within 20 min, while the peak action of NPY (1.25 nmol) in the awake animals is sustained. Means ± s.e.m. are shown. Area calculations comparing the hypotensive responses of these maximal doses of NPY and A are significantly different ($p<0.05$; Mann-Whitney U-test) in the awake animals.

analyzed in the awake animals. It is found that the combined administration of A and NPY significantly counteracts the cardiovascular actions of either A or NPY as evaluated from the analysis of peak effects or vasodepressor areas. No antagonistic interaction is found for a combined administration of A and NPY when heart rate is considered.

DISCUSSION

The present paper gives physiological evidence for the existence of an antagonistic interaction between α_2-adrenergic and NPY receptors in cardiovascular centers of the medulla oblongata. The receptor autoradiographical evidence of Fuxe et al. (see this volume) suggests that this integrative mechanism in arterial blood pressure regulation may take place within the nTS. The

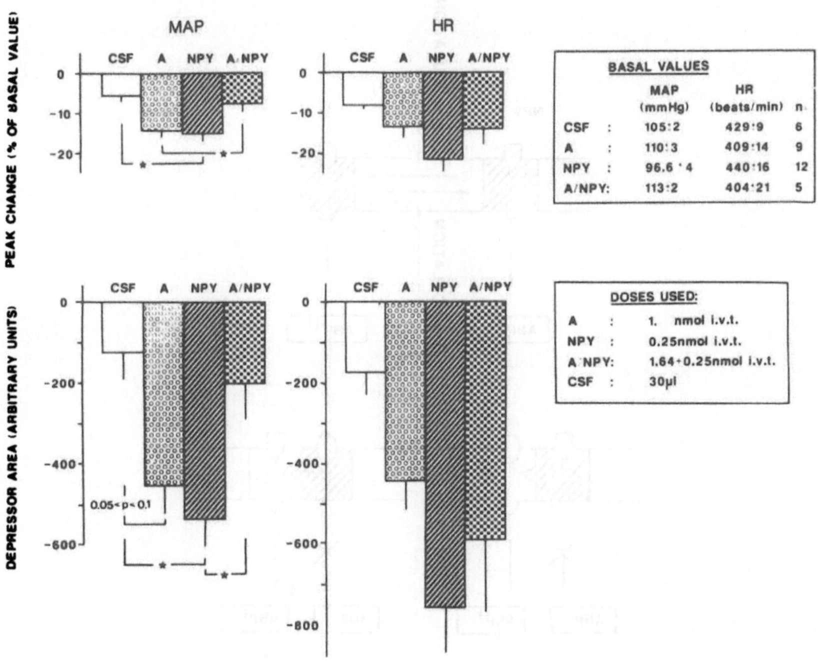

Fig. 5. The antagonistic effect of combined central administration of NPY (0.025 nmol) and A (1.0 nmol) on mean arterial blood pressure (MAP) but not on heart rate (HR) is shown in the awake freely moving male rat. Both maximal responses (peak) and the overall effect (vasodepressor area) are given. Note the antagonistic effect of combined administration with regard to the hypotensive effect. * = $p<0.05$.

hypothesis on an antagonistic interaction is mainly based on two observations.

1. Neuropeptide Y and adrenaline when given together in the lateral ventricle antagonize the vasodepressor activity of one another.
2. The administration of an α_2-adrenergic receptor antagonist in the lateral ventricle significantly enhances the vasodepressor actions of a low dose of NPY.

As indicated in the paper of Fuxe et al. (see this volume) biochemic and receptor autoradiographic studies on the receptor-receptor interaction are also compatible with an antagonistic interaction. Thus, NPY and CL reduce the binding of one another and NPY counteracts the α_2-agonist promoted coupling of the Ni protein to the adenylate cyclase. Our present view of the molecular mechanism involved and on the interaction between NPY and

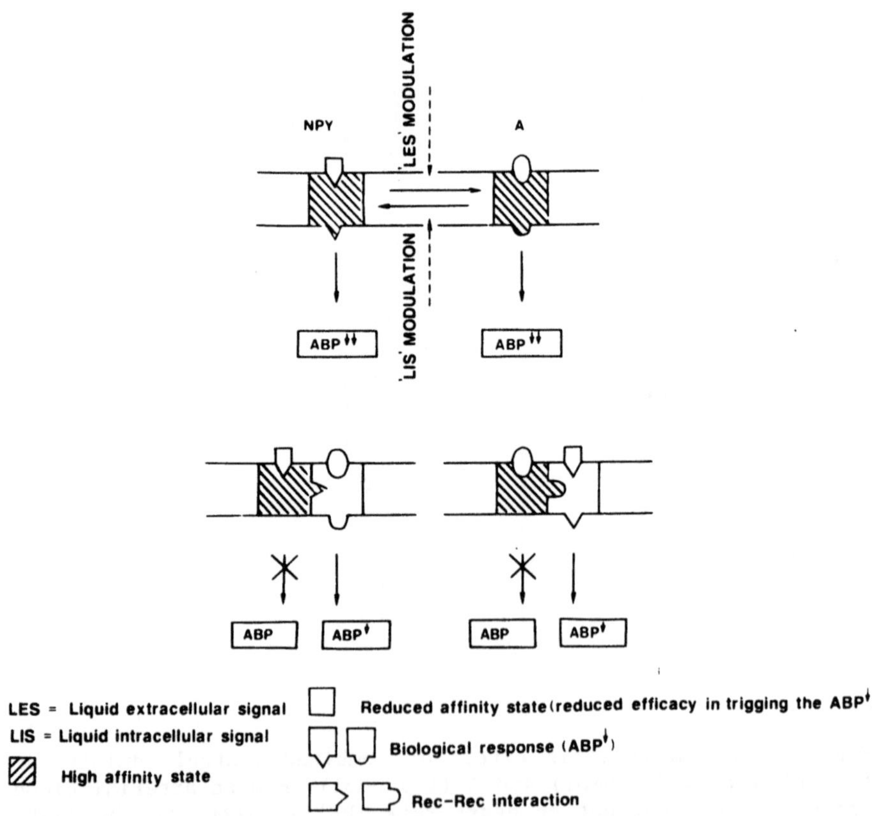

Fig. 6. Present hypothesis on the receptor-receptor interactions underlying the antagonistic interactions of NPY and adrenaline in central vasodepressor responses. See also Fuxe et al., this symposium.

α_2-adrenergic receptors is presented in fig. 6. It is illustrated that the receptor-receptor interaction leads to a reduced affinity state of the NPY and α_2-adrenergic receptors having a reduced efficacy in triggering the biological response, i.e., the lowering of arterial blood pressure. Furthermore, it is assumed that one of the interacting receptors becomes uncoupled from the biological effector. The intramembrane nature of the receptor-receptor interaction is also illustrated as well as its dependency upon extra- and intracellular signals. Controlling e.g. membrane fluidity and the phosphorylated state of the membrane proteins (see Greengard, this symposium).

The present study shows that NPY and A mimic one another with regard to the cardiovascular effects. However, one important difference exists and that is the prolonged lowering of arterial blood pressure obtained with NPY compared with A in the awake unrestrained male rat. A prolonged action of the NPY may reflect the fact that it is not as rapidly broken down as A. Based on the mismatch between the distribution of the NPY IR nerve terminals and the high affinity ^{125}I-NPY binding sites in the nTS, a role of NPY in volume transmission (Agnati et al., 1986, Agnati et al., this volume) has also been suggested. Thus, NPY may diffuse from the dmnX to reach neuronal targets in the nTS to coordinate the activity in visceral afferent/efferent systems in the dorsal medulla.

In conclusion, the present findings provide physiological evidence for an antagonistic interaction between NPY and α_2-adrenergic receptors in the central regulation of arterial blood pressure.

ACKNOWLEDGEMENTS

This work has been supported by a grant (04X-715) from the Swedish Medical Research Council. For excellent technical assistance we are grateful to Bettan Andbjer, Lars Rosén and Anders Jansson. For excellent secreterial assistance we are grateful to Elisabeth Sandqvist, Monia Särd and Anne Edgren.

REFERENCES

Agnati, L.F., Fuxe, K., Zoli, M. Merlo Pich, E., Benfenati, F., Zini, I. & Goldstein, M. (1986). Aspects on the information handling by the central nervous system: focus on the aged rat brain. In: Coexistence of neuronal messengers: a new principle in chemical transmission. (Hökfelt, T., Fuxe, K. & Pernow, B. eds.). Progress in Brain Res. 68, 291-301.

Agnati, L.F., Fuxe, K., Zoli, M., Zini, I., Toffano, G. & Ferraguti, F. (1986). A correlation analysis of the regional distribution of central enkephalin and β-endorphin immunoreactive terminals and of opiate receptors in adult and old male rats. Evidence for the exxistence of two main types of communication in the central nervous system: the volume transmission and the wiring transmission. Acta Physiol. Scand. 128, 201-207.

Bolme, P. and Fuxe, K. (1971). Pharmacological studies on the hypotensive effects of clonidine. Eur. J. Pharmacol. 13, 168-174.

Brazenor, R.M. and Bentley, G.A. (1981). Central effects of α-adrenoceptor agonist drugs on a cardiovascular reflex evoked by stimulation of somatic afferents. Europ. J. Pharmacol. 73, 213-215.

Burkowski, K.R. and Finch, L. (1980). The hypotensive actions of centrally administered adrenaline. In: Central adrenaline neurons, basic aspects and their role in cardiovascular functions. (Fuxe K., Goldstein M., Hökfelt B. and Hökfelt T. eds.). Pergamon Press Oxford and New York, pp. 225-234.

Chronwall, B.M., DiMaggio, D.A., Massari, V.J., Pickel, V.M., Ruggiero, D.A. and O'Donohue, T.L. (1985). The anatomy of neuropeptide-Y containing neurons in rat brain, Neuroscience Vol. 15, No. 4, pp. 1159-1181.

Doxey, J.C.,Roach, A.G.& Smith, C.F.C. (1983). Studies on RX 781094: a selective,potent and specific antagonist of α_2-adrenoceptors. Br. J. Pharmac. 78, 489-505.

Everitt, B.J., Hökfelt, T., Terenius, L., Tatemoto, K., Mutt, V. and Goldstein, M. (1984). Differential co-existence of neuropeptide Y (NPY)-like immunoreactivity with catecholamines in the central nervous system of the rat. Neuroscience 11, 443-462.

Fuxe, K. Agnati, L., Härfstrand, A., Zini, I., Tatemoto, K., Merlo Pick, E., Hökfelt, T., Mutt, V. and Terenius, L. (1983). Central administration of neuropeptide Y induces hypotension, bradypnea and EEG synchronization in the rat. Acta Physiol. Scand. 118, 189-192.

Fuxe, K., Agnati, L., Härfstrand, A., Zoli, M. and Janson, A.M. (1985a). Image analysis and determination of codistribution and coexistence of neuroactive substances in nerve terminal populations. Acta Stereol. 4:2, 181-186.

Fuxe, K., Agnati, L.F., Härfstrand, A., Janson, A.M., Neumeyer, A., Andersson, K., Ruggeri, M., Zoli, M. and Goldstein, M. (1986). Morphofunctional studies on the neuropeptide Y/adrenaline co-storing nerve terminal systems in the dorsal cardiovascular region of the medulla oblongata. Focus on receptor-receptor interactions in cotransmission. Prog. Brain Res. 68, 303-320.

Haeusler, G. (1973). Activation of the central pathway of the baroreceptor reflex, a possible mechanism of the hypotensive action of clonidine. N-S Arch Pharmacol. 278, 231-346.

Hollander, M. and Wolfe, D.A. (1973). Nonparametric statistical methods. John Wiley & Sons, New York, U.S.A.

Härfstrand, A. (1986d). Intraventricular administration of neuropeptide Y (NPY) induces hypotension, bradycardia and bradypnea in the awake unrestrained male rat. Counteraction by NPY-induced feeding behaviour. Acta Physiol. Scand. 128, 121-123.

Härfstrand, A., Fuxe, K., Agnati, L.F., Ganten, D., Eneroth, P., Tatemoto, K. and Mutt, V. (1984). Studies on Neuropeptide-Y catecholamine interactions in central cardiovascular regulation in the α-chloralose anaesthetized rat. Evidence for a possible new way of activating the μ-2 adrenergic transmission line. Clin. Exp. Hypertens. Theor. and Pract. A6 (10&11), 1947-1950.

Härfstrand, A., Fuxe, K., Agnati, L.F., Eneroth, P., Zini, I., Zoli, M., Andersson, K., von Euler, G., Terenius, L., Mutt, V. and Goldstein, M. (1986c). Studies on neuropeptide Y-catecholamine interactions in the hypothalamus and in the forebrain of the male rat. Relationship to neuroendocrine function. Neurochem. Int. Vol 8, No. 3, 355-376.

Härfstrand, A., Kalia, M., Terenius, L. and Fuxe, K. (1986b). Neuropeptide Y immunoreactive perikarya and nerve terminals in the rat medulla oblongata. Relationship to cytoarchitecture and catecholaminergic cell groups. J. Comp. Neurol. In press.

Härfstrand, A., Fuxe, K., Agnati, L.F., Benfenati, F. and Goldstein, M. (1986a). Receptor autoradiographical evidence for high densities of ^{125}I-Neuropeptide Y binding sites in the nucleus tractus solitarius of the normal male rat. Acta Physiol. Scand. 128, 195-200.

Härfstrand, A., Fredholm, B.B. and Fuxe, K. (1986e). Inhibitory effects on cyclic AMP accumulation in slices of the nucleus tractus solitarius region of the rat. Neurosci. Lett., in press.

Härfstrand, A., Fuxe, K., Agnati, L.F., Kitayama, I., Cintra, A., Janson, A.M., Kalia, M., Vanderhaegen, J-J., Goldstein, M. and Terenius, L. (1986). Intracisternal administration of cholecystokinin-8 counteracts the central cardiovascular effects of adrenaline and NPY. A study based on the existence of cholecystokinin, phenylethanolamine N-methyltransferase and neuropeptide Y immunoreactive neurons in the nucleus tractus solitarius. Neurochem. Int., in press.

Hökfelt, T., Fuxe, K., Goldstein, M. and Johansson, O. (1973). Evidence for adrenaline neurons in the rat brain. Acta Physiol. Scand. 89, 286-288.

Hökfelt, T., Fuxe, K., Goldstein, M., Johansson, O. (1974). Immunohistochemical evidence for the existence of adrenaline neurons in the rat brain. Brain Res. 66, 235251.

Hökfelt, T., Lundberg, J.M., Tatemoto, K., Mutt, V., Terenius, L., Polak, J., Bloom, S., Sasek, C., Elde, R. and Goldstein, M. (1983a). Neuropeptide Y (NPY)-and FMRF amide neuropeptide-like immunoreactivities in catecholamine neurons of the rat medulla oblongata. Acta Physiol. Scand. 117, 315-318.

Hökfelt, T, Lundberg, J.M., Lagerkrantz, H., Tatemoto, K., Mutt, V., Terenius, L., Everitt, B.J., Fuxe, K., Agnati, L.F. and Goldstein, M. (1983b). Occurence of neuropeptide Y (NPY)-like immunoreactivity in catecholamine neurons in the human medulla oblongata. Neurosci. Lett. 36, 217-222.

Hökfelt, T., Everitt, B.J., Fuxe, K., Kalia, M., Agnati, L.F., Johansson, O., Härfstrand, A., Lundberg, J.M., Terenius, L., Theodorsson-Norheim, E. and Goldstein, M. (1984a). Transmitter and peptide systems in areas involved in the control of blood pressure. Clin, Exp. Hypertens. Theor. Pract. A6 (1&2), 23-41.

Hökfelt, T., Johansson, O. and Goldstein, M. (1984b). Central catecholamine neurons as revealed by immunohistochemistry with special reference to adrenaline neurons. In: Handbook of Chemical Neuroanatomy. Vol. 2: Classical Transmitters in the CNS. (Björklund A. and Hökfelt T. eds.), Elsevier Scinece Publishers B.V., pp. 157-276.

Kalia, M. (1981). Localization of Aortic and Carotid Baroreceptor and Chemoreceptor primary Afferents in the Brain Stem. In: Central Nervous System Mechanisms in Hypertension. (Buckley J.P and Ferrerio C.M. eds.) Raven Press, New York. Vol 6, 9-24.

Kubo, T. and Misu, Y. (1981). Pharmacological characterisation of the α-adrenoreptors responsible for a decrease of blood pressure in the nucleus tractus solitarii of the rat. Naunyn-Schmiedeberg's Arch Pharmacol. 317, 120-125.

Lundberg, J.M., Tatemoto, K. (1982). Pancreatic polypeptide family (APP, BPP NPY and PYY) in relation to sympathetic vasoconstriction resistant to α-adrenoreceptor blockade. Acta Physiol. Scand. 116, 393-402.

Zandberg, P., DeJong, W. and DeWied, D. (1979). Effect of catecholamine-receptor stimulating agents on blood pressure after local application in the nucleus tractus solitarii of the medulla oblongata. Europ. J. Pharmacol. 55, 43-56.

42
Behavioral Analysis of the Functional Significance of Peptide–Transmitter Coexistences

Jacqueline N. Crawley

Immunocytochemical techniques have revealed a growing number of anatomical pathways in mammalian brain in which two or more neurotransmitters coexist within the same neuron. At first iconoclastic, the concept of coexisting transmitters is becoming more the rule than the exception, expanding our thinking about a whole new level of neuronal regulatory processes. Many examples of coexistences involving neuropeptides have been identified in brain regions which are involved in primary behavioral functions and in a variety of neuropsychiatric disorders. Our laboratory is using a behavioral approach to investigate the functional significance of peptide-transmitter coexistences. Postsynaptic sites of brain nuclei identified immunocytochemically as containing two or more neuroactive substances are implanted with indwelling cannulae for discrete microinjections in awake, behaving rats. Coexisting neurotransmitters are microinjected separately and in combination over a wide dose range. Behavioral paradigms are chosen from the literature or newly developed which are specific to each neurotransmitter at that postsynaptic site. Functional relationships are then evaluated in terms of synergistic, additive, or antagonistic interactions between the coexisting transmitters. Three cases of coexistence have been studied to date. In two cases neuropeptides coexisting with "classical" transmitters appear to function as limited modulators of the function of the primary transmitter, while in the case of two neuropeptides coexisting with each other, each peptide appears to function independently and competitively.

Cholecystokinin (CCK) is an eight amino acid peptide found in both gut (Mutt, 1978) and brain (Vanderhaeghen, 1975). Hokfelt and coworkers discovered that CCK coexists with dopamine (DA) in the ventral tegmental neurons of rats, monkeys and humans (Hokfelt et al, 1980a,b; Skirboll et al, 1981). The CCK-DA coexistence is limited to mesolimbic neurons projecting to the nucleus accumbens and olfactory tubercles, as it is not found in nigrostriatal neurons projecting to the caudate nucleus (Hokfelt et al, 1980a,b;

Studler et al, 1981; Fallon et al, 1983). The mesolimbic dopamine pathway has been shown to mediate locomotor behaviors in rats (Iversen, 1974), and has been implicated in the pathophysiology of human schizophrenia (Stevens, 1973). In this first case of one peptide coexisting with one "classical" transmitter, CCK was microinjected into the nucleus accumbens, alone and in combination with DA, to investigate its role in mediating locomotor behaviors. Male Sprague-Dawley rats, 200 grams starting weight, were stereotaxically implanted under sodium pentobarbital anesthesia with bilateral 24 gauge hypodermic stainless steel guide cannulae, within the medial posterior region of the nucleus accumbens, 1.2 mm anterior to bregma, $\pm$ 1.2 mm lateral to the saggital suture, and 5.7 mm ventral to the surface of the skull at bregma. One week later, each rat was microinjected through the guide cannulae, using a 31 gauge injection tube lowered to a site 1.0 mm below the ventral surface of the guide cannula, with either 0.9% physiological saline, CCK (Bachem, Inc., Torrance, CA), DA (Sigma Chemical Co., St. Louis, MO), or CCK + DA, in a fixed volume of 0.2 µl over a one minute period, using a Sage microinfusion pump (Orion Research, Inc. Cambridge, MA). Microinjection parameters were chosen to minimize tissue damage and spread of compounds away from the intended injection site. Immediately after microinjection, rats were tested for locomotor activity in a Digiscan photocell-equipped animal activity monitor (Omnitech, Inc., Columbus, OH) for ten minutes. At the end of each experiment, all animals were microinjected with fast green dye in a manner identical to the standard microinjections, and sacrificed for histological analysis of the microinjection site, using frozen sections and cresyl violet staining. Data from animals with incorrect microinjection sites were eliminated from the statistical analysis. CCK had no effect on locomotor activity at doses from 20 pg to 200 ng (Figure 1, Crawley, et al, 1985c). However, CCK significantly potentiated DA-induced hyperlocomotion at doses from 20 pg to 200 ng (Figure 1, Crawley et al, 1985c). A trend toward a biphasic response can be seen, since 2 µg of CCK did not potentiate DA-induced hyperlocomotion. Previous studies have reported that microgram-range doses of CCK inhibited amphetamine and apomorpine-induced behaviors (Schneider et al, 1983; Widerlov et al, 1983; Ellinwood et al, 1983; Van Ree et al, 1983; Hamilton et al, 1984). CCK may therefore act to potentiate dopaminergic function at lower doses, in the more physiological picogram to nanogram range, while CCK may act to inhibit dopaminergic function at higher doses, in the more pharmacological microgram range. These results suggested that CCK acts as a modulator of DA, having no behavioral function alone but facilitating the behavioral activity of DA. Further studies with analogs of CCK found that neither unsulfated CCK octapeptide nor the C-terminal tetrapeptide had any effect on DA-induced hyperlocomotion, suggesting a pharmacological specificity to the sulfated octapeptide (Crawley et al, 1985c). Topographical analysis of the terminal regions within the nucleus accumbens and the caudate nucleus found that the ability of CCK to potentiate DA-induced hyperlocomotion was confined to the region of coexisting CCK-DA terminals, in the medial and posterior areas of the nucleus accumbens. CCK microinjection sites in the caudate nucleus (Crawley et al, 1985c), and in the lateral and anterior portions of the nucleus accumbens (Crawley et al, 1985a), had no effect on DA-induced behaviors.

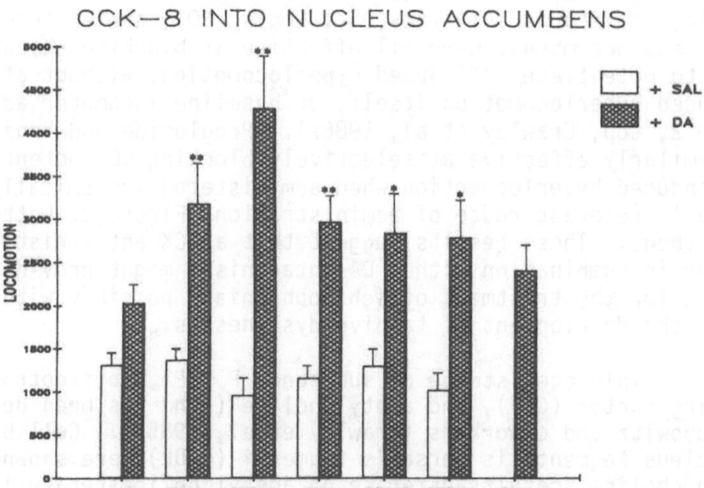

Figure 1. Coexistence of cholecystokinin (CCK) and dopamine (DA) in the mesolimbic pathway. Bilateral microinjection of CCK or DA (20 ug) into the medial posterior nucleus accumbens of awake, behaving rats, was immediately followed by a ten minute test session for locomotion in a photocell-equipped Digiscan optical activity monitor. CCK had no effect on locomotion when microinjected alone into the nucleus accumbens, over a dose range of 20 pg to 200 ng, as compared to saline controls. CCK significantly potentiated DA-induced hyperlocomotion when microinjected in combination with DA (20 ug) into the nucleus accumbens, at doses of 20 pg, 200 pg, 2 ng, 20 ng, and 200 ng CCK. ANOVA $F(6,48) = 4.46$, $p < 0.01$, Newman-Keuls $p < 0.05$ (*), $p < 0.01$ (**) as compared to DA + saline. Data are presented as mean + standard error of the mean. N = 6-10 rats for each dose of each treatment. (Crawley et al, 1985c).

These results highlighted a facilitatory role for CCK in dopaminergic behaviors. Because of the implication of the mesolimbic pathway in the treatment of schizophrenia, several clinical research groups have tested CCK as a novel antipsychotic treatment, with controversial but generally negative results to date (Nair et al, 1982; Moroji et al, 1982; Lotstra et al, 1984; Tamminga et al, 1984; Hommer et al, 1985). Our data suggest that a CCK antagonist might be a better candidate for an antipsychotic, either alone or in combination with a DA antagonist, as previously suggested (Innis, et al, 1986). CCK antagonists were therefore tested alone, in combination with DA, and in combination with DA + CCK, after microinjection into the nucleus accumbens or after intraperitoneal injection, in the ten minute rat locomotion session. Proglumide (Rotta Laboratorium, Milano, Italy), benzotript (Rotta), and antiserum to CCK (gift from Dr. M. Beinfeld, St. Louis Univerity, St. Louis, MO), administered into the nucleus accumbens, were all effective at blocking the ability of CCK to potentiate DA-induced hyperlocomotion, without affecting DA-induced hyperlocomotion itself, or baseline locomotor activity (Figure 2, top, Crawley et al, 1986a). Proglumide and benzotript were similarly effective at selectively blocking CCK potentiation of DA-induced hyperlocomotion when administered systemically, the clinically relevant route of administration (Figure 2, bottom, unpublished). These results suggest that a CCK antagonist, either alone or in combination with a DA antagonist, might provide a new approach for the treatment of schizophrenia, possibly with less risk of the development of tardive dyskinesias.

A triple coexistence of substance P (SP), corticotropin releasing factor (CRF), and acetylcholine (Ach) has been described by Jacobowitz and coworkers (Crawley et al, 1985b). Cell bodies in the nucleus tegmentalis dorsalis tegmenti (NTDL) were shown to contain choline acetyltransferase or acetylcholinesterase in combination with SP (Vincent et al, 1983) and with CRF (Olschowka et al, 1983; Crawley et al, 1985b). These NTDL neurons, which are adjacent and medial to the locus coeruleus, project to the thalamus, septum, and to the medial anterior portion of the cortex in the rat. To investigate the functional significance of this coexistence between two peptides and one "classical" transmitter, we employed the same strategy as for the CCK-DA coexistence. Rats were similarly cannulated bilaterally in the postsynaptic region receiving the AchE-SP-CRF terminals, the medial anterior cortex, 3.8 mm anterior to bregma, $\pm$ 0.8 mm lateral to the saggital suture, and 3.5 mm ventral to the surface of the skull at bregma. One week after surgery, SP (Peninsula Laboratories, Belmont, CA), CRF (Peninsula Laboratories, Belmont, CA), saline, or the non-selective cholinergic agonist, carbachol (carbamylcholine, Sigma Chemical Co., St. Louis, MO) was microinjected over a wide dose range. No observable behavioral effects of either SP or CRF were observed, over a dose range from 10 pg to 5 μg, including no effects on locomotion, vertical activity, grooming, sniffing, exploration,

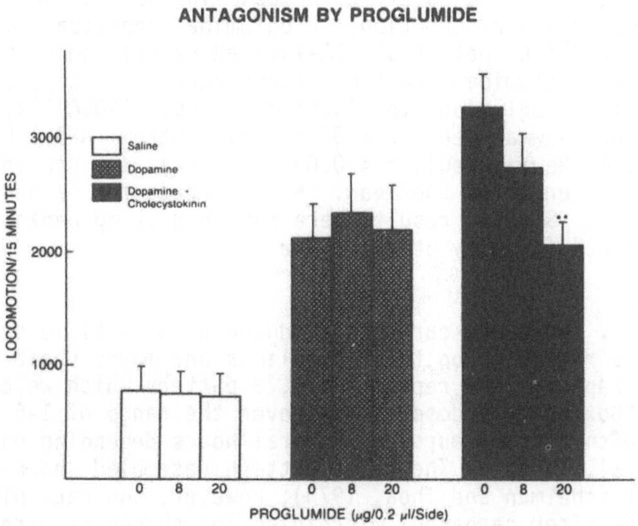

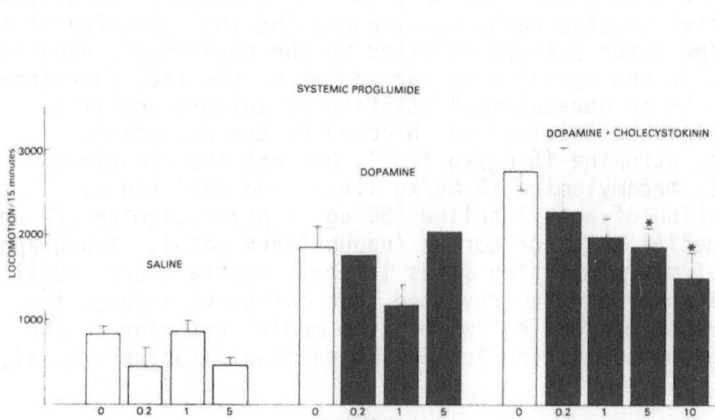

Figure 2. CCK antagonists. Proglumide was microinjected

bilaterally into the nucleus accumbens (top) or intraperitoneally (bottom), five minutes before bilateral microinjection of saline, DA (20 ug), or CCK (1 ng) + DA (20 ug) into the nucleus accumbens. Rats were then tested for locomotor activity over a ten minute session. Proglumide had no effect on baseline locomotion or on dopamine-induced hyperlocomotion. Proglumide significantly blocked the ability of CCK to potentiate DA-induced hyperlocomotion at a dose of 20 ug proglumide into the nucleus accumbens (top) and at a dose of 5 mg/kg proglumide i.p. (bottom). Top: ANOVA $F(2,15) = 5.21$, $p < 0.05$, Newman-Keuls $p < 0.05$ (*). Bottom: ANOVA $F(4,37) = 2.64$, $p < 0.05$, Newman-Keuls $p < 0.05$ (*). Data are presented as mean + standard error of the mean. N = 6-8 rats for each dose of each treatment. Similar results were obtained using benzotript and antiserum to CCK (Crawley et al, 1986a).

postures, etc. However, carbachol induced an unusual motor pattern in which rats reared up on their hindlimbs and moved their forepaws in the air, rapidly and repetitively, a pattern which we call "boxing". "Boxing" was dose related over the range of 1-6 µg carbachol, with a time course of several hours depending on dose (Crawley et al, 1985b). The motor pattern resembled shock-elicited aggression (Eichelman and Thoa, 1973); however, two rats placed in the same cage after carbachol microinjection showed no directed aggression but went to opposite ends of the cage and "boxed" independently. Electroencephalographic analysis from frontal and fronto-occiptal cortical electrodes in rats microinjected with carbachol revealed high amplitude bursts of spiking, characteristic of seizures, concommittant to "boxing" behavior (Crawley et al 1986b). Standard anticonvulsants such as diazepam, clonazepam, and pentobarbital blocked carbachol-induced "boxing" (Crawley et al, 1986b). The motor pattern elicited by the cholinergic agonist, carbachol, in the medial anterior cortex of the rat, therefore, appears to be an unusual manifestation of seizure activity. Carbachol-induced "boxing" was blocked by the muscarinic antagonist, atropine (5 mg/kg i.p.), but not the nicotinic antagonist, mecamylamine (3 mg/kg i.p.), and mimicked by microinjection of acetylcholine (80 µg) + physostigmine (20 µg) into the medial anterior cortex (unpublished data). Topographical analysis of microinjection sites lateral, ventral, and caudal to the medial anterior site revealed that carbachol induces its maximal amount of "boxing" within the medial anterior region containing the coexisting Ach-SP-CRF terminals (Stivers et al, 1987).

Coinjection of SP and carbachol revealed that SP potentiated carbachol-induced "boxing" at a dose of 1 µg (Figure 3, Crawley et al, 1985b). A long-lasting analog of SP, [pGlu^5, MePhe^8, Sar^9]-SP (5-11), similarly potentiated carbachol-induced "boxing," at doses of 500 ng and 1 µg (Stivers and Crawley, 1987). CRF significantly inhibited carbachol-induced "boxing," at doses of 1 ng and above (Crawley et al, 1985b). CRF did not completely

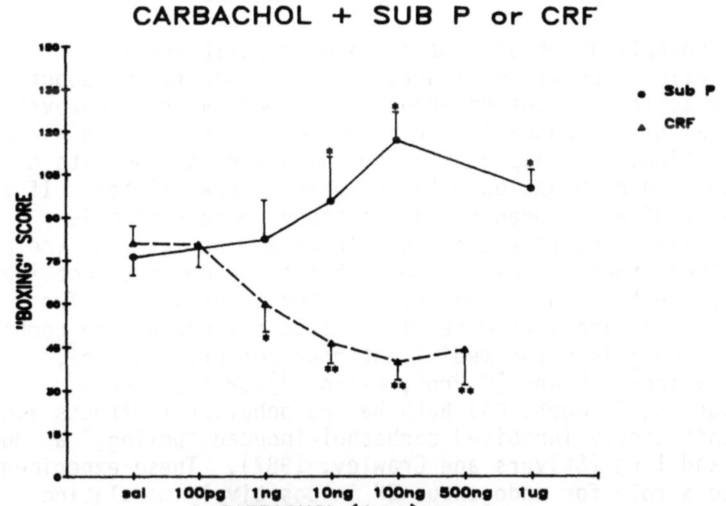

Figure 3. Coexistence of substance P (SP), corticotropin releasing factor (CRF), and acetylcholinesterase in the dorsolateral tegmental nucleus projecting to the medial anterior cortex. Carbachol (4 ug) bilaterally microinjected into the medial anterior cortex of awake, behaving rats, induced a motor manifestation of seizures resembling human "boxing." The characteristic upright posture and rapid, repetitive forepaw treading were quantitated by an observer, uninformed of the treatment condition, over a fifteen minute session. Neither SP nor CRF induced "boxing" or any other observable spontaneous behavior when microinjected alone at doses up to 5 ug SP and 200 ng CRF. SP significantly potentiated carbachol-induced "boxing" at doses of 10 ng, 100 ng, and 1 ug, ANOVA $F(4,26) = 3.67$, $p < 0.05$, * $p < 0.05$ by Newman-Keuls post-hoc analysis of individual means. CRF significantly decreased carbachol-induced "boxing" at doses of 1 ng, 10 ng, 100 ng and 500 ng, ANOVA $F(5,43) = 6.63$, $p < 0.01$, * $p < 0.05$, ** $p < 0.01$ by Newman-Keuls analysis. Data are presented as mean + standard error of the mean. N = 6-10 for each dose of each treatment. (Crawley et al, 1985b).

inhibit carbachol-induced "boxing", at doses up to 400 ng. In this case of a triple coexistence, it appears that both peptides act as modulators of the primary, "classical" transmitter, having no effect alone but influencing the level of effect of the primary transmitter. One peptide, SP, appears to act as a positive, facilitatory modulator, while the other peptide, CRF, appears to act as a negative, inhibitory modulator. Combinations of SP+CRF+carbachol tested in pilot studies gave "boxing" scores significantly below those for carbachol alone, suggesting that the CRF influence may be more efficacious than the SP influence.

Microapplication of peptides and transmitters into postsynaptic sites yields a great deal of information about the possible actions of the endogenous neuromodulators. However, the functional significance of the endogenous peptides remains open to investigation. One approach is to administer antagonists of the modulators in combination with the primary transmitter. If the antagonist of an endogenous facilitatory neuropeptide is behaviorally inactive alone, but blocks the activity of the primary transmitter, then it seems likely that the endogenous peptide is normally functioning in concert with the transmitter. Two substance P antagonists were microinjected alone and in combination with carbachol into the medial anterior cortex. [D-Pro^2, D-Phe^7, D-Trp^9]-SP and [D-Pro^2, D-Trp^{7,9}]-SP (Peninsula Laboratories, Belmont, CA) both had no behavioral effects alone, but significantly inhibited carbachol-induced "boxing," at doses of 500 ng and 1 μg (Stivers and Crawley, 1987). These experiments point to a role for endogenous SP in positively modulating cholinergic function in the medial anterior cortex. An antagonist of CRF, a-helical CRF [9-41] (Peninsula Laboratories, Belmont, CA), had no effect on carbachol-induced "boxing", although this compound reportedly has a low potency (Britton, et al, 1986).

To test for the pharmacological specificity of SP and CRF in modulating carbachol-induced "boxing," other peptides localized in the medial anterior cortex of the rat were similarly microinjected alone and in combination with carbachol. As seen in Figure 4, there was no effect of a wide dose range of calcitonin gene related peptide, galanin, or met-enkephalin, on carbachol-induced "boxing." Somatostatin was the only peptide tested to date which significantly potentiated carbachol-induced "boxing," at a dose of 200 ng. Somatostatin may be acting on cholinergic mechanisms originating from the nucleus basalis of Meynert, rather than the NTDL (Stivers et al, 1987). The lack of activity of the other three peptides supports the possibility that modulation of carbachol-induced "boxing" is specific to peptides coexisting with acetylcholine in the NTDL terminal field.

Another type of coexistence involves two neuropeptides without a "classical" neurotransmitter. Oxytocin (OXY) and cholecystokinin (CCK) have been reported to coexist in the supraoptic and paraventricular nuclei of the hypothalamus (Vanderhaeghen, et al, 1981). The projections of these neurons has not been described; however, OXY-immunoreactive neurons in the paraventricular nuclei have been shown to project to the lateral hypothalamus, via the ventral tegmentum, descending through the medulla to the spinal cord (Buijs et al, 1981; Sofroniew, 1985). To investigate the possible actions and interactions of CCK and OXY in this type of coexistence, rats were bilaterally cannulated in the caudal ventral tegmental area, 6.3 mm posterior to bregma, $\pm$ 0.5 mm lateral to the

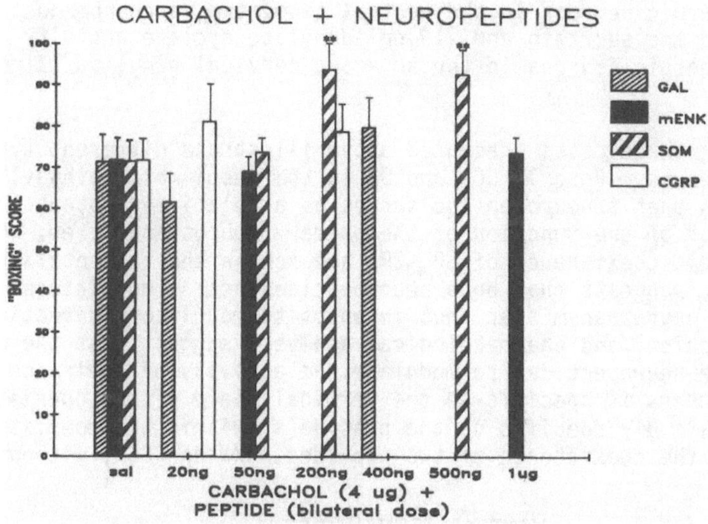

Figure 4. Peptides localized in the medial anterior cortex but not coexisting with acetylcholine. Calcitonin gene related peptide (CGRP), met-enkephalin (mENK), and galanin (GAL) had no effect on carbachol-induced "boxing" at doses from 15 ng to 1 ug. Somatostatin significantly potentiated carbachol-induced "boxing" at doses of 200 ng and 500 ng. ANOVA $(F3,25) = 8.34$, $p < 0.01$, ** $p < 0.01$ by Newman-Keuls analysis. Data are presented as mean + standard error of the mean. N = 6-8 for each dose of each treatment. (unpublished data).

saggital suture, and 8.0 mm ventral to the surface of the skull at bregma. One week later, rats were microinjected with saline, OXY, or CCK, over a dose range of 10 pg to 400 ng. The most striking behavior elicited by OXY was grooming, which has been previously reported for OXY administered into the lateral ventricle (Drago et al, 1986a,b; Fahrbach et al, 1984). OXY induced significant increases in normal patterns of grooming activity at doses of 20 ng and 200 ng (ANOVA $F(4,18) = 8.01$, $p < 0.01$, Figure 5, Kaltwasser and Crawley, 1987). CCK induced increases in grooming which were significant at the 10 ng dose of CCK $(F4,31) = 4.60$, $p < 0.01$. Combinations of CCK and OXY produced grooming scores which were equal to or less than the grooming scores obtained with each peptide alone at that dose (Figure 6, Kaltwasser and Crawley, 1987). The maximum amount of grooming which could be elicited by each neuropeptide appeared to be considerably lower than the amount of grooming elicited by other peptides such as ACTH and bombesin (Crawley and Moody, 1983; Dunn et al, 1979; Gispen and Isaacson, 1981). Combination treatments of OXY and CCK yielded grooming scores which appeared to stay below this "ceiling," presenting a profile characteristic of a competitive interaction between these two peptides. The mechanism of this competition is unknown, but could involve competition for the same second messenger of postsynatpic neurons in the ventral tegmentum, a phenomenon reported for secretin and VIP on adenylate cyclase activity in noradrenergic neurons in the superior cervical ganglion (Zigmond, 1985).

The three cases presented above illustrate different types of coexistences. Case 1, CCK and DA in the mesolimbic pathway, suggests that a neuropeptide serves as a relatively potent modulator of the function of the primary neurotransmitter. Case 2, the triple coexistence of SP, CRF and Ach in the tegmental-cortical pathway, suggests that both neuropeptides act as modulators of the primary neurotransmitter, but in opposite modulatory directions. Topographical and pharmacological analyses suggest that the ability of these neuropeptides to modulate the activity of their coexisting transmitters is specific to the terminal field of the coexistence and relatively specific to the peptide(s) within the coexistence. Case 3, the coexistence of two peptides, OXY and CCK, without a

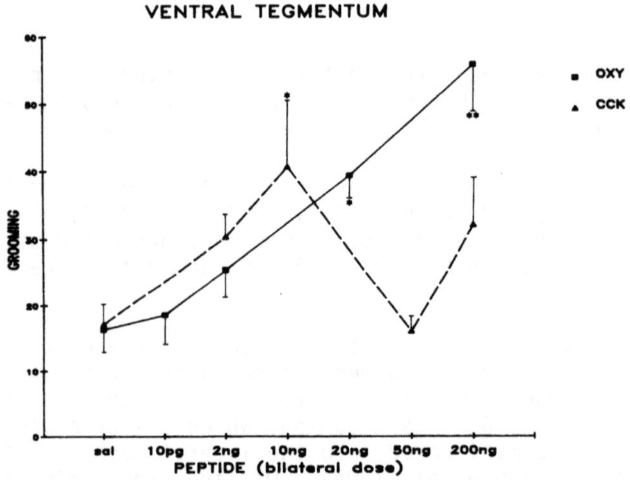

Figure 5. Coexistence of oxytocin (OXY) and cholecystokinin (CCK) in the supraoptic and paraventricular nuclei of the hypothalamus. Bilateral microinjection of saline, OXY, or CCK into the caudal portion of the ventral tegmentum, one possible terminal regions of this coexistence, was immediately followed by a thirty minute session in which all forms of grooming were scored as present or absent, every fifteen seconds, by an observer uninformed of the treatment condition. OXY significantly elevated baseline grooming at does of 20 ng and 200 ng, ANOVA $F(4,18) = 8.01$, $p < 0.01$, * $p < 0.05$, ** $p < 0.01$ by Newman-Keuls post-hoc analysis of individual means. CCK significantly elevated baseline grooming at a dose of 10 ng, ANOVA $F(4,31) = 4.60$, $p < 0.01$, * $p < 0.05$ by Newman-Keuls analysis. Data are presented as mean + standard error of the mean. N = 6-10 for each dose of each treatment. (Kaltwasser and Crawley, 1987).

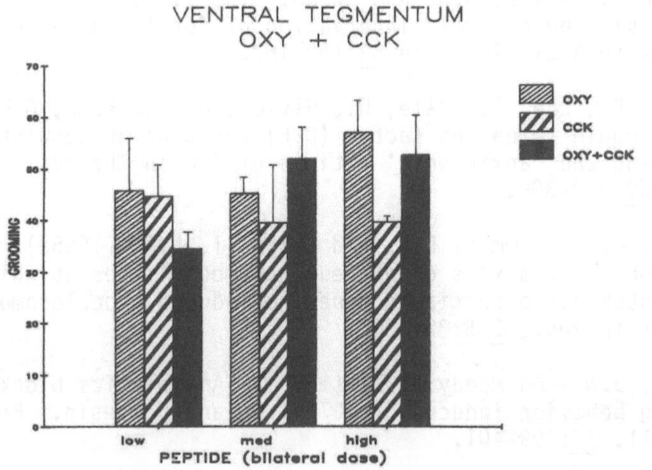

Figure 6. Interaction between OXY and CCK. Coinjection of OXY and CCK yielded grooming scores which were not significantly different than grooming scores obtained for each peptide microinjected alone at that dose. Low doses: OXY 100 pg, CCK 10 ng; Medium doses: OXY 2 ng, CCK 40 ng; High doses: OXY 200 ng, CCK 200 ng. Data are presented as mean + standard error of the mean. N = 6-8 for each dose of each treatment. (Kaltwasser and Crawley, 1987).

primary neurotransmitter, suggests that each peptide can function independently, but that the two peptides together have a competitive interaction. A great deal of work remains to be initiated on the mechanism of functional interactions between coexisting neurotransmitters, including studies of synthesis, release (Markstein and Hokfelt, 1984); turnover (Fuxe et al, 1980); allosteric receptor sites (Agnati et al, 1983); and second messengers (this volume). However, from the three cases described to date, it is interesting to note that the peptides do have functional significance in awake, behaving rats, which is primarily of a modulatory nature. Coexisting neuropeptides may serve as the "fine tuning" system of neuropharmacology, providing new avenues for the development of neuropsychiatric drugs with great selectivity and reduced toxicity.

References

Agnati, L.F., Fuxe, K., Benfenati, F., Cleani, M.F., Battistini, N, Mutt, V., Cavicchioli, L., Galli, G. and Hokfelt, T. (1983).

Differential modulation of CCK-8 and CCK-4 of ^{3}H-spiperone binding sites linked to dopamine and 5-hydroxytryptamine receptors in the brain of the rat. Neurosci. Lett. 35:179-183.

Buijs, R.M., Velis, D.N. and Swaab, D.F. (1981). Extrahypothalamic vasopressin and oxytocin innervation of fetal and adult brain. Progress in Brain Research 53:159-167.

Britton, K., Lee, G., Vale, W., Rivier, J. and Koob, G.F. (1986). Corticotropin releasing factor (CRF) receptor antagonist blocks activating and 'anxiogenic' actions of CRF in the rat. Brain Res., 369:303-306.

Crawley, J.N., Hommer, D.W. and Skirboll, L.R. (1985a). Topographical analysis of nucleus accumbens sites at which cholecystokinin potentiates dopamine-induced hyperlocomotion in the rat. Brain Res., 335:337-341.

Crawley, J.N. and Moody, T.W. (1983). Anxiolytics block excessive grooming behavior induced by ACTH 1-24 and bombesin. Brain Res. Bull. 10:399-401.

Crawley, J.N., Olschowka, J.A., Diz, D.I. and Jacobowitz, D.M. (1985b). Behavioral investigation of the coexistence of substance P, corticotropin releasing factor, and acetylcholinesterase in lateral dorsal tegmental neurons projecting to the medial frontal cortex of the rat. Peptides 6:891-901.

Crawley, J.N, Stivers, J.A., Blumstein, L.K. and Paul, S.M. (1985c). Cholecystokinin potentiates dopamine-mediated behaviors: evidence for modulation specific to a site of coexistence. J. Neurosci. 5:8:1972-1983.

Crawley, J.N., Stivers, J.A., Hommer, D.W., Skirboll, L.R. and Paul, S.M. (1986a). Antagonists of central and peripheral behavioral actions of cholecystokinin octapeptide. J. Pharmacol. Exp. Ther. 236:2:320-330.

Crawley, J.N., Stivers, J.A., Martin, J.V. and Mendelson, W.B., (1986b). Cholinergic induction of seizures in the rat prefrontal cortex. Life Sci. 38:2347-2354.

Drago, F., Caldwell, J.D., Pedersen, C.A., Continella, G., Scapagnini, U. and Prange, A.J. (1986a). Dopamine neurotransmission in the nucleus accumbens may be involved in oxytocin-enhanced grooming behavior of the rat. Pharmacol. Biochem. Behav. 24:1185-1188.

Drago, F., Pedersen, C.A., Caldwell, J.D. and Prange, A.J. (1986b). Oxytocin potently enhances novelty-induced grooming behavior in the rat. Brain Res. 368:287-295.

Iversen, S.D. (1977). Brain dopamine systems and behavior. In Handbook of Psychopharmacology, volume 8, pp. 333-384. (eds. L.L. Iversen, S.D. Iversen, and S.H. Snyder). Plenum Press, New York.

Kaltwasser, M.T. and Crawley, J.N. (1987). Interactions of central oxytocin and cholecystokinin in rat grooming behavior. Submitted.

Lotstra, F., Verbanck, P., Mendlewicz, J. and Vanderhaeghen, J.J., (1984). No evidence of antipsychotic effect of caerulein in schizophrenic patients free of neuroleptics: a double-blind cross-over study. Biol. Psychiatry 19:877-882.

Markstein, R. and Hokfelt, T. (1984). Effect of cholecystokinin octapeptide on dopamine release from slices of cat caudate nucleus. J. Neurosci. 4:570-575.

Moroji, T., Watanabe, N., Aoki, N. and Itoh, S., 1982, Antipsychotic effects of ceruletide (caerulein) on chronic schizophrenia. Arch. Gen. Psychiatry 39:485-486.

Mutt, V., 1978, Hormone isolation. In Gut Hormones. pp. 21-27. (eds. S.R. Bloom). Churchill, London.

Nair, N.P.V., Bloom, D.M. and Nestoros, J.N. (1982). Cholecystokinin appears to have antipsychotic properties. Prog. Neuro-Psychopharmacol. Biol. Psychiatry 6:509-512.

Olschowka, J.A., Diz, D.I. and Jacobowitz, D.M. (1983). Coexistence of substance P, CRF and AChE in neurons of the nucleus tegmenti dorsalis lateralis. Society for Neuroscience Abstract 9:167.5

Schneider, L.H., Alpert, J.E. and Iversen, S.D. (1983). CCK-8 modulation of mesolimbic dopamine: antagonism of amphetamine-stimulated behaviors. Peptides 4:749-753.

Skirboll, L.R., Grace, A.A., Hommer, D.W., Rehfeld, J., Goldstein, M., Hokfelt, T. and Bunney, B.S. (1981). Peptide-monoamine coexistence: studies of the actions of cholecystokinin-like peptide on the electrical activity of midbrain dopamine neurons. Neuroscience 6:2111-2124.

Sofroniew, M.V. (1985). Vasopressin, oxytocin and their related neurophysins. In Handbook of Chemical Neuroanatomy, volume 4, pp. 93-265. GABA and Neuropeptides in the CNS. (eds. A. Bjorklund and T. Hokfelt). Elsevier Science Publishers, North Holland.

Stevens, J.R. (1973). An anatomy of schizophrenia. Arch. Gen. Psychiatry 29:177.

Stivers, J.A. and Crawley, J.N. (1987). Substance P antagonists block carbachol-induced "boxing" behavior at a site of coexistence in the rat prefrontal cortex. Submitted.

Stivers, J.A., Skirboll, L.R., Long, R. and Crawley, J.N. (1987). Anatomical analysis of frontal cortex sites at which carbachol induces "boxing"-like seizures in the rat. Submitted.

Studler, J.M., Simon, H., Cesselin, F., Legrand, J.C., Glowinski, J. and Tassin, J.P. (1981). Biochemical investigation on the localization of cholecystokinin octapeptide in dopaminergic neurons originating from the ventral tegmental area of the rat. Neuropeptides 2:131-139.

Tamminga, C.A., Lucignani, G., Porrino, L.J., Littman, B., Thaker, G.K. and Alphs, L. (1984). Cholecystokinin: potential peptidergic antipsychotic. Clin. Neuropharmacol. 7:Suppl 1:556-557.

Vanderhaeghen, J.J., Lotstra, F., Vandesande, F. and Dierickx, K. (1981). Coexistence of cholecystokinin and oxytocin-neurophysin in some magnocellular hypothalamo-hypophyseal neurons. Cell Tissue Res. 221:227-231.

Vanderhaeghen, J.J., Signeau, J.C., and Gepts, W. (1975). New peptide in the vertebrate CNS reacting with gastrin antibodies. Nature 221:557-559.

Van Ree, J.M., Gaffori, O. and De Wied, D. (1983). In rats, the behavioral profile of CCK-8-related peptides resembles that of antipsychotic agents. Eur. J. Pharmacol. 93:63-78.

Vincent, S.R., Satoh, K., Armstrong, D.M. and Fibiger, H.C. (1983). Substance P in the ascending cholinergic reticular system. Nature 306:688-691.

Widerlov, E., Kalivas, P.W., Lewis, M.H., Prange, A.J. and Breese, G.R. (1983). Influence of cholecystokinin on central monominergic pathways. Regul. Pept. 6:99-109.

Zigmond, R.E. (1985). Biochemical consequences of synaptic stimulation: the regulation of tyrosine hydroxylase activity by multiple transmitters. Trends in NeuroSciences 8:63-69.

Overview

43
Receptor Regulation and Receptor–Receptor Communication

M.D. Hollenberg

INTRODUCTION

As recently as five years ago, what one might term the "grind and bind era" of receptor studies was still in full bloom; and a large number of studies were focussed on determining receptor affinities and numbers under a variety of physiologic situations. It became clear from such studies that receptors are not static elements, fixed in the cell membrane at a single site. Rather, it is now fully appreciated that receptors for neurotransmitters, hormones and other messenger molecules are dynamic elements both in the plasma membrane and in the cytoplasmic compartment, as outlined in Fig. 1.

Progress in receptor research over the past decade has yielded an intriguing picture of these cell surface signal transduction elements. Much information is now available for the detailed physicochemical and chemical (e.g. amino acid sequence) properties of a number of receptors; and reliable data are available concerning the cell surface mobility and aggregation properties of many receptors. Structural and functional relationships between receptors of different types (e.g. receptors for growth factors like insulin, insulin-like growth factor-I (IGF-I), platelet-derived growth factor (PDGF) and epidermal growth factor-urogastrone (EGF-URO) are now recognized; and a number of distinct receptor-triggered transmembrane signalling reactions have now been described. All of this information provides a rich perspective for considering the question: how is receptor function regulated; and by what mechanism(s) might the activation of one receptor regulate the behaviour of a second receptor? In the following sections, I will try to provide a partial answer to this question, along with some specific examples of receptor-receptor interactions.

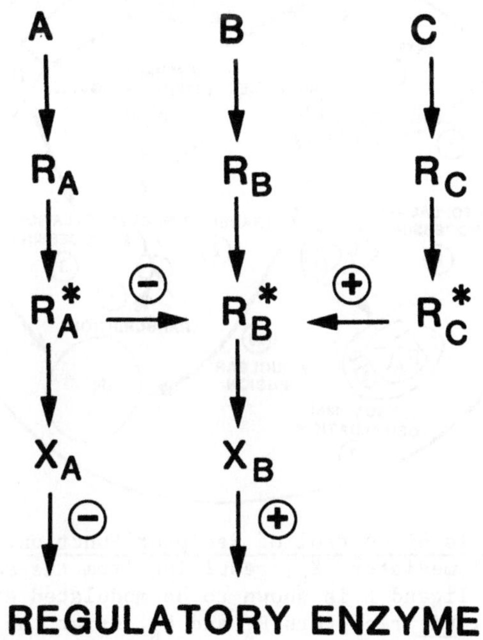

Figure 1. Receptor Regulation. The numbers indicate potential sites at which receptor regulation (e.g. biosynthesis, turnover and function) may occur.

LEVELS OF CONTROL OF RECEPTOR FUNCTION

There are many levels at which receptors may be controlled, both with respect to receptor number and with respect to overall receptor activity. In principle, as outlined in Fig. 2, control could be exerted at the level of the receptor, at the level of mediators generated by receptor activity, or at some point farther along in the reaction pathway. For this discussion, attention will be focussed on the receptor *per se* and not on control points that might occur distal to the receptor-generated signal.

Receptor control may occur either at the level of the gene, where the stage of the cell cycle, the state of cell differentiation, or the influence of agents like glucocorticoids may change receptor expression; or, receptor regulation may occur at an epigenetic level in response both to non-specific (e.g. change in cell substratum; cell-cell contact; chemical insult) and to receptor-triggered stimuli. When thinking about the possible

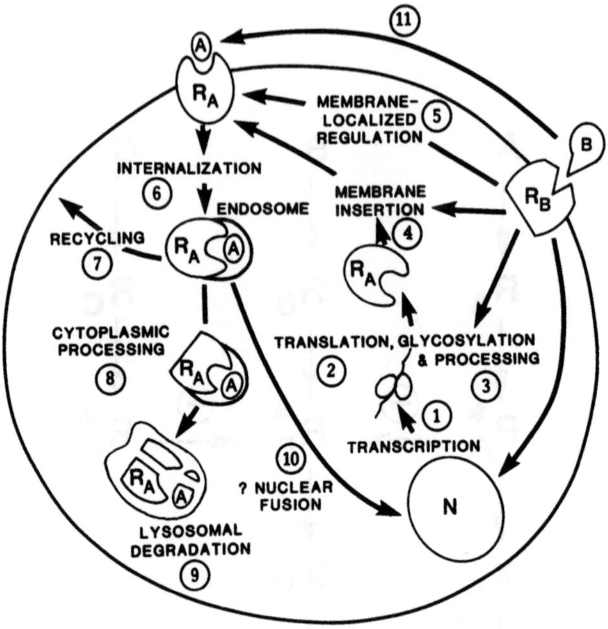

Figure 2. Levels of Control of Receptor Function. The positive cellular signal mediator, X_B, resulting from the activation of receptor R_B by ligand B is shown to be modulated either at the receptor level, via receptors R_A and R_B, or at the cellular level by the negative cellular mediator, X_A, generated by the triggering of receptor R_A by ligand A. The asterisks denote the ligand-activated states of the receptor.

mechanisms that may be involved in regulating receptors, one must consider not only the ligand binding domain of the receptor but also other receptor domains that may be involved in overall receptor function. Receptor domains apart from the ligand binding region that may be targets for receptor-regulatory processes could include: (1) a receptor catalytic site (e.g. tyrosine kinase); (2) a specific ion binding site (e.g. for sodium, calcium or chloride); (3) phosphate acceptor sites (serine, threonine or tyrosine) that control overall receptor conformation; (4) oligosaccharide attachment sites; (5) sites involved in clustering and internalization and; (6) a cryptic peptide-messenger domain that could be liberated proteolytically from the receptor. In summary, irrespective of the stimulus that triggers receptor regulation (i.e. non-specific or ligand-triggered events), mechanisms that involve the receptor *per se* can be thought of at the level of specific receptor domains that can be targets for the reactions to be discussed in the following sections.

MECHANISMS OF RECEPTOR-MEDIATED SIGNALLING

In order to fulfill their recognition/cell activation function, membrane receptors must be able to generate a cellular signal that can be greatly amplified. It is reasonable to assume that the same reactions involved in generating and amplifying a transmembrane signal may also be involved in regulating the very elements (i.e. the receptors) that initially trigger the signal. As summarized in greater detail elsewhere (Hollenberg, 1985, 1986), three basic processes related to transmembrane signalling can be singled out: (1) ligand-modulated ion channel activity; (2) ligand-regulated enzymatic activity, with the receptor catalytic domain located at the inner aspect of the plasma membrane; and (3) ligand-directed liberation of cryptic mediators via direct interactions of intracellular receptor domains with other membrane-associated constituents like the so-called G- or N-regulatory oligomers involved in the regulation of the adenylate cyclase system (Gilman, 1984; Northup, 1985; Rodbell, 1980); it is also possible that cryptic mediators may be liberated directly from the membrane as a consequence of the action of the receptor-enzyme activity (process 2). Each of the three ligand-triggered processes can lead to a change in receptor activity by reactions that may or may not involve the formation of a covalent bond.

MECHANISMS OF RECEPTOR REGULATION

Non-Covalent Interactions

Non-covalent mechanisms that might regulate receptor function could include: (1) changes in membrane potential; (2) changes in receptor distribution (patching/capping etc.); (3) allosteric interactions of the receptor with other membrane proteins (e.g. G-proteins) or with smaller ligands (cations, anions, nucleotides); and (4) changes in the membrane lipid environment. For instance, the change in membrane potential generated by receptor-mediated events causes profound changes in the behaviour of the voltage-sensitive sodium channel (Catterall, 1982; Noda et al., 1984). As repeatedly pointed out by Zierler and colleagues (Zierler, 1985; Zierler et al., 1981), changes in membrane potential could also have a profound effect on the function of many membrane proteins. Thus a change in membrane potential caused by one membrane receptor could readily change the function of other receptors on the same cell surface. Amongst the other mechanisms outlined above, the interaction of receptors with G-protein complexes can be singled out as potentially playing a critical role in regulating receptor activity. As envisioned by the mobile or floating receptor model, developed by a number of investigators (e.g. Jacobs and Cuatrecasas, 1985) and summarized in some detail elsewhere (Hollenberg, 1986; Hollenberg and Goren, 1985), the noncovalent interaction of the ligand-occupied receptor with a membrane effector like a G-protein would be expected to change the receptor's affinity for the ligand. In summary, all four of the non-covalent mechanisms outlined above could have broad implications for the control of receptor function.

Covalent Modification of Receptors

Several covalent reactions can be singled out for their potential to play a role in the rapid membrane-localized regulation of receptor function (sites 5 and 6, Fig. 1). In principle, the potential reactions very likely reflect the same ones used for the regulation of cytoplasmic enzymes. Thus, it comes as no surprise that phosphorylation-dephosphorylation reactions, involving the receptor either as enzyme or substrate, are prime candidates for the control of receptor activity. Two other covalent reactions that merit attention are: (1) disulfide-sulfhydryl interchange and (2) limited proteolysis.

Examples of all three of the above mechanisms can now be cited in the context of receptor studies. For instance, the phosphorylation of the EGF-URO receptor caused by the activation of the phorbol ester receptor (i.e. kinase C) is thought to regulate both the binding activity and kinase activity of the EGF-URO receptor (Downward *et al.*, 1985). Alternatively, desensitization of the beta-adrenergic receptor is thought to involve phosphorylation mediated either *via* a cyclic-AMP-dependent kinase or *via* a novel kinase that can phosphorylate the beta-receptor primarily when the receptor is occupied by an agonist (Benovic *et al.*, 1986). Disulfide-sulfhydryl exchange reactions may be important not only for the regulation of the structure and oligomeric forms of receptors like the one for insulin (Maturo *et al.*, 1983) but also for the control of the sulfhydryl-reagent-sensitive tyrosine protein kinase activity of growth factor receptors (EGF-URO, insulin, IGF-I, PDGF). Finally, proteolytic mechanisms might not only control receptor binding (Cuatrecasas, 1971) but may also liberate limited receptor domains that could regulate intracellular processes (Kiechle and Jarett, 1985). Potentially, many of the rapid membrane-localized receptor-receptor interactions that have now been documented (reviewed by Hollenberg, 1985; 1986) could be rationalized in terms of the non-covalent or covalent reactions outlined above.

RECEPTOR CASCADES

Although most of this chapter is focussed on mechanisms situated at the inner aspect of the plasma membrane, it is instructive to consider examples whereby the activation of one receptor can, *via* the secretion of an extracellular messenger, activate a second cell-surface receptor. This process can be thought of as a cascade mechanism leading to a positive amplification via a matrix of receptor-mediated signals, triggered by the activation of a single receptor. In a sense this kind of cascade can be thought of as an autocrine system that can function within a selected microenvironment for somatic cells. In the nervous system, co-release of a variety of transmitters might play a similar role. Drawing from work by my colleagues and myself, one can point to the generation of IGF-I-like material by the action of EGF-URO on fibroblasts (Atkison et al., 1984a, b), to the likely participation of prostaglandins in the contractile response of rat arterial strips to EGF-URO (Muramatsu et al., 1985), and to the possible participation of surface-localized interferon-related constituents in the course of a mitogenic response to EGF-URO (Lin et al., 1983). Other, and perhaps better examples of this kind of receptor cascade situation are to be found in the literature; it is as yet not certain as to how wide-spread this mechanism of receptor-receptor communication may be.

POTENTIAL RECEPTOR CODES FOR COORDINATE REGULATION

Since it is now recognized that quite distinct receptors can have homologous receptor domains (e.g. the tyrosine kinase domains in receptors for growth factors like insulin and EGF-URO; or the protein kinase target domain of the beta-receptor and rhodopsin); and since the basic transmembrane signalling mechanisms may be few in number, so as to be used by several distinct receptors on a given cell type, it is reasonable to suspect that the activation of a single receptor may lead to the coordinate regulation of many receptors on a given cell. For instance, the elevation of cyclic AMP caused by activating the beta adrenergic receptor might lead to a concerted control of all receptors that possess a regulatory cyclic AMP kinase substrate domain. The code for receptor regulation would be the same, irrespective of the agonist that caused the initial elevation of cyclic AMP. Whether or not the code for coordinated control would be bi-directional or uni-directional would depend on the rates and reversibility of the specific phosphorylation reactions involved. In theory, most of the specific receptor-regulatory reactions outlined above could be used for coordinated receptor control. These reactions comprising a unique "receptor code", in principle, might be either unidirectional or reversible. In practice, the challenge is to identify those receptor-targeted reactions that can be seen to play a role in the actual physiologic setting of an intact cell.

SUMMARY

Using the same set of reactions that are designed for communicating with intracellular reaction pathways, receptors can also communicate with each other. This communication can take place via a variety of intramembranous interactions that may or may not result in a covalent modification of the receptors themselves. In some cases, receptors may also communicate with each other via a cascade of secreted messenger molecules that act at the cell surface in an autocrine fashion. The existence of common receptor polypeptide domains and a limited number of cell activation pathways raises the possibility that certain classes of receptors (e.g. those activating adenylate cyclase) may modulate overall cell function by using a common code of receptor-regulatory reactions.

References

Atkison, P.R., Bala, R.M., and Hollenberg, M.D. (1984a). Somatomedin-like activity from cultured embryo-derived cells: Partial characterization and stimulation of production by epidermal growth factor (urogastrone). Can. J. Biochem. Cell Biol., 62, 1335-1342.

Atkison, P.R., Hayden, L.J., Bala, R.M., and Hollenberg, M.D. (1984b). Production of somatomedin-like activity by human adult tumor-derived transformed and normal cell cultures and by cultured rat hepatocytes: Effects of culture conditions and of epidermal growth factor (urogastrone). Can. J. Biochem. Cell. Biol., 62, 1343-1350.

Benovic, J.L., Strasser, R.H., Caron, M.G., and Lefkowitz, R.J. (1986). β-Adrenergic receptor kinase: Identification of a novel protein kinase that phosphorylates the agonist-occupied form of the receptor. Proc. Natl. Acad. Sci. USA, 83, 2797-2801.

Catterall, W.A. (1982). The emerging molecular view of the sodium channel. Trends Neurosci., 5, 303-306.

Cuatrecasas, P. (1971). Perturbation of the insulin receptor of isolated fat cells with proteolytic enzymes. J. Biol. Chem., 246, 6522-6531.

Downward, J., Waterfield, M.D., and Parker, P.J. (1985). Autophosphorylation and protein kinase C phosphorylation of the epidermal growth factor receptor. J. Biol. Chem., 260, 14538-14546.

Gilman, A.G. (1984). Guanine nucleotide-binding regulatory proteins and dual control of adenylate cyclase. J. Clin. Invest., 73, 1-4.

Hollenberg, M.D. (1985). Control of receptor function by homologous and heterologous ligands. In Mechanisms of Receptor Regulation. (eds. G. Poste and S.T. Crooke). Plenum Press, New York. pp. 295-322.

Hollenberg, M.D. (1986). Mechanisms of receptor-mediated

transmembrane signalling. Experientia, 42, 718-727.

Hollenberg, M.D. and Goren, H.J. (1985). Ligand-receptor interactions at the cell surface. In Mechanisms of Receptor Regulation. (eds. G. Poste and S.T. Crooke). Plenum Press, New York. pp. 323-373.

Jacobs, S. and Cuatrecasas, P. (1985). The mobile receptor hypothesis and polypeptide hormone action. In Polypeptide Hormone Receptors. (ed. B. Posner). Marcel Dekker, New York. pp. 39-60.

Kiechle, F.L. and Jarett, L. (1985). The molecular basis of insulin action: Membrane-associated reactions and intracellular mediators. In Insulin, Its Receptor and Diabetes. (ed. M.D. Hollenberg). Marcel Dekker, New York. pp. 181-204.

Lin, S., Ts'o, P.O.P., and Hollenberg, M.D. (1983). Epidermal growth factor-urogastrone action: Induction of 2'5'oligoadenylate synthetase activity and enhancement of the mitogenic effect by anti-interferon antibody. Life Sci., 32, 1479-1488.

Maturo, J.M. III, Hollenberg, M.D., and Aglio, L.S. (1983). Insulin receptor: Insulin-modulated interconversion between distinct molecular forms involving disulfide-sulfhydryl exchange. Biochemistry, 22, 2579-2586.

Muramatsu, I., Hollenberg, M.D., and Lederis, K. (1985). Vascular actions of epidermal growth factor-urogastrone: Possible relationships to prostaglanding production. Can. J. Physiol. Pharmacol., 63, 994-999.

Noda, M., Shimizu, S., Tanabe, T., Takai, T., Kayano, T., Ikeda, T., Takahashi, H., Nakayama, H., Kanaoka, Y., Minamino, N., Kanagawa, K., Matsuo, H., Raftery, M.A., Hirose, T., Inayama, S., Hayashida, H., Miyata, T., and Numa, S. (1984). Primary structure of Electrophorus electricus sodium channel deduced from cDNA sequence. Nature, 312, 121-127.

Northup, J.K. (1985). Overview of the guanine nucleotide regulatory protein systems, N_s and N_i, which regulate adenylate cyclase activity in plasma membranes. In Molecular Mechanisms of Transmembrane Signalling. (eds. P. Cohen and M. Houslay). Elsevier, Amsterdam. pp. 91-116.

Rodbell, M. (1980). The role of hormone receptors and GTP-regulatory proteins in membrane transduction. Nature, 284, 17-22.

Zierler, K. (1985). Membrane polarization and insulin action. In Insulin: Its Receptor and Diabetes. (ed. M.D. Hollenberg). Marcel Dekker, New York. pp. 141-179.

Zierler, K. and Rogus, E.M. (1981). Effects of peptide hormones and adrenergic agents on membrane potentials of target cells. Fedn. Proc., 40, 121-124.

Acknowledgements

Work in the author's laboratory is supported in large part by the Canadian Medical Research Council, the Alberta Heritage Foundation for Medical Research and the Alberta Heritage Savings Trust Fund-Cancer.

Index

MIX
Papier aus verantwortungsvollen Quellen
Paper from responsible sources
FSC® C105338

If you have any concerns about our products,
you can contact us on
ProductSafety@springernature.com

In case Publisher is established outside the EU,
the EU authorized representative is:
Springer Nature Customer Service Center GmbH
Europaplatz 3, 69115 Heidelberg, Germany

Printed by Libri Plureos GmbH
in Hamburg, Germany